IET ENERGY ENGINEERING SERIES 158

Advanced Dielectric Materials for Electrostatic Capacitors

Other volumes in this series:

Volume 1	**Power Circuit Breaker Theory and Design** C.H. Flurscheim (Editor)
Volume 4	**Industrial Microwave Heating** A.C. Metaxas and R.J. Meredith
Volume 7	**Insulators for High Voltages** J.S.T. Looms
Volume 8	**Variable Frequency AC Motor Drive Systems** D. Finney
Volume 10	**SF_6 Switchgear** H.M. Ryan and G.R. Jones
Volume 11	**Conduction and Induction Heating** E.J. Davies
Volume 13	**Statistical Techniques for High Voltage Engineering** W. Hauschild and W. Mosch
Volume 14	**Uninterruptible Power Supplies** J. Platts and J.D. St Aubyn (Editors)
Volume 15	**Digital Protection for Power Systems** A.T. Johns and S.K. Salman
Volume 16	**Electricity Economics and Planning** T.W. Berrie
Volume 18	**Vacuum Switchgear** A. Greenwood
Volume 19	**Electrical Safety: A guide to causes and prevention of hazards** J. Maxwell Adams
Volume 21	**Electricity Distribution Network Design, 2nd Edition** E. Lakervi and E.J. Holmes
Volume 22	**Artificial Intelligence Techniques in Power Systems** K. Warwick, A.O. Ekwue and R. Aggarwal (Editors)
Volume 24	**Power System Commissioning and Maintenance Practice** K. Harker
Volume 25	**Engineers' Handbook of Industrial Microwave Heating** R.J. Meredith
Volume 26	**Small Electric Motors** H. Moczala *et al.*
Volume 27	**AC-DC Power System Analysis** J. Arrillaga and B.C. Smith
Volume 29	**High Voltage Direct Current Transmission, 2nd Edition** J. Arrillaga
Volume 30	**Flexible AC Transmission Systems (FACTS)** Y.-H. Song (Editor)
Volume 31	**Embedded Generation** N. Jenkins *et al.*
Volume 32	**High Voltage Engineering and Testing, 2nd Edition** H.M. Ryan (Editor)
Volume 33	**Overvoltage Protection of Low-Voltage Systems, Revised Edition** P. Hasse
Volume 36	**Voltage Quality in Electrical Power Systems** J. Schlabbach *et al.*
Volume 37	**Electrical Steels for Rotating Machines** P. Beckley
Volume 38	**The Electric Car: Development and future of battery, hybrid and fuel-cell cars** M. Westbrook
Volume 39	**Power Systems Electromagnetic Transients Simulation** J. Arrillaga and N. Watson
Volume 40	**Advances in High Voltage Engineering** M. Haddad and D. Warne
Volume 41	**Electrical Operation of Electrostatic Precipitators** K. Parker
Volume 43	**Thermal Power Plant Simulation and Control** D. Flynn
Volume 44	**Economic Evaluation of Projects in the Electricity Supply Industry** H. Khatib
Volume 45	**Propulsion Systems for Hybrid Vehicles** J. Miller
Volume 46	**Distribution Switchgear** S. Stewart
Volume 47	**Protection of Electricity Distribution Networks, 2nd Edition** J. Gers and E. Holmes
Volume 48	**Wood Pole Overhead Lines** B. Wareing
Volume 49	**Electric Fuses, 3rd Edition** A. Wright and G. Newbery
Volume 50	**Wind Power Integration: Connection and system operational aspects** B. Fox *et al.*
Volume 51	**Short Circuit Currents** J. Schlabbach
Volume 52	**Nuclear Power** J. Wood
Volume 53	**Condition Assessment of High Voltage Insulation in Power System Equipment** R.E. James and Q. Su
Volume 55	**Local Energy: Distributed generation of heat and power** J. Wood
Volume 56	**Condition Monitoring of Rotating Electrical Machines** P. Tavner, L. Ran, J. Penman and H. Sedding
Volume 57	**The Control Techniques Drives and Controls Handbook, 2nd Edition** B. Drury
Volume 58	**Lightning Protection** V. Cooray (Editor)
Volume 59	**Ultracapacitor Applications** J.M. Miller

Volume 62	**Lightning Electromagnetics** V. Cooray
Volume 63	**Energy Storage for Power Systems, 2nd Edition** A. Ter-Gazarian
Volume 65	**Protection of Electricity Distribution Networks, 3rd Edition** J. Gers
Volume 66	**High Voltage Engineering Testing, 3rd Edition** H. Ryan (Editor)
Volume 67	**Multicore Simulation of Power System Transients** F.M. Uriate
Volume 68	**Distribution System Analysis and Automation** J. Gers
Volume 69	**The Lightening Flash, 2nd Edition** V. Cooray (Editor)
Volume 70	**Economic Evaluation of Projects in the Electricity Supply Industry, 3rd Edition** H. Khatib
Volume 72	**Control Circuits in Power Electronics: Practical issues in design and implementation** M. Castilla (Editor)
Volume 73	**Wide Area Monitoring, Protection and Control Systems: The enabler for Smarter Grids** A. Vaccaro and A. Zobaa (Editors)
Volume 74	**Power Electronic Converters and Systems: Frontiers and applications** A.M. Trzynadlowski (Editor)
Volume 75	**Power Distribution Automation** B. Das (Editor)
Volume 76	**Power System Stability: Modelling, analysis and control** Abdelhay A., Malik and Om P. Malik
Volume 78	**Numerical Analysis of Power System Transients and Dynamics** A. Ametani (Editor)
Volume 79	**Vehicle-to-Grid: Linking electric vehicles to the smart grid** J. Lu and J. Hossain (Editors)
Volume 81	**Cyber-Physical-Social Systems and Constructs in Electric Power Engineering** S. Suryanarayanan, R. Roche and T.M. Hansen (Editors)
Volume 82	**Periodic Control of Power Electronic Converters** F. Blaabjerg, K. Zhou, D. Wang and Y. Yang
Volume 86	**Advances in Power System Modelling, Control and Stability Analysis** F. Milano (Editor)
Volume 87	**Cogeneration: Technologies, optimisation and implementation** C.A. Frangopoulos (Editor)
Volume 88	**Smarter Energy: From smart metering to the smart grid** H. Sun, N. Hatziargyriou, H.V. Poor, L. Carpanini and M.A. Sánchez Fornié (Editors)
Volume 89	**Hydrogen Production, Separation and Purification for Energy** A. Basile, F. Dalena, J. Tong and T.N.Veziroğlu (Editors)
Volume 90	**Clean Energy Microgrids** S. Obara and J. Morel (Editors)
Volume 91	**Fuzzy Logic Control in Energy Systems with Design Applications in MATLAB®/Simulink®** İ.H. Altaş
Volume 92	**Power Quality in Future Electrical Power Systems** A.F. Zobaa and S.H.E.A. Aleem (Editors)
Volume 93	**Cogeneration and District Energy Systems: Modelling, analysis and optimization** M.A. Rosen and S. Koohi-Fayegh
Volume 94	**Introduction to the Smart Grid: Concepts, technologies and evolution** S.K. Salman
Volume 95	**Communication, Control and Security Challenges for the Smart Grid** S.M. Muyeen and S. Rahman (Editors)
Volume 96	**Industrial Power Systems with Distributed and Embedded Generation** R. Belu
Volume 97	**Synchronized Phasor Measurements for Smart Grids** M.J.B. Reddy and D.K. Mohanta (Editors)
Volume 98	**Large Scale Grid Integration of Renewable Energy Sources** A. Moreno-Munoz (Editor)
Volume 100	**Modeling and Dynamic Behaviour of Hydropower Plants** N. Kishor and J. Fraile-Ardanuy (Editors)
Volume 101	**Methane and Hydrogen for Energy Storage** R. Carriveau and D.S.-K. Ting
Volume 104	**Power Transformer Condition Monitoring and Diagnosis** A. Abu-Siada (Editor)
Volume 106	**Surface Passivation of Industrial Crystalline Silicon Solar Cells** J. John (Editor)

Volume 107 **Bifacial Photovoltaics: Technology, applications and economics** J. Libal and R. Kopecek (Editors)
Volume 108 **Fault Diagnosis of Induction Motors** J. Faiz, V. Ghorbanian and G. Joksimović
Volume 110 **High Voltage Power Network Construction** K. Harker
Volume 111 **Energy Storage at Different Voltage Levels: Technology, integration, and market aspects** A.F. Zobaa, P.F. Ribeiro, S.H.A. Aleem and S.N. Afifi (Editors)
Volume 112 **Wireless Power Transfer: Theory, technology and application** N. Shinohara
Volume 115 **DC Distribution Systems and Microgrids** T. Dragičević, F.Blaabjerg and P. Wheeler
Volume 117 **Structural Control and Fault Detection of Wind Turbine Systems** H.R. Karimi
Volume 119 **Thermal Power Plant Control and Instrumentation: The control of boilers and HRSGs, 2nd Edition** D. Lindsley, J. Grist and D. Parker
Volume 120 **Fault Diagnosis for Robust Inverter Power Drives** A. Ginart (Editor)
Volume 121 **Monitoring and Control using Synchrophasors in Power Systems with Renewables** I. Kamwa and C. Lu (Editors)
Volume 123 **Power Systems Electromagnetic Transients Simulation, 2nd Edition** N. Watson and J. Arrillaga
Volume 124 **Power Market Transformation** B. Murray
Volume 125 **Wind Energy Modeling and Simulation, Volume 1: Atmosphere and plant** P. Veers (Editor)
Volume 126 **Diagnosis and Fault Tolerance of Electrical Machines, Power Electronics and Drives** A.J.M. Cardoso
Volume 128 **Characterization of Wide Bandgap Power Semiconductor Devices** F. Wang, Z. Zhang and E.A. Jones
Volume 129 **Renewable Energy from the Oceans: From wave, tidal and gradient systems to offshore wind and solar** D. Coiro and T. Sant (Editors)
Volume 130 **Wind and Solar Based Energy Systems for Communities** R. Carriveau and D.S.-K. Ting (Editors)
Volume 131 **Metaheuristic Optimization in Power Engineering** J. Radosavljević
Volume 132 **Power Line Communication Systems for Smart Grids** I.R.S Casella and A. Anpalagan
Volume 139 **Variability, Scalability and Stability of Microgrids** S.M. Muyeen, S.M. Islam and F. Blaabjerg (Editors)
Volume 145 **Condition Monitoring of Rotating Electrical Machines** P. Tavner, L. Ran, C. Crabtree
Volume 146 **Energy Storage for Power Systems, 3rd Edition** A.G. Ter-Gazarian
Volume 147 **Distribution Systems Analysis and Automation, 2nd Edition** J. Gers
Volume 155 **Energy Generation and Efficiency Technologies for Green Residential Buildings** D. Ting and R. Carriveau (Editors)
Volume 157 **Electrical Steels, 2 volumes** A. Moses, K. Jenkins, Philip Anderson and H. Stanbury
Volume 160 **Microgrids for Rural Areas: Research and case studies** R.K. Chauhan, K. Chauhan and S.N. Singh (Editors)
Volume 172 **Lighting Interaction with Power Systems, 2 volumes** A. Piantini (Editor)
Volume 905 **Power system protection, 4 volumes**

Advanced Dielectric Materials for Electrostatic Capacitors

Edited by
Qi Li

The Institution of Engineering and Technology

Published by The Institution of Engineering and Technology, London, United Kingdom

The Institution of Engineering and Technology is registered as a Charity in England & Wales (no. 211014) and Scotland (no. SC038698).

First published 2020

The Institution of Engineering and Technology
Michael Faraday House
Six Hills Way, Stevenage
Herts, SG1 2AY, United Kingdom

www.theiet.org

British Library Cataloguing in Publication Data
A catalogue record for this product is available from the British Library

ISBN 978-1-78561-988-5 (hardback)
ISBN 978-1-78561-989-2 (PDF)

Typeset in India by MPS Limited

Contents

About the editor

Qi Li is an Associate Professor of Electrical Engineering at Tsinghua University, Beijing, China. His research interests focus on nanodielectrics for energy applications. He was the recipient of the MRS (Materials Research Society) Postdoctoral Award in 2016, the IEEE Caixin Sun and Stan Grzybowski Young Professional Achievement Award in 2018, and the National Natural Science Fund for Excellent Young Scholars in 2019. He serves as an editorial member of IET Nanodielectrics. He is a member of the Professional Committee of Dielectric Polymer Composite Materials and Applications, Composite Materials Association of China, and a member of the Professional Committee of High Voltage, Beijing Electrical Engineering Association. He is an IEEE Senior Member, a Senior Member of the China Electrotechnical Society and he is also a member of the IET and MRS.

Foreword

I am delighted to write the foreword for the *Advanced Dielectric Materials for Electrostatic Capacitors*. The electrostatic capacitor technology has a long history that dates back to mid-18th century when Ewald Georg von Kleist first stored charge with a hand-held glass jar. After centuries' development, the electrostatic capacitors based on synthetic dielectric materials such as polymers and ceramics are now the indispensable passive electrical apparatus for charge storage and regulation in a multitude of applications ranging from modern electronics to electrical power systems. Recently, the development of novel capacitor dielectrics with greater storage capacitance, temperature-invariant performance and environmentally benign nature has been spurred by the numerous emerging needs, e.g., electrified transportation, next-generation switching technology, grid-tied renewable energy, advanced propulsion and pulse forming systems. In this context, the book is at the perfect timing to meet its readership.

I personally share the research interest in capacitor dielectric materials for over 15 years. Dr. Qi Li, the editor and one of the primary authors of this book, is an active and established professional in this area who is the author to many of the pioneering literatures on this research topic. I appreciate the organization of the whole book in that it not only presents a comprehensive updates of the recent progress in advanced capacitor dielectrics, but also contains almost all the important contents and fundamental aspects for a beginner to learn this specific field. I am also very impressed by the writing team composed of fellow engineers, experimentalists, computational scientists from multiple disciplines including materials science, electrical engineering and information technology. For instance, the first chapter, an introduction to the capacitor technology, is contributed by a highly experienced research scientist specialized in capacitors who used to work at Honeywell, CTS Corporation, GE Global Research and W.L. Gore for 20 years, and now is a professor at university. The second chapter offers a detailed overview of the techniques for capacitor dielectrics characterization, written by a group of experts running a company specialized in high voltage dielectric materials and instruments. The rest of chapters focusing on polymer and ceramic materials are all authored by renowned scholars that have long been active in the technical field of capacitor dielectrics. They in the respective chapters provide insightful comments to the current trends and future prospects in the research and development of capacitor dielectrics. From a scientific point of view, the quality of the book amazes me a lot. I would like to thank all the contributors for presenting such wonderful chapters that constitute the book.

I expect that the book will provide an effective learning experience and referenced resource for all graduate students, electronic and electrical engineers, and academic professionals that are relevant to the electrostatic capacitors community, and will inspire their future study and research on capacitor dielectrics.

C. P. Wong

Regents' Professor
Charles Smithgall Institute Endowed Chair
Member of the National Academy of Engineering of the USA
Foreign academician of the Chinese Academy of Engineering
Georgia Institute of Technology
Materials Science and Engineering
Atlanta, GA

Preface

The rapidly growing electricity demand and increased electrification worldwide entail high-performance electronic components for the development and innovation of power systems in the industrial sector, transport, large appliances and others. Electrostatic capacitors that capitalize on the characteristic of electrical polarization in solid-state dielectric materials to store, control and regulate charges are one of the most important passive elements in modern electronic devices and electrical power systems. For instance, the development of high power electronics, grid-tied renewable energy, hybrid electric vehicles, electrified aircraft, information technology and electromagnetic repulsion systems all pose an urgent need for high-performance electrostatic capacitors. The capacitive performance of electrostatic capacitors is largely determined by the dielectric materials thereof, which are the core material ingredient of the capacitor technology. Advanced dielectric materials are thus an enabling and most promising approach to high-performance electrostatic capacitors.

This book presents the current progress in searching and developing advanced dielectric materials towards high-performance electrostatic capacitors, as well as the basic principle in designing these materials. The focus of material category in the book is polymer- and ceramic-based systems because they are currently the most pursued capacitor dielectrics on the market, and the main content of the book revolves around these two topics. The subject is elucidated from a materials science viewpoint by concentrating the discussion on the synthetic method and property–structure correlation. In addition to the experimental studies, chapters on the computational modelling are also included based on the consideration that this emerging method has proved efficient in the design and prediction of new dielectric materials, especially those occurring over the last decade. A comprehensive set of characterization techniques are introduced as a separate chapter to guide the evaluation of new dielectric materials prior to the commercialization efforts towards packaged capacitors.

The book is designed to be self-contained for the purpose of guiding the development of novel capacitor dielectrics, which should be of use to graduate students, as well as scientists in both industry and academia. All the chapters are contributed by renowned scholars who have long been active in the relevant research field. In order to keep the book a manageable size, there are important and deliberate omissions. I hope that the book will give the reader sufficient grounding to study those content elsewhere.

I am grateful to the Institution of Engineering and Technology for the kind invitation to edit and contribute to this wonderful book. I would like to thank all the chapter authors for their efforts in preparing the book content. I would like to also thank the National Natural Science Foundation of China, Beijing Natural Science Foundation and the State Grid Corporation of China for the financial support on my research in this field.

Qi Li
Beijing, China
2020

Chapter 1

Introduction to electrostatic capacitor technology

Daniel Q. Tan[1]

1.1 Capacitor construction

1.1.1 Capacitor basic attributes

A capacitor is a device that stores electric energy between a pair of electrodes on which electric charges (*Q* in Coulomb) accumulate. Historically, capacitors have taken the form of a pair of thin metal plates, which are either flat or tightly wound up in a cylinder having capacitance (*C*). This is a measure of the potential difference or voltage (*V*), which appears across the plates for a given amount of energy stored on each plate (*Q*/*V* in a unit of Farad). A traditional parallel-plate capacitor stores the amount of energy that is proportional to the surface area (*A*) of the conducting plate and inversely proportional to the distance (*d*) between the plates. It is also proportional to the permittivity of the dielectric substance that separates the plates, whether vacuum, air, or electrically insulating materials chosen for their special dielectric characteristics [1].

The capacitance of a parallel-plate capacitor is equal to the following:

$$C = \varepsilon_0 \varepsilon_r A/d \tag{1.1}$$

where ε_0 is the electrostatic permittivity of vacuum or free space, and ε_r is the relative permittivity of the dielectric substance between metal plates or dielectric constant. Since Farad is a very large unit, values of capacitors are usually expressed in millifarads (mF), microfarads (μF), nanofarads (nF), or picofarads (pF).

The selection of a dielectric substance decides the relative permittivity, voltage withstanding capability, or breakdown strength and hence the volumetric energy storage capability. A capacitor is a key passive component in electrical circuit such as a direct current (DC) to alternating current (AC) inverter as shown in Figure 1.1(a) [2]. The bus link capacitor is used in DC to AC inverters to decouple the effects of the inductance from the DC voltage source to the power bridge because of its low impedance. In a non-ideal or practical DC voltage supply, such

[1]Department of Materials Science and Engineering, Technion Israel Institute of Technology and Guangdong Technion Israel Institute of Technology, Shantou, Guangdong, China

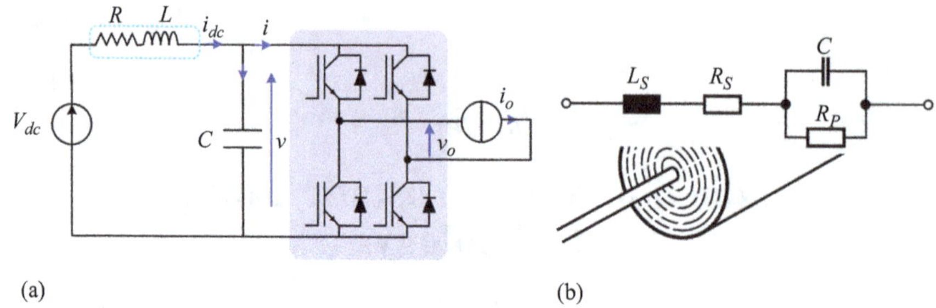

Figure 1.1 An inverter circuit design containing a DC-link capacitor and the equivalent circuit of the capacitor. (a) Basic circuit scheme of a single-phase full-bridge inverter to determine the DC-link voltage ripple [2]. (b) Equivalent circuit of the capacitor showing the basic nature

as a battery, photovoltaic module, fuel cell, or even the DC–DC converter or pulse-width modulation (PWM) rectifier, due to the presence of a series DC impedance, a voltage ripple appears on the DC-link if a switching current ripple circulates through the capacitor. And, a DC-link capacitor in the design can suppress the voltage ripple.

A capacitor responds differently to the applied electric voltage and frequency. In a DC circuit, a capacitor acts like an open circuit where the voltage impressed across it serves as an energy source for the circuit. In an AC circuit, a capacitor cyclically stores and releases energy at twice the frequency of the forcing function, since stored energy varies as the square of the voltage. The amount of "resistance" of a capacitor to AC is known as capacitive reactance and varies depending on the AC frequency. Capacitive reactance (*Xc*) and impedance (*Zc*) are given by this formula [1]:

$$Xc = -\frac{1}{2\pi fC} \tag{1.2}$$

$$Zc = -\frac{j}{2\pi fC} \tag{1.3}$$

Thus, the reactance is inversely proportional to the frequency. For DC, the capacitors completely block direct current. For high-frequency alternating currents, the reactance is small enough to be considered as zero in approximate analyses. At frequencies high enough, various unusual behaviors are exhibited to render the lumped circuit model inapplicable in favor of the distributed circuit model.

Real capacitors have imperfections within the capacitor's material that create resistance in addition to the *Xc*. This is specified as the equivalent series resistance (ESR) of the capacitor. Its equivalent circuit consists of resistance in series (R_s), and in parallel (R_p), inductance (L_s), and the ideal capacitance (C) as shown in Figure 1.1(b). Because of the ESR, the ripple current causes heat to be generated within the capacitor due to the dielectric losses caused by the changing field strength together with the current flow across the slightly resistive supply lines or the electrolyte in the capacitor. If capacitors have low ESRs, they have the capacity to deliver huge currents into short circuits, which can be dangerous. For board-level capacitors, this is done by placing a high-power 1 to 10 Ω resistor across the terminals.

Equivalent series inductance (ESL) is also important for signal capacitors. For any real-world capacitor, there is a frequency above DC at which it ceases to behave as a pure capacitance. This is called the (first) resonant frequency. This is also critically important with local supply decoupling for high-speed logic circuits. The local decoupler (a 0.1–1.0 μF capacitor placed near a power pin of a digital IC) must look like a capacitor at 10× the highest (usually, a clock) frequency associated with the logic chip, down to DC. This capacitor supplies transient current to the chip. Without decouplers, the IC demands current faster than the connection to the power supply can supply it, as parts of the circuit rapidly switch on and off.

1.1.2 Energy storage capacity

The choice of dielectric and the construction of the component will determine the capacitor either to store energy effectively in a wide temperature and frequency range or to leak energy adversely in terms of heat generation. The energy stored in a capacitor is equal to the work done to charge it up. Moving a small element of charge *dq* from one plate to the other against the potential difference requires the work (*W*). The energy stored in a capacitor is then [1],

$$W = \int_0^Q V(q)dq \tag{1.4}$$

In the simple case like linear dielectric between uniform electrode plates, a capacitor can store the energy in the following expression.

$$W = \frac{1}{2}CV^2 \tag{1.5}$$

The maximum voltage a capacitor can withstand is determined by the dielectric strength (E_d) of the dielectric material. Therefore, the capacitor's dielectric properties are critical important for its component performance. The dielectric also subjects to certain electrical leakage, resistance, and temperature dependence. Due to the intrinsic dielectric loss and resistance, the electrical energy lost in the dielectric and is converted into heat via the ESR. A typical ESR for most capacitors is between 0.0001 and 0.01 Ω, low values being preferred for high current, high power, or long-term integration applications to avoid power loss. One method to depict loss factor from dielectric and electrode separately is to measure the ESR over frequency of interest. Based on the ESR or loss factor at the frequency of use, power loss due to heat generation can be figured out.

$$ESR = R_D + R_\Omega \tag{1.6}$$

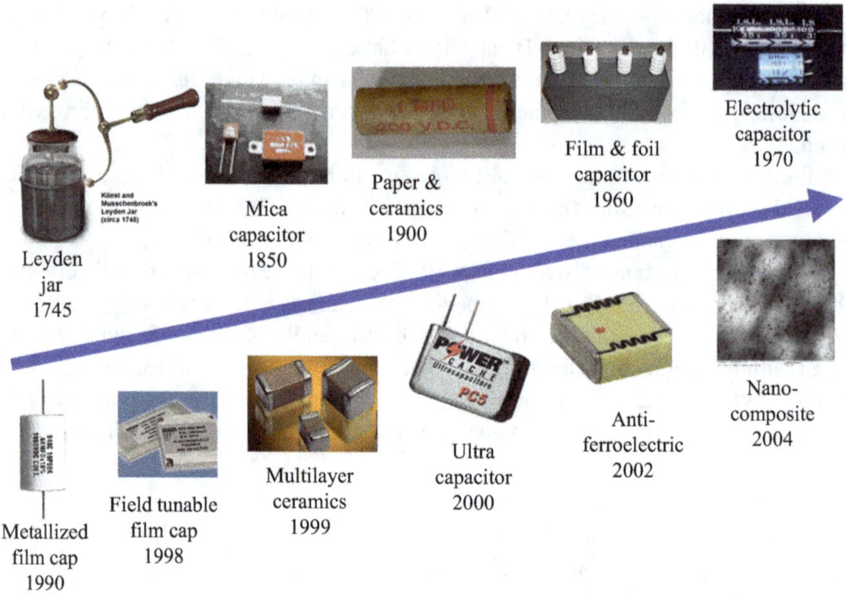

Figure 1.2 Historical evolution of capacitor technologies. © 2006 The Institute of Electrical Engineers of Japan. Reproduced, with permission, from [3]

$$R_D = \frac{\tan(\delta_D)}{C \cdot \omega} \tag{1.7}$$

where

$\tan(\delta_D)$ = Dissipation factor of the dielectric at frequency f [Hz]
ω = Frequency $[\text{c/s}^{-1}] = 2\pi f$
R_D = Dielectric losses resistance [Ω]
R_Ω = Electrode and capacitor plates losses resistance [Ω]

In order to maximize the charge that a capacitor can hold, the dielectric material needs to have as low a loss with frequency as possible. Leakage is equivalent to a resistor in parallel with the capacitor. Constant exposure to heat can cause dielectric breakdown and excessive leakage.

1.1.3 Capacitor technology classification

Capacitor technology has existed over two centuries and its evolutionary history is shown in Figure 1.2 [3,4]. Various types of capacitors besides film/foil and metallized film capacitors have rapidly advanced since 1950. Capacitors can be divided into electrostatic (high voltage), electrolytic (high capacitance), and electrochemical supercapacitor (high energy density). The electrostatic capacitor is the focal point of this book. The electrostatic capacitors can be further classified into three categories; polymeric films, ceramics (MLCC)/mica, and thin film as shown in Figure 1.3. They currently comprise billions of USD capacitor market annually

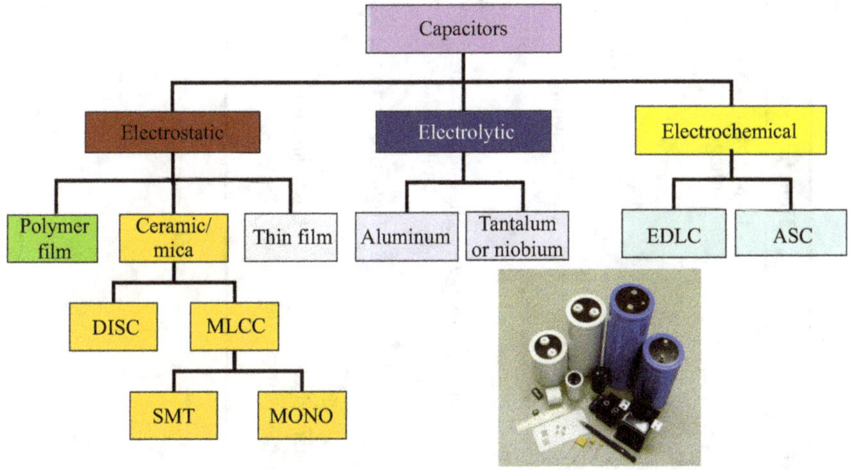

Figure 1.3 Classification of capacitor technologies and their relation with dielectric materials

worldwide [5]. In comparison with electrolytic capacitors, electrostatic capacitors are less lossy and withstand various voltages dominating the capacitor markets. Ceramic type can operate at higher temperatures and frequencies, but suffer from non-graceful aging failure and require lower operating electric field. Polymer capacitors have greater advantages for their higher power storage, lower voltage ripple, and greater reliability, but are limited to 105 °C.

High-temperature capacitor is an emerging technology designed to meet various high-temperature applications. Table 1.1 summarizes their characteristics and main issues. Polymer film capacitors have played an important role in utility, industry, transportation, etc. In the late 1980s and early 1990s, a significant effort was made to improve the energy density of film capacitors for military application where the pulse width is typically in the millisecond ranges [6–8]. The state-of-the-art metallized film capacitors can offer an energy density of 2–4 J/cm^3 using biaxially-oriented polypropylene (BOPP) dielectric film. In order to further increase the energy density of a capacitor, nanodielectric materials and engineering strategy occurring around year 2000 brings a hope to a new era of capacitor technology and electrical insulation [9,10].

On the other hand, thin film approach of ceramic dielectrics was conducted to achieve higher temperature and energy density with a support of proper substrate. Diamond-like carbon film, aluminum nitride film, and SiO_2 film of sub-micron films received early attention, which were deposited on metal foil or metallized polymer film carrier using vapor-phase deposition techniques (e.g., ion beam, sputtering) [11]. The thin film dielectric devices can work at high temperatures, yet they are limited by low capacitance and processing difficulties due to the slow deposition rate and the production of large-scale pin-hole free films. Metal-organic chemical solution deposition of barium titanate thin film has been studied for more than 10 years mainly for field tunable filter and memory applications. The

Table 1.1 Existing capacitor technology

Capacitor type	Advantage	Disadvantage	Cryogenics	High heat
Polymer capacitor (linear dielectric)	High voltages, low ESR, graceful failure, temperature stability	High volumetric efficiency, limited temperature range, high dielectric loss for high-K polymer films	Brittle, reliability issue	Lower breakdown voltage
X7R capacitor (nonlinear dielectric)	Good volumetric efficiency, very low ESR, ESL	Poor temp./voltage stability, mechanical crack/thermal shock	Reduced ESR, decreased efficiency	Capacitance drift, lower ripple current
NPO/COG (linear dielectric)	Excellent temperature stability and ripple current, very low loss, high reliability	Low volumetric efficiency, low energy density	Reduced ESR, temperature stability	Low capacitance
Antiferroelectric (nonlinear dielectric)	High energy density and volumetric efficiencies, very low ESR	Lack of understanding of aging/failure mechanisms, cyclic crack	Reduced energy storage	High-temperature stability
Thin film (linear or nonlinear dielectric)	High breakdown strength, small case, temperature stability	Low voltage, reduced energy density due to substrate, scalability	Lower capacitance	Higher dielectric loss
Electrolytic (linear dielectric)	High capacitance, medium-high voltages, high ripple current	High ESR, frequency dependent	Substantial electrolyte degradation	Lossy and lower voltage

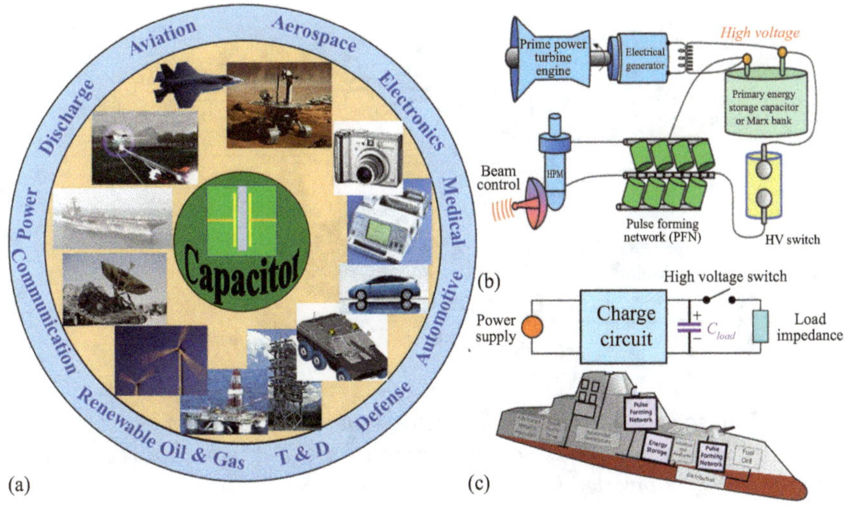

Figure 1.4 Various application of capacitor components. (a) Capacitor needs in various applications. (b) Schematic of a pulsed electric power system consisting of primary capacitor bank and pulse forming network. © 2019 Wiley-VCH Verlag GmbH & Co. KGaA, Weinheim. Reproduced, with permission, from [13]. (c) Power capacitor roles for future war ship

development of the inorganic-based capacitor level component remains to be challenged by manufacturing yield.

1.2 Perceived needs for advanced capacitors

1.2.1 Emerging capacitor applications

Next generation military applications are moving toward more electrical system, pulsed power weapons, and platforms. Current electrical system size and weight are limited primarily by passive components rather than the active components; this is especially true in pulse power applications where the capacitor comprises up to 50% of the system volume. To address the increasing demand, the US Department of Defense (DOD) initiated several programs such as the Combat Hybrid Power System AFRL Film Capacitor programs and the MURI program to develop a full-scale hybrid electric power system for advanced combat vehicles [12]. The advance in high-temperature and high-frequency SiC switches demanded a corresponding capacitor technology with high-temperature and high-frequency performance. Fulfilling these requirements has been identified as one of the most difficult

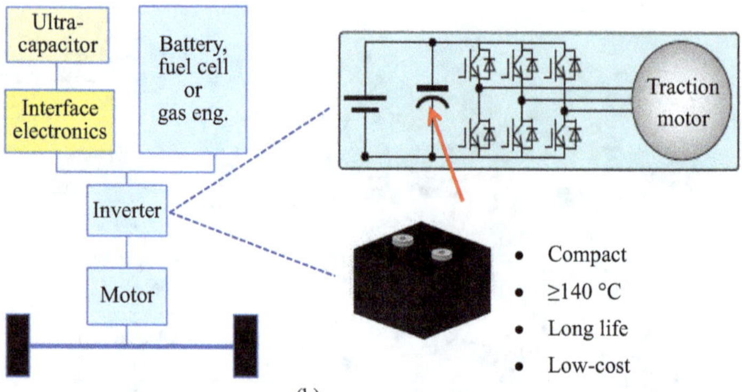

(a) (b)

Figure 1.5 Capacitor role in hybrid electric vehicle. (a) Model of electric vehicle driven by various power sources. (b) DC-link capacitor requirements for an inverter [14]

technological barriers. High voltage, high energy, and fast discharge capacitors are important elements in the high power microwave (HPM) and laser pulse forming network (Figure 1.4) [12,13]. The HPM requires regulated high voltage, typically >10 kV, and lasers require current regulation at a load voltage <1 kV. Therefore, high energy density and smart capacitors are essential to developing a more compact pulse-forming network (PFN) for a pulsed HPM, railgun, and laser source.

The all-electric warship is a major component of the Navy's vision for advancing its future combat capability, including the lethality and survivability of the surface Navy. A reduction in capacitor size would, in turn, allow for more compact primary energy storage systems. The US Army also has great interest in the energy density and high temperature capability of capacitors for the serviceability and power of ground vehicles, and lower mass of the equipment ground troops carry. The US Department of Energy (DOE) funded several programs developing cost-effective, high-performance capacitors that meet hybrid electric vehicle application [14]. The advanced capacitors are required to replace polypropylene film capacitors that are the key component in the inverter of hybrid electric vehicles as shown in Figure 1.5(a) and (b). They should be able to operate at or above 140 °C and to meet compact design requirements because today's capacitors suffer from either low temperature, low capacitance, or low voltage ratings and low volumetric energy density.

Advanced capacitor packaging technology is required to fully utilize the dielectric properties and to tolerate pulsed current with reduced heat generation. Current intelligent film capacitors either use fuses, thyristors, or ripple voltage diagnosis; however, they are slow-switching devices. Further optimization is needed for impregnation to improve voltage stand-off and to reduce the free margin width. Under a very large current pulse such as 12.5 kA and a rating voltage of 1.6 kV, poor end metal termination could cause a significant temperature rise.

Research is required to design metal end termination that enables the entire end spray surface to be used as a current carrier and capacitor seal.

In addition, thick metallization and heavy edge metallization on films are required to handle high current pulse power operation. Another important consideration for the packaging of the capacitor is low inductance. A careful consideration of volumetric efficiency and a strategy to keep the inductance low should be constructed.

Reliable, high power systems also require integration of "intelligence" into high energy density capacitors. This is because of the following reasons:

1. Availability of mission critical pulsed weapons, electrical propulsion and distribution systems hinges on reliable energy storage.
2. Without effective ways to predict, detect, and isolate faults internal or external to the capacitor, catastrophic failures, plasma events, and propagated faults will damage the system due to extremely high stored energy.
3. Early prediction, detection, and isolation of faults will allow current-carrying structures, such as fuses, bus work, interconnects, and thermal management, to be size optimized and not be excessively derated to handle extreme stresses during fault conditions.

The integrated "Smart" capacitor assembly with these functions may lead to system energy density improvement if capacitor assemblies can be functionally integrated with the power switches in the pulsed power or power conditioning circuits.

High-temperature capacitors are directly applicable for hybrid electric vehicle, oil and gas drilling, and aircraft power electronics as well. They are fundamental building blocks providing electrical charge build-up barriers and electrical current handling capability. Many modules in real-time measurements also require electronics integrated with a high-voltage power supply, amplifier and discriminator, and high-power batteries that can provide the necessary energy to operate the many devices. Moreover, capacitors as the critical ubiquitous passive components are used for power filters, supply pumping and bypass, AC coupling, analog filters, and timing (Figure 1.6). Many other applications exist in additional energy markets such as geothermal, automotive, nuclear power, power turbines, and the military.

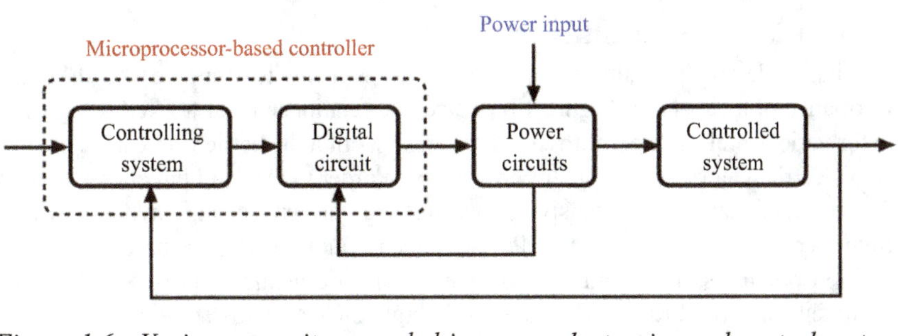

Figure 1.6 Various capacitors needed in power electronics and control systems

Old technical concept is now called to exhibit a new life and function to meet various advanced applications. Future electronic and electrical power systems require the high performance and smart capacitors to show the following capabilities:

1. High capacitance range (μF–mF)
2. Capacitor breakdown strength of >200 V/μm
3. Capacitor energy density of 5–20 J/cm^3 (J/cc)
4. Working temperature of 100–150, 150–250, >300 °C
5. Long service time without failure or pulsed discharge cycle of >1,000 cycles
6. External control electronics for capacitors with activation, deactivation, and discharge functions
7. Large-scale cost-effective processing of capacitor grade films (< 3 μm) with manufacturing reliability.

1.2.2 Dielectric material requirements

The desirable advanced capacitors require advanced dielectric materials that possess such properties as high dielectric constant, high dielectric strength, high resistivity/low-leakage current, high-temperature stability, and reliable electrical and mechanical characteristics under temperature cycling. Minimizing the dielectric losses, and thus reducing heat, can promote greater stability, lower distortion, and lower mean-time-before-failure (MTBF) of a capacitor. In comparison with organic polymers, inorganic dielectrics possess the advantage of high dielectric permittivity that leads to high capacitance and miniaturization of multilayer ceramic capacitor (MLCC) construction. Organic dielectrics, on the other hand, offer a self-healing feature, contributed a grace failure mechanism that enabled reliable operation and higher capacitance stability. Ideally, the improved energy density is achievable by increasing both dielectric permittivity and the dielectric strengths of dielectric or its composites. In fact, the dielectric material having high dielectric permittivity generally possess a narrower energy band gap and thus a lower dielectric strength (Figure 1.7) [13,15]. Therefore, the advanced dielectric material should be carefully inspected and wisely designed to render a synergistic dielectric permittivity and strength, low dielectric loss, high temperature rating, thinner film, and higher energy density.

1.2.2.1 Dielectric permittivity

Increasing dielectric permittivity is an obvious pathway to improve the dielectric performance of a capacitor. Figure 1.8 depicts the relationship for a fixed design voltage for various dielectric permittivity. The thickness of a dielectric film can be smaller for a dielectric material with higher permittivity in order to reach a high energy density [4]. Researchers have been extensively investigating various dielectric materials with a permittivity higher than 10 such as Polyvinylidene fluoride (PVDF)-based polymers, nanofilled polymers, and nonlinear inorganic dielectric ceramics [16–18]. However, this strategy always brings up issues with high dielectric loss, dielectric strength reduction due to filler addition and limited permittivity increase.

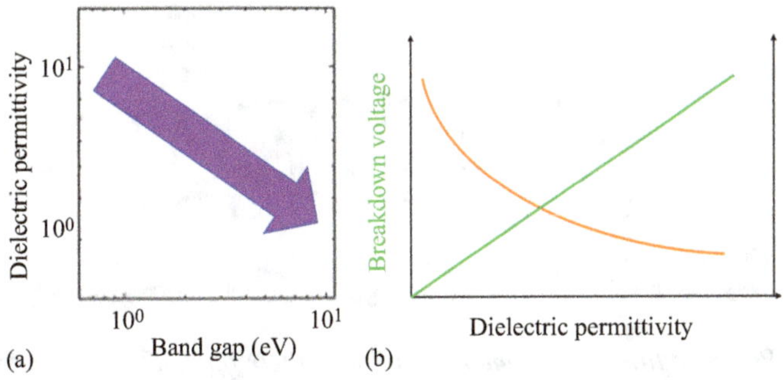

Figure 1.7 Relationship between dielectric permittivity, breakdown strength, and energy band gap. (a) Dependence of dielectric permittivity on energy band gap. © 2019 Wiley-VCH Verlag GmbH & Co. KGaA, Weinheim. Reproduced, with permission, from [13]. (b) Breakdown strength and capacitance of dielectric films as a function of dielectric permittivity

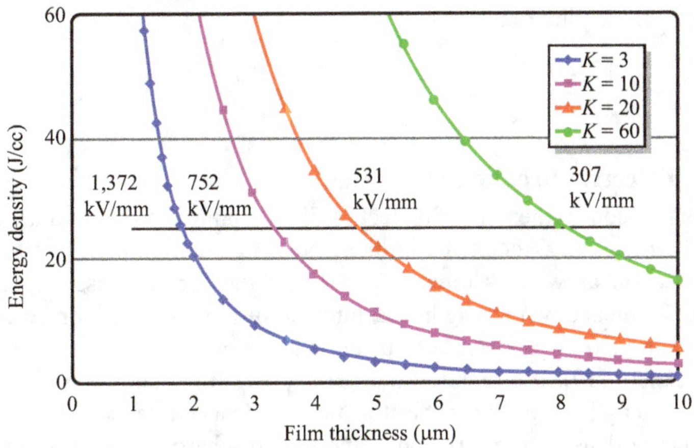

Figure 1.8 Theoretical design rule for dielectric films illustrating the relationship of energy density, film thickness, dielectric constant, and breakdown strength to meet the requirement for energy density of 25 J/cm³ [4]

1.2.2.2 Dielectric thickness

Capacitor volume, capacitance density, and energy density are all dielectric thickness dependent. Equation (1.8) shows the relationship of volumetric energy density with the dielectric thickness (d). It has been a trend to develop thinner gauge dielectric materials or films over the last two decades as shown in various dielectric polymer films in Figure 1.9(a). The effort to create thin films, however, is not

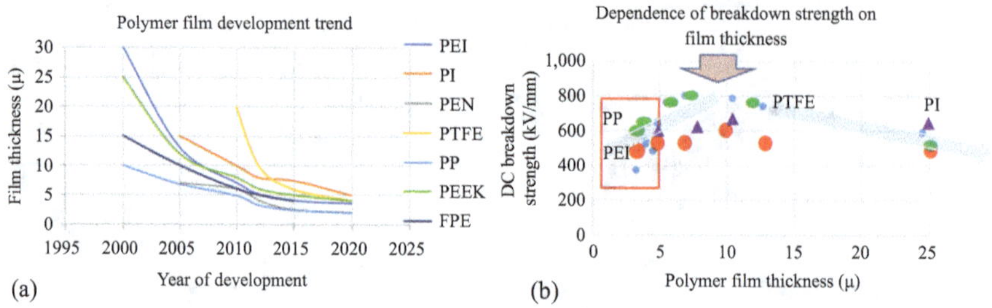

Figure 1.9 Polymer film development trend and their thickness dependence. (a) Thickness reduction trend for various polymers. (b) Thickness dependence of breakdown strength of thin films (metallized film electrodes for polypropylene (PP), Polytetrafluoroethylene (PTFE) and ball-plan electrodes for polyetherimide (PEI) and polyimide (PI)) [19]

always suitable for high-performance capacitor technology because of various material and processing requirements.

$$\frac{W}{Volume} = \frac{1}{2}\frac{\varepsilon_0 \varepsilon_r V^2}{d^2} \tag{1.8}$$

1.2.2.3 Dielectric breakdown strength

One big concern down the path of dielectric film thinning is the loss of its dielectric strength. This may be associated with the increasing contribution of surface effect and imperfection as well as other defects from the film process. Figure 1.9(b) shows the DC breakdown strength as a function of the thickness for several polymers. Dielectric strength decrease turns out when a polymer film is thinner than a certain thickness [19]. It's required that a dielectric film should provide a higher breakdown strength so as to enable the future advanced capacitors to offer kV voltage, energy density of 5–30 J/cm^3, and <μs discharge time. Incorporation of appropriate fillers and improvement of formulation has been revealed to be a possible method [16,19]. A thorough understanding of the nanocomposite engineering, improvement of materials' properties, and quality film processing is critical for achievement of this type of capacitor with reliable service life.

1.2.2.4 Dielectric loss

Dielectric materials usually exhibit higher dielectric loss at higher temperatures. The intrinsic energy leakage subjects the dielectric materials to the self-generated heat when energizing a capacitor. This additional dielectric loss can be higher than the ambient temperature and lead to a degradation and premature failure. Heat generated within the capacitor can be calculated from any one of the following three formulas, the choice of which depends on available information about the capacitor. It is obvious that dielectric loss is mainly responsible for the power loss in a capacitor.

$$P_{loss} = 2 \cdot \pi \cdot f \cdot C \cdot V_{RMS}^2 \cdot \tan(\delta) \tag{1.9}$$

$$P_{loss} = \frac{I_{RMS}^2 \cdot \tan(\delta)}{2 \cdot \pi \cdot f \cdot C} \tag{1.10}$$

$$P_{loss} = I_{RMS}^2 \cdot ESR \tag{1.11}$$

where V_{RMS} is room mean square (RMS) voltage across the capacitor and I_{RMS} is RMS current through the capacitor.

In the case that the current or voltage waveform is sinusoidal, the above equations apply directly. For non-sinusoidal waveforms such as which would be present in the pulsed power circuit, one must first analyze the harmonic content of the waveform to determine V_{RMS} and I_{RMS} for each harmonic. The temperature rise in the core of the capacitor is the sum of the ambient temperature, the temperature rise from the outer skin to the inner core, and the temperature rise of the inner core. The rise in temperature in the core of the capacitor causes an increase in losses and can lead to premature failure. The thermal characteristics of the capacitor include operating ambient temperature, self-generated heat, and cooling abilities of the package, which makes up the capacitor, play an important role. Often, the self-generated heat can exceed 50 °C and overshadow the ambient temperature. The mechanisms that are responsible for heat generation and the flow of heat from the capacitor include the following sources and should be taken care of:

- The dielectric material and its inherent power losses and heat-retention properties
- Electrode ohmic losses (I^2R, where the resistivity can vary from 1 Ω/sq to 30 Ω/sq)
- Interfacial spraying (Schooping, e.g. arc spray or flame spray) materials
- Termination method (wire, tabs, or terminals).

1.2.2.5 High temperature requirement

Increasing the temperature rating and energy density of a capacitor has been identified as one of the most difficult technological barriers [19]. Capable capacitors are greatly desirable to operate in such environments as pulse power systems, deep oil & gas exploration, hybrid vehicles, commercial pulsed forming networks, military vehicles, and electric ships. Figure 1.10(a) shows the temperature rating for different kinds of capacitors. Ceramic- and polymer film-based dielectrics may present high-temperature candidate materials. A number of candidate polymers are currently assessed and improved accordingly as shown in Figure 1.10(b). Some of them are being converted into capacitors in response to some customers' calls. Some of them are believed to be adopted by customers and capacitor manufacturers in no longer time.

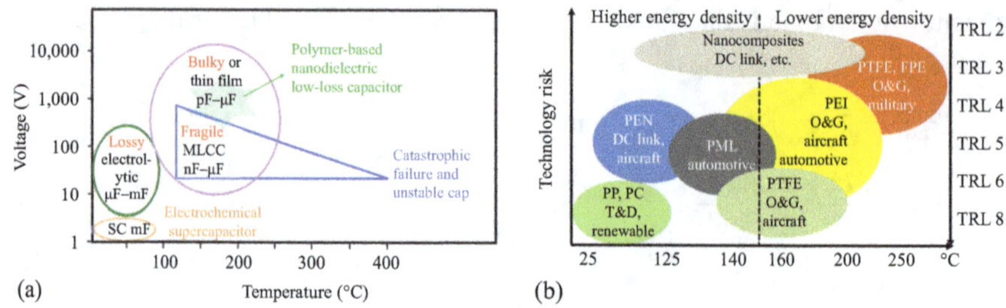

Figure 1.10 *Capacitor and dielectric properties as a function of temperature. (a) Temperature and voltage ranges for various capacitors. © 2019 Wiley-VCH Verlag GmbH & Co. KGaA, Weinheim. Reproduced, with permission, from [17]. (b) Polymer dielectric candidates for various temperature application*

1.2.2.6 Dielectric film scalability

Capacitors are basically constructed in two forms: lamination and winding of dielectric layers. This construction poses additional requirements on mechanical strength and processability. In consideration of various requirements of advanced capacitor, a desirable dielectric film of the following property metrics is postulated according to the author's practical experiences working with capacitor manufacturers.

- Tg or operation temperature >125 °C
- Dielectric strength Eb: >350 kV/mm
- Dielectric constant: >2
- Loss factor: <0.005
- Thickness: <10 μm thick (K relevant)
- Mechanical strength: > 5,000 psi
- Film processing: extrusion or solution
- Film flexibility: bendable >θ
- Self-healing: capable
- Moisture absorption: <1%

1.3 Recent advancement in dielectric researches

As a figure-of-merit, energy storage density and energy storage efficiency are concerned targets with dielectric material development. The energy density (W_{rec}) and the efficiency (η) of dielectric material is governed by the applied electric field (E) and the polarization (P), which can be expressed below [1].

$$W_{rec} = \int_{P_r}^{P_m} dP \tag{1.12}$$

$$\eta = \frac{W_{rec}}{W_{rec} + W_{loss}} \tag{1.13}$$

These equations are often used for nonlinear ferroelectric or relaxor-ferroelectric or antiferroelectric in addition to linear dielectrics (Eq. 1.9). Electrical charging and discharging capacity of a dielectric material can be correctly assessed based on these equations. Among various investigations toward advanced dielectric materials, researches have been emphasized on polymer dielectric films, ceramic dielectric layers, and thin film dielectrics so far.

1.3.1 Polymer dielectric films

In order to produce high-performance, favorable polymeric dielectrics, scientists worldwide have conducted an assessment of various commercial polymer films, syntheses of new polymer dielectrics, and engineering of nanodielectric composites. Figure 1.11(a) outlines the historical effort conducted by dielectric and insulation community in the last two decades. The central idea is to wisely leverage linear and nonlinear dielectric materials to increase dielectric breakdown strength (DBV) and dielectric constant (*K*) [20,21]. Great obstacle and difficulty were encountered in enhancing DBV and *K* through the nanodielectric strategy in the first decade (before 2010). Significant progresses were accomplished in the second decade (after 2010), however, limited to polymer matrices exhibiting high dielectric loss. Scientists and engineers from General Atomics, Lithium Power, and PennState pioneered the development of nonlinear PVDF film capacitors claiming an energy density in the range of 4.3–12 J/cm^3 [8,22,23]. However, the ferroelectric property of PVDF results in significant dielectric loss, high dielectric absorption under high voltage, with a slow release over time. These capacitors hardly find nowhere to survive for high-performance energy storage applications.

As a result, various polymers and nanodielectric composites are investigated as summarized in Table 1.2. Films such as PTFE, PEI, PEN, FPE, PPS, and PI are very attractive for their high temperature rating, high dielectric strength, low dielectric loss, and commercial maturity [3,13,18,20]. However, some of them are lack of self-clearing capability and expensive. A second type of dielectrics under development hold promise for greater dielectric properties and high temperature

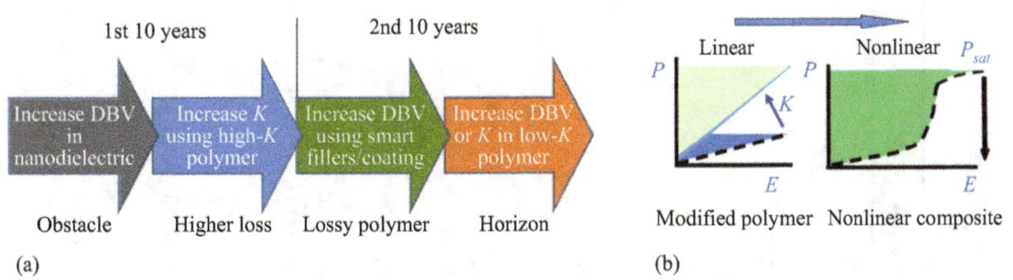

Figure 1.11 Dielectric polymer development focus in two decades. (a) General description of research focus in dielectric researches. (b) Schematic of relationship between polarization and electric field for linear and nonlinear dielectric materials. © 2019 Wiley-VCH Verlag GmbH & Co. KGaA, Weinheim. Reproduced, with permission, from [13]

Table 1.2 High-temperature and high-performance polymer films (cf. [19])

Type	Polymer films	Advantage	Challenges
I – Commercial film	PEI	Overall property and film scalability (0.1% shrinkage)	High cost (>\$200/kg) and lower dielectric strength comparing with BOPP (800kV/mm)
	PEN	High dielectric strength and thinner gauge	Dielectric loss (>0.001), Cost (>\$200/kg)
	PEEK	Scalability in manufacturing of pure film and inorganic filled composites (6 μm)	Dimensional change and lower dielectric strength, and dielectric loss (>0.001)
	PPS	Low dielectric loss (0.0003) and thin gauge (<1 μm)	Lack of self-clearing and higher dielectric loss (0.002)
	PI	Scalability in manufacturing and high temperature (300 °C)	
	PTFE	Low dielectric loss, high dielectric strength, and high temperature	High cost, lower mechanical strength, and low dielectric constant
	FPE	High temperature (275 °C)	Solution coating process and high cost
II – Lab research dielectrics	Polyurea	High dielectric strength (1000kV/mm), High dielectric constant (4.5)	
	Polythiourea	High dielectric strength (800kV/mm), High temperature (180 °C)	High dielectric loss (0.005), Film process scalability
	Imide based BTDA-HAD or BTDA-HK511	Medium high dielectric constant (7.8), high dielectric strength (>670kV/mm)	Higher loss (0.007), film process scalability High dielectric loss (0.005), cost, and film process scalability
	CN-PEI	High temperature (220 °C), high dielectric constant (4.7)	Film process scalability
	Organometallic polymer	High dielectric constant (10-30)	Higher dielectric loss (>0.01), low breakdown strength (<400kV/mm), film scalability
	Modified PP	Higher dielectric constant (3.3-4.5) and high dielectric strength (>600kV/mm)	Film process scalability and induced polarization loss
	Acrylate PML	High dielectric strength (1000kV/mm)	Too thin, no stand-alone film

(Continues)

	PHONDI	Medium dielectric constant (3.2) and temperature (150 °C)	High dielectric loss (0.007) and medium dielectric strength (350kV/mm)
III – Nano-composite	PEEK	Increased dielectric constant and proven composite films	Lower dielectric strength (<300kV/mm)
	PVDF or PVDF copolymer	Higher dielectric constant (10-80) and mechanical strength	High dielectric loss (0.02), low temperature (125 °C), film process scalability
	FPE	High temperature (275°C), and dielectric constant after adding fillers	Low dielectric and mechanical strength, film process scalability
	PEI	Higher dielectric constant after adding low filler concentration and high dielectric strength	Lower mechanical strength containing fillers above 10wt%, film process scalability
	PEEU (poly(arylene ether urea)	Higher dielectric constant (>7), high dielectric strength	High dielectric loss (0.01)
	PI	Higher dielectric constant (>5) after adding fillers	Low dielectric and mechanical strength
	Silicone	Higher dielectric constant (>5)	Low mechanical strength and thick film
	c-BCB (cross-linked divinyl-tetramethyldi-siloxane-bis (benzo-cyclobutene)	Lower dielectric loss (0.001) and electrical conduction, and stable breakdown strength (400kV/mm) up to 250 °C	Low mechanical strength and film process scalability
	PMMA (poly(methyl methacrylate))	High dielectric constant (5-20)	High dielectric loss (0.02-0.1), low dielectric strength (20kV/mm), low temperature (105 °C)
	PE	Reduced conductivity and high voltage endurance	Film process scalability and low filler uniformity
	Cellulose acetate, cyanoethyl	High dielectric constant (>20), organic field-effect transistor (OFET) dielectric	High dielectric loss (>0.02) and low mechanical strength

ratings; however, they are facing the challenges in dielectric strength, dielectric loss, and high-quality film processing. Nanocomposite dielectric is the third type that leverages a high-performance polymer matrix to which proper nanofillers are incorporated through ingenuous processing methods. The challenges are basically associated with low dielectric strength, mechanical strength, and scale-up of composite films. Therefore, the dielectric polymers with good overall properties and processes capable of replacing BOPP film is very limited because of the requirements in dielectric loss, dielectric strength, harmful solvents, heat shrinkage, film-forming ability, etc.

Among these endeavors, fabricating nanocomposites and thinning polymer films received a great deal of attentions. Figure 1.12(a) shows the reasoning of synergizing inorganic (high-*K*) and organic materials (higher E_d); however, this combination has taken enormous efforts of the scientific community [24]. Specifically, researches were directed to add nanoparticles in polymers, engineering filler–polymer interfaces, and modifying film surfaces through single layer or multilayer construction.

1.3.1.1 Single layer vs multilayer polymer

Dielectric strength as an intrinsic material property is sensitive to its surface condition, especially in its thinner gauge. Minimizing the surface defects and contamination and their adverse contributions becomes more important for polymer films thinner than 10 μm. A conventional approach is to deposit a layer of organic material on the surface of a polymer film to reduce the nonuniformity of the film surface and the adverse effects of surface defects [25]. The organic coating concept was extended to multilayer lamination using two different polymers [26].

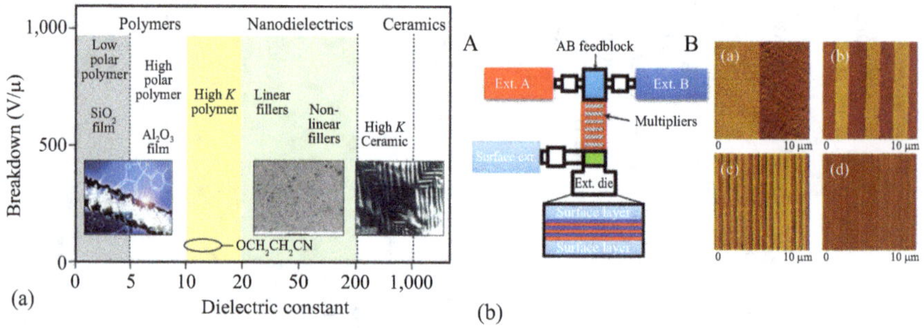

Figure 1.12 Nanodielectric composites through polymer–filler and polymer–polymer engineering. (a) Synergy between polymer and inorganic dielectric fillers. (b) Schematic representation of the multilayer polymer film coextrusion process (A) and AFM phase images of PC/PVDF 50/50 (vol/vol) multilayer films (B): (a) 2-layer, (b) 8-layer, (c) 32-layer, and (d) 256-layer. The total film thicknesses are around 12 μm. © 2017 American Chemical Society. Reproduced, with permission, from [26]

Figure 1.12(b) shows a schematic principle and co-extruded multilayer film images. The 1D system, with tailored material choices and interfaces, is believed to minimize the disadvantageous polarizations and maximize the dielectric strength for polymer films. As such, one can have more design space by incorporating a high-*K* polymer and a low-*K* polymer with synergistic properties.

An inorganic oxide coating effect on dielectric strength of single polymer film was started 10 years ago [27]. Different deposition techniques (sputtering, reactive sputtering, plasma-enhanced chemical vapor deposition (PECVD), e-beam, and atomic layer deposition (ALD)) were acceptable for the surface modification and dielectric strength increase were achieved [28]. Recently, BN coating on polymers like PEI, BOPP, FPE, PI, and c-BCB resulted in higher dielectric strength for a given charge–discharge efficiency at above 100 °C. This further demonstrated the apparent enhancement of dielectric strength at higher temperature when coating with higher thermal conductivity *h*-BN [29].

1.3.1.2 Core-shell filler vs junction effect

Among various fillers, ceramic (SiO_2, Al_2O_3, TiO_2, $BaTiO_3$, lead zirconate titanate (PZT), lead mangnesium niobate-lead titanate (PMN-PT), lead lanthanum zirconate titanate (PLZT), MgO, BN), semiconductors (SiC, ZnO, ZnS), and conductors (Al, Ag, graphene, carbon nanotubes (CNTs)) [3] are mostly used. It was observed that low-dimensional fillers such as 1D and 2D morphology are preferred to maintain a lower loading fraction in a polymer matrix. In any cases, the distribution and dispersion of fillers themselves have been emphasized. In recent years, researchers paid more attention to the role and ingenious design of filler–polymer interfaces. The interfacial phase can occupy >50% by its volume ratio and thus can have a significant impact on the overall performance of nanocomposites and their devices [30,31]. Computer simulation points out the necessity of ingenious design of core-shell configuration as shown in Figure 1.13 [32]. Various experimental designs were executed leading to positive and the new wave of intensive dielectric composite investigation [32–36]. The role of the local polarization at the interface was directly detected using the modified Kelvin probe force microscope (KPFM) [37]. As a result, the properties of the interface region must be controlled to achieve desirable capabilities of nanodielectric composites via core-shell or semiconductor junction configuration. Details about the interfacial contribution can be seen in several articles and review papers [38–41].

Besides inorganic shell coating, the organic shell strategy was also taken in the recent years [42]. Table 1.3 shows the combination of inorganic core and organic shell. Organic dielectric shell was introduced onto inorganic fillers by chemical treatment, grafting shell layers, using multiple hierarchical shells on the surface of the fillers, or multi-layered structures as reported [43,44]. Currently, the roles of interfaces reside in their function as charge transport barrier and charge traps. In the core-shell case, the interface comprises an inorganic shell or surface coating, which raises the energy barrier height and blocks the charge injection and movement. Prior experiments on *h*-BN-coated PEI films also showed increased electrical resistivity and energy barrier charges demonstrating more difficulty for the

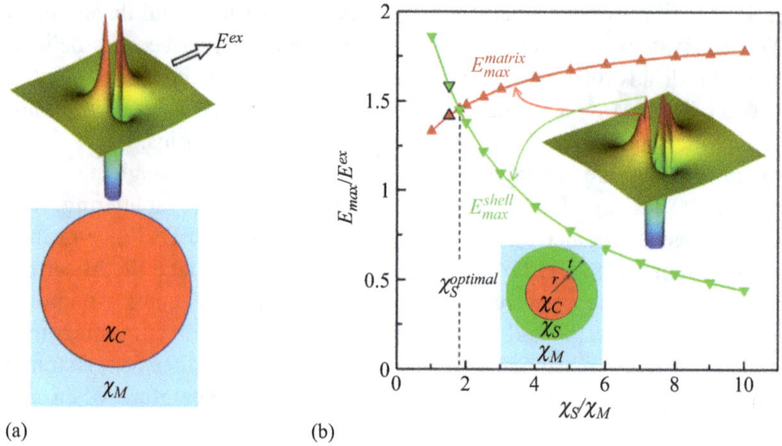

Figure 1.13 (a) Filler particle without core-shell structure embedded in matrix with χ_M:χ_C = 1:100. Without shell, the field is more concentrated at the interface (more spiky), (b) Effect of shell dielectric susceptibility on local electric field concentration in core-shell filled composite with χ_M:χ_C = 1:100 and t/r = 2:3. Maximum local electric field magnitudes concentrated in the matrix (red) and shell (green) as a function of χ_S. Inset corresponds to data point χ_M:χ_C = 1.5:1 as highlighted by black triangles. © 2011 AIP Publishing. Reproduced, with permission, from [32]

electrons to be injected into the samples. The suppression and delay of charge transport lead to a higher energized voltage.

Recently, the semiconductor ZnO surface structure was further engineered to form a non-conductive junction, and resulted in substantially enhanced dielectric breakdown strength and energy density. It was conjectured that charge carriers were captured and confined in the composite through the potential barrier formed at the heterojunctions of ZnO–ZnS nanoparticles. The space charge accumulation at the filler–matrix interface is relieved and the concentration of a local electric field is weakened [45]. When the interface dielectrics are very thin (nm), Fowler–Nordheim tunneling may be considered, and higher breakdown strength and low dielectric loss may be expected. When dielectric films are exposed to high electric fields and temperatures, a hopping mechanism may play a role by releasing the trapped charges and causing higher electric conductivity and failure [46]. Nevertheless, the dielectric breakdown mechanism remains complicated and unclear with respect to nanocomposite engineering and should be an interesting topic of future investigation.

1.3.2 Ceramic dielectric layers

Ceramic capacitors are known for their high temperature rating, small case size, low inductance (ESL), and high frequency of operation. Investigation of ceramic

Table 1.3 Summary of the range of modifiers used for filler surface modification in dielectric nanocomposites [19,42]

Core filler	Dielectric constant	Organic modifier/shell (lower-*K*)	Polymer matrix	Coating method	Inorganic shell
Particle	High-*K* $BaTiO_3$	Phosphonic acid, Carboxylic acids, Poly-vinylprrolidone, Dopamine/Polydopa-mine, Hydantoin epoxy, Hyperbranched polyamide, Poly(trifluoroethyl acrylate)	PVDF-base	Surface absorption, Grafting, Polycondensation for organic shell.	TiO_2, Al_2O_3, SiO_2, Fe_3O_4, C, Ag
	High-*K* $BaTiO_3$	PHEMA@PMMA, HBP@PMMA	PMMA-base	Sol–gel, hydro-thermal, hydrolysis reaction, coaxial electro-spinning, CVD for inorganic shell.	
	High-*K* $BaTiO_3$	Gallic acid	PVA		
	High-*K* $BaTiO_3$	Polystyrene	Polystyrene		
	High-*K* $BaTiO_3$	Amino-terminated silane molecules	Glycidyl methacrylate		
	Low-*K* Al_2O_3	Phosphonic acid-terminated poly(ethylene-co-1-butene)	Polypropylene		
	Low-*K* SiO_2	Fluoride 1H,1H,2H,2Hper-fluorooctyl-triethoxy-silane	P(VDF–HFP)		C
Fiber	High-*K* $BaTiO_3$	Fluoro-polydopamine, PVP	PVDF base	Surface absorption, Grafting, Polycon-densation	TiO_2, Al_2O_3, SiO_2
	High-*K* $BaSrTiO_3$	H_2O_2, Ethylenediamine	PVDF base		
	High-*K* $NaNbO_3$	Polydopamine	PVDF base		
Platelet	Graphene	Hyperbranched aromatic polyamide, Polydopamine	PVDF-base, PBO	Grafting	
	Graphene oxide	PVA, p-Phenylenediamine	Nitrile butadiene rubber, Polyimide, PVDF-base		
	Boron nitride	g-Aminopropyl triethoxysilane	PVDF		

PBO: Poly(*p*-phenylene benzobisoxazole)
PMMA: Poly(methyl methacrylate)
PVA: Poly(vinyl alcohol)
PVP: Polyvinylprrolidone
PVDF-base: PVDF, P(VDF–HFP), Poly(vinylidene fluoride-trifluoroethylene-chlorofluoroethylene), Poly(vinylidene fluoride-trifluoroethylene-chlorotrifluoroethylene)

dielectric materials was directed to linear dielectric, ferroelectric, antiferroelectric, lead-based, and lead-free dielectric ceramics as described in the following section.

1.3.2.1 High voltage disc capacitors

For very low capacitance values and high voltage, one can use a single ceramic disc such as $SrTiO_3$, TiO_2, and $BaTiO_3$-$CaZrO_3$ with both sides coated with electrodes [47]. Increasing the dielectric strength is one of the challenges because the presence of grain boundaries, porosity, impurities, surface defects, and chemical deterioration are mainly responsible for the low-field failure. For high capacitance, a high dielectric constant ceramic has to be utilized. The most wanted ceramic capacitors with higher capacitance remain to be MLCCs because they can be made small size and high capacitance.

1.3.2.2 Miniaturized MLCC

MLCC showing higher capacitance, thinner layers, and smaller case sizes are replacing aluminum film capacitors, promising a higher market occupation. Its growth trend continues with increasing market needs for higher capacitance, higher voltage, and smaller size [48]. Murata successfully developed smaller and more uniform ceramic grains and advanced thin-film techniques allowing the development of more versatile and stable capacitors as well as higher capacitance density in both high- and low-*K* dielectrics. For example, the small packaged MLCC with values up to 1 μF in a tiny EIS0402 case size (1.0 × 0.5 mm), and 100 μF in a EIA1210 package became available [49]. MLCCs can withstand a temperature of 270 °C for up to 60 s in a reflow environment. Ceramic dielectric materials are processed at much higher temperatures during their manufacture and assembly into monolithic components. Therefore, the fundamental materials are not affected at soldering temperatures, even when using higher temperature lead free solder. Today, the world's smallest MLCC EIA008004 (0.25 × 0.125 mm), largest capacitance (1,000 μF), and high-voltage MLCC have been manufactured by Taiyo Yuden (Figure 1.14) [50]. One of the key enablers is the development of ultrafine ceramic and nickel powders that are used for internal electrodes in MLCC

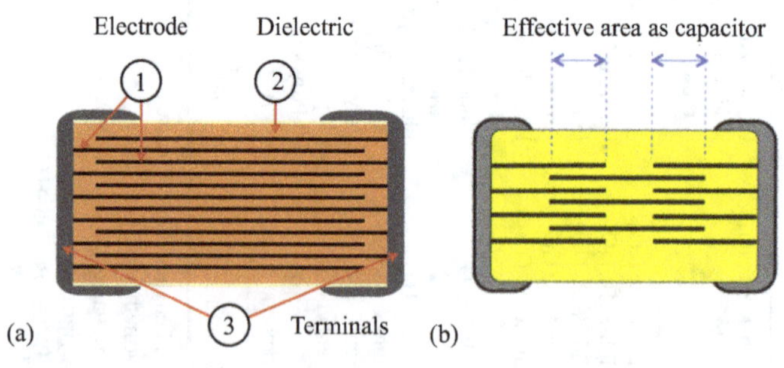

Figure 1.14 Schematic of MLCC construction involving inner electrode and termination [50]. (a) Standard MLCC capacitor. (b) High voltage MLCC capacitor

construction, as shown in Figure 1.14 [51]. Owing to the advantageous characteristics, the nickel ultrafine powder of JEF Mineral easily forms an electrically continuous and homogeneous thin film electrode after firing, and the drop-in capacitance of MLCC with the reduction of nickel weight per layer is small [51]. JFE Mineral has developed NFP101 having an average particle size of 0.1 μm by adjusting the CVD reaction conditions to realize the requirement of 0.5 μm thick dielectric layer.

1.3.2.3 High-temperature compositions

Morgan Electro Ceramics has developed a ceramic material based on Strontium Titanate dielectric that has high dielectric strength and low dissipation [52]. Further modifications of X7R ceramics attempted for a higher operation temperature (>210 °C), such as using relaxor ferroelectric compositions by TRS Ceramics and modified $BaTiO_3$ by Kemet and Presidio Components for >250 °C [53,54]. Eclipse NanoMed has developed surface mount MLCCs and leaded ceramic capacitor assemblies of 0.68 μF that are utilized in the high-temperature oil exploration and +300 °C geothermal markets, as well as military, aerospace, medical, and specialty telecom applications [55]. Multilayer fabrication of BN dielectrics for even >600 °C was attempted by MicroCoating Technologies [56]. Many researchers investigated lead-free relaxor ferroelectric ceramics. For example, the $0.6BaTiO_3$–$0.4Bi(Mg_{1/2}Ti_{1/2})O_3$ (0.6BT–0.4BMT) ceramic was shown to exhibit very high energy storage performance with a η of 93% and a W_{rec} of 4.49 J/cm^3 [57]. Two main types of perovskite ceramics $BaTiO_3$ and/or $Na_{0.5}Bi_{0.5}TiO_3$ have advantages of temperature-stable dielectric relaxor characteristics with consistency in permittivity of ±15% over wide ranges of temperature extending to 200–500 °C [58]. The representative lead-free relaxor ceramic systems with mixed cation site occupancy on the perovskite lattice that exhibit high dielectric are shown in Figure 1.15. One can see more perspective on recent researches

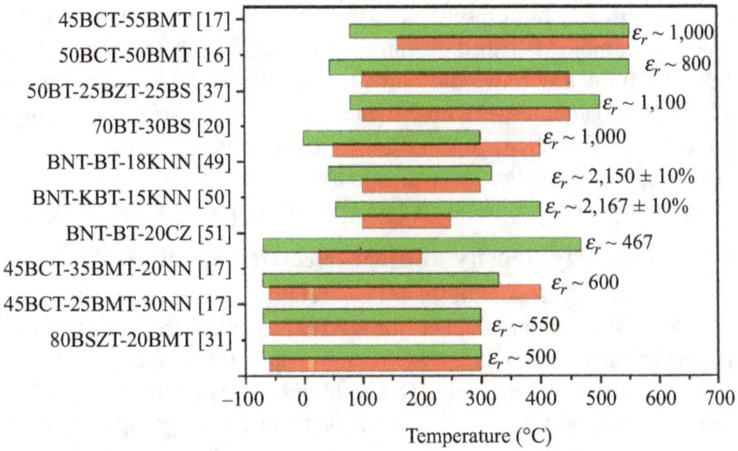

Figure 1.15 Bar plots representing temperature range of ±15% or better stability of ε_r (green bars): range of low tan $\delta < 0.02$ represented by red bars. © 2015 Springer Science Business Media New York. Reproduced, with permission, from [58]

into lead-free dielectric ceramics with high and stable relative permittivity well beyond the operating limit of traditional high-*K* dielectrics [58]. However, converting the advanced dielectric materials into high energy density capacitor products require more engineering efforts.

1.3.2.4 Nanoceramic oxide capacitors

Scientists at University of Missouri at Rolla, under the US DOD funding since 2008, systematically studied TiO_2 nanoceramic with high dielectric constant, dielectric strength, and high temperature. It was claimed that high dielectric strength was achieved as opposed to the normal micron sized TiO_2 [59], which promised a great energy density because of the high dielectric constant (~100). Presidio Components took over the ceramic component processing, demonstrating the feasibility of nano TiO_2 capacitors. However, the scale-up production was not announced till now.

1.3.2.5 High dielectric strength glass-based capacitors

Typically, glass ingredients are applicable for high-temperature energy storage applications due to their excellent thermal stability ($Tg > 550$ °C) and dielectric strength, particularly of alkali-free glasses [60]. It was reported that composites of glass ceramics containing PZT and nanodispersoids of silver shows high dielectric constant. However, the packaged capacitor can only offer an energy density of 3–4 J/cm^3 because of processing defects and packaging issues [61]. Defect-free antiferroelectric-to-ferroelectric phase-change materials containing 10%–40% glass are attempted. By growing antiferroelectric ceramic (AFE) ceramic particles in the glass matrix, TRS developed glass–ceramic system in 2005 to exhibit dielectric constant of ~10^3 and breakdown strength on the order of 1,000 kV/cm, potentially resulting in energy densities up to 40 J/cm^3 [62]. Glass–ceramics with ultra-high energy density up to 14.58 $\pm$ 1.14 J/cm^3 with high breakdown strength of 2,382 $\pm$ 92 kV/cm and high discharge efficiency have been prepared through the melt-quenching combined with the controlled crystallization technique [63]. Glass sheet of 5–10 μm thick commercial boroaluminosilicate glasses were produced by NEG and Shiga, Japan, and were investigated by PennState scientists giving rise to an energy density of 11 J/cm^3 up to 200 °C. Even though polymeric coating was applied to the glass sheet, the mechanical strength and flexibility remains to be a challenge in winding capacitors in a small volume [64].

1.3.2.6 High energy density antiferroelectric capacitors

Nevertheless, the energy density of the MLCC ceramic capacitors remains low due to lower breakdown strength and thicker films. AFE is another great candidate for high dielectric performance due to the phase transition upon the application of an electric field greater than the switching field. This field-induced phase transition releases the high power in a very short time. AFE was divided into several groups, e.g. perovskite, liquids crystal and pyrochlore group. Among them are $PbZrO_3$, $PbHfO_3$, $NaNbO_3$, and $(Na_{0.5}Bi_{0.5})TiO_3$–$BaTiO_3$-based composites. Novacap firstly developed PZT AFE capacitor claiming a packaged energy density of 5–8 J/cm^3, but carried risk factors that include electromechanical cracking, low breakdown strength, and low production yield [65]. Compositional modification was conducted to realize the favorable AFE-FE transition behavior and slimmer FE loop to avoid mechanical crack. Doping of Ag_2O-enabled PLZT4/90/10 exhibiting slim loop antiferroelectric characteristics and an energy density of 2.25 J/cm^3 and the efficiency of 78.7% [66]. In 2015, spark plasma sintering was employed to suppress the diffusion behavior between tetragonal phase and orthorhombic phase in $(Pb_{0.858}Ba_{0.1}La_{0.02}Y_{0.008})(Zr_{0.65}Sn_{0.3}Ti_{0.05})O_3$–$(Pb_{0.97}La_{0.02})(Zr_{0.9}Sn_{0.05}Ti_{0.05})O_3$ AFE bulk ceramics [21]. As a result, the increased FE–AFE phase transition field for an enhanced energy density of 6.4 J/cm^3 was obtained at 30.6 kV/mm. Therefore, compositional modulation of AFEs is able to dramatically suppress the large strain associated with field-induced phase transformation promising a new horizon for AFE capacitors.

1.3.3 Thin film dielectrics

The limited breakdown strength of bulk ceramic layers can be changed by reducing the ceramic thickness and carefully controlling film-growing conditions. When the thickness decreases to a critical value, the dielectric strength becomes dimension independent and reaches the intrinsic one if surface defects are excluded. Figure 1.16 shows the dielectric strength versus dielectric constant for various thin-film dielectric materials [67,68]. The figure suggests an inverse relationship between the dielectric constant and dielectric strength. The required thickness of the dielectric layer depends on the dielectric strength of the dielectric material and is around 200 nm [68].

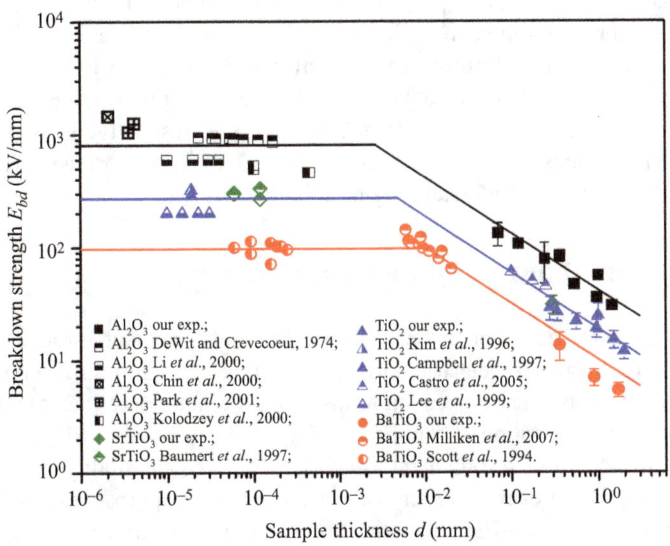

Figure 1.16 Correlation between dielectric strengths and dielectric constant of thin-film dielectric materials. *Reproduced, with permission, from [67]*

Higher breakdown strength (470 V/μm) and energy density were achieved by increasing the number of interfaces in the $Ba_{0.3}Ca_{0.3}TiO_3/BaZr_{0.2}Ti_{0.8}O_3$ multilayer systems [69]. In addition, doping effect has been employed to effectively improve the breakdown strength of lead-based AFE thin films. For example, $Pb_{0.92}La_{30.08}Zr_{0.52}Ti_{0.48}O_3$ thin film always exhibits better energy storage performances than that of bulk ceramic layers due to the ultrahigh values of Pm–Pr and breakdown strength [70]. The striking results are that a giant energy density was obtained in crack-free AFE PLZT films grown on $LaNiO_3$-buffered nickel (Ni) foils. AFE capacitors have high energy density due to the double electric-hysteresis loops caused by electric field-induced phase transition from AFE phase to FE phase, 17.4 J/cm^3 was obtained in Fe_2O_3 doped-$PbZrO_3$ AFE films [71]. Y doping in $Pb_{0.97}Y_{0.02}[(Zr_{0.6}Sn_{0.4})_{0.925}Ti_{30.075}]O_3$ AFE thin films resulted in a high recoverable energy density of 21.0 J/cm^3 with a high energy storage efficiency of $\eta = 91.9\%$ under an electric field of 130 V/μm [72]. Proper doping even results in 53 J/cm^3 in $Pb_{0.92}La_{0.02}(Zr_{0.95}Ti_{0.05})O_3$ at 350 MV/m at room temperature. Furthermore, outstanding temperature stability was found in compositionally graded multilayer $Pb_{1-3x/2}La_xZr_{0.85}Ti_{0.15}O_3$ AFE thick films [73]. Moreover, scientists at Argonne National Lab developed La-doped PLZT ferroelectric films deposited on base metal foils using the so-called "film-on-foil" technology. Using PLZT aerosol precursors and low deposition temperatures, they demonstrated excellent dielectric properties of flexible ceramic films. PLZT films that were fabricated on Kapton sheet via this process possessed a high dielectric constant, low dielectric loss, weak dependence on the applied field, and high recoverable energy density, and were suitable for high field and high-temperature operation. However, scaling-up for multilayer capacitors is not easy and failed to be rolled into manufacturing [74].

Alternatively, Zhang *et al.* propose an effective way to improve the breakdown strength and energy storage density by constructing opposite double-heterojunction ferroelectricity–insulator–ferroelectricity configuration. For the PZT/AO/PZT thin films annealed at 550 °C, the obtained dielectric strength, storage density, and efficiency are 571.1 V/μm, 63.7 J/cm^3, and 81.3%, respectively [75]. Various material performances were summarized in Figure 1.17. More extensive review can be accessed for a better understanding on this matter [76].

1.4 Horizon of advanced dielectrics and capacitors

Advanced capacitor development has been seeking its way out through extensive investigation of advanced dielectric polymers, ceramics, and thin films. Polymer film development exhibits great success in terms of high temperature increase, thickness decrease, and miniaturization of capacitors using PTFE, PEN, and PEI film in spite of the high cost. Nanofilled polymer composites see the feasibility of enhancing dielectric strength and energy density in spite of different approaches, challenges in mechanism, and scale-up. The alternative approach of film surface coating is underway to reveal coating effect on dielectric films and capacitor performance, which may become less development period. Although the fundamental mechanisms for the

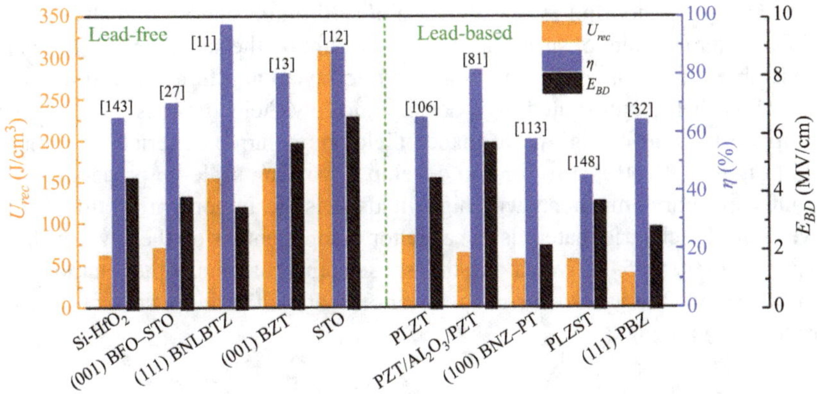

Figure 1.17 The comparison of energy density, efficiency, and breakdown field of various Pb-based and Pb-free dielectric ceramic films (BFO–STO: $0.4BiFeO_3$–$0.6SrTiO_3$; BNLBTZ: $(Bi_{0.5}Na_{0.5})0.9118La_{0.02}Ba_{0.0582}(Ti_{0.97}Zr_{0.03})O_3)$; BZT: $Ba(Zr_{0.2}Ti_{0.8})O_3$; STO: $SrTiO_3$; PLZT: $Pb_{0.92}La_{0.08}(Zr_{0.52}Ti_{0.48})O_3$; PZT: $Pb(Zr,Ti)O_3$; BNZ–PT: $0.44Bi(Ni_{0.5}Zr_{0.5})O_3$–$0.6PbTiO_3$; PLZST: $(Pb_{0.97}La_{0.02})(Zr_{0.91}Sn_{0.04}Ti_{0.05})O_3$; PBZ: $(Pb_{0.8}Ba_{0.2}ZrO_3)$. © 2018 Wiley-VCH Verlag GmbH & Co. KGaA, Weinheim. Reproduced, with permission, from [76]

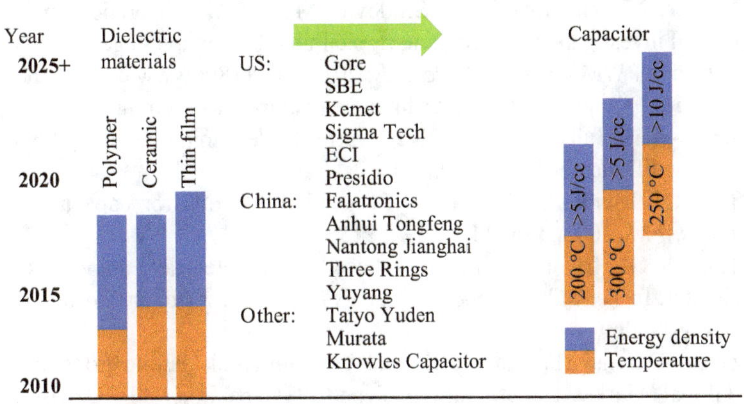

Figure 1.18 The projected roadmap for the advanced dielectric and converted capacitor technologies

dielectric nanocomposite engineering are to be further investigated, it is optimistic to see the market insertion of high-temperature film capacitors compared to high energy density capacitors based on high breakdown strength composite films.

Ceramic layers witness the energy density increase in AFE layer and thin films, as well as significant improvement in miniaturized MLCC through fine dielectric and nickel powders. Thin film dielectrics demonstrate significantly higher energy density and temperature stability through compositional doping in the PZT system in spite of suffering from a discount by the substrate. There remain considerable efforts in converting both ceramic layers and thin films to capacitor components that demonstrated high density and low dielectric loss. Nevertheless, the horizon is seen for high-performance dielectric ceramic materials as summarized in Figure 1.18. It is also time to develop a working scale-up process to manufacture capacitor components with high yields. It is the author's projection that the conversion of dielectric materials to capacitor components is on their way of being acceptable in next 5–10 years. In the subsequent chapters, other authors will describe in depth the various technical advancements of the dielectric materials and capacitor technologies.

Acknowledgments

This work was financially supported by Guangdong Technion Israel Institute of Technology, and Guangdong Basic and Applied Basic Research Foundation - 2019A1515012056.

References

[1] 'Capacitor', https://en.wikipedia.org/wiki/Capacitor, accessed Dec. 15, 2019.

[2] Vujajic, M., Hammami, M., Srndovic, M., *et al.*, 'Theoretical and experimental investigation of switching ripple in the DC-link voltage of single-phase H-bridge PWM inverters', *Energies*, 2017, **10**, 1189; www.mdpi.com/journal/energies, https://www.researchgate.net/figure/Basic-circuit-scheme-of-the-single-phase-full-bridge-H-bridge-inverter-to-determine-the_fig1_319071311, accessed December 15, 2019.

[3] Tan, Q., Irwin, P., Cao, Y., 'Advanced dielectrics for capacitors', *IEEE Trans. FM*, 2006, **126**, 11, 1209–1215.

[4] Tan D. Q., Irwin P., 'Polymer based nanodielectric composites', Book chapter, In *Tech Online book Advances in Ceramics* Ed. By Costas Sikalidis, 2011–09–06.

[5] Zogbi Dennis M., Market Report by Paumanok Publications Inc 2004, 'Global Market Update for 16 Types of Capacitor', 11/06/2018. https://www.ttieurope.com/content/ttieurope/en/resources/marketeye/categories/passives/me-zogbi-20181106.html, accessed December 15, 2019.

[6] Reed, C. W., DeVre, W., Gasworth, M. W., 'Progress with advanced materials for high-energy-density capacitors', ARL-CR-150, Contract DAAL01–91-C-0149 1994, https://apps.dtic.mil/dtic/tr/fulltext/u2/a281443.pdf, accessed Dec. 15, 2019.

[7] Sarjeant, W. J., Zirnheld, J., Mcdougall, F. W., 'Capacitors', *IEEE Trans. Plasma Sci.*, 1998, **26**, 5, 1368–1392.

[8] Adams, S., Macdougall, F. W., Ellwangar, R., Yializis, A., 'Advanced capacitors for airborne pulsed high power microwave'. AIAA-2003–5915,

1st International Energy Conversion Engineering Conference, Portsmouth, Virginia, August 2003.

[9] Lewis, T. J., 'Interfaces are the dominant feature of dielectrics at the nanometric level', *IEEE Trans. Dielectr. Electr. Insulation*, 2004, **11**, 739.

[10] Nelson, J. K., *Dielectric polymer nanocomposites*, Springer, Berlin, Germany, 2010.

[11] Fries-Carr, S., Adams, S., Weimer, J., *et al.*, 'Advanced dielectrics for pulsed power capacitor devices'. *SAE Technical Paper Series 2004–02–3216*, Power Systems Conference, Reno, Nevada, Nov. 2–4, 2004.

[12] Lanagan, M., FY2005 Multidisciplinary University Research Initiative (MURI) Program Award Winners 2005, https://studylib.net/doc/12891252/, accessed February, 2019.

[13] Tan, D. Q., 'Review of polymer-based nanodielectric exploration and film scale-up for advanced capacitors', *Adv. Funct. Mater.*, 2019, **1808567**, 1–23.

[14] Taylor, R. S., Silvi, N., Chen, Q., *et al.*, 'Development, test and demonstration of a cost-effective, compact, light-weight, and scalable high temperature inverter for HEVs, PHEVs, and FCVs', *US DOE report DE-FC26–07NT43121*, 2012.

[15] Tan, D., 'Differentiation of roughness and surface defect impact on dielectric strength of polymeric thin films', *IET Communications*, 2020, **3**, 28–31

[16] Dang, Z.-M., Yuan, J.-K., Zha, J.-W., *et al.*, 'Fundamentals, processes and applications of high-permittivity polymer–matrix composites', *Prog. Mater. Sci.*, 2012, **57**, 660–723.

[17] Zhang, X., Shen, Y., Shen, Z. H., 'Achieving high energy density in PVDF-based polymer blends: Suppression of early polarization saturation and enhancement of breakdown strength', *ACS Appl. Mater. Interfaces*, 2016, **8**, 27236−27242.

[18] Prateek, Thakur, V. K., Gupta, K., 'Recent progress on ferroelectric polymer-based nanocomposites for high energy density capacitors: Synthesis, dielectric properties, and future aspects', *Chem. Rev.*, 2016, **116**, 4260.

[19] Tan, D. Q., The search for enhanced dielectric strength of polymer-based dielectrics: A focused review on polymer nanocomposites. *J. Appl. Polym. Sci.* 2020, e49379. https://doi.org/10.1002/app.49379

[20] Tan, D. Q., Zhang, L. L., Chen, Q., *et al.*, 'High temperature capacitor polymer films', *J. Elect. Mater.*, 2014, **43**, 4569.

[21] Yao, Z. H., Song, Z., Hao, H., *et al.*, 'Homogeneous/inhomogeneous-structured dielectrics and their energy-storage performances', *Adv. Mater.*, 2017, **29**, 1601727.

[22] Munshi, M., Zafar, A., WO 00/79550 A1, 2000.

[23] Chu, B. J., Zhou, X., Ren, K. L., *et al.*, 'A dielectric polymer with high electric energy density and fast discharge speed', *Sci.*, 2006, **313**, 334.

[24] Kobayashi, S., Mullen, K., *Encyclopedia of Polymeric Nanomaterials*, Springer, Berlin, Heidelberg, 2014.

[25] Yializis, A., Powers, G. L., Shaw, D. G., 'A new high temperature multilayer capacitor with acrylate dielectrics', *IEEE Trans. Comp. Hybrids Manuf. Technol.*, 1990, **13**, 611.

[26] Baer, E., Zhu, L., '50th anniversary perspective: Dielectric phenomena in polymers and multilayered dielectric films', *Macromolecules*, 2017, **50**, 2239.
[27] Tan, D. Q., Dalakos, G. T., Cao, Y., *et al.*, 'Coated polymer dielectric film', *US Patent 8354166*, 2011.
[28] Zhou, X., Chen, Q., Zhang, Q. M., *et al.*, 'Dielectric behavior of bilayer films of P(VDF-CTFE) and low temperature PECVD fabricated Si3N4', *IEEE Trans. Dielectr. Electr. Insul.*, 2011, **18**, 2, 463–470.
[29] Aziz, A., Gadinski, M. R., Li, Q., *et al.*, 'High-performance polymers sandwiched with chemical vapor deposited hexagonal boron nitrides as scalable high-temperature dielectric materials', *Ad. Mater.*, 2017, 1701864.
[30] Kumar, S. K., Benicewicz, B. C., Via, R. A., *et al.*, '50th anniversary perspective: Are polymer nanocomposites practical for applications?' *Macromolecules*, 2017, **50**, 3, 714–731.
[31] Raetzke, S., Kindersberger, J., 'The effect of interphase structures in nanodielectrics', *IEEJ Trans. Fundam. Mater.*, 2006, **126**, 1044.
[32] Wang, Y. U., Tan, D. Q., Krahn, J., 'Computational study of dielectric composites with core-shell filler particles', *J. Appl. Phys.*, 2011, **110**, 044103.
[33] Huang, X. Y., Sun, B., Zhu, Y., *et al.*, 'High-k polymer nanocomposites with 1D filler for dielectric and energy storage applications', *Prog. Mater. Sci.*, 2018, **100**, 187.
[34] Tan, D., Cao, Y., Tuncer, E., *et al.*, 'Nanofiller dispersion in polymer dielectrics', *Mater. Sci. Appl.*, 2013, **04**, 6.
[35] Plesa, Il., Notingher, V. P., Schlögl, S., *et al.*, 'Properties of polymer composites used in high-voltage applications', *Polymers*, 2016, **8**, 173.
[36] Luo, S. B., Yu, J. Y., Yu, S. H., *et al.*, 'Significantly enhanced electrostatic energy storage performance of flexible polymer composites by introducing highly insulating ferroelectric micro hybrids as fillers', *Adv. Energy Mater.*, 2019, **9**, 1803204.
[37] Peng, S. M., Zeng, Q. B., Yang, X., *et al.*, 'Local dielectric property detection of the interface between nanoparticle and polymer in nanocomposite dielectrics', *Sci. Rep.*, 2016, **6**, 38978.
[38] Pitsa, D., Danikas, M., 'Interfaces features in polymer nanocomposites: A review of proposed models', *NANO: Brief Rep. Rev.*, 2011, **6**, 497–508, World Scientific Publishing Company.
[39] Pan, Z., Zhai, J., Shen, B., 'Multilayer hierarchical interfaces with high energy density in polymer nanocomposites composed of $BaTiO_3$@TiO_2@Al2O_3 nanofibers', *J. Mater. Chem. A*, 2017, **5**, 15217.
[40] Mendes, S. F., Costa, C. M., Caparrós, C., *et al.*, 'Effect of filler size and concentration on the structure and properties of poly(vinylidene fluoride)/$BaTiO_3$ nanocomposites', *J. Mater. Sci.*, 2012, **47**, 1378.
[41] Prateek, Bhunia, R., Siddiqui, S., Garg, A., Gupta, R. K., 'Significantly enhanced energy density by tailoring the interface in hierarchically structured TiO_2-$BaTiO_3$-TiO_2 nanofillers in PVDF-based thin-film polymer nanocomposites', *ACS Appl. Mater. Interfaces*, 2019, **11**, 14329–14339.
[42] Luo, H., Zhou, H. F., Ellingford, *et al.*, 'Interface design for high energy density polymer nanocomposites', *RSC Chem. Soc. Rev.*, 2019, **48**, 4424–4465.

[43] Mallakpour, S., Madani, M., 'A review of current coupling agents for modification of metal oxide nanoparticles', *Prog. Org. Coat.*, 2015, **86**, 194–207.

[44] Kim, Y., Kathaperumal, M., Chen, M., *et al.*, 'Bilayer structure with ultrahigh energy/power density using hybrid sol–gel dielectric and charge-blocking monolayer', *Adv. Energy Mater.*, 2015, **5**, 1500767.

[45] Yu, S. H., Ding, S. J., Tan, D. Q., *et al.*, 'Heterostructured nanoparticles as fillers: A new strategy to enhance energy density of polymer-based composites', submitted to *Ad. Funct. Mat.*, 2019.

[46] Huang, X., Jiang, P., Tanaka, T. A., 'A review of dielectric polymer composites with high thermal conductivity', *IEEE Electr. Insul. Mag.*, 2011, **27**, 8–16.

[47] Buchanan, R. C., *Ceramic materials for electronics*, CRC Press, Third Edition, 2004.

[48] Dennis, M., Zogbi, 'As 2019 Ends, How Healthy is the Global MLCC Market?' 11/12/2019 in Passives, https://www.ttieurope.com/content/ttieurope/en/resources/marketeye/categories/passives/me-zogbi-20191112.html.

[49] Kubota, K., Nishiyama, S., Malhotra, K., 'Ceramic capacitor aid high-voltage designs', *Power Elect. Technol.*, May 2004, 14–23; 'Chip Multilayer Ceramic Capacitors for General', Murata catalogue, November 27, 2017, Izumo, Japan, www.murata.com.

[50] Yuden, T., https://www.mouser.com/pdfdocs/MLCCWhitePaper2016_WA_final-2.pdf, https://www.yuden.co.jp/eu/solutions/mlcc/product/; https://en.wikipedia.org/wiki/Ceramic_capacitor#/media/File:MLCC-Principle.svg, accessed December 15, 2019.

[51] 'Improved Nickel Powder for Small Case Size MLCC with High Capacitance', *JFE TECHNICAL REPORT* No. 6, Oct. 2005, https://www.jfe-steel.co.jp/.

[52] Morgan Electroceramics Inc., http://www.morganelectroceramics.com/cap1.html, accessed December 15, 2019.

[53] Kemet, www.kemet.com, 2019.

[54] Presidio Components, www.Presidio.com, 2019.

[55] http://www.eclipsenanomed.com/

[56] Microcoating Technologies, http://www.microcoating.com/

[57] Hu, Q. Y., Tian, Y., Zhu, Q. S., 'Achieve ultrahigh energy storage performance in $BaTiO_3$–Bi(Mg1/2Ti1/2)O_3 relaxor ferroelectric ceramics via nano-scale polarization mismatch and reconstruction', *Nano Energy*, https://doi.org/10.1016/j.nanoen.2019.104264.

[58] Zeb, A., Milne, S. J., 'High temperature dielectric ceramics: A review of temperature-stable high-permittivity perovskites', *J. Mater. Sci.: Mater. Electron*, 2015, **26**, 9243–9255.

[59] Dogan, F., Devoe, A., Burn, I., 'Nanostructured dielectric materials for high energy density multilayer ceramic capacitors', US Patent US 8,644,000 B2, February 4, 2014.

[60] Fan, B. Y., Liu, F. H., Yang, G., *et al.*, 'Dielectric materials for high-temperature capacitors', *IET Nanodielectrics*, 2018, **1**, 32–40.

[61] Kundu, T. K., Chakravorty, D., 'Nanocomposites of lead–zirconate–titanate glass ceramics and metallic silver', *Appl. Phys. Lett.*, 1995, **67**, 18, 2732.

[62] Hackenburger, W., TRS Technologies Inc., http://www.trsceramics.com/, “Very High Energy Density Capacitors for Miniaturized, Modular Power Electronics”. 2005, White Papers.

[63] Wang, H. T., Liu, J. H., Zhai, J. W., *et al.*, ‘Ultra high energy-storage density in the barium potassium niobate-based glass-ceramics for energy-storage applications’, *J. Am. Ceram. Soc.*, 2016, **99**, 9, 2909–2912.

[64] Manoharan, M. P., Zou, C., Furman, E., *et al.*, ‘Flexible glass for high temperature energy storage capacitors’, *Energy Technol.*, 2013, **1**, 313–318.

[65] Novacap, A. Dova Companym, http://www.novacap.com/product_page.php?Pk=PR11.

[66] Zhao, T., Zhu, Q. S., Xu, R., Tian, J. J., Feng, Y. J., ‘Effects of Ag_2O doping on dielectric properties of (Pb0.96La0.04) (Zr0.9Ti0.1)0.99O3 antiferroelectric ceramics’, *Ceramics Inter.*, 2019, **45**, 1887–1892.

[67] Neusel, C., Schneider, G. A., ‘Size-dependence of the dielectric breakdown strength from nano to millimeter scale’, *J. Mech. Phys. Solids*, 2014, **63**, 201–213.

[68] Imanaka, Y., Shioga, T., Baniecki, J. D., ‘Decoupling capacitor with low inductance for high-frequency digital applications’, *FUJITSU Sci. Tech. J.*, 2002, **38**, 1, 22–30 (June).

[69] Sun, Z. X., Ma, C. R., Liu, M., *et al.*, ‘Ultrahigh energy storage performance of lead-free oxide multilayer film capacitors via interface engineering’, *Adv. Mater.*, 2017, **29**, 1604427.

[70] Hu, G. L., Ma, C. R., Wei, W., ‘Enhanced energy density with a wide thermal stability in epitaxial $Pb_{0.92}La_{0.08}Zr_{0.52}Ti_{0.48}O_3$ thin films’, *Appl. Phys. Lett.*, 2016, **109**, 193904.

[71] Sa, T. L., Cao, Z. P., Wang, Y. J., *et al.*, ‘Enhancement of charge and energy storage in $PbZrO_3$ thin films by local field engineering’, *Appl. Phys. Lett.*, 2014, **105**, 043902.

[72] Ahn, C. W., Amarsanaa, G., Won, S. S., *et al.*, ‘Antiferroelectric thin-film capacitors with high energy-storage densities, low energy losses, and fast discharge times’, *ACS Appl. Mater. Interfaces* 2015, **7**, 26381.

[73] Wang, Y., Hao, X., Yang, J., *et al.*, ‘Fabrication and energy-storage performance of $(Pb,La)(Zr,Ti)O_3$ antiferroelectric thick films derived from polyvinylpyrrolidone-modified chemical solution’, *J. Appl. Phys.*, 2012, **112**, 3, 034105.

[74] Balachandran, U., Kwon, D. K., Narayanan, M., *et al.*, ‘Development of PLZT dielectrics on base metal foils for embedded capacitors’, *J. Eur. Ceram. Soc.*, 2010, **30**, 365.

[75] Zhang, T. D., Li, W. L., Zhao, Y., *et al.*, ‘High energy storage performance of opposite double-heterojunction ferroelectricity–insulators’, *Adv. Funct. Mater.*, 2018, 1706211, 1–9.

[76] Palneedi, H., Peddigari, M., Hwang, G.-T., *et al.*, ‘High-performance dielectric ceramic films for energy storage capacitors: Progress and outlook’, *Adv. Funct. Mater.*, 2018, 1803665, 1–33.

Chapter 2

Techniques for capacitor dielectrics characterization

Bo Li[1], Haijuan Zhang[1] and Shihai Zhang[1]

2.1 Introduction

Electrostatic capacitors are indispensable components in high voltage pulsed power systems and power electronics [1,2]. They are widely employed in applications such as pulse-forming networks, switched-mode power supplies, medical defibrillators, and power electronics in hybrid electric vehicles (HEV), grid-tied wind turbine generators, high-speed trains, photovoltaics, etc. They perform the conversion of prime electric energy to energize large loads, and regulate electrical inputs into stable outputs over a wide range of load conditions. In most applications, the electrostatic capacitors are not the primary energy storage device; rather, their function is more likely conditioning primary electrical energy to certain pulse form, AC/DC, frequency, or voltage level to drive the final electrical load [2,3].

Biaxially oriented polypropylene (BOPP) still dominates the film capacitor market because of the ease of manufacturing (BOPP can be biaxially oriented to high-quality thin films with thickness from 1.9 to 10 μm, at line speed higher than 1,000 m/min and film width wider than 10 m), superior high breakdown strength (>700 V/μm in kJ packaged capacitors), and extremely low dielectric loss. However, the low dielectric constant of BOPP (ca. 2.2) greatly limits its energy density, and a large capacitor size is needed to provide adequate energy and power outputs [4–6]. In addition, the low thermal stability of BOPP imposes stringent constrains on working conditions. Additional air or liquid cooling systems are usually required for BOPP capacitors to maintain their normal operation under continuous cycles [7].

Numerous efforts have been devoted in search of new materials and processing technologies that can replace BOPP in advanced power electronics and energy storage systems. The success in such endeavors will definitely depend on the clear and comprehensive understanding of application environments, accurate characterization of material performance, and evaluation of the feasibility of large-scale manufacturing (Figure 2.1) [8].

[1]PolyK Technologies, State College, PA, USA

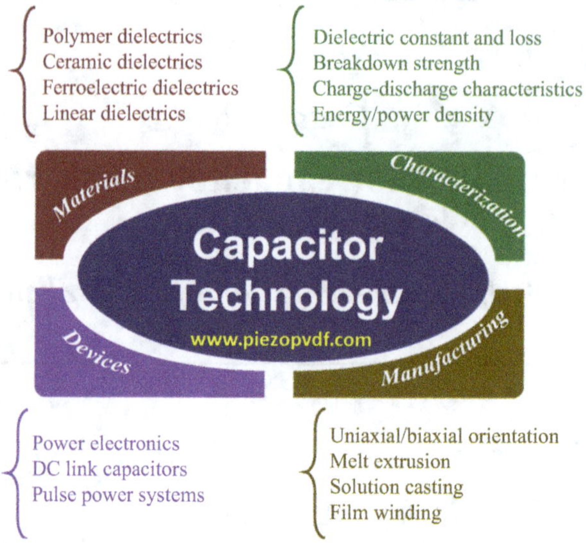

Figure 2.1 Key elements and concepts involved in capacitor technology. More information can be referred to www.piezopvdf.com

In this chapter, we will discuss the techniques for dielectric and high voltage characterization, as well as the methods to determine manufacturing feasibility. Particularly, in addition to general characterization tools that are widely used in many publications, we focus on new test protocols that can simulate practical capacitor operation scenario and the test results are more relevant for capacitor applications. As it usually costs over \$10 million US dollars to commercialize a new dielectric materials to large size capacitors (>10 m^2 film, >1,000 g mass, or >1,000 J energy), it is critical to ensure accurate, comprehensive, and relevant measurement at small area samples (millimeter to centimeter) such that they can be confidently used to make investment decision for future commercialization [6]. Figure 2.2 illustrates the scale up process from <10 mm^2 dielectric sample to capacitors using >10 m^2 pinhole-free film with uniform thickness.

2.2 Regular dielectric measurement

Several parameters are used to evaluate the performance of dielectric materials for the capacitor applications. The energy density indicates the amount of electric energy that can be stored in a unit volume or weight of capacitors or dielectric materials. A higher energy density is desirable to reduce the weight or volume of the capacitor, which often occupies a significant volume or weight of a system. Since the energy density of dielectrics is proportional to the dielectric constant and the square of the applied electric field, the materials with a high dielectric constant and large breakdown strength are highly appreciated. The energy stored in a

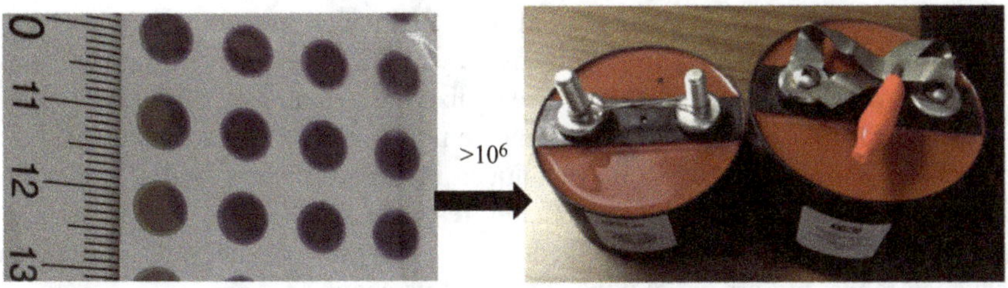

1. Weight: milligram to gram
2. Film: mm^2–cm^2, single layer
3. Capacitance: 100 pF–1 nF
4. Energy: <mJ
5. No thermal runaway issue

1. Weight: >1 kg
2. Film: 10–1,000 m^2, (<5 μm thick)
3. Capacitance: >100 μF
4. Energy: >1,000 J
5. Thermal runaway becomes critical

Figure 2.2 From lab-scale dielectric film to packaged large size capacitors, the challenges to commercialization

capacitor or dielectric material during the charging process cannot be completely released to the external loads during the discharging process because of the dielectric loss or conduction of the dielectric material used in the capacitor. The energy that cannot be discharged normally dissipates as heat, which has a detrimental effect on the performance of capacitors by raising the temperature of the capacitors. Therefore, the dielectric loss under the targeted operation scenario is very critical parameter to evaluate the merit of a dielectric material, even more important than the energy density or power density [9,10].

The following characterization methods illustrate the measurement of dielectric constant and loss, breakdown strength, discharge energy density and efficiency, and leakage current. This information is commonly used as guidance to predict the dielectric behavior of materials (preliminary screening of candidate materials).

2.2.1 Low field dielectric constant and dielectric loss

The dielectric constant and dielectric loss of dielectric material can be very easily measured by an inductance-capacitance-resistance (LCR) meter. This technique involves the application of a small AC voltage (usually less than 1 V) and the recording of current signals passing through the sample. The capacitance (C) of the dielectric sample can be obtained via $C = \frac{1}{2\pi f X}$, where X is the calculated impedance of the sample and f is the measuring frequency. The dissipation factor (tan δ) can be acquired by the phase shift (δ) between the measured current with respect to the applied voltage.

The conventional LCR meter can provide accurate measurements from 10^2 to 10^6 Hz. With increasing frequency, the impedance of dielectric materials becomes smaller and the inductance of the cables connecting samples to instruments becomes larger. When the sample impedance approaches the level of the cable inductance, the measured capacitance could deviate from the actual value because of the influence of the cable inductive resistance. Another issue regarding the high-frequency measurement is the insufficient power of the LCR meters. The reduction in the sample impedance increases the current level under the same driving voltage. As a result, the electrical power delivered by the LCR mete will increase with increasing measuring frequency. It can be realized that to obtain good measurement results at high frequency, short cables and samples with small capacitance should be used. Advanced measurement techniques include transmission/reflection line method, open-ended coaxial probe method, resonant method, etc. These measurement methods extend the measurement range to radiofrequency and microwave frequency range, but are beyond the scope of this chapter.

In addition to the problems at high frequencies, there are also difficulties associated with the low-frequency measurement. At low frequencies, the sample impedance is increased and the current level becomes small even under the maximum applied voltage (usually 1 V). The low current level imposes great challenges to the accurate measurement, especially for the phase angle, due to the interference of noise signals. Similarly, it can be obtained that samples with large capacitance and a high driving voltage should be used for low-frequency measurement. Advanced measurement protocol involves the use of a lock-in amplifier to filter the noise from the environment and allow the precise measurement down to 0.001 Hz.

The low field AC dielectric measurement includes frequency and temperature scan modes. In the frequency scan mode, the sample is placed in an environment with a constant temperature. The dielectric properties of a sample are recorded as a function of frequency. This mode is commonly used in fundamental studies and delineates the polarization mechanisms. In the temperature scan mode, the temperature is ramped at a constant heating or cooling rate, and the dielectric properties of the sample are measured. The dependence of dielectric properties at a specific frequency on temperature can provide practical guidance in quickly determining material performance for particular applications.

Figure 2.3 shows the typical temperature scanning results of a high dielectric constant (*K*) polymer, P(VDF-TrFE-CFE). P(VDF-TrFE-CFE) terpolymer with appropriate compositions is a semicrystalline polymer with melting temperature around 120–140 °C. It exhibits ferroelectric relaxor (ferrorelaxor) behavior as

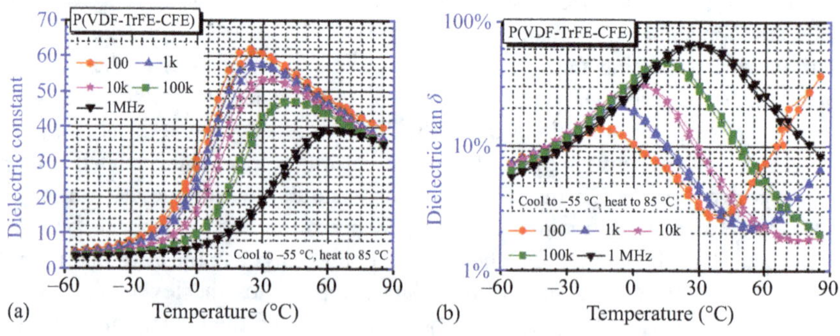

Figure 2.3 Dielectric constant (a) and dielectric loss (b) of P(VDF-TrFE-CFE) as a function of temperature

certain ceramic materials, i.e., no ferroelectric hysteresis. Their dielectric constant can be above 50, which is about five-time higher than polyvinylidene fluoride (PVDF), and their dielectric breakdown strength can be above 400 MV/m. The sample is first cooled to –55 °C then heated to 85 °C at a ramp rate of 5 °C/min, while the dielectric properties are recorded at five discrete frequencies. As shown in Figure 2.3(a), the dielectric constant of the terpolymer increases from 5 at low temperatures to above 50 at room temperature, followed by a decline with further increasing temperature. Accompanied by the strong increase in dielectric constant is an associated peak in the dielectric loss spectrum (Figure 2.3(b)). Regardless of the mechanisms underlying the high dielectric constant for PVDF-TrFE-CFE, the temperature scanning results at least suggest that these terpolymers are potential candidate materials for high energy density film capacitors. In fact, high energy density >10 J/cc can be easily obtained in without adding any nanoparticles.

While the temperature scan mode can provide the dielectric constant and loss of a dielectric material at a particular temperature and frequency, the frequency sweep mode can reveal the polarization mechanism and the correlation of dielectric properties with structural development. Figure 2.4 shows the frequency scanning results of a PVDF-co-hexafluoropropylene (PVDF-HFP) copolymer before and after stretching. Stretching is commonly adopted in the manufacturing of high-quality polymer capacitor films with thin thickness.

All films exhibit a strong dielectric relaxation at the high-frequency end (~10^6 Hz), which is evidenced by a decrease in dielectric constant and a corresponding loss peak. The relaxation is related to the glass transition of PVDF-HFP (also known as α_a or T_g), arising from the segmental motion of polymer chains. The dielectric relaxation strength is defined as the amplitude of loss peak or the step-change in dielectric constant, which increases with stretching. The increased relaxation strength can be attributed the crystal orientation. As the PVDF-HFP crystals gradually align with the stretching direction, the amorphous chains in the interphase regions should also be oriented. This orientation facilitates dipole

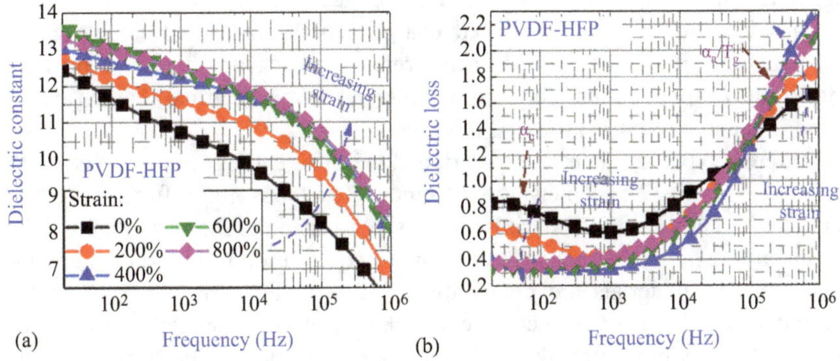

Figure 2.4 Dielectric constant (a) and dielectric loss (b) of PVDF-HFP as a function of frequency and degree of film orientation

rotation, as the applied field is now parallel to the dipole moment of the CH_2–CF_2 dipoles [11]. It has been reported that the oriented interphase PVDF can exhibit a dielectric constant of ~21, much higher than the isotropic counterpart [12]. Apart from the glass transition, another relaxation mechanism can be observed in the unstrained films at the low frequency end (~1 Hz), which is absent in the stretched films. This relaxation is associated with the switching of slightly tilted CF_2 dipoles along the chain direction in α crystals (also known as α_c relaxation) [13]. With increasing stretching, more α crystals are conversed into β phase and the polymer chains are gradually oriented perpendicular to the electric field. Consequently, it becomes more difficult for the CF_2 to switch with the electric field and the α_c relaxation is suppressed in the stretched films.

The low field dielectric measurement with LCR meter or impedance analyzer provides a quick overview of dielectric materials. Although the dielectric behavior at low electric fields is not necessarily equivalent to the high field performance, materials with poor dielectric performance at low field will certainly become worse at high field and they shall be immediately excluded from future study. On the other hand, material compositions with good low-field dielectric properties can be moved to the next phase high field investigation.

2.2.2 Polarization-electric field hysteresis loop

The dielectric properties of materials could exhibit strong field dependence, resulting in substantially different behaviors at low field and high fields. This is especially true for ferroelectric materials and lossy materials with high conduction loss. As a result, the polarization-electric field hysteresis loop technique (P-E loop) is developed to evaluate the high field dielectric performance of the materials.

For a typical test, a sinusoidal or triangular ac field is applied to a sample, and the induced electric displacement or polarization is recorded. The P-E loop measurement circuit follows the traditional design of the Sawyer–Tower circuit. The tested dielectric material is connected in series with a reference capacitor with known capacitance. The capacitance of the reference capacitor is required to be independent of the applied voltage. Since the electrical charges are the same for two capacitors in series, the induced charges on the measured sample can be evaluated from the voltage signal on the reference capacitor, which is then converted to the charge outputs. The applied voltage and the electrical charges are then plotted in a curve to obtain the P-E hysteresis loop.

Different types of dielectric materials have different signature shapes of P-E loops [14]. Figure 2.5 depicts the common P-E loops for different dielectric materials [9]. Ferroelectric materials would show a shape as shown in Figure 2.5(a). Ferroelectric materials normally have a higher polarization response than common dielectrics. The materials are potentially a class of dielectric materials of high energy density. However, ferroelectric switching causes substantial dielectric loss, which should be prevented during capacitor applications. The antiferroelectric materials exhibit a typical P-E loop shown in Figure 2.5(b). In antiferroelectrics, the spontaneous polarization in the neighboring unit cells is anti-parallel and

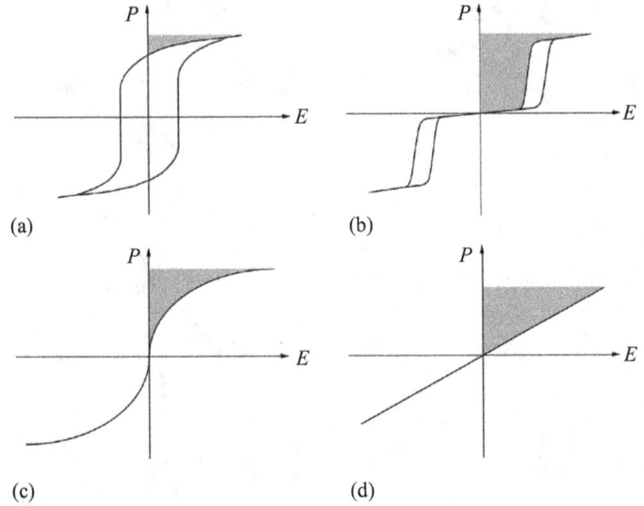

Figure 2.5 Different polarization response under high electric fields for linear and nonlinear dielectric materials. Reprinted from [9] with permission from Elsevier

cancels each other. At a low electric field, the polarization response varies linearly with the applied field, and the materials can be converted into a ferroelectric state at a high electric field. Figure 2.5(c) shows the polarization response of the relaxor ferroelectric or paraelectric materials. Compared with normal ferroelectric materials, the hysteresis behavior vanishes on the P-E loops of the relaxor ferroelectric or paraelectric materials. All the aforementioned materials exhibit nonlinear field-dependent dielectric response under a high electric field. The dielectric properties of some materials including many polymers do not change with electric field and their polarization response varies linearly with the electric field, as shown in Figure 2.5(d). It should be pointed out that the polarization response of the materials shown in Figure 10.3(c) and (d) is the ideal case. In reality, these materials may have hysteresis-like dielectric responses under high electric field due to the inevitable conduction loss [9].

In addition to distinguishing different types of dielectric materials, the P-E loop is widely used in energy storage applications for determining discharged energy density, energy loss, and discharging efficiency of a capacitor material. Figure 2.6(a) and (b) shows the unipolar P-E loops for an unstretched PVDF-HFP and 800% stretched PVDF-HFP, respectively.

The recoverable energy density (U_R) measures the energy that can be released per poling cycle, which is obtained by integrating the area between the discharging curve and the charge-density ordinate. As shown in Figure 2.6(c), with increasing applied field, the energy density is increased and shows a parabolic dependence with field. The effective dielectric constant (ε_{eff}) can be derived via $U_R = \frac{1}{2}\varepsilon_0\varepsilon_{eff}E^2$ (U_R is the discharged energy density, E is the applied field, ε_0 the vacuum

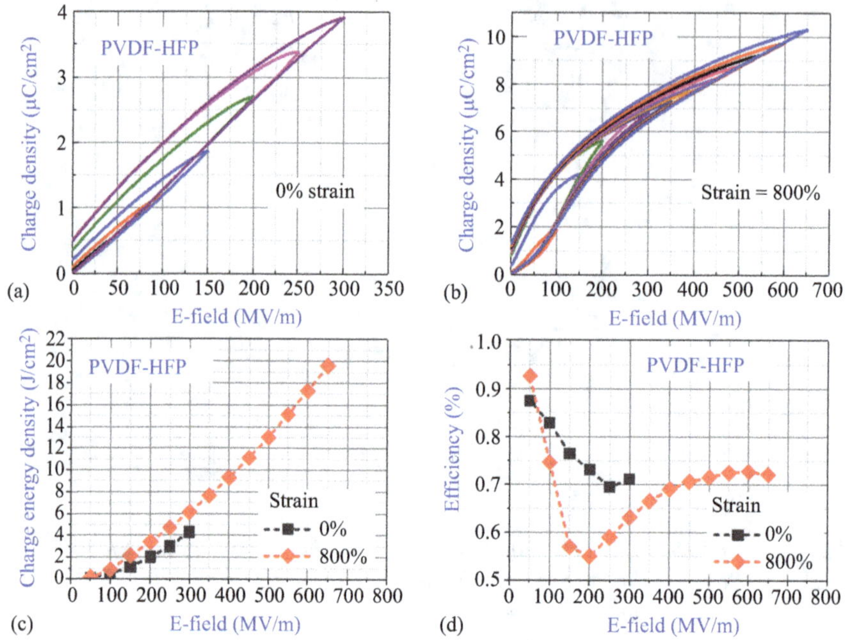

Figure 2.6 Unipolar P-E loops of (a) unstretched PVDF-HFP and (b) stretched PVDF-HFP. (c) The discharged energy density and (d) efficiency derived from the P-E loops

permittivity). Note that ε_{eff} is not equal to the low field dielectric constant (ε_r) measured by the LCR meter (as will be detailed in "**High field dielectric constant and dielectric loss**"). In some publications, the energy density of dielectric material was frequently calculated using the weak-field dielectric constant ε_r and breakdown electric field in the above equation. Sometimes, it can be quite misleading as ferroelectric materials will demonstrate a strong nonlinear dielectric behavior with e-field; moreover, unpredictable energy loss further causes significant deviation in the calculation. Also from Figure 2.6(c), it can be observed that the stretched film always has a higher energy density than the unstretched film. It is well known that the stretching can transform the α phase PVDF-HFP into the β phase. The enhanced energy density is not attributed to the large content of β crystals, because once polarized, the ferroelectric domains cannot switch back and participate in the next charging-discharging cycles [15]. However, film stretching can also promote the formation of nanoconfined ferroelectric domains when the ferroelectric crystals are fragmented into smaller crystallites. These decoupled crystal dipoles can exhibit reversible polarization and provide an additional increase to the discharged energy density [16].

The discharging efficiency (η) is calculated by $\eta = U_R/(U_R + U_L)$, where U_R is the recoverable energy density as mentioned before and U_L is the energy loss defined by the area inside the loop. Figure 2.6(d) presents the discharging efficiency of the two films. The unstrained film exhibits a monotonic decrease in efficiency with increasing field, arising from the intensified charge migration and the injection of charge carriers at high fields. The stretched films show significantly different behaviors. With increasing applied field, the efficiency shows an abrupt reduction at low fields, followed by a gradual increase at higher fields. The minimum efficiency is obtained at ~200 MV/m, which coincides well with the coercive field for ferroelectric switching in PVDF-based polymers, indicating that the substantial efficiency decline is caused by the irreversible ferroelectric switching [15]. After the ferroelectric crystalline dipoles are gradually aligned by the applied field, the efficiency begins to recover with further increasing field. The efficiency is also an important parameter to characterize the energy storage performance of a dielectric material. The low efficiency not only reduces the effectiveness of energy storage but also implies large dielectric loss at high fields, which can cause serious heating issues during applications especially for high repetition rate operations.

Furthermore, the bipolar P-E loop is commonly used to measure the ferroelectric switching in ferroelectric materials, a mature technique in ferroelectric community. For ferroelectric materials, the spontaneous polarization exhibits bistable states and an electric field higher than their coercive field (E_c) is required to switch the polarization between the two states. Figure 2.7(a) shows the bipolar P-E hysteresis loops for an α phase dominated PVDF-HFP [15]. It exhibits a typical ferroelectric behavior at high fields, as evidenced by the propeller-shaped double hysteresis loops. The ferroelectric properties in the α PVDF-HFP arise from the transition from nonpolar α phase to ferroelectric γ phase at high electric fields. The remnant polarization and maximum polarization as a function of poling field are plotted in Figure 2.7(b). The shaded area represents the region of ferroelectric domain switching. Before the region, the remnant polarization is negligible, and the

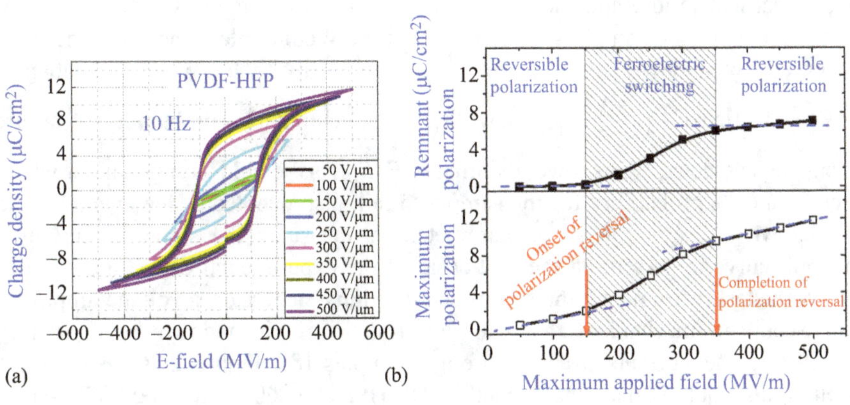

Figure 2.7 (a) Bipolar P-E loops of PVDF-HFP. (b) Remnant polarization and maximum polarization derived from the P-E loops. Reprinted from [15] with permission of AIP Publishing

maximum polarization increases linearly with electric field. It is consistent with the linear dielectric response, indicating the main contribution coming from the amorphous dipoles. After the region, the remnant polarization reaches a maximum value, which is almost independent on electric field, while the maximum polarization resumes the linear increase with field, with the same slope as before the region. It suggests that after all ferroelectric domains are polarized, the extra contribution to the polarization still comes from the amorphous dipolar orientation.

2.2.3 Dielectric breakdown

Apart from high dielectric constant and low dielectric loss, high voltage endurance is as important to dielectric materials as the other dielectric parameters. Materials with low breakdown strength cannot be operated under high electric fields, and the corresponding energy density would be greatly reduced because of the quadratic dependence of discharge energy on applied field. Several mechanisms have been developed to explain the electrical failures in the dielectric materials, including electron avalanche breakdown, electromechanical breakdown, thermal breakdown, and partial discharge breakdown. However, it is usually very difficult, if not impossible, to pinpoint the exact breakdown mechanism, as the electrical breakdown is a stochastic process, which is determined by the weakest regions in a sample [17]. The experimentally observed breakdown fields are often analyzed within the framework of a statistic method, e.g., the Weibull method, to characterize the statistic distribution. It shall be pointed out that the dielectric breakdown in polymeric materials has been studied for over 70 years for high voltage electrical insulation applications, even though the dielectric breakdown mechanism in capacitor film (<5 μm thick) and high voltage insulation materials (> mm thick) may be significantly different [18].

There are several types of breakdown measurements. The most common one is the DC breakdown, where a constant ramped voltage is applied to the test sample until electrical failure and the breakdown voltage is recorded as a data point. At least ten to twenty data points are required for the Weibull plot. The experimentally observed breakdown fields are then fitted by the Weibull statistics: $P(E) = 1 - \exp\left(-\left(\frac{E}{E_{BD}}\right)^{\beta_w}\right)$, where $P(E)$ is the cumulative probability of failure occurring at electric fields lower or equal to E. E_{BD} is the field strength under which there is a 63% probability for the sample to fail, also denoted as the characteristic breakdown strength; the shape parameter β_w measures the slope of the fitted Weibull curve, representing the scattering of the experimental data. Figure 2.8(a) shows the test instrument, which consists of a signal generator, a voltage amplifier, and a built-in mechanism that can detect dielectric breakdown by constantly monitoring the currents passing through samples [8]. Figure 2.8(b) shows the Weibull distribution of an unstretched PVDF-HFP and 800% stretched PVDF-HFP. The characteristic breakdown strength E_{BD} is significantly improved by stretching, from 480 MV/m for the unstretched sample to 770 MV/m for the stretched sample. The Weibull modulus (shape parameter) β_w also shows a mild increase with

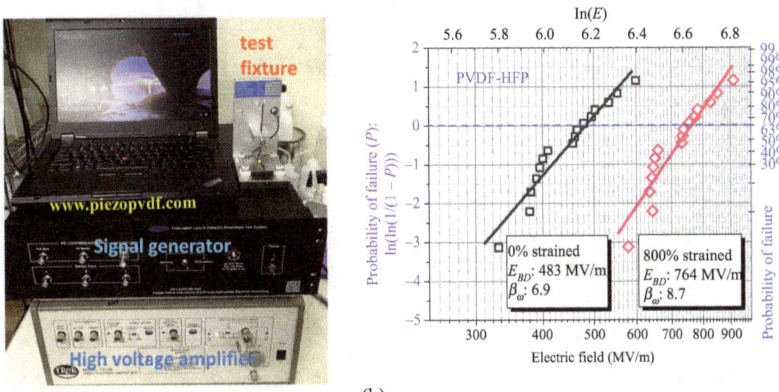

(a) (b)

Figure 2.8 (a) Dielectric breakdown measurement instrument [8]. (b) Weibull plots of breakdown fields for unstretched and stretched PVDF-HFP

stretching. The retention of the shape parameter suggests that the improved breakdown properties are also reflected in the low field failures. For example, the entire population of failures for the strained film lies at fields higher than the field corresponding to the 95% failure probability for the unstretched film. In other words, the stretched film does not exhibit any dielectric breakdown for fields well above 550 MV/m, where most of the respective unstrained films have already failed. This consideration has important practical implications, as it is often the low field breakdown strength, rather than the characteristic Weibull strength, which determines the operation field in applications [19].

In order to meet practical requirements and facilitate the communication between the materialists and engineers, various breakdown test modes have also been developed. The AC breakdown test mode is designed to examine the insulation properties of cable jackets, and it follows the international standard IEC 60243–1 [20] and ASTM D149 [21]. During the measurement, sinusoidal AC voltages with stepwise increasing amplitude are applied to a dielectric sample. Each AC voltage is maintained for a certain time period before ramping to the next higher voltage. The ac breakdown strength is defined as the electric field under which the sample cannot survive for the prescribed time duration. A schematic diagram showing the voltage profile is demonstrated in Figure 2.9(a), where an initial 100 V AC voltage is applied to the test sample for 2 s, and then the voltage amplitude is increased by 50 V each step until the electrical failure.

Another breakdown mode is called lifetime test, which applies a constant DC or AC voltage across the measured dielectric and records the time when the sample is failed. This mode is further developed to investigate the self-healing behavior of polymer dielectrics. As shown in Figure 2.9(b), polyimide (PI) and polyimide nanocomposites filled with zirconium dioxide (ZrO_2) are subjected to continuous cycles of 10 Hz and 500 MV/m AC electric field [22]. While the unfilled PI polymer fails after less than 7,000 cycles, the 2 wt% PI nanocomposites can sustain

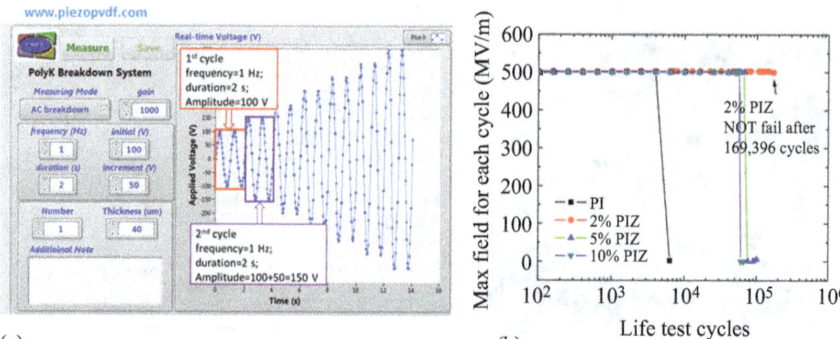

(a) (b)

Figure 2.9 (a) Voltage profile in AC breakdown measurement [8]. (b) Lifetime test of dielectric breakdown. Reprinted from [22] with permission of AIP Publishing

almost 170,000 cycles. The greatly improved lifetime in the nanocomposites can be partially attributed to the high thermal conductivity of ZrO_2, which helps to dissipate the heat generated during discharging, and partially to the evaporation of ZrO_2 upon electrical breakdown. The latter can consume a significant amount of heat, thus substantially impeding the thermal aging of the PI.

2.2.4 Leakage current

Leakage current experiment is another critical measurement in capacitor and electrical insulation materials. In fact, accurate measurement of leakage current (electrical resistance) is one of the screening parameters to exclude poor dielectric materials, as a majority of commercial DC link capacitors are operated under a continuous and constant DC bias voltage (e.g., 200 V/μm) with a small ripple voltage ($<\pm 10\%$ of DC voltage) for over 10,000 h. In many applications, the leakage current dominates the energy loss, rather than the charge-discharge process in the DC link capacitors [23].

Dielectrics with a high leakage current level would be more prone to electrical breakdown because of the increased possibility of electron avalanche, which can be further exacerbated by the conduction-loss induced heating. Moreover, the large electrical conductivity reduces the maximum voltage that can be obtained in high energy/power density capacitors, because stored charges will continue to dissipate as strong conduction currents. For certain applications such as railgun or high power microwave weapons, electrical power depends on the instantaneous release of huge voltages stored on the capacitors and the limited voltage level would put a cap on the maximum power output that can be realized in these applications.

As reflected by the name, leakage current measures the current passing through the measured sample under applied electric fields. However, it is not as easy as it seems to be. The leakage current usually follows an exponential decay with measuring time. As shown in Figure 2.10(a), leakage current is measured as a function of test time [24]. The initial current density is large, because it evaluates the overall

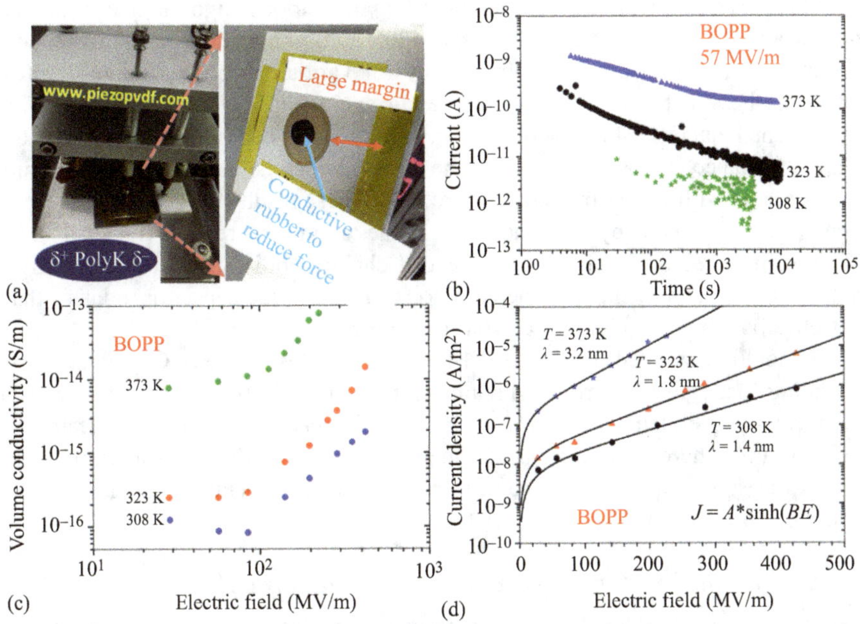

Figure 2.10 (a) Leakage current test fixture developed by PolyK Technologies [8]. (b) Leakage current of BOPP measured with time duration. (c) Volume conductivity of BOPP at different temperatures and electric fields. (d) Leakage current densities of BOPP fitted to hopping conduction model [24]

effects of ion transport, dipolar orientation, and electron conduction. With the lapse of time, the dipolar orientation begins to saturate and the ions gradually accumulate near the sample-electrode interface, causing the decay of leakage current. The current will become more and more stable over time, and then only the actual electron conduction is being measured, which is defined as steady leakage current. From this perspective, the leakage current measurement is a long time process, during which the premature breakdown is easy to occur. As a result, the electrode clamping force needs to be adjusted just enough to maintain the electrical contact in order to prevent premature breakdown, and the sample margin area (the surface area without electrode) should be kept large enough to minimize the surface current (Figure 2.10(b)).

In addition to the long stabilization time, the steady current also follows a nonlinear increase with field when the applied field exceeds some critical value. As shown in Figure 2.10(c), the BOPP shows Ohmic conduction until 100 MV/m, above which the leakage current exhibits an exponential field dependence, resulting in a strongly increased volume conductivity. The enhanced conductivity arises from the detrapping of charge carriers from localized traps, which increases the apparent charge density [24].

The steady current density can be fitted to various high field conduction models and the fitting results provide insights into conduction mechanisms and guidance on tailoring polymer morphology for various purposes. Figure 2.10(d) shows the best fit to the BOPP's leakage currents using the hopping model. The experimental data can be described very well by the hopping mechanism. The hopping distance can be derived from the fitting, which is on the order of 1–3 nm and increases with increasing temperature [24]. These results suggest the electrical currents are dominated by the conduction in the amorphous regions (rather than in the crystalline regions), where the enhanced chain mobility facilitates the charge motion. This is consistent with the common practice in the capacitor industry that dielectric polymers with higher crystallinity are more preferred.

It should be pointed out that ASTM D257 "DC Resistance or Conductance of Insulating Materials1" is proposed mainly for electrical insulation materials which are normally operated at <10 V/μm [25], while most capacitors are operated at >100 V/μm, where nonlinear dielectric behavior is significant. Therefore, it is critical measure the DC resistance at relevant capacitor operation conditions, i.e., higher electric field and longer time.

2.2.5 *Thermally stimulated depolarization current*

Thermally stimulated depolarization current (TSDC) measures the depolarization current as a function of temperature. It consists of a high voltage source, a high precision electrometer, and a temperature-controlled oven (Figure 2.11(a)). Prior to the measurement, a sample is polarized at a high temperature. The polarization time should be at least comparable to the relaxation time of the polarizable units in the sample so that these units could respond to the applied field and be polarized. Then, the temperature is decreased while maintaining the applied field across the sample.

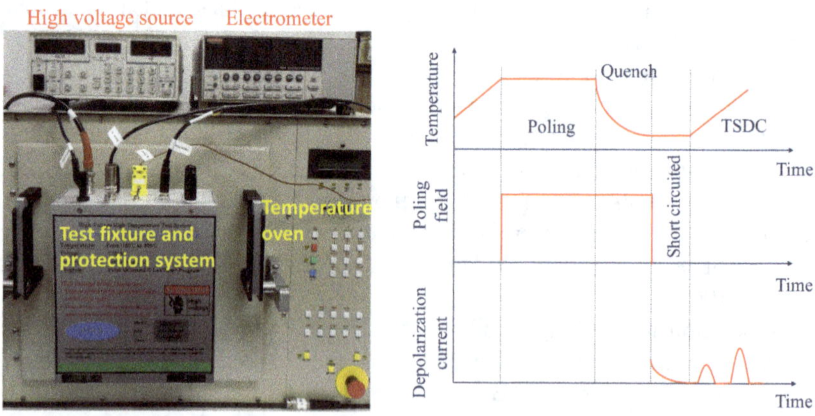

(a) (b)

Figure 2.11 (a) TSDC measurement instrument from PolyK Technologies [8]. (b) Schematic diagram of TSDC measurement

This step is to freeze the polarizations in the sample, which can be retained even after the removal of the electric field. When the temperature is cooled down to the setpoint, usually below the glass transition temperature to prevent depolarization, the sample is short-circuited for a few minutes to remove the surface charges. Then, the sample is heated at a constant rate and the depolarization current is measured as a function of temperature. A schematic diagram is shown in Figure 2.11(b) to illustrate the entire process.

From the description, it is known that the depolarization current is strictly related to dipolar depolarization and charge detrapping, processes that can be activated by both temperature and applied field. The TSDC results contain information on such relaxation processes, which can be deconvoluted. Since this technique does not require external high voltages during the measurement, the electrical conduction through the sample is absent and will not obscure the output signals. Besides, the electrometer connected to the sample can measure currents as low as 10^{-15} A, enabling the detection of very mild relaxation processes.

Figure 2.12(a) shows TSDC results of a PVDF thin film, which is polarized at various fields for under 50 °C for 20 min [12]. Three major depolarization processes can be identified, located around –45 °C, 60 °C, and 90 °C. To delineate the origins of these relaxation processes, the charge densities of the three discharge peaks are obtained by integrating the depolarization current underlying each peak, as summarized in Figure 2.12(b). The low-temperature peak (at –45 °C) is apparently associated with the glass transition of PVDF. The depolarization of amorphous dipoles gives rise to a peak in the TSDC spectrum. From its nature, the charge density for this peak should be proportional to the applied field with the slope being the dielectric constant of amorphous PVDF. As shown in Figure 2.12(b), the best linear fit yields a dielectric constant of 10, which is in good agreement to the low field dielectric constant of PVDF. The 60 °C peak originates from the injected charges. Under high field poling, the homocharges could be injected into the sample-electrode interface. The depolarization of the injected homocharges

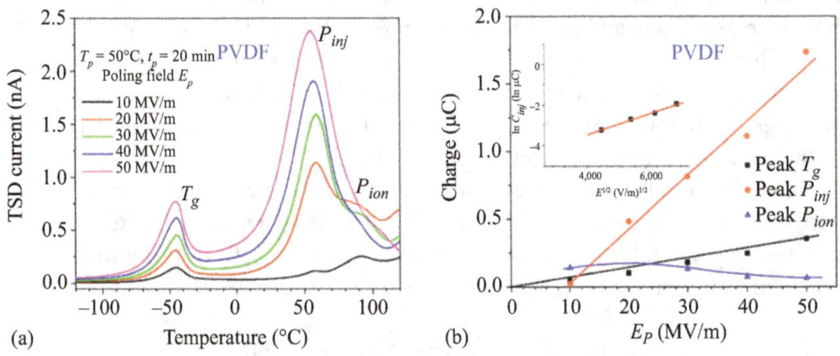

Figure 2.12 (a) TSDC measurement of PVDF poled at different fields. (b) Integrated charge density from TSDC peaks. © 2015 American Chemical Society. Reprinted, with permission, from [12]

generates this 60 °C TSDC peak. If this is the case, the injected charges should follow Schottky emission, and the total charges being injected would scale as $\ln C_{inj} \propto E_p^{1/2}$, where C_{inj} is the charge density for the 60 °C P_{inj} peak and E_p is the poling field. This is exactly observed in this measurement, as depicted in the inset of Figure 2.12(b). With the assignment of the two relaxation processes, the 90 °C high-temperature peak can only be attributed to the depolarization of trapped space charges within the PVDF. Note the low intensity of 90 °C current peak is not due to the small population of ions, but instead should be related with the slow dynamics of ion polarization, i.e., the poling time is not long enough to fully polarize the mobile charges. The deconvoluted three relaxation processes provide comprehensive information about what happened during the application of high voltages. This information is very useful to understand the high field dielectric behavior of polymer dielectrics [12].

2.3 Advanced dielectric measurement

The above characterization methods do not consider the application environment that the capacitors are practically used in. Since many dielectrics exhibit nonlinear dielectric properties (with electric field, applied frequency, DC bias, etc.), the actual dielectric performance of these dielectric materials during practical operation cannot be easily extrapolated based on the measurement results obtained under completely different conditions.

For instance, in many applications, the stored energy is required to be released to loads in a certain period. The discharging speed of a capacitor is not only related to the intrinsic properties of the dielectric materials but is also largely determined by the external loads connected to the capacitor. Very often, the power density, the ratio of the energy density to discharging time, is used to characterize the capacitors or dielectric materials. The power density is determined by the maximum energy density and the quickest discharging speed that can be achieved by the dielectric materials. Another example is the DC link capacitors, which usually work under a large DC bias voltage and provide electric powers by continuous low-peak-to-peak voltage fluctuation (ripple). The electric powers can be systematically increased by increasing switching frequency, which is limited by the dielectric loss of materials (i.e., large operation frequency will increase the inner temperature of capacitors) and the dynamics of polarization relaxation [10].

Since the **regular dielectric measurement** as mentioned before cannot characterize the actual dielectric properties of capacitors when used in actual devices, many new test modules are developed to simulate the practical operation environments and to provide a more accurate evaluation of material performance.

2.3.1 High-frequency polarization-electric field hysteresis loop

Currently, most literature still use 10 Hz P-E loops to evaluate the energy/power storage performance of dielectric materials. In fact, this methodology is borrowed from the ferroelectric community, where researchers use low-speed AC voltage to polarize large ferroelectric domains and study ferroelectric switching, which are relatively slow. In practical applications, especially for high energy/power density requirements, the capacitors are always used at high repetition rates, enabled by high rate switches (e.g., silicon insulated-gate bipolar transistor (IGBT) semiconductor), in order to enhance the electrical power delivered by the capacitors. The switching speed of the current IGBT switches is rated at 1–2 kHz, which determines the operation frequency of capacitors in these applications. With the development of advanced wide bandgap (WBG) devices such as SiC and GaN, the switching rate can be further increased to over 10 kHz, thus greatly improving electrical power conversion efficiency and power density. It was found that SiC-enabled power inverter could cut the energy loss by 33% comparing these with Si-switches.

It can be seen that in any case, the capacitors are not used at low speeds. Since the dielectric properties of dielectrics are strongly dependent on frequency, the evaluation of potential capacitor materials should be carried out at the same environment for better accuracy. Besides, for polymer nanocomposites, heavy filler loadings are usually required to generate considerable improvement in properties, as predicted by the logarithmic mixing rule. The high-frequency P-E loop measurement can exclude electrical conduction effects and reveal the general trend in the high field dielectric properties of nanocomposites upon filler doping. In the following, the dielectric performance of alumina/polyetherimide (Al_2O_3/PEI) nanocomposites are studied as a function of charge-discharge frequency using a high-speed P-E loop test instrument developed by the team (up to 300 kHz). Two concentrations of fillers are used here to represent the dielectric materials with low loss (5 wt% nanocomposite) and high loss (60 wt% nanocomposite), respectively.

Figure 2.13 shows the dielectric properties of 5 wt% Al_2O_3/PEI measured at 1 to 1,000 Hz. As shown in Figure 2.13(a), the loop becomes narrow with increasing frequency, indicating a reduced dielectric loss. As a result, the discharge efficiency is substantially increased from ~77% at 1 Hz to above 97% at 1,000 Hz (Figure 2.13(b)). In a charge-discharge cycle, the charged energy evaluates the contribution from dipolar orientation (fast and reversible), charge displacement (slower and reversible), and charge conduction (slowest and irreversible), whereas the discharged energy only evaluates the reversible contribution. As shown in Figure 2.13(c), the discharged energy remains almost the same while a marked decrease in the charged energy is observed with increasing frequency. It confirms that the improved efficiency and the narrowed P-E loops are attributed to the restricted charge conduction. It also implies that the charge displacement is not a major contribution to the measured polarization, in contrast to the high filler-loading composite (Figure 2.14). The absence of charge displacement could be due to the low amount of impurity ions that are introduced along with the fillers. Although it requires additional experiments to determine whether this sample is feasible for energy storage applications, the preliminary results suggest that the dielectric loss under practical application setting is not as bad as it seems to be in the low-frequency regular measurement.

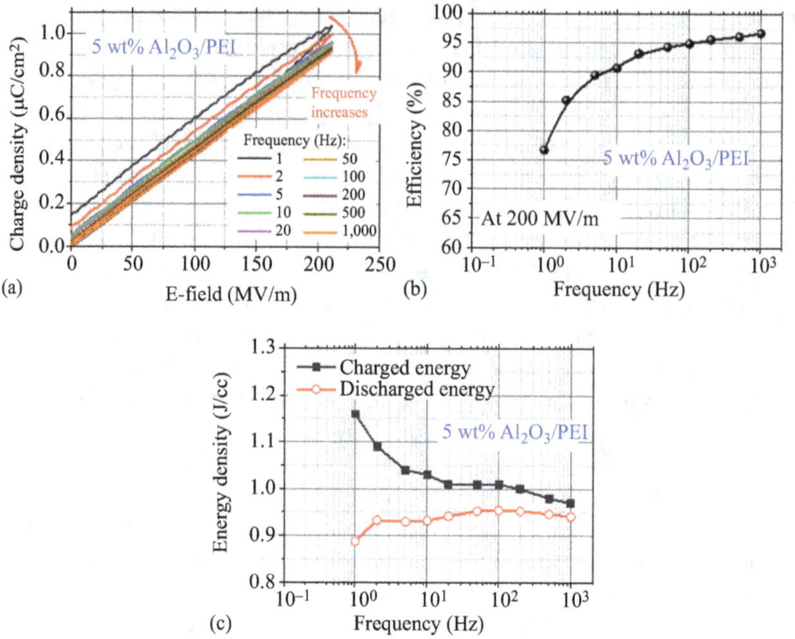

Figure 2.13 *(a) P-E loops of 5 wt% Al_2O_3/PEI at different frequencies. (b) Discharge efficiency and (c) discharge energy density as a function of frequency. The test was performed with a high-speed P-E loop test instrument and a high power amplifier developed by the team (up to 300 kHz)*

Figure 2.14 shows the dielectric properties of 60 wt% Al_2O_3/PEI from 200 to 10,000 Hz. A similar trend is observed with gradually narrowed P-E loops at increasing frequencies. The P-E loops at 200 Hz are highly distorted and cannot be used in the evaluation of material performance, e.g., the efficiency at 140 MV/m is calculated close to 0 (Figure 2.14(g)). However, the linear dielectric behavior is recovered in the P-E loops when the measuring frequency is increased to 10,000 Hz (Figure 2.14(f)). The discharge efficiency of the same sample is substantially enhanced to ~75% at 10,000 Hz, compared to ~5% at 200 Hz (Figure 2.14(g)). Figure 2.14(h) and (i) presents the charged and discharged energy densities at different frequencies, respectively. The discharged energy density shows an abrupt decrease with increasing frequency, consistent with the frustrated charge conduction. The charged energy density also shows a systematic increase with increasing frequency, e.g., the energy density is enhanced from 0.2 J/cc at 500 Hz to 0.5 J/cc at 10,000 Hz. It reflects the contribution from charge displacement. With increasing frequency, more and more charges cannot exhibit long-range migration across the sample (i.e., charge conduction); instead, they would be separated over a finite distance by the applied field. The short-range oscillating movement of charges

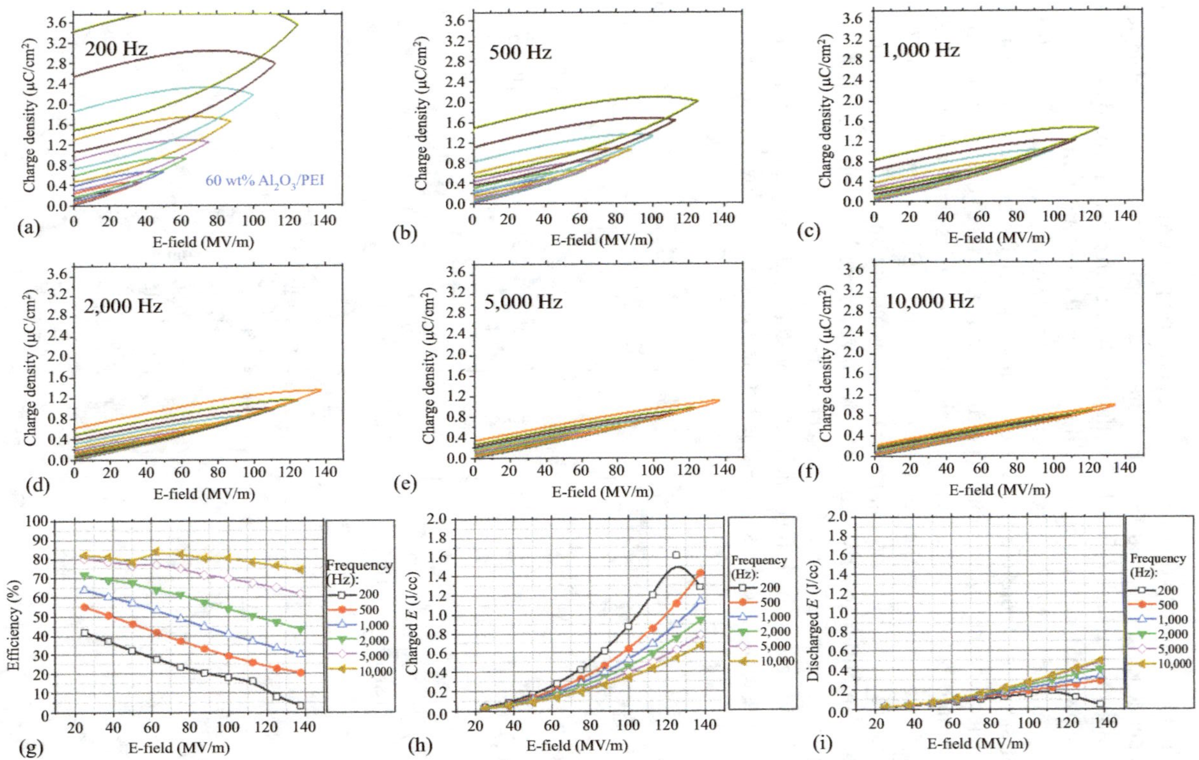

Figure 2.14 *(a–f) P-E loops of 60 wt% Al_2O_3/PEI with increasing frequency (up to 10,000 Hz). (g) Discharge efficiency, (h) charged energy density, and (i) discharged energy density as a function of frequency*

provides addition polarization effect, adding to the instantons dipolar polarization of the polymer nanocomposite.

2.3.2 *High field dielectric constant and dielectric loss*

In addition to the frequency dependence, the dielectric properties could also exhibit a strong dependence on applied field. For practical applications where dielectrics are always subjected to high AC or DC voltages, their performance would deviate a lot from the low field measurement. In this regard, the test instrument should be capable of simulating working conditions and evaluating the "real" dielectric response of samples. There are two different test modes for the measurement of high field dielectric constant and loss. The first one applies a large and constant DC bias on a sample before measuring its dielectric properties with a normal LCR meter. The driving signal is the combination of a large DC bias and a small AC voltage in this case. The second test mode does not require any DC sources. It uses a high power amplifier to drive large AC voltages directly to the samples, and acquires their dielectric constant and dielectric loss.

Figure 2.15 presents high field dielectric constant and loss for a PVDF with a constant DC bias applied to the sample (the first test mode). The low field dielectric

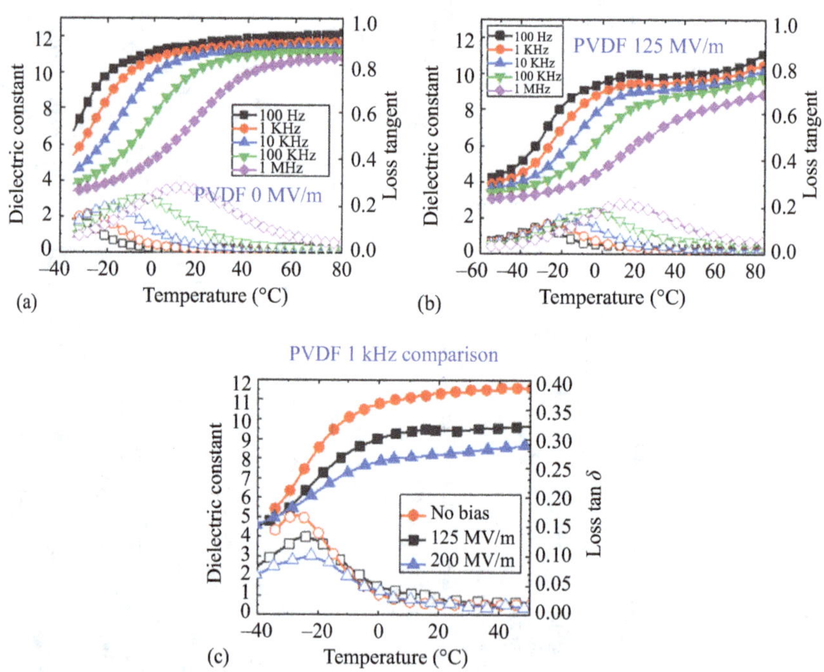

Figure 2.15 (a) Dielectric properties of PVDF without DC bias. (b) Dielectric properties of PVDF at 125 MV/m DC bias field. (c) Comparison of dielectric properties of PVDF at different DC bias voltages

measurement of the same sample is shown in Figure 2.15(a) for comparison. In the low field measurement, the dielectric constant shows a step increase at the glass transition temperature (T_g), accompanied by an associated loss peak. With the application of a 125 MV/m DC bias, both dielectric constant and dielectric loss are reduced at every frequency, but the relaxation temperature is retained regardless of the applied bias (Figure 2.15(b)). A more direct comparison is demonstrated in Figure 2.15(c), where the measurement results under three different DC bias voltages are compared. A similar trend is observed. During the measurement, the randomly aligned dipoles without external DC bias should be easier to rotate with the measuring AC field than the dipoles that are already pre-aligned by a large bias field. Since the dielectric constant and loss measure the dynamics of dipole orientation, the measurement with DC bias will generate a lower dielectric constant and dielectric loss than the measurement without DC bias. The relaxation temperature, on the other hand, signifies the onset of dipole orientation. The applied DC bias should only reduce the dipolar orientation rate, but will not affect the temperature under which the dipoles are mobilized.

Figure 2.16 shows the measured dielectric constant and dielectric loss of the same PVDF under large AC voltages (the second test mode). Figure 2.16(a) and (b) presents the 3D graphs of the measurement results as a function of temperature and frequency. The corresponding contour plots are shown in Figure 2.16(c) and (d). The dielectric performance measured in this mode is significantly different from that measured in the first mode. As shown in Figure 2.16(c), at every measuring frequency, the dielectric constant shows an initial increase with increasing electric field, followed by a decline with further increasing field. A maximum dielectric constant is obtained at intermediate electric fields (critical field). The critical field shifts to higher values with increasing frequency (as marked by a yellow dotted line). The exact same trend is observed in the dissipation factor, which assumes the maximum values at the same frequency and temperate range where the maximum dielectric constants are realized (Figure 2.16(d), marked also by a yellow line).

Unlike the low field dielectric measurement and the high field measurement with a small AC voltage (the first test mode), samples measured in the second test mode are subjected to a large AC signal, which could switch the dipoles in the ferroelectric crystalline phase. With increasing driving voltage, more and more crystalline dipoles participate in the polarization process, providing an extra contribution to the measured dielectric constant. After the majority of ferroelectric dipoles have responded to the applied field, further increasing voltage will not generate as much polarization as expected; and the dielectric constant will experience a decrease due to the saturation of crystalline dipolar orientation [26]. From this perspective, the critical field to obtain the maximum dielectric constant (or dissipation factor) should correspond to the coercive field for ferroelectric switching. It can be understood that a larger coercive field is required to polarize ferroelectric crystals when poling frequency increases. Similarly, the critical field will also shift to higher values with increasing measuring frequency.

The two different high field measurements yield strikingly different results, even for the same sample and with the same amplitude in the applied voltage. The

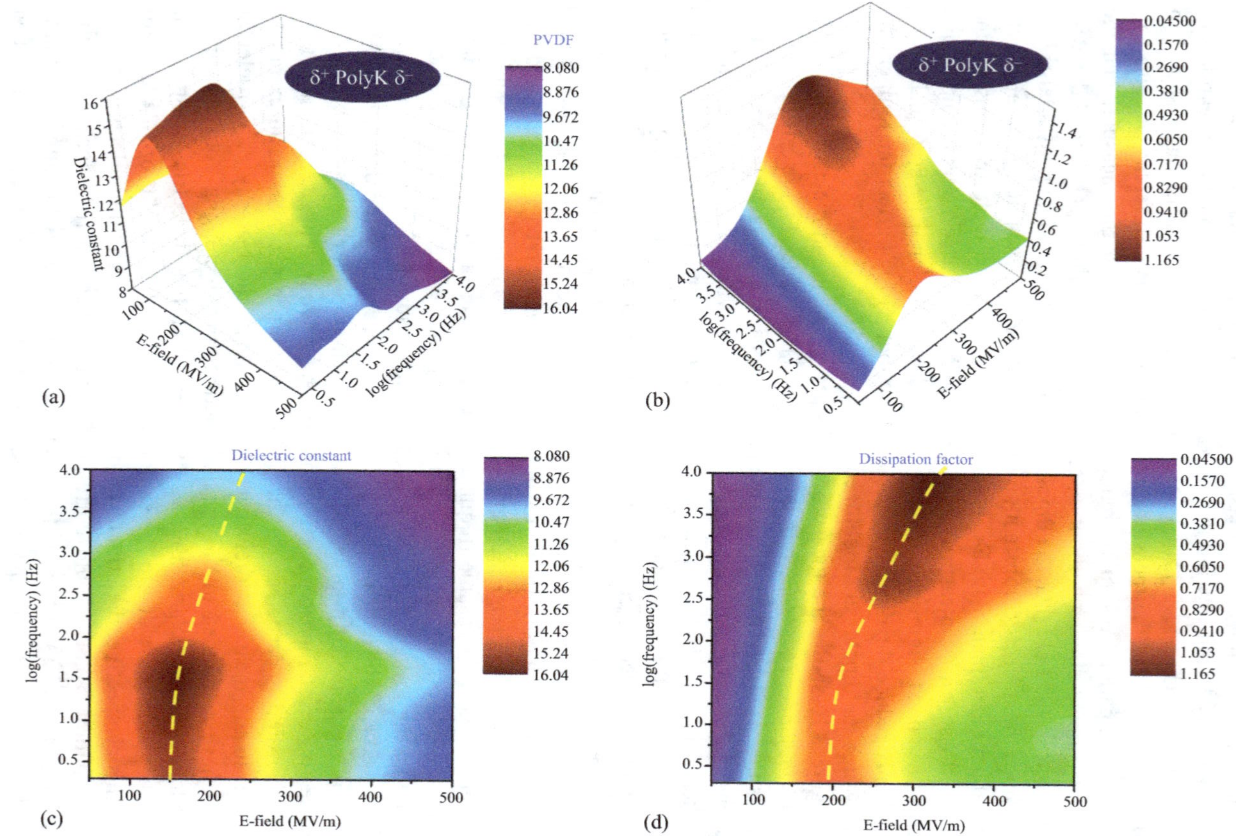

Figure 2.16 (a), (c) Dielectric constant of PVDF as a function of applied field and frequency. (b), (d) Dielectric loss of PVDF as a function of applied field and frequency. Measurement was performed with a high-speed high-power P-E loop test setup developed by the team [8]

measurement results not only provide an accurate prediction for practical applications but also deep insights into the molecular origins of dielectric properties.

2.3.3 Ripple measurement with DC bias for electrical vehicles

Most of high energy/power capacitors are used as DC link capacitors in power electronics for wind energy and electrical vehicles. They are replacing the so far used electrolytic capacitors due to more stringent electrical requirements. In these applications, the DC link capacitors converse electrical powers and couple different electrical grids to one DC voltage level. One the one hand, these capacitors act as an output filter, absorbing switching currents for minimum voltage fluctuation (ripple). On the other hand, they can also provide high-frequency currents to energize large loads while retaining the constant DC voltage. The DC link capacitors can improve system energy density and address the challenges of ripples during rapid switching power conversion. With the development of advanced wide-bandgap switches, the switching rate can increase tens or hundreds of times faster than that of the traditional technology, thus greatly increasing electrical out powers. New materials need to be developed and old materials need to be re-evaluated to take better advantage of the fast switch technology.

It is seen that the environment where these capacitors are operated is very different from the traditional P-E loops. It is well established in the dielectric community that PVDF-based materials are very lossy materials. As shown in Figures 2.6(a) and 2.7(a), PVDF-HFP shows a broad opening in the P-E loop at high electric fields, resulting in a large dielectric loss and a low efficiency. However, at the high field range, the loop front actually is very narrow. It can be envisaged that if a constant DC bias is applied to the sample and the dielectric response is only measured with a small AC signal, the dielectric loss would be much less than expected.

Based on this consideration, a special high-speed instrument for evaluating capacitor performance under a constant DC bias and a small AC ripple voltage is developed by the team [8]. This measurement can accurately characterize dielectric materials when used as DC link capacitors. Figure 2.17 shows the dielectric behaviors of PVDF-HFP driven by the same small ripple voltage but under different DC bias. Substantially different behaviors are observed, implying that the dielectric performance of this dielectric material is strongly affected by the bias voltage. Roughly, three stages can be distinguished. In the first stage when the DC voltage is below 400 V (Figure 2.17(a)), the P-E loop becomes broader and the slope of the P-E loop also increases with increasing DC bias. It reflects the polarization of ferroelectric crystals. With increasing field, more and more ferroelectric crystals can rotate along the electric field, providing an additional increase in dielectric constant (increased slope of P-E loop); meanwhile, these crystals cannot be depolarized, leaving enhanced remnant polarization and dielectric loss. The second state is above 400 V and below 1,100 V (Figure 2.17(b)), where the P-E loop becomes narrower with field. When the maximum switching rate of ferroelectric

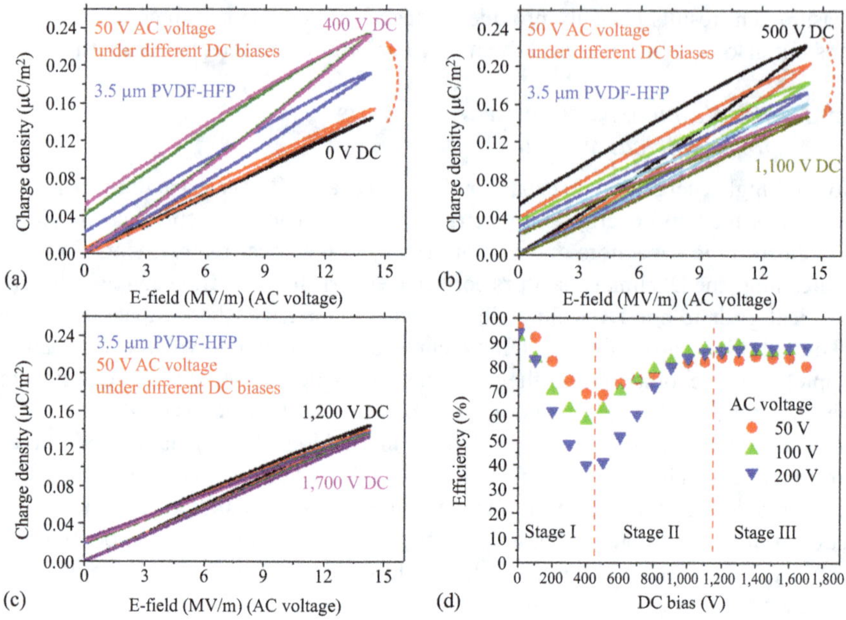

Figure 2.17 (a–c) P-E loop of 3.5 μm thick PVDF-HFP driven by the same ripple voltage but under different DC voltages. (d) Discharge efficiency as a function of DC voltage and ripple voltage. The PVDF-HFP film is free standing and it was manufactured using a melt-extrusion and biaxial orientation process

dipoles is realized (at ~400 V), further increasing DC bias will gradually saturate the ferroelectric orientation and therefore reduce energy loss. The last stage is when most of the ferroelectric domains have already been polarized which occurs at electric fields larger than 1,100 V (Figure 2.17(c)). The PVDF-HFP reassumes the linear dielectric behavior, with dielectric response determined by the small AC signals regardless of DC bias. It is also worth mentioning that at this stage the remnant polarization comes from the conduction loss, instead of ferroelectric loss, which explains the slightly lower efficiency of the sample at this stage than at the first stage without DC bias (Figure 2.17(d)), although both stages evaluate only the contribution from amorphous dipolar orientation.

The similar trend is also reflected in other ripple voltages (Figure 2.17(d)). The minimum efficiency is always obtained at 400–500 V, i.e., ~130 MV/m. This electric field just corresponds to the critical field for ferroelectric switching. From this measurement, it can be obtained that the notorious ferroelectric loss can be well suppressed at high electric fields, and the efficiency of PVDF-HFP is not as low as expected (it is well above 80% at high fields). It can also be seen that a high crystallinity is preferred in these applications, as conduction loss dominate the dielectric loss at high DC bias and polymers with high crystallinity could exhibit a

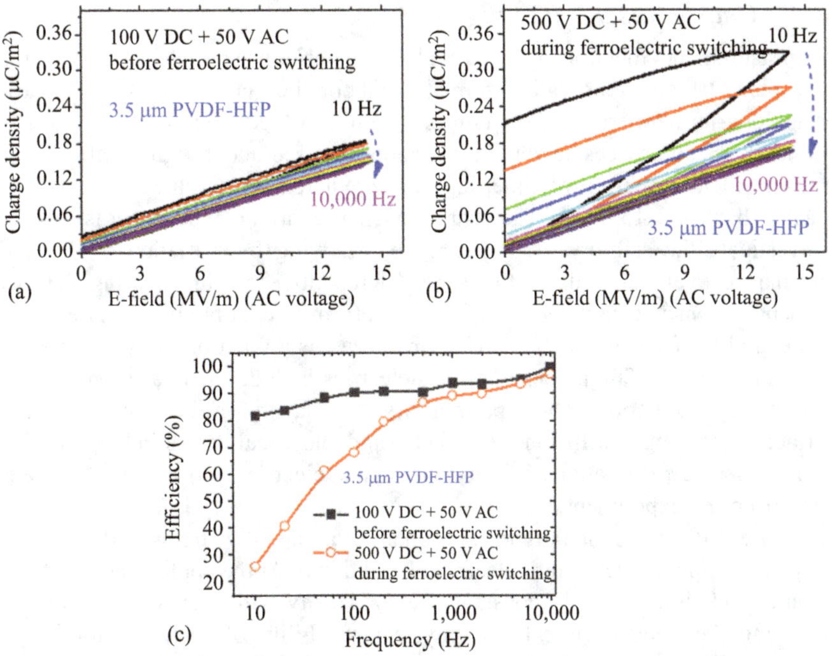

Figure 2.18 (a) Effects of frequency on ripple P-E loop of PVDF-HFP at a small DC bias. (b) Effects of frequency on ripple P-E loop of PVDF-HFP at a large DC bias. (c) Discharge efficiency as a function of frequency for the above two cases

low leakage current. As mentioned before, the DC-link capacitors work at much higher frequencies than the current measurement frequency (100 Hz). The frequency dependence of ripple dielectric performance is further examined.

Figure 2.18 shows the ripple dielectric properties of PVDF-HFP as a function of measuring frequency. Two DC biases are employed, with one below the critical field of ferroelectric switching and the other close to the critical field. At a low bias voltage, the frequency does not have significant impacts on efficiency, because the dielectric response comes from the dipolar orientation in the amorphous phase and the relaxation speed is too fast to be affected by the measurement frequency. At a high bias voltage, the frequency change has substantial effects on efficiency. Since ferroelectric domain switching is a slower process and the switching rates fall in the same range of measurement frequency, increasing frequency would impede ferroelectric switching, resulting in a narrowed P-E loop and higher discharging efficiency. The efficiency at 10,000 Hz is above 95% for both DC bias voltages, justifying the feasibility of PVDF-HFP in these application conditions. These measurement results also provide guidance in selecting appropriate film thickness to achieve low dielectric losses, when the application frequency, input/output bias voltage, and power density are considered together.

2.3.4 *Charge-discharge fatigue*

The measurements mentioned above evaluate material properties using a one-time experiment. It cannot provide information about the stability of the dielectrics during long-term operation. Such information is of equal importance to other parameters, as all devices need to be continuously used under high electrical stress for a considerable period of time, e.g., the predicted service time for a DC bus capacitor bank is >15,000 h. The charge-discharge fatigue experiment is designed to investigate the stability of dielectrics by studying the variation of dielectric performance over hundreds of thousands of repetitive cycling. During the measurement, a constant unipolar or bipolar waveform is continuously applied to the sample, and the measurement results are recorded as a function of cycle time. The possible reasons for fatigue behavior in dielectrics include the relaxation of lattice strain energy, formation of low permittivity dead layers on the sample-electrode interface, long-range diffusion of defects, and electrical stress-induced decomposition. Besides, the fatigue life at a particular electric field can also be determined from the experiment.

Figure 2.19 shows the fatigue experiment of a biaxially stretched PVDF under continuous unipolar cycles until electrical breakdown. At the applied electric fields, the sample exhibits a nearly constant energy density with increasing cycle time, confirming the good chemical and physical reliability of PVDF. Although the Weibull breakdown strength of PVDF can reach 700–800 MV/m, the sample can only withstand tens of thousands cycles even at half of the Weibull strength (i.e., ~400 MV/m). The maximum cycle time is substantially increased when the applied field is reduced to 350 MV/m (data not shown here). It confirms what was mentioned in "**Dielectric breakdown**": it is the low field breakdown strength, rather

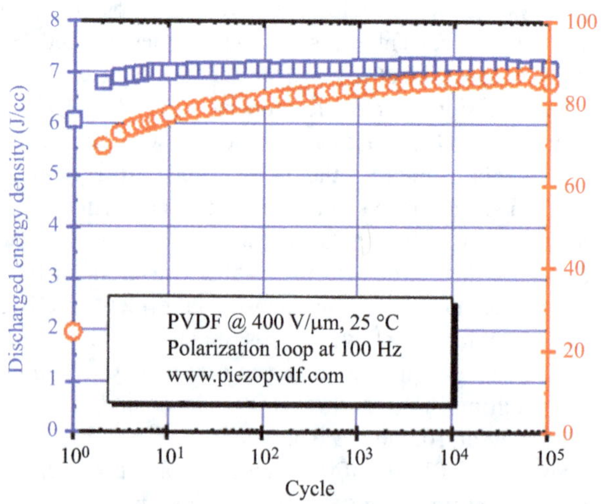

Figure 2.19 Discharged energy density and discharge efficiency as a function of fatigue cycles for PVDF

than the Weibull breakdown strength, which determines the operation field during practical applications.

In contrast to the stable performance of discharged energy density, the efficiency shows a continuous increase with increasing cycle time (Figure 2.19). It reflects the gradual alignment of ferroelectric domains with the applied field. The ferroelectric switching is an irreversible process that will substantially reduce energy storage efficiency. The initial low efficiency is related to the polarization of ferroelectric crystals. With the lapse of time, more and more ferroelectric domains can be polarized, resulting in an increase in efficiency. When most of the ferroelectric dipoles are aligned with the electric field, the efficiency reaches the stable and maximum value, as now it only evaluates the reversible polarization from the amorphous dipoles. The stabilization time can be shortened at a larger electric field, e.g., it needs hundreds of thousands of cycles to stabilize efficiency at 350 MV/m (data not shown here), but only needs thousands of cycles when the field is increased to 450 MV/m (Figure 2.19). This indicates that ferroelectric switching is a field-activated dynamic process, where a higher electric field can speed up dipolar orientation.

2.3.5 Capacitor discharge test

The energy storage performance of dielectric material is usually evaluated by the P-E loop, as mentioned before. During the measurement of P-E loop, the stored energy is discharged in a controlled way determined by the waveform of the applied voltage, e.g., using a triangular waveform, the voltage on the test sample will be released in a linear manner with respect to discharge time. However, for practical applications, capacitors are connected to the external loads that are about to be powered. Discharging of stored energy follows an exponential decay with time, which is determined by the entire circuit (including the loads, electrical interconnections, and capacitors). The discharge conditions and the discharged energy density, in this case, are significantly different from those in the P-E loop measurement [9].

To address this issue, the capacitor discharge test method is developed to evaluate the actual energy storage performance of dielectric materials (Figure 2.20(a)). During the measurement, the dielectric materials are first charged to a given voltage and after that, the stored energy in the capacitor is discharged to an external load. The discharging process is recorded by an oscilloscope, which is used to calculate the energy/power density of the measured sample. In a typical discharging curve (Figure 2.20(b)), the voltage drop shows an exponential dependence on discharging time and the discharging rate can be evaluated by the time constant $R_L C$ (R_L is the resistance of the external load, C is the capacitance of the sample). From the profile of voltage decay ($V(t)$), the discharge power (P) can be calculated by $P = \frac{V(t)^2}{R_L}$, whereas the energy density (U) under this charging voltage can be calculated by $U = \frac{\int_0^\infty P dt}{Volume} = \frac{\int_0^\infty \frac{V(t)^2}{R_L} dt}{Volume}$.

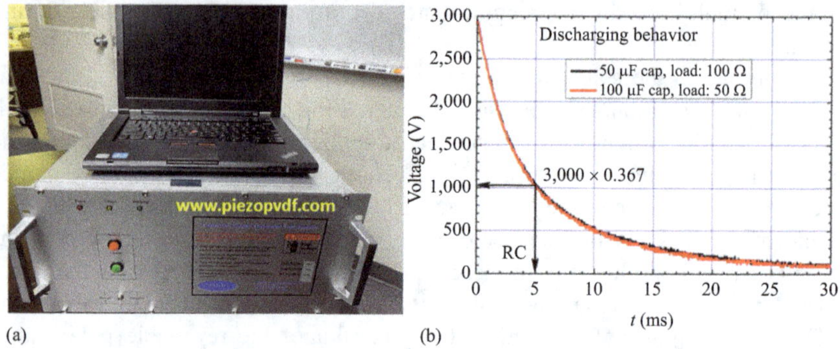

Figure 2.20 (a) Capacitor discharge test system developed by PolyK Technologies. (b) Typical discharge curve at different external loads [8]

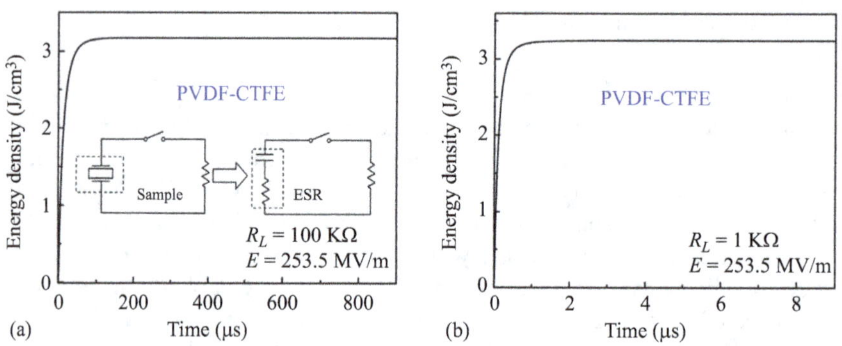

Figure 2.21 Discharge energy density as a function of time measured from the direct discharge of the PVDF-CTFE copolymer to a resistor load R_L. (A) R_L=100 kΩ (B) R_L=1 kΩ. The charging field is 253.5 MV/m in both cases. Reprinted from [9] with permission from Elsevier

Figure 2.21 present the time dependence of the discharged energy density for a PVDF-co-chlorotrifluoroethylene (PVDF-CTFE) copolymer when connected to various resistor loads. With increasing resistance of external load, the discharging rate is reduced accordingly. It should be mentioned that the increase in discharge time would not be strictly proportional to the reduction in load resistance. This is because the test sample also has a certain dielectric loss, which can be modeled as an equivalent series resistor (ESR) and should be considered in the calculation of time constant, as shown in the inset of Figure 2.21(a). The ESR is estimated from the dissipation factor (tan δ) by $ESR = \frac{\tan\delta}{wC}$ (w is the angular frequency, C the capacitance of the sample). As can be seen, the ESR becomes more obvious at a lower discharging rate, resulting in a more-than-expected increase in the time constant. Besides, the capacitance of the sample also has frequent dependence.

With reducing frequency, the capacitance would increase, resulting in an even larger time constant. Both effects shift the time constant to higher values when a larger external load is connected.

The total discharged energy densities of the copolymer to different load resistors R_L can also be calculated from Figure 2.21, with 3.5 J/cm^3 for 100 kΩ resistor and 3.3 J/cm^3 for 1 kΩ resistor. With reducing R_L, the discharged energy density is decreased, due to the energy loss in the ESR. When the external load is increased, the time constant will be increased, resulting in a higher ESR value [9].

It shall be pointed out that the power density measured from the discharge test is actually misleading. While the power density certainly increases with short discharge time (smaller load) for almost all dielectric materials, in a large size packaged capacitor, the practical power density is always limited by the heat dissipation capability; thus, the actual power density may be millions time smaller than the value measured with a millimeter size sample, where the heat can be immediately dissipated to the surrounding air or insulation oil. Therefore, power density measured from a small piece of film is meaningless.

2.3.6 Positive up and negative down

Ferroelectric polymers are promising dielectric materials for energy storage because of their high dielectric constant. The increased energy density can reduce capacitor size for better integration into electronic devices with continuous miniaturization requirements. However, these ferroelectric materials also suffer large dielectric loss, which prevents the utilization at high repetition rates due to serious heating issues, therefore greatly impeding the power density that can be realized in these capacitors. The development of new ferroelectric materials first requires a clear understanding of ferroelectric switching in these materials. However, proper characterization of ferroelectric polarization is challenging, as dielectric loss generally measured from other technologies is an overall effect, combing the contribution from ferroelectric loss and conduction loss. Positive up and negative down (PUND) is a new technology enabling the deconvolution of different polarization events [27,28], which can be used to study ferroelectric switching and reveal the underlying structural correlation.

As shown in Figure 2.22, the PUND measurement consists of two consecutive positive pulses and two subsequent negative pulses. Before data acquisition, a negative poling pulse is applied to the sample to "preset" the ferroelectric domains. During the measurement, the ferroelectric component (FC) abruptly rises when an electric field is applied and remains a constant value even upon the removal of the field. The paraelectric component (PC) also abruptly rises when the electric field is applied and then disappears after the electric field is removed. The leakage current component (LC) exhibits a linear increase with electric field, and is the only one of the three components that depends on the pulse width (the duration of the applied field). As a result, in the first positive pulse, the abrupt increment in electric displacement (FP) should be a summation of FC and PC, and the polarization of ferroelectric reversal can be calculated by $FC = FP - PC$ [29]. Besides, the relative permittivity can also be estimated from the decrement in electric displacement

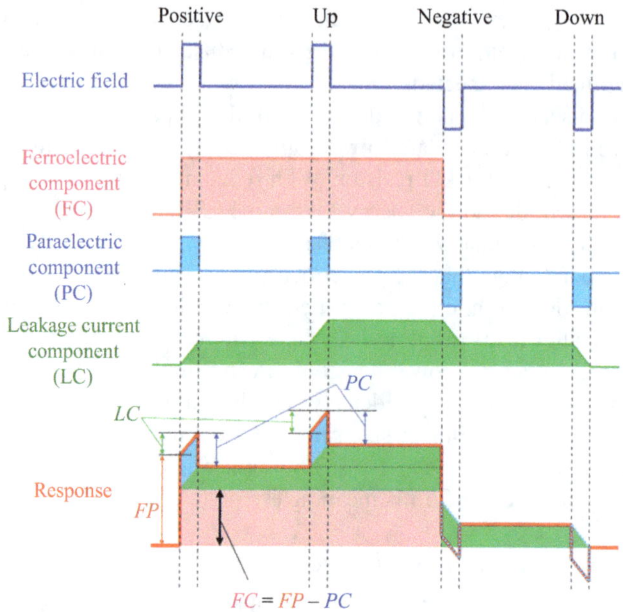

Figure 2.22 Schematic illustration of pulse responses in the PUND measurement. © 2008 The Japan Society of Applied Physics. Reprinted, with permission, from [29]

when the pulse is removed (PC), and the leakage current density can be calculated from the gradient of the pulse response (LC) when the pulse is applied.

Figure 2.23(a) shows an example of PUND measurement on a ferroelectric thin film of bismuth ferrite ($BiFeO_3$) [29]. The pulse field is 1.73 MV/cm and the pulse width is 5 μs. The magnified pulse responses under the first two consecutive positive pulses are shown in Figure 2.23(b). In Figure 2.23(b), the leakage current densities can be evaluated from the slope of LC_1 and LC_2, which is on the order to 1 A/cm^2 for the applied pulse field. The paraelectric components are estimated to be ~35 μC/cm^2 from both PC_1 and PC_2, and the total electric displacement upon the application of first pulse (*FP*) is measured to be ~170 μC/cm^2. As a result, the actual polarization of ferroelectric reversal should be ~135 μC/cm^2 (170–35 μC/cm^2). The measured sample is a leaky specimen, and the conduction loss will significantly obscure the ferroelectric polarization. The deconvolution of each contribution from total polarization would be very difficult, if not impossible, by other dielectric measurement techniques. PUND measurement, on the other hand, provides a feasible approach to separate out irreversible polarization (e.g., ferroelectric polarization), reversible polarization (e.g., paraelectric polarization), and dissipative term (such as leakage current).

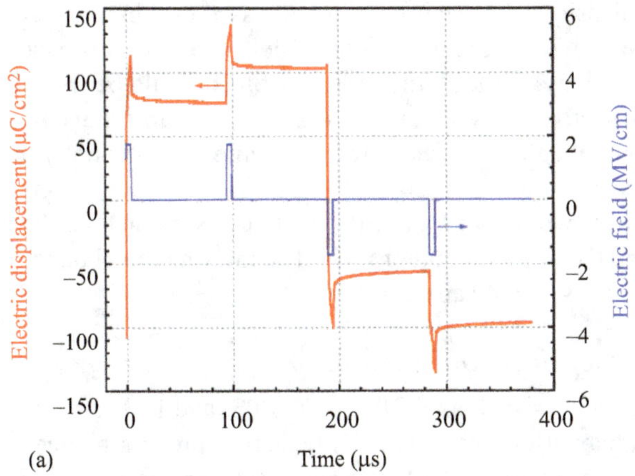

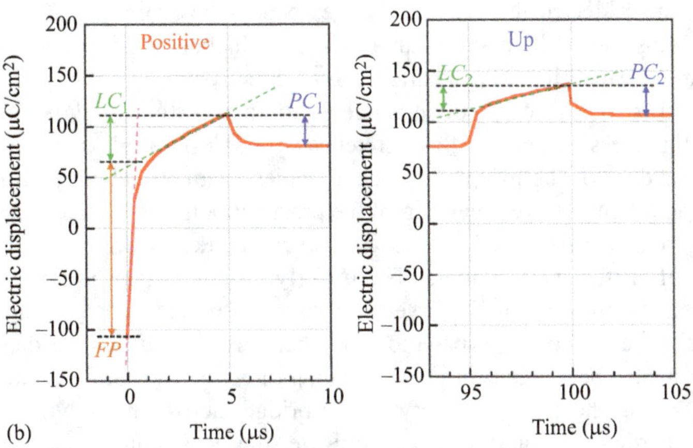

Figure 2.23 Magnified pulse responses for (a) the first positive pulse and (b) the second positive pulse.

2.4 Manufacturing related test

When materials with attractive dielectric properties are developed, the next step is to evaluate their feasibility for use in capacitors. In addition to good film quality and uniformity, the capacitor films need to meet certain thickness requirements. It turns out to be a determinant step that impedes the development of new capacitors. For instance, polyetherimide (PEI) is a high-temperature polymer, which can exhibit excellent dielectric properties, including low dielectric loss, high breakdown strength, and low leakage current. However, it is very challenging to manufacture PEI films with a thickness below 4 μm. Because of the voltage restriction in power electronics, any thickness above the critical thickness will only reduce the energy/power density of capacitors. To address the thickness issue of PEI, GE global research and Sabic spent millions of dollars and tens of years in improving processing technology. Besides, the capacitor films should be flexible enough to be wound into large rollers. Many inorganic dielectric materials fail for this consideration even though their dielectric properties substantially outperform the polymer dielectrics.

In this section, we will introduce two techniques mimicking the manufacturing of thin films and the film winding process. The information obtained can be used as guidance in the subsequent capacitor design.

2.4.1 Machine direction orientation

Commercial capacitors such as BOPP, PET, PPS, and PEN are all semicrystalline polymers with biaxial orientation. The orientation process reduces film thickness and improves film quality, which is critical for capacitor films. Because of the voltage limit in electrical devices (voltages cannot be arbitrarily increased), the capacitor films should be manufactured into thin thickness to increase the energy/power density. Since it is very challenging to directly produce thin polymer films by melt extrusion, the common practice is to first extrude polymer films with moderate thickness, which are then stretched to the desired thickness. As a result, the understanding of the morphology development with stretching and the effects on dielectric properties are crucial from the practical point of view. Furthermore, in addition to reducing film thickness, stretching can markedly improve the dielectric properties of crystalline polymers. The underlying mechanism is associated with the oriented spherulite, which serve as barriers to impede charge transport and frustrate electrical treeing propagation (see below for details). The thermal conductivity can also be increased in the stretching direction as heat is conducted more effectively along the established crystalline bridges across the polymer film. The enhanced thermal conductivity can mitigate the loss-induced heating issues, prolonging the service life of capacitors. In this sense, the classical film stretching is not unlike the recent popular approaches in polymer nanocomposites, where high aspect ratio and high thermal conductivity nanofillers are introduced in polymers to obtain excellent dielectric performance [30,31].

A prototype-stretching machine is built by the team [8], as schematically shown in Figure 2.24. It consists of two stepper motors, both moving in the same direction but at different speeds, and a heating wire which only locally heats a narrow region just above the wire. While the stretching ratio can be adjusted by varying the speed ratio of the two motors, the stretching of polymer films is always limited within the narrow heating zone. The narrow stretching zone minimizes the necking in stretched films. The unstrained portions of films are continuously sent to the heating zone for further stretching. In this manner, stretched polymer films with low necking and uniform stretching ratios can be prepared, with film properties very similar to those produced in the industrial orientation machines.

The structural development of a stretched PVDF-HFP is measured by 2DXRD, as also shown in Figure 2.24. The films are always placed with the drawing

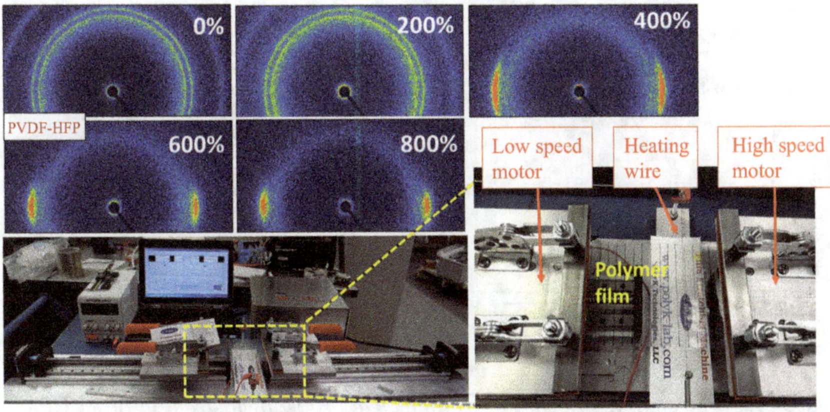

Figure 2.24 (Up) XRD measurement of stretching-induced crystal orientation in PVDF-HFP and (down) schematic illustration of a zone-stretching machine [8]

direction up. With stretching, isotropic circular rings are azimuthally narrowed into pairs of arcs, reflecting the straining-induced crystal orientation. These diffraction patterns, from the inner to the outer bright rings, correspond to the scatterings of $(100)_\alpha$, $(020)_\alpha$, $(110)_\alpha$, and $(110)_\beta/(200)_\beta$, respectively. Since all these scatterings come from the crystalline planes (*hk0*), i.e., the planes parallel to the chain direction, the accumulation of these patterns around the equatorial regions indicates that the polymer chains are aligned along the stretching direction. At high strains, the appearance of β phase scattering is accompanied by the suppression of α phase reflection, indicating the stretching-induced phase transformation.

The effects of stretching on dielectric properties are presented in Figure 2.25. Using the Weibull distribution statistics (more on "**Dielectric breakdown**"), the breakdown fields of the PVDF-HFP with different strains are plotted in Figure 2.25(a). The Weibull curve is shifted to higher electric fields in a parallel manner with stretching, indicating that the uniaxial straining can increase the E_{BD} of the film while still retaining the shape parameter β_w. It has been reported that electrical treeing would be initiated at and propagate through the defective amorphous regions under high electric fields. The randomly oriented crystals in the unstretched film allow a short path for charge transport across the sample. After stretching, the oriented crystals with the lamellae perpendicular to e-field can substantially increase the path tortuosity of discharging and frustrate the treeing-induced breakdown [32]. The same reason can also explain the systematical reduction in leakage current, measured from a series of stretched polyethylene (Figure 2.25(b)) [33].

2.4.2 Folding endurance test

Polymer films are wound into rollers before assembling into capacitors [34]. The flexibility of polymers offers their remarkable advantages over the ceramic

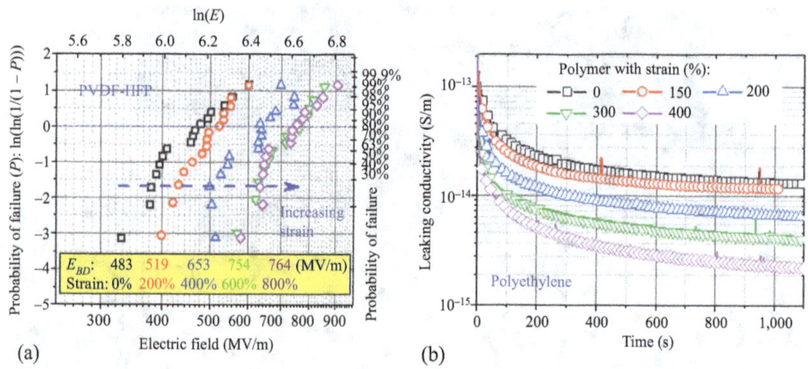

Figure 2.25 Stretching effects on (a) dielectric breakdown strength of PVDF-HFP and (b) leakage current of PE. © 2018 American Chemical Society. Reprinted, with permission, from [30]

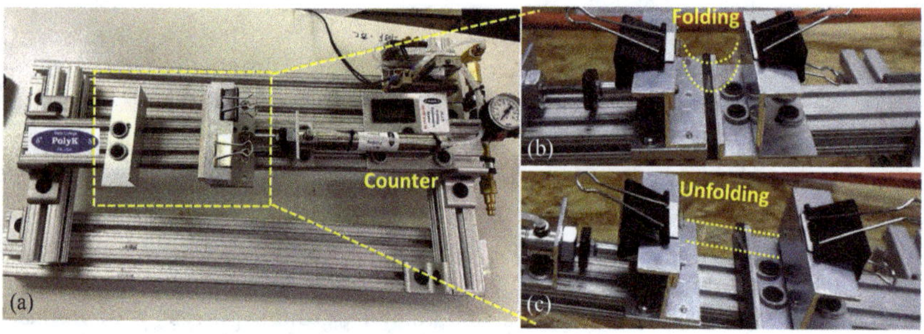

Figure 2.26 (a) M.I.T. folding endurance tester. (b–c) Schematic illustration of folding and unfolding of a polymer thin film

counterparts, because large size capacitors with high energy/power density can be made out of polymer dielectric films but would be very challenging for ceramic dielectric materials. Owing to the winding requirements in the manufacturing of capacitors, it is imperative to fabricate thin-film capacitors resistive to mechanical curvatures.

To test whether polymer dielectrics can withstand mechanical folding without deteriorating dielectric properties, a folding endurance tester is built by PolyK Technologies following the ASTM D 2176 (Standard Test Method for Folding Endurance of Paper and Plastics Film) [35]. As shown in Figure 2.26(a), it has a movable shaft driven by an air pump, and an electronic counter which can record up to 9,999,999 foldings. Both folding distance (curvature) and folding speed can be adjusted. During the measurement, the film is clamped to the fixture stages, with one stage fixed in place and the other one connected to the movable shaft. The folding endurance of polymer films can be readily examined over this machine by repetitive folding and unfolding hundreds of thousands of times (Figure 2.26(b–c)).

2.5 Conclusion and outlook

In this chapter, different experimental methods for the characterization of dielectric materials are summarized and discussed. The characterization techniques cover not only the basic dielectric measurement (such as dielectric constant and loss, breakdown strength, leakage current, charge-discharge characteristics, etc.) but also the measurement instruments developed by the team based on practical application environments (such as high-frequency P-E loop, dielectric properties under DC bias, ripple measurement, capacitor discharge test, etc.). In most cases, PVDF-based polymers are used as example materials to show how to extract useful information from measurement, how to analyze the experimental results, and how to correlate the dielectric properties with the material morphologies. These characterization tools can provide comprehensive information regarding the dielectric performance of candidate materials and their feasibility for use in capacitors.

While dielectric materials performing well in all above tests will still not necessarily work in practical capacitors, dielectric materials with poor performance in any of the tests will definitely perform poorly in packaged capacitors and they may be immediately excluded from future commercialization efforts.

References

[1] Sarjeant, W.J., Zirnheld, J., and MacDougall, F.W. 'Capacitors', *IEEE Transactions on Plasma Science*, 1998, **26**, (5): 1368–1392.

[2] Sarjeant, W.J., Clelland, I.W., and Price, R.A. 'Capacitive Components for Power Electronics', *Proceedings of the IEEE*, 2001, **89**, (6): 846–855.

[3] Ho, J., Jow, T.R., and Boggs, S. 'Historical Introduction to Capacitor Technology', *IEEE Electrical Insulation Magazine*, 2010, **26**, (1): 20–25.

[4] Rabuffi, M. and Picci, G. 'Status Quo and Future Prospects for Metallized Polypropylene Energy Storage Capacitors', *IEEE Transactions on Plasma Science*, 2002, **30**, (5): 1939–1942.

[5] Wang, Y., Zhou, X., Chen, Q., Chu, B.J., and Zhang, Q.M. 'Recent Development of High Energy Density Polymers for Dielectric Capacitors', *IEEE Transactions on Dielectrics and Electrical Insulation*, 2010, **17**, (4): 1036–1042.

[6] Zhang, S., Zou, C., Kushner, D.I., *et al.* 'Semicrystalline Polymers with High Dielectric Constant, Melting Temperature, and Charge-Discharge Efficiency', *IEEE Transactions on Dielectrics and Electrical Insulation*, 2012, **19**, (4): 1158–1166.

[7] Ho, J. and Greenbaum, S.G. 'Polymer Capacitor Dielectrics for High Temperature Applications', *ACS Applied Materials & Interfaces*, 2018, **10**, (35): 29189–29218.

[8] PolyK Technologies. Available from https://piezopvdf.com [Accessed Dec 2019].

[9] Chu, B., Zhou, Y., and Zhang, S. 'Charging and Discharging Characteristics of Dielectric Polymer Materials' in Dang, Z.-M. (ed.). *Dielectric Polymer*

Materials for High-Density Energy Storage, William Andrew Publishing, 2018, pp. 351–382.

[10] Zhu, L. 'Exploring Strategies for High Dielectric Constant and Low Loss Polymer Dielectrics', *The Journal of Physical Chemistry Letters*, 2014, **5**, (21): 3677–3687.

[11] Guan, F., Pan, J., Wang, J., Wang, Q., and Zhu, L. 'Crystal Orientation Effect on Electric Energy Storage in Poly(Vinylidene Fluoride-Co-Hexafluoropropylene) Copolymers', *Macromolecules*, 2010, **43**, (1): 384–392.

[12] Yang, L., Ho, J., Allahyarov, E., Mu, R., and Zhu, L. 'Semicrystalline Structure-Dielectric Property Relationship and Electrical Conduction in a Biaxially Oriented Poly(Vinylidene Fluoride) Film under High Electric Fields and High Temperatures', *ACS Applied Materials & Interfaces*, 2015, **7**, (36): 19894–19905.

[13] Mijovic, J., Sy, J.-W., and Kwei, T.K. 'Reorientational Dynamics of Dipoles in Poly(Vinylidene Fluoride)/Poly(Methyl Methacrylate) (Pvdf/Pmma) Blends by Dielectric Spectroscopy', *Macromolecules*, 1997, **30**, (10): 3042–3050.

[14] Yang, L.Y., Li, X.Y., Allahyarov, E., Taylor, P.L., Zhang, Q.M., and Zhu, L. 'Novel Polymer Ferroelectric Behavior Via Crystal Isomorphism and the Nanoconfinement Effect', *Polymer*, 2013, **54**, (7): 1709–1728.

[15] Li, B., Yuan, M., Zhang, S., Rajagopalan, R., and Lanagan, M.T. 'Abnormal High Voltage Resistivity of Polyvinylidene Fluoride and Implications for Applications in High Energy Density Film Capacitors', *Applied Physics Letters*, 2018, **113**, (19): 193903.

[16] Guan, F.X., Wang, J., Pan, J.L., Wang, Q., and Zhu, L. 'Time and Poling History Dependent Energy Storage and Discharge Behaviors in Poly (Vinylidene Fluoride-Co-Hexafluoropropylene) Random Copolymers', *Chinese Journal of Polymer Science*, 2011, **29**, (1): 65–80.

[17] Dissado, L.A. and Fothergill, J.C. *Electrical Degradation and Breakdown in Polymers*. London: IET; 2008.

[18] Diaham, S., Zelmat, S., Locatelli, M.L., Dinculescu, S., Decup, M., and Lebey, T. 'Dielectric Breakdown of Polyimide Films: Area, Thickness and Temperature Dependence', *IEEE Transactions on Dielectrics and Electrical Insulation*, 2010, **17**, (1): 18–27.

[19] Tomer, V., Polizos, G., Randall, C.A., and Manias, E. 'Polyethylene Nanocomposite Dielectrics: Implications of Nanofiller Orientation on High Field Properties and Energy Storage', *Journal of Applied Physics*, 2011, **109**, (7): 074113.

[20] *Electric strength of insulating materials - Test methods - Part 1: Tests at power frequencies*. IEC 60243–1, 2013.

[21] *Standard Test Method for Dielectric Breakdown Voltage and Dielectric Strength of Solid Electrical Insulating Materials at Commercial Power Frequencies*. ASTM D149–09, 2013.

[22] Zou, C., Kushner, D., and Zhang, S. 'Wide Temperature Polyimide/ZrO_2 Nanodielectric Capacitor Film with Excellent Electrical Performance', *Applied Physics Letters*, 2011, **98**, (8): 082905.

[23] Ho, J. and Jow, T.R. 'High Field Conduction in Biaxially Oriented Polypropylene at Elevated Temperature', *IEEE Transactions on Dielectrics and Electrical Insulation*, 2012, **19**, (3): 990–995.

[24] Ho, J. and Jow, T.R. *Electrical Conductivity of Biaxially Oriented Polypropylene Under High Field at Elevated Temperature*. ARL-TR-5720, 2011.

[25] *Standard Test Methods for DC Resistance or Conductance of Insulating Materials*. ASTM D257–14, 2014.

[26] Guan, F.X., Wang, J., Pan, J.L., Wang, Q., and Zhu, L. 'Effects of Polymorphism and Crystallite Size on Dipole Reorientation in Poly (Vinylidene Fluoride) and Its Random Copolymers', *Macromolecules*, 2010, **43**, (16): 6739–6748.

[27] Furukawa, T., Nakajima, T., and Takahashi, Y. 'Factors Governing Ferroelectric Switching Characteristics of Thin Vdf/Trfe Copolymer Films', *IEEE Transactions on Dielectrics and Electrical Insulation*, 2006, **13**, (5): 1120–1131.

[28] Hu, W.J., Juo, D.M., You, L., *et al*. 'Universal Ferroelectric Switching Dynamics of Vinylidene Fluoride-Trifluoroethylene Copolymer Films', *Scientific Reports*, 2014, **4**: 4772.

[29] Naganuma, H., Inoue, Y., and Okamura, S. 'Evaluation of Electrical Properties of Leaky $BiFeO_3$ Films in High Electric Field Region by High-Speed Positive-up–Negative-Down Measurement', *Applied Physics Express*, 2008, **1**: 061601.

[30] Li, B., Salcedo-Galan, F., Xidas, P.I., and Manias, E. 'Improving Electrical Breakdown Strength of Polymer Nanocomposites by Tailoring Hybrid-Filler Structure for High-Voltage Dielectric Applications', *ACS Applied Nano Materials*, 2018, **1**, (9): 4401–4407.

[31] Li, Q., Han, K., Gadinski, M.R., Zhang, G., and Wang, Q. 'High Energy and Power Density Capacitors from Solution-Processed Ternary Ferroelectric Polymer Nanocomposites', *Advanced Materials*, 2014, **26**, (36): 6244–6249.

[32] Li, B., Xidas, P.I., Triantafyllidis, K.S., and Manias, E. 'Effect of Crystal Orientation and Nanofiller Alignment on Dielectric Breakdown of Polyethylene/Montmorillonite Nanocomposites', *Applied Physics Letters*, 2017, **111**, (8): 082906.

[33] Li, B., Xidas, P.I., and Manias, E. 'High Breakdown Strength Polymer Nanocomposites Based on the Synergy of Nanofiller Orientation and Crystal Orientation for Insulation and Dielectric Applications', *ACS Applied Nano Materials*, 2018, **1**, (7): 3520–3530.

[34] Boggs, S.A., Ho, J., and Jow, T.R. 'Overview of Laminar Dielectric Capacitors', *IEEE Electrical Insulation Magazine*, 2010, **26**, (2): 7–13.

[35] *Standard Test Method for Folding Endurance of Paper and Plastics Film by the M.I.T. Tester*. ASTM D2176–16, 2016.

Chapter 3

Dielectric polymers and dielectric metamaterials for high-energy capacitors

Xin Chen[1,2]*, Tian Zhang*[1,3]*, Yash Thakur*[1,3,*]*, Qiyan Zhang*[1,4] *and Q.M. Zhang*[1,2,3]

3.1 Introduction

Dielectric materials store and regulate energy electrostatically through various polarization mechanisms [1]. Dielectric capacitors are unparalleled in flexibility, adaptability, and efficiency for electrical energy storage, filtering, and power conditioning [2–16]. The demand for capacitive energy storage has increased due to continuing electrification of land and sea transportation, as well as military and civilian systems [17–20]. These applications require capacitors with high energy density, low loss, high efficiency, and high operating temperature. Compared to ceramics and electrolytic capacitors, polymer-based capacitors are attractive because they feature low manufacturing cost and low dielectric loss, can be used under high voltage due to high breakdown strength, and fail gracefully with an open circuit [2–4]. In many of these devices and systems, capacitors constitute a substantial fraction of volume and weight (>30% volume and weight) [21–23]. To meet the demand of continued miniaturization and increased functionalities with a given volume of modern electrical and electronic systems, the energy density of dielectric polymers must be improved. In general, the energy stored in a capacitor is proportional to the dielectric constant K and the square of the electric field E, e.g.,

$$U = \frac{1}{2}\varepsilon_0 KE^2 = \frac{1}{2}DE, \tag{3.1}$$

[1]Materials Research Institute, The Pennsylvania State University, University Park, PA, USA
[2]Materials Science and Engineering, The Pennsylvania State University, University Park, PA, USA
[3]School of Electrical Engineering and Computer Science, The Pennsylvania State University, University Park, PA, USA
[4]Department of Chemical Engineering, Tsinghua University, Beijing, China

[*]Present address: Center for Physics Sciences and Technology, Intel Corporation, PTD, Oregon

where ε_0 is the vacuum permittivity. Therefore, the materials of interest should display high dielectric constant and high breakdown strength. In addition, the dielectric loss and the conduction loss at high electric fields can generate large Joule-heating in many applications. Hence, it is also critical to develop low-cost approaches in polymer dielectrics to reduce significantly these losses over a broad temperature range.

The state-of-the-art high energy density film capacitors use biaxially oriented polypropylene films (BOPP). It is attractive for energy storage and regulation applications, such as capacitors in hybrid electric vehicles (HEV) and power grids due to its high dielectric breakdown strength (>700 MV/m) and low dielectric loss (<0.018%, when measured at low electric field). However, the low dielectric constant ($K \approx 2.2$) of BOPP limits its energy density [8,21]. In addition, the low operation temperature of BOPP (<80 °C) requires external cooling to maintain safe operation temperature of these capacitors, since many applications are in high temperature environments (to 150 °C) [22,23]. This external cooling increases system size and cost. Also, for many widely used dielectric polymers, including BOPP, it is found that the conduction loss becomes significant at high-applied fields. In general, these conduction losses increase exponentially with the electric field and cause Joule heating [24–26]. This also results in the need to have a cooling system to avoid thermal runaway of the film capacitors.

Extensive materials development efforts have led to several alternative dielectric polymers, including polycarbonate (PC), poly(ethylene terephthalate) (PET), and poly(phenylene sulfide) (PPS) with high operating temperatures (>125 °C). However, the dielectric constant of these polymers is still below 3.3 [8,27]. Table 3.1 summarizes the dielectric properties of polymers used in industry and some recently developed dielectric polymers in research labs.

On the other hand, the strong coupling among dipoles have led to high dielectric constants of polyvinylidene fluoride (PVDF)-based ferroelectric polymers ($K > 10$) [3–5,29]. By proper defect modifications of PVDF-based polymers, it has been shown that these polymers can achieve either a high dielectric constant at room temperature ($K > 50$) or a very high energy density (>25 J/cm^3) [30–32]. However, the strong dipolar coupling in these ferroelectric polymers causes large polarization hysteresis loss. The operating temperature is still limited to below 100 °C due to low T_m (<140 °C). Nevertheless, the results demonstrate the potential of tailoring nano and meso-structures of dielectric polymers to achieve high dielectric performance.

Many studies have been conducted in the past to improve the dielectric constant [4,8,29,33,34]. In dielectric polymers, a necessary condition for achieving high dielectric constant is that they contain dipoles in the polymer chains. Strongly dipolar materials, in which dipole moment is larger than 3 Debye with weak dipolar coupling have the potential to reach relatively high dielectric constant than the widely used non-polar polymers including the state-of-the-art BOPP, and exhibit a lower loss. During the past several years, we have been devoting efforts to (i) develop high glass transition ($T_g > 200$ °C) dipolar polymers with improved K (>5) and low loss; (ii) develop novel (and low cost) strategies to enhance the

Table 3.1 Summary of commercial and recent developed dielectric polymers for capacitor applications [8,28]

Material	Dielectric constant (25 °C)	Dielectric loss at 1 kHz (10^{-3})	Working[a] temperature (°C)	Dielectric strength[b] (MV/m)	Energy density at breakdown strength[c] (J/cm^3)
BOPP	2.2	0.2	90	820	6.2
PTFE	2.1	0.5	260	300	2.1
Solvent-cast PC	3.1	1.3	125	<820	9.2
Cyano-PC	3.2	3	180	710	7.1
PhONDI	3.2	7.8	150	350	1.7
PPS	3.1	0.5	150	470	3
PEEK	3.2	4	150	320	1.4
PEI	3.2	2	200	460	3
PI	3.3	2	300	300	1.3
PEN/Si_3N_4	3.7	7	125	578	5.4
Nanolayer PCPVDF-HFP	4.6	10	125	750	14
Cyano-PEI	4.7	3	220	745	11.5
P(TFE-co-VDF)	10.2	30	270	>225	>2
Alkali-free barium boroaluminosilicate glass	6	5	180	>1,000	>30

PTFE, polytetrafluoroethylene; PC, polycarbonate; PhONDI, exo-N-phenyl-7-oxanorbornene-5,6-dicarboximide; PPS, polyphenylene sulfide; PEEK, polyether ether ketone; PEI, polyetherimide; PI, polyimide; PEN, polyethylene naphthalate; PVDF-HFP, poly(vinylidene fluoride-hexafluoropropylene).
[a]Melting temperatures for polymers with high crystallinity, glass transition temperatures for amorphous polymers, or semi-crystal polymers with low crystallinity.
[b,c]Values at room temperature.

dielectric performance, including K, breakdown strength, and conduction losses in high T_g amorphous as well as semi-crystalline polymers. One significant progress made through these studies is the discovery and development of a class of highly scalable dielectric metamaterials in which very low volume loading of nanofillers, through interfacial effects, can lead to marked increase in K, breakdown field, and reduction in the high field conduction loss. This chapter will present some of these works.

3.2 Polyurea- and polythiourea-based high temperature dipolar polymers

In this pursuit, we have developed a class of high T_g amorphous polymers, containing high density dipoles possessing high dipole moment p, i.e., urea and thiourea, which have p of 4.56 and 4.89 Debye, respectively [35–38]. It has been shown that by increasing the dipole moment and the dipole density, the dielectric constant in this series of polymers increases from 4.1 to 5.7. The high dipole moments in these amorphous polymers provide strong polar-scattering centers and traps, which significantly reduce the conduction loss at high electric fields. As a result, these polymers exhibit an improved electrical energy density than BOPP [35]. The high glass transition temperature also leads to a higher operating temperature. In strongly dipolar polymer materials, such as the polyurea and polythiourea, the orientation polarization is the dominant polarization mechanism compared to the electronic, atomic, or ionic polarization. Orientation polarization occurs by the reorientation of the permanent dipole moment toward the direction of the electric field.

Many dielectric models have been developed in the past to describe the dipolar responses of polymers [39]. In general, a high dipole moment and dipolar density in polymer chains may lead to a high dielectric constant. Four dielectric polymers based on polyurea and polythiourea have been developed, which include aromatic polyurea (ArPU), aromatic polythiourea (ArPTU), *meta*-phenylene polyurea (*m*-PhPU/*meta*-PU), and methylene polythiourea (MePTU), to study the influence of dipole moment and dipole density on the dielectric properties [40]. Figure 3.1 shows the chemical structure of these polymers and Table 3.2 summarizes their dielectric properties.

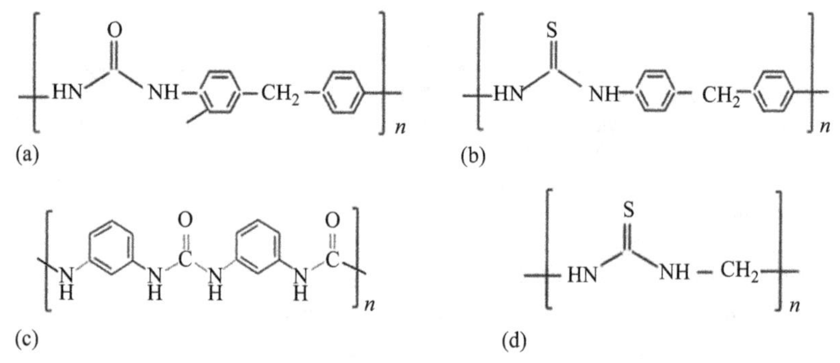

Figure 3.1 Schematics of (a) aromatic polyurea (ArPU), (b) aromatic polythiourea (ArPTU), (c) meta-phenylene polyurea (m-PhPU/ meta-PU), and (d) methylene polythiourea (MePTU)

Table 3.2 Summary of dielectric properties of ArPU, ArPTU, m-PhPU, and MePTU [40]

Polymer	Dielectric constant (1 kHz)	Loss tangent (1 kHz)	Breakdown strength E_b (MV/m)	Energy density at E_b (J/cm^3)
ArPU	4.1	0.87%	800	13.5
ArPTU	4.4	0.64%	>1,000	20.1
meta-PU	5.7	1.71%	670	13
MePTU	5.7	1.55%	500	7.5

Presented in Figure 3.2(a) is the dielectric constant and loss measured at room temperature of ArPU, ArPTU, *m*-PhPU, and MePTU as a function of frequency. With a high dipole moment of 4.56 Debye in the urea unit, and 4.89 Debye in the thiourea unit, the dielectric constants of ArPU and ArPTU are 4.1 and 4.4, respectively, with a low dielectric loss (<1%). The dielectric constants of ArPU and ArPTU are higher than most of the polymer dielectrics, which are non-polar or weakly dipolar materials with dielectric constant around 2–3.5 [25]. As shown in Figure 3.1(a) and (b), ArPU and ArPTU have nearly the same volumetric dipole density as they contain the same methylene diphenyl group between the two dipolar units. The dielectric constant of ArPTU is slightly higher than ArPU because of higher dipole moment of thiourea than urea. However, the increase in dielectric constant is not proportional to the square of dipole moment as predicted by many existing dielectric models [39].

In ArPU and ArPTU (Figure 3.1(a), (b)), methylene diphenyl (two aromatic rings), as one blocking unit, separates neighboring dipoles along the polymer chain, leading to relatively low volumetric dipole density. In general, the dielectric constant can be enhanced by increasing the volumetric dipole density [39]. Therefore, a new type of polymers, *m*-PhPU and MePTU (Figure 3.1(c) and (d)), with higher volumetric dipole density, were synthesized by modifying the block structures between dipoles in the polymer chain. An aromatic ring in *m*-PhPU separates the two dipolar units, while methylene is the separating unit in MePTU. The dielectric constant of both *m*-PhPU and MePTU is 5.7, as shown in Figure 3.2(a). As predicted, these two polymers have a higher dielectric constant than ArPU and ArPTU.

The distance between two dipolar units can be estimated by bond length and bond angle [41]. The bond length for ArPU and ArPTU is 1,251 *pm*, while it is 754 *pm* and 512 *pm* for *m*-PhPU and MePTU, respectively. Comparing *m*-PhPU with ArPTU, the dipole density along the chain increases by 66% (distance reduced from 1,251 *pm* to 754 *pm*), which leads to an increase of dielectric constant by more than 40%. This deviates from the results predicted by the Frohlich model as

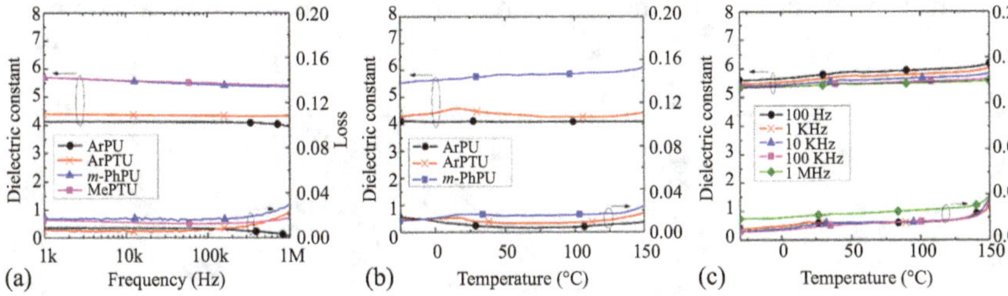

Figure 3.2 (a) Dielectric constant and loss as a function of frequency for ArPU, ArPTU, m-PhPU, and MePTU. (b) Dielectric constant and loss at 1 kHz as a function of temperature for ArPU, ArPTU, and m-PhPU. (c) Dielectric constant and loss at different frequency as a function of temperature for m-PhPU. © 2015 AIP. Reproduced, with permission, from [40]

well as other existing dielectric models [39]. We presume that there might be changes in the interchain spaces due to the disordered amorphous structure of these polyurea and polythiourea based films. The amorphous structure of these polyurea and polythiourea based films will be discussed in the later part of this section.

Figure 3.2(b) and (c) shows the dielectric properties at 1 kHz as function of temperatures for all four polymers, showing relatively high stability at high temperature, up to 150 °C, of the dielectric constant and losses. Besides higher dielectric constants, these polyureas or polythioureas also show higher thermal stability, compared to BOPP [25].

It should be pointed out that for most dielectric polymers, even though their dielectric constant is independent of the applied field, the losses due to the conduction may increase at high electric fields [42,43]. The high field efficiency η is defined as the ratio of the discharged energy density to the total input energy density in the charge–discharge cycle [44]. 1-η is defined as the high field loss in the equation:

$$\eta = \frac{U_r}{U_s} = 1 - \frac{U_l}{U_s}, \tag{3.2}$$

where U_s, U_r, and U_l are the total input, discharged, and electric loss energy densities in the charge–discharge cycle, respectively.

Figure 3.3(a)–(c) presents the charge–discharge curves (P vs. E) measured under different applied electric fields at 10 Hz and at room temperature for ArPU, ArPTU, and *m*-PhPU. Due to the high dielectric constant of *m*-PhPU, a high polarization of 3.75 μC/cm^2 was achieved at 670 MV/m. The data reveal that the dielectric loss, which is measured by the enclosed area between the charge and discharge curves, does not increase significantly at high fields, leading to high charge–discharge efficiency. The high field measurements were not performed on MePTU due to the quality of films with low breakdown field.

The energy densities are calculated from the charge–discharge curves in Figure 3.3(a)–(c) and presented in Figure 3.4(a). Overall, high energy densities >13 J/cm^3 can be achieved in these polyureas- and polythioureas-based polymers

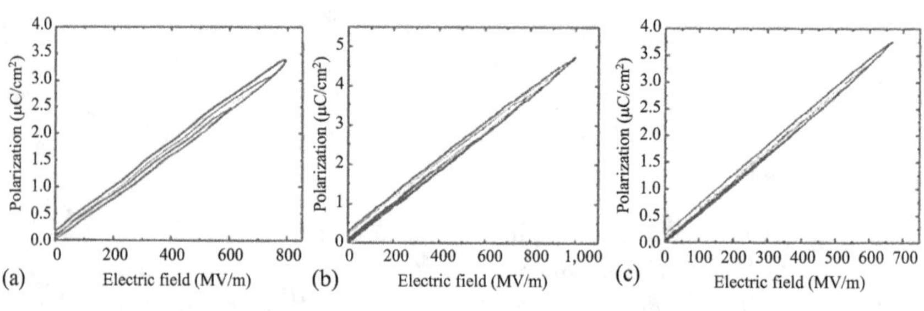

Figure 3.3 Charge–discharge curves of (a) ArPU, (b) ArPTU, and (c) m-PhPU measured under different electric fields at 10 Hz. © 2015 AIP. Reproduced, with permission, from [40]

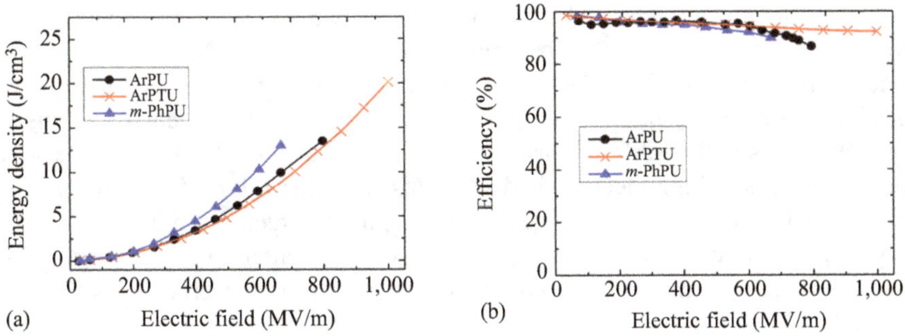

Figure 3.4 (a) Discharged energy densities vs. applied field and (b) the efficiency of ArPU, ArPTU, and m-PhPU. © 2015 AIP. Reproduced, with permission, from [40]

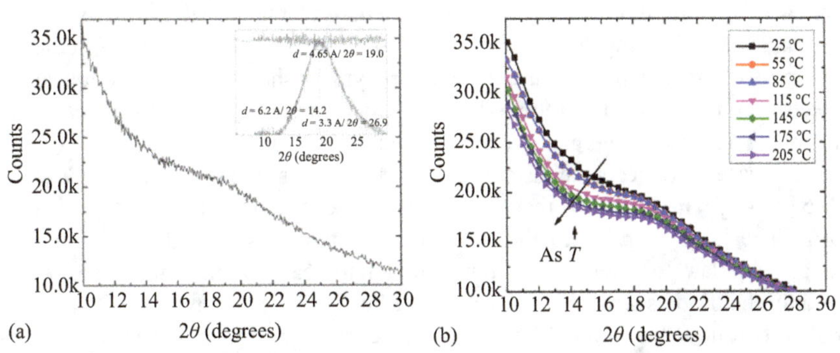

Figure 3.5 (a) XRD pattern of ArPTU at room temperature and the inset is the background normalized pattern and (b) XRD patterns of ArPTU at varied temperatures from 25 to 205 °C. © 2015 AIP. Reproduced, with permission, from [40]

with charge–discharge efficiency >90% (Figure 3.4(b)). The high efficiency under high electric field is due to extremely low conductivity of these polymers, leading to very low current leakage at high electric fields. As Shan Wu *et al.* [36] have sown it, the very smooth surface and high field quality lead to a high breakdown field of 1.1 GV/m for the ArPTU films. The high dielectric constant of *m*-PhPU due to the higher dipole density leads to an energy density of 13 J/cm^3 at 670 MV/m, while ArPTU and ArPU with lower volume dipole density lead to lower energy density at the same electric field of 670 MV/m (Figure 3.4(a)).

X-ray diffraction measurements on these films were carried out and the results reveal the amorphous structures of the films. Presented in Figure 3.5(a) is an example of the ArPTU films which show a broad X-ray peak around $2\theta = 19°$, indicating that the ArPTU films do not possess long-range crystalline ordering and are amorphous. Structure analysis suggests that the X-ray peak around $2\theta = 19°$ is likely due to the interchain diffraction. It was also found that the amorphous structure of ArPTU does not change much from 25 to 205 °C (Figure 3.5(b)). This is quite different from BOPP, which is a semi-crystalline polymer and has a glass transition temperature below –30 °C [3,45]. Above T_g, the segmental motions of polymer chains will facilitate charge hopping and lead to the observed increase in conduction loss at high electric fields [25,26,46]. While the random dipoles with high dipolar moment and amorphous glass-phase structure in polyurea- and polythiourea-based polymers provide much stronger scattering to the charge carriers, resulting in a much lower conduction loss under high electric fields [25].

3.3 Effect of dipole motion on the permittivity of strongly dipolar polymers with high T_g

In addition to the dipole density and dipole moment, the "dipole mobility," e.g., how easy or hard it is for dipoles to reorient under external fields, is equally important for achieving high permittivity. It is noted that in earlier studies of dipolar polymers, such as polyurethane, the polymers show a low loss, but also low dielectric constant at temperatures below T_g, as the motion of the dipoles is significantly restricted in the glassy state. On the other hand, these polymers exhibit a large increase in dielectric constant after undergoing a glass transition at T_g, e.g., $K > 6$ [47,48]. The penalty is that the dielectric loss is also increased markedly (loss >5%). The large increase in the dielectric constant observed above T_g in these strongly dipolar polymers is attributed to an increase of the empty space surrounding the dipoles termed "free-volume effect," (FVE) which makes it easier for dipoles to follow the applied field and hence reach a higher dielectric constant. However, large-chain-segment motions above T_g, which have long relaxation times, also cause high dielectric loss [48]. Now, the question is whether one can design a dipolar polymer that can generate the "FVE" at nanoscale at temperatures below T_g, thus leading to high K while avoiding large-chain-segment motion causing dielectric loss.

3.3.1 FVE in dipolar polymers with high T_g

Through a combined theoretical and experimental investigation, we show that it is the FVE at temperatures below T_g ($T_g > 200$ °C) that leads to a high dielectric constant ($K > 5.6$) in *meta*-phenylene polyurea (*meta*-PU) [35–37,40,49]. The computer simulation study of *meta*-PU reveals significantly larger ionic (dipolar) permittivities in the disordered structures compared to ordered structure. Figure 3.6 presents a comparison of the ionic contributions of the disordered and the ordered structures. The simulation results show that the ionic (dipolar) permittivity is proportional to the variation of the dipole moment. Therefore, a larger dipole motion could lead to a larger permittivity. The specific volume is ~12% larger in the disordered structures compared to the ordered structure. The larger volume gives urea units more free space to vibrate, see Figure 3.7. The disordered structures have the variation 70~100% larger than the ordered structure.

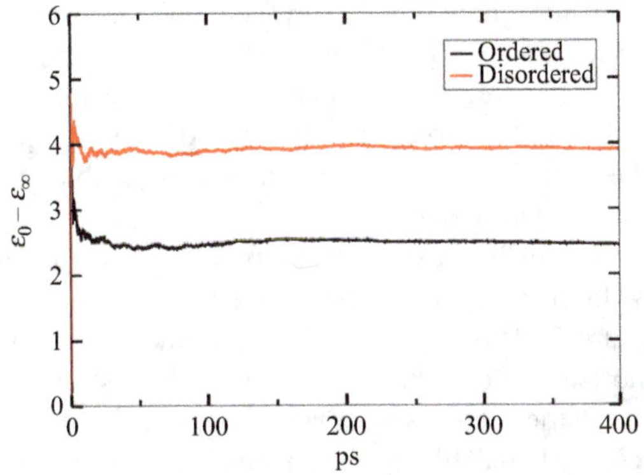

Figure 3.6 The disorder effect on the ionic dielectric response. The plot shows the convergence of the ionic contribution to permittivity with simulation time. The black curve is the ordered structure, while the red curve is the average of the three disordered structures. The disordered structures have much higher ionic contributions. © 2015 Elsevier. Reproduced, with permission, from [37]

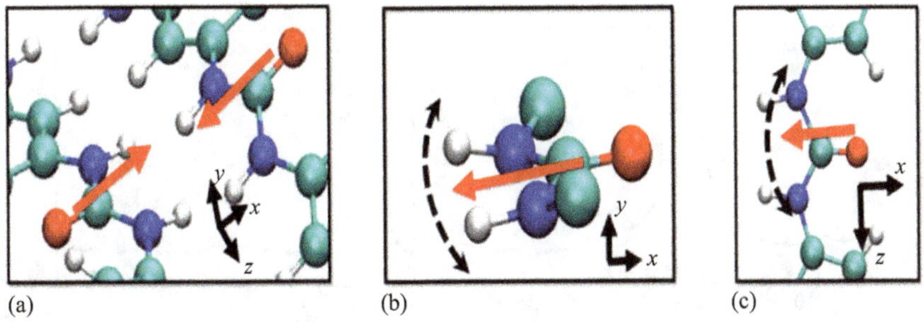

Figure 3.7 Single-urea motion contributions to dipole moment variation in the ordered meta-PU structure. (a) Antiparallel dipole pairs aligned along the x-axis. (b) Rotation of a single urea unit in xy-plane. (c) Wiggling of a single urea unit in xz-plane. © 2015 Elsevier. Reproduced, with permission, from [37]

The increase in free volume in the disordered structure gives rise to an enhancement in permittivity. More importantly, this mechanism does not introduce high loss, as shown by the experimental data on *meta*-PU. To further confirm that disorder is responsible for the high dielectric constant, simulations were also performed for aromatic polyurea (ArPU). The simulation results show that the disordered structure, which has larger free volume locally for dipoles, has a higher dielectric constant than the ordered one.

The simulation results show that the presence of large free volume in the disordered structure can lead to higher ionic permittivity, compared to that of the ordered structure. A large free volume enables easier reorientation of dipoles in response to an electric field. Moreover, the lack of long-range order removes a lower bound on the energies of optical phonons and enables longer wavelength vibrations, which also increase permittivity.

Experimentally, *meta*-PU with ordered (crystalline) and disordered phases were prepared. Synthesized *meta*-PU powder shows relatively sharp X-ray diffraction peaks (Figure 3.8(a)), indicating the presence of a crystalline phase in addition to an amorphous component. The crystalline structure from computer simulations and its lattice constants match well with the lattice spacing in Figure 3.8(a). In contrast, the films made from solution casting display a broad X-ray peak centered at $2\theta = 9.5°$, shown in Figure 3.8(b), indicating an amorphous (disordered) structure.

The dielectric properties of meta-PU films have been presented in Figure 3.2(a) and Figure 3.2(b), showing a high dielectric constant and low loss over-a-broad temperature range, which is due to the high glass transition temperature. Even more importantly, the *meta*-PU films exhibit a linear dielectric response and very low loss even at very high electric field, as has been shown in Figure 3.3(c) (refer to figure in last part, for the charge/discharge curves).

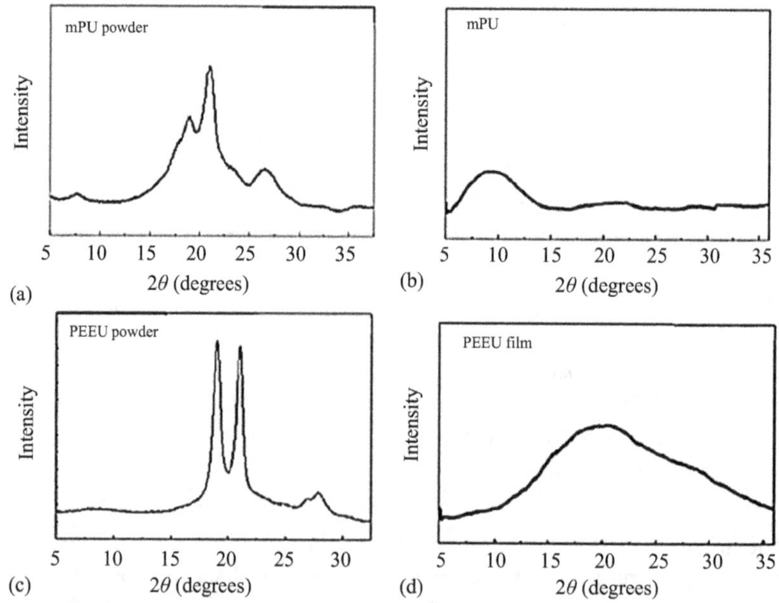

Figure 3.8 X-ray data for (a) ordered meta-PU structures; (b) films of disordered meta-PU structure; (c) ordered PEEU structures, powder; (d) films of disordered PEEU structure. © 2015 Elsevier. Reproduced, with permission, from [37]

The dielectric properties of the *meta*-PU powder (ordered structure) were measured which has a $K_1 = 3.8$, smaller than $K = 5.7$ measured in the disordered (amorphous) structures. The results indicate that *meta*-PU in the disordered structure possesses a higher dielectric constant than that in the ordered phase.

To further confirm that the results obtained here are not an isolated case, we also synthesized poly(arylene ether urea) (PEEU) polymer (see Figure 3.9), which has an ordered structure in the as synthesized polymer, and see sharp X-ray peaks for the as synthesized PEEU (Figure 3.9(c)). In contrast, the solution-cast films display a broad X-ray peak (Figure 3.9(d)), indicating an amorphous structure. The dielectric constant of PEEU films is 4.7, while the dielectric constant of as synthesized PEEU is 3.65, which is smaller than that of the disordered structure.

3.3.2 Enhancing the FVE in high T_g polymers enable by polymer blending

In Section 3.3.1, we show, through combing theoretical and experimental studies, that in *meta*-PU and several other high T_g dipolar polymers, disordered phases exhibit higher dielectric constants compared to those of ordered phases, due to the large built-in free volumes in the disordered structures. The presence of free volumes in disordered phases of these dipolar polymers at temperatures far below the glass transition enables easier reorientation of dipoles in response to an electric field. The "volume effect" reduces barriers for dipole reorientations to the applied electric field, leading to high dielectric constants while preserving low dielectric loss.

Here, we present an innovative mechanism, which through nanostructure engineering of dipolar polymers, further enhances the FVE. Blending two polymers may create additional free volume due to partial mismatches between two dissimilar polymer chains, reducing constraints of the glassy structure on the dipoles in the polymers, and enhancing the dielectric constant with low dielectric loss. An aromatic polythiourea (ArPTU, $K = 4.4$) and a poly(ether ether urea) (PEEU, $K = 4.7$) were chosen as the blend components. The T_g of both polymers are above 200 °C. As shown in Figure 3.10(a), by blending two dipolar polymers, a higher dielectric constant, $K = 7.5$ can be obtained while maintaining low dielectric loss (<1%) [50]. Figure 3.10(b) presents the dielectric properties vs. temperature (at 10 kHz), showing that the blend can maintain a high dielectric constant and low loss up to 120 °C.

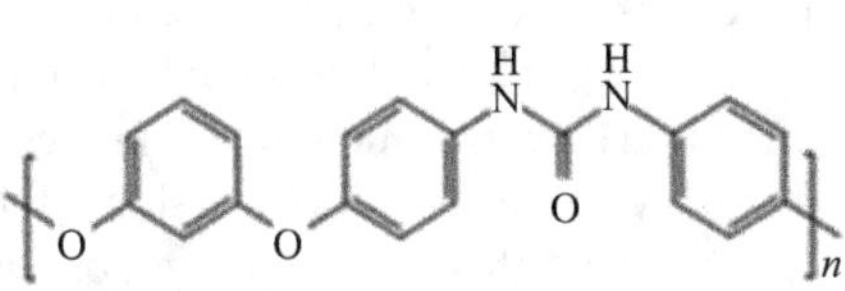

Poly(arylene ether urea) (PEEU)

Figure 3.9 Schematic of molecule structure of PEEU

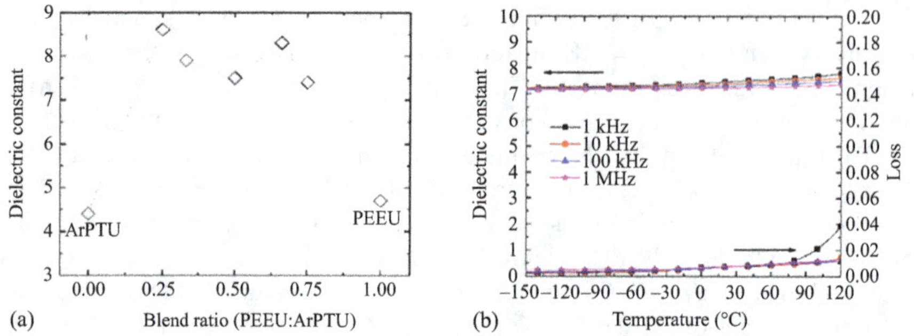

Figure 3.10 (a) The dielectric constant of blends of PEEU:ArPTU vs. the ratio between the two. (b) Dielectric data as a function of temperature at different frequencies. © 2017 Elsevier. Reproduced, with permission, from [50]

Structure data (see XRD in Figure 3.11) reveals that the nano (molecular)-scale uniform mixing of the two dipolar polymers (PEEU and ArPTU) causes an increase of the interchain spacing (~9%) in the blends in the glassy state. The larger interchain spacing in the blend enables easier dipole reorientation to the applied field, leading to a higher dielectric constant compared to those of the neat polymers while maintaining low dielectric loss.

Computer simulation results presented in Figure 3.12 indicate that this "free space" generates a higher dielectric response without compromising the dielectric loss. However, it was found that the breakdown field of the blends is reduced, a feature not desired, possibly due to a 9% (large) increase in the interchain spacing. Nevertheless, the results indicate the potential of nanostructure engineering in raising the dielectric constant of polymers without relying on high dielectric constant components.

3.3.3 Enhancing the dielectric response in high T_g polymers by weakening intermolecular hydrogen bonds

Many dipolar polymers possess strong hydrogen bonds. In these dipolar polymers, beside the free volume, strong hydrogen bonds also significantly restrict the motion of dipoles. Hence, disrupting hydrogen bonding locally can improve dipole mobility under electric field. For instance, quenched nylon films, in which hydrogen bonds between dipoles were broken, exhibited obvious ferroelectric switching; whereas the annealed nylon, where the motion of dipoles was constrained by densely packed hydrogen-bonded sheets, did not show ferroelectric switching [51,52]. Recently, we demonstrated that the hydrogen bonding in high T_g dipolar polymers can be disrupted by blending it with another misciable high T_g dipolar polymer, which does not have hydrogen bonds [53]. In details, two high T_g polymers, i.e., commercial PEI and poly(ether methyl ether urea) (PEMEU), see

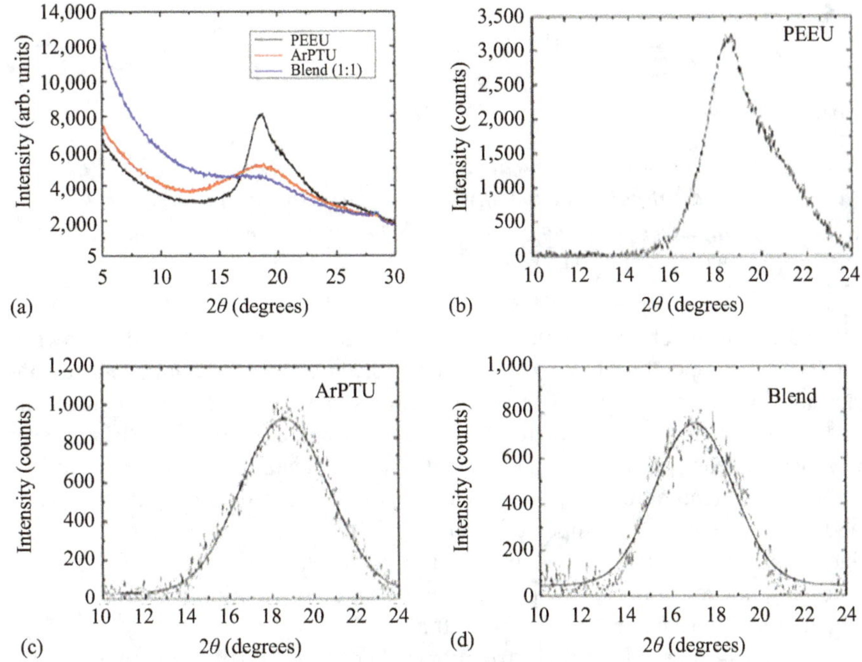

Figure 3.11 (a) XRD data of ArPTU and PEEU, and their 1:1 blend. (b), (c), (d) Background subtracted data for PEEU, ArPTU, and the 1:1 blend, respectively. © 2017 Elsevier. Reproduced, with permission, from [50]

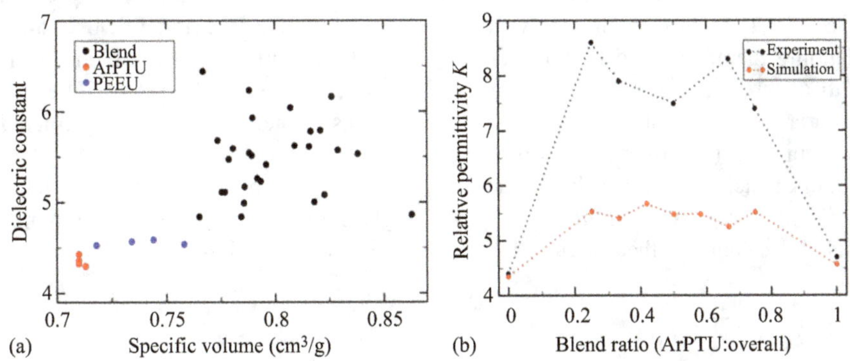

Figure 3.12 (a) Computational results of dielectric constant vs. specific volume for PEEU, ArPTU, and blends for various supercells with different PEEU:ArPTU weight ratios. (b) Comparison of simulation and experimental data of dielectric constant vs. blend composition (PEEU:ArPTU weight ratio) at room temperature. © 2017 Elsevier. Reproduced, with permission, from [50]

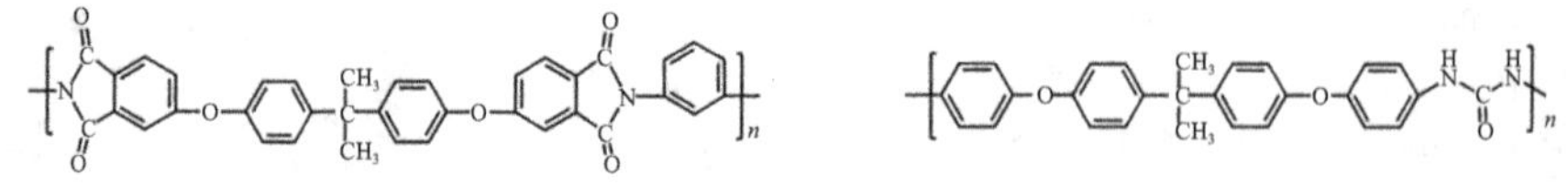

(a) (b)

Figure 3.13 Schematic of chemical structures of (a) poly(ether imide) (PEI) and (b) poly(ether methyl ether urea) (PEMEU) used to prepare dielectric polymer blends. © 2019 Elsevier. Reproduced, with permission, from [53]

Figure 3.13 for the chemical structures, are blended with different ratio in order to prepare high-temperature (>150 °C) polymer dielectrics with high dielectric constant and low dielectric loss. The results demonstrate a large enhancement in the dielectric constant, i.e., more than 80% increase compared to pristine PEI and near 50% increase compared to pristine PEMEU, while maintaining low dielectric loss (<1%) at 1:1 blend ratio. Moreover, the dielectric constant and dielectric loss show good thermal stability at temperature ranging from –70 to 150 °C. To our knowledge, this is the first report on improving the intrinsic dielectric properties by decreasing the constraint of hydrogen bonding on dipole motion in polymer chain through a highly scalable and low-cost strategy.

Figure 3.14 presents the dielectric properties of PEI, PEMEU, and their blends. As can be seen, the blend with ratio of 1:1 exhibits remarkably high dielectric constant ($K = 5.8$ at 1 KHz) under ambient temperature while maintaining low loss (below 1%). By measuring the 1:1 blend film under various temperature conditions, thermally stable dielectric properties are observed in a broad temperature range from –70 to 150 °C (Figure 3.14(c)).

The improvement of dielectric constant is believed to be due to the increased "free" local motion of dipoles of polymer chains in the PEI and PEMEU after blending them. PEI and PEMEU show good compatibility, as all the blend films with different ratio were transparent and homogenous (Figure 3.14(d)). The differential scanning calorimetry (DSC) thermograms of the blend films are depicted in Figure 3.15(a). All the blend films show single T_g that all fall in the midst of T_g values of each component. In addition, the T_g value of the blends increases with the PEI content due to the higher T_g of PEI. The glass transition temperature of blends was also calculated theoretically by using the Fox equation [54]:

$$\frac{1}{T_g} = \frac{w_1}{T_{g1}} + \frac{w_2}{T_{g2}} \tag{3.3}$$

where T_{g1} and T_{g2} are glass transition temperatures (K) of individual polymers, and w_1 and w_2 are the weight fractions of each polymer in the blend. The experimental and theoretically calculated data of T_g are compared and shown in Figure 3.15(b). The results confirm that PEI and PEMEU are miscible in the molecular level. The AFM images of the blends with 3:1, 1:1, and 1:3 PEI/PEMEU ratios all show single

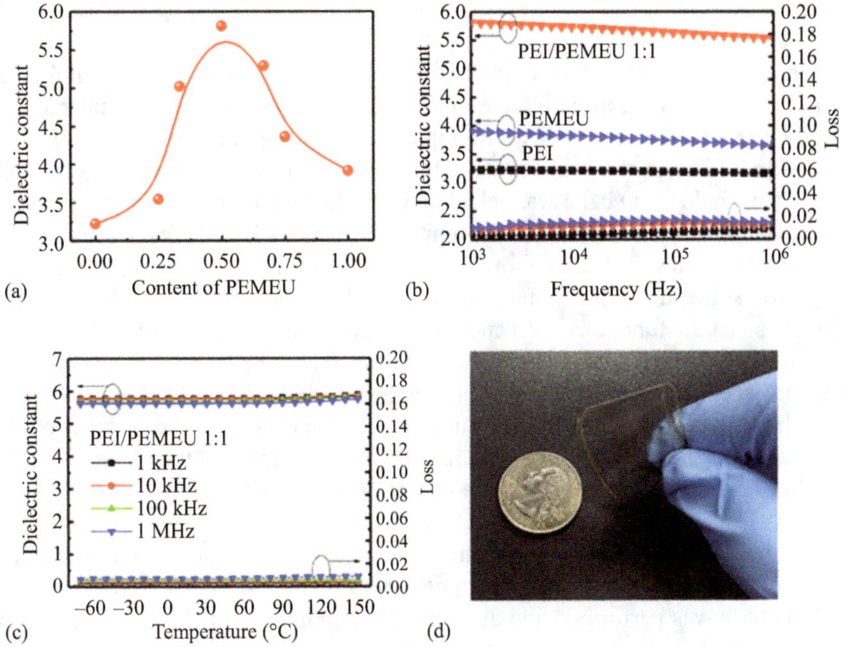

Figure 3.14 *(a) Dielectric constant vs. blend composition at ambient temperature and 1 KHz; (b) dielectric properties of the PEI, PEMEU, and 1:1 blend of PEI and PEMEU as a function of frequency at ambient temperature; (c) dielectric properties of 1:1 blend of PEI and PEMEU as a function of temperature at different frequencies; and (d) sample photo image of the blend film with ratio of 1:1. © 2019 Elsevier. Reproduced, with permission, from [53]*

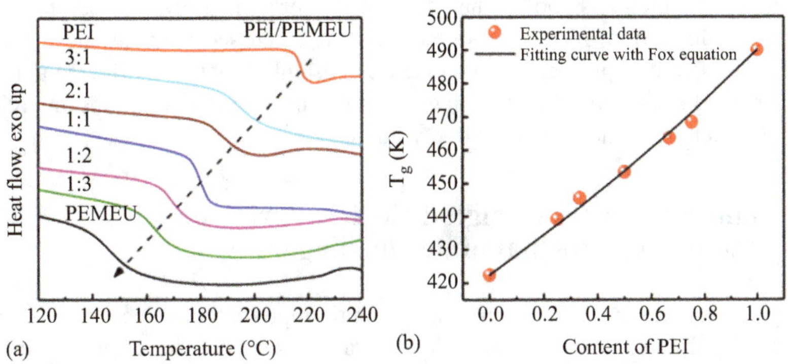

Figure 3.15 *(a) DSC heating curves and (b) glass transition temperatures (T_gs) of PEI, PEMEU, and their blends with different composition. © 2019 Elsevier. Reproduced, with permission, from [53]*

phase, suggesting the uniform mixing of the two polymers in the blend at the nanoscale.

On the other hand, the strength of intermolecular interaction between polymer chains is a major factor affecting the motion of dipoles and the specific interaction should be changed when these two miscible polymers (i.e., PEI and PEMEU) were blended [54–57]. As show in Figure 3.16(f), the ratio of hydrogen-bonded C=O band to the whole C=O band was calculated to estimate the changes in the fractions of free and hydrogen-bonded C=O groups with various blend compositions. The stretching vibration frequencies of N–H and C–N peaks were also plotted with respect to the blend compositions (Figure 3.16(g) and (h)). Figure 3.16(f), (g), and (h) reveals that the functions of fraction of free C=O groups, stretching vibration of N–H and C–N peaks vs. composition are nonlinear, and the large variations can be observed when the PEI content is below 0.5, while nearly no changes observed when the PEI content is over 0.5. These results indicate that the destruction of hydrogen bonds between PEMEU chains reach the maximum at the composition of 0.5. This explains why the dielectric constant of the blends exhibited a peak at the ratio of 0.5.

The hydrogen bonding would restrain the motion of polymer chains and hence play an important role in the interchain packing structure of polymers [58,59]. The XRD analysis was performed and as shown in the Figure 3.17, PEI, PEMEU, and their 1:1 ratio blend all exhibited broad and low intensity amorphous scattering peak at 16.6°, 17.6°, and 16.9°, respectively. The corresponding average interchain space of the films, deduced from the peak positions, are 5.33, 5.03, and 5.24 Å for PEI, PEMEU, and the blend, respectively. The slightly smaller interchain spacing of PEMEU than that of PEI suggests that the free-volume distribution is much tighter in PEMEU. This mainly attributed to the strong hydrogen bonds between adjacent urea units in the PEMEU chains, which can align to form a geometrically dense two-dimensional sheet [60,61]. The blend shows higher interchain spacing than that of PEMEU and even close to the value of PEI, which implies the chain packing in the blend is looser than that in the pristine PEMEU resulted from remarkable disruption of the hydrogen bonds as discussed above. This indicates that the significant improvement of dielectric constant of blends is affected not only by the hydrogen bonding between the polymer chains but also by the interchain packing structure or the degree of free volume.

3.4 Nanocomposites of high dielectric performance with low volume nanofiller loading

In Sections 3.1–3.3, a "free volume" approach to improve the dielectric constant in both pure polymers and polymer blends was presented. The approach entails designing "new" dielectric materials which utilize free volume in the glass phase to enhance dielectric constant of high temperature (high T_g) dipolar polymers. One interesting question is whether the similar strategy can be further applied to nanocomposites for a broad range of high T_g dipolar polymers, which are

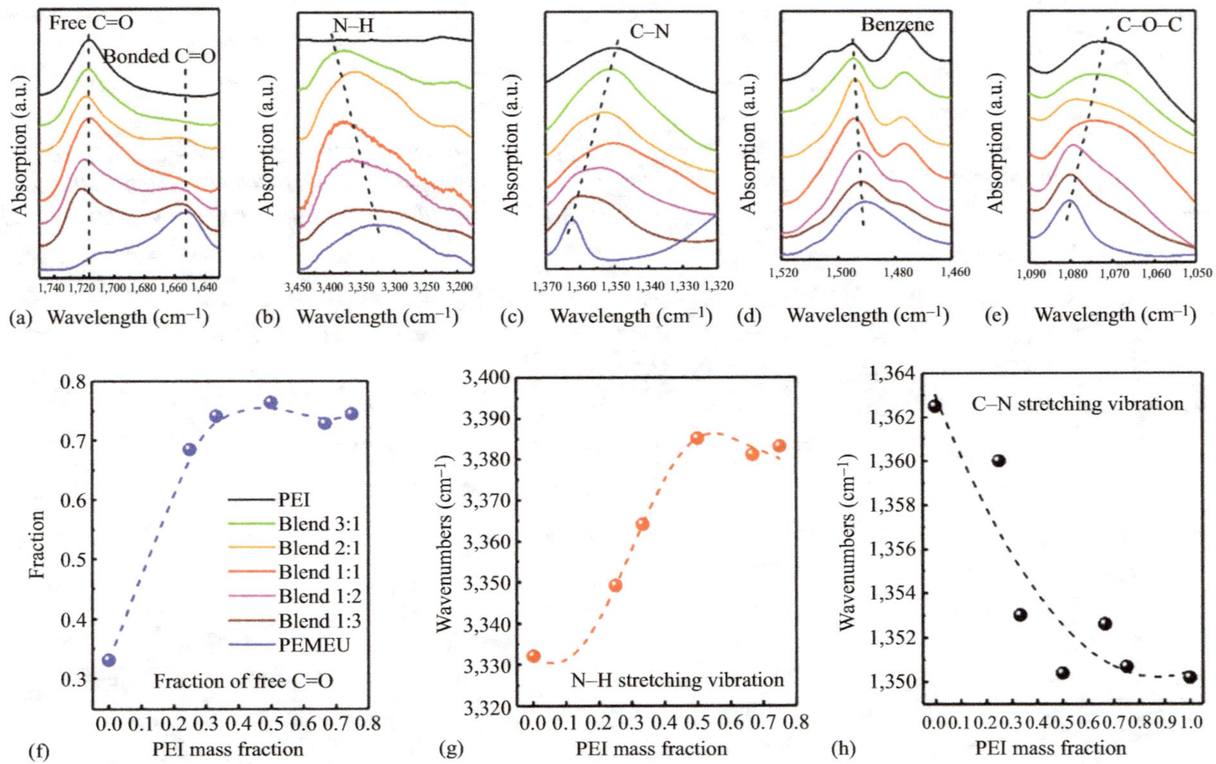

Figure 3.16 *Fourier-transform infrared spectroscopy (FTIR) spectra for PEI, PEMEU, and their blends at room temperature in the range of (a) 1,750–1,630 cm^{-1}, (b) 3,450–3,150 cm^{-1}, (c) 1,370–1,320 cm^{-1}, (d) 1,520–1,460 cm^{-1}, and (e) 1,090–1,050 cm^{-1}. Composition dependence of (f) the fraction of free carbonyl (C=O) groups, (g) the frequency of N–H stretching vibration, and (h) the frequency of C–N stretching vibration. © 2019 Elsevier. Reproduced, with permission, from [53]*

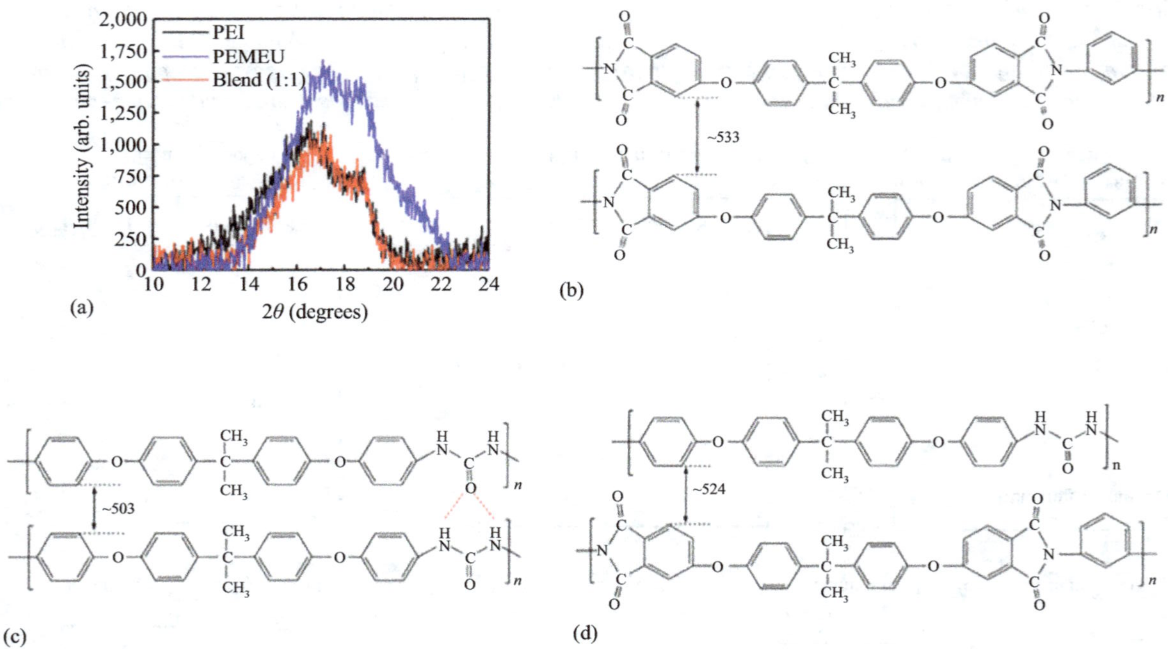

Figure 3.17 *(a) Wide angle X-ray diffraction data of PEI, PEMEU, and their 1:1 blend. D-spacing of two parallel molecule chains with different intermolecular interactions in (b) pristine PEI, (c) pristine PEMEU, and (d) the blend (1:1). © 2019 Elsevier. Reproduced, with permission, from [53]*

commercially available, to enhance and improve their dielectric response in the glass state without affecting other properties of these polymers. Nanocomposite strategies have been widely investigated in the past decades to raise the dielectric constant of polymers. In the traditional nanocomposite approach, nanoparticles with dielectric constant much higher (e.g., 100×) than that of the polymer are mixed with the polymer matrix [3,4,33,62–85]. The increase in the dielectric constant of such composites is largely attributed to the high dielectric constant of the inorganic fillers, since the composite dielectric constant is a volume average of those of the constituents. However, this approach suffers from several shortcomings. A major problem is that the inclusion of high dielectric constant fillers at high volume loading (>10 vol.%) results in a large reduction of the electric breakdown strength and consequently lowering the energy density [31,67,86–89]. This is caused by highly inhomogeneous electric fields at the interface of polymer and nanoparticle due to the large difference in permittivity between the two phases, causing intensification of local electric fields in the polymer matrix near the filler particles, leading to a large reduction of the electric breakdown strength. In addition, high volume inorganic filler loading in polymers significantly change the mechanical properties of the products, making it a challenge to employ low-cost roll-to-roll polymer film fabrication techniques to produce dielectric films over a broad thickness range.

In contrast to the traditional polymer nanocomposites, which rely on the high dielectric constant fillers in high volume loading to raise the dielectric constant of polymer composites, in this new approach, we explore the large interfacial areas and volumes in nanocomposites and the interfacial associated effects on dielectric response. We show that in high T_g dipolar polymers investigated, very low volume loading of inorganic nanoparticles (<0.5 vol.%) can lead to significantly enhanced dielectric properties. We note that dielectric metamaterials have been studied quite extensively in the past decades for high-frequency applications (from microwave to optical frequencies). In these dielectric metamaterials, local structures and interfaces, rather than averaged structures, play a significant role in enhancing the material performance and generating new material responses, which are not found in natural materials. The results presented in this section demonstrate that such a dielectric metamaterial strategy can also be explored at low frequencies for controlling and storing charges and electric energy in dielectric nanocomposites.

3.4.1 Nanocomposites with low content of nanofillers in high T_g amorphous dipolar polymers

3.4.1.1 Nanofiller size effect in polyetherimide (PEI) nanocomposites

Compared to semi-crystalline polymers, amorphous polymers present simpler dielectric polymer structures for the study. Hence, the nanocomposites with low content of nanofillers in high T_g amorphous dipolar polymers were first studied. In [90], PEI, a high temperature polymer (T_g ~ 217 °C), was selected as the matrix. PEI contains dipolar units in the polymer chains [91] and exhibits a dielectric

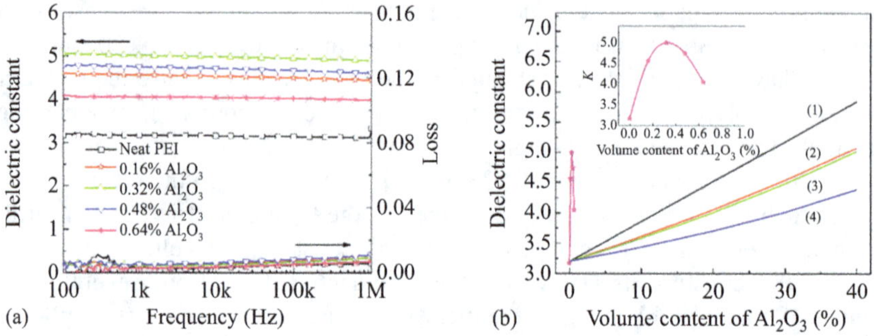

Figure 3.18 (a) Room temperature dielectric properties of PEI/alumina (20 nm particle size) nanocomposites at different alumina nanoparticle loading (in vol.%) vs. frequency. (b) Dielectric constant of nanocomposite films of PEI/alumina (20 nm particle size) vs. nanofiller volume content and comparison with several widely used dielectric models of diphasic dielectric composites (lines with no data points): curve (1) parallel model, (2) Maxwell model, (3) Lichtenecker model, and (4) series model. Inset shows an expanded view of the dielectric constants of the composite films vs. alumina loading. © 2017 The Royal Society of Chemistry. Reproduced, with permission, from [90]

constant of 3.2 and low dielectric loss (0.29%). Due to its high operating temperature, modest *K*, and low loss, PEI has been considered for the next generation of high-performance capacitor dielectric material [8,27]. Alumina nanoparticles were used as the nanofillers. As presented in Figure 3.18(a) and (b), films with a very low volume content of alumina nanoparticles (20 nm size, 0.32 vol.%) can achieve a dielectric constant >5, a more than 55% increase over that of PEI, while maintaining low loss. Figure 3.18(b) reveals this large increase in the dielectric constant occurs in a narrow composition range. With additional nanofiller, the dielectric constant decreases. There have been several studies of PEI nanocomposites in the past [27,72,83]. These earlier reports focused on high nanofiller content (>2 vol.%). It is startling that the large enhancement in the dielectric response of PEI/alumina (20 nm size) nanocomposites occurs at such a low filler volume content. Figure 3.18(b) also presents a comparison of the experimental data with several classical dielectric composite models using the dielectric properties of PEI and alumina. None of these models can describe the observed phenomenon, since they are based on the geometrical volume average of the dielectric responses of the constituents.

The temperature dependence of the dielectric properties of the neat PEI and its 0.32 vol.% alumina (20 nm) nanocomposite film are shown in Figure 3.19. The data show that the nanocomposite maintains low loss at temperatures in excess of 150 °C, the same as that of PEI.

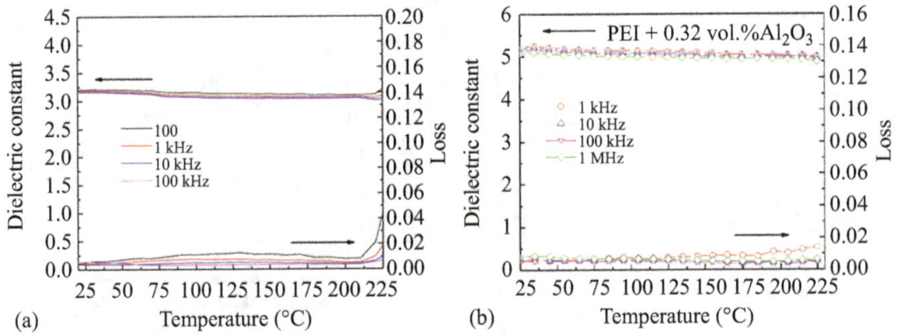

Figure 3.19 Dielectric properties at different frequencies of (a) neat PEI and (b) PEI/alumina (20 nm size) nanocomposite with 0.32 vol.% alumina loading as a function of temperature. © 2017 The Royal Society of Chemistry Reproduced, with permission, from [90]

For the nanocomposites with low volume content of nanofillers, the breakdown field is not compromised. Figure 3.20(a) presents the charge–discharge curves of the 0.32 vol.% alumina (20 nm) nanocomposite films at room temperature. These films exhibit a high breakdown field of 525 MV/m, similar to that of pure PEI [27], achieving a discharged energy density of 5.25 J/cm^3 (see inset in Figure 3.20(a)). The increased discharged energy density at different temperatures is shown in Figure 3.20(b), (c), and (d), the room temperature discharged energy density is 2.9 J/cm^3 under 350 MV/m, compared to that of pure PEI (1.9 J/cm^3) under the same field. At 150 °C (Figure 3.20(d)), there is increased conduction loss in both PEI and nanocomposite films at high field, causing a reduction in the discharged energy density. Nevertheless, the discharge energy density of the nanocomposite is still 12% higher than that of PEI, i.e., PEI films with only 0.32 vol.% alumina nanofiller exhibit enhanced discharge energy density and higher dielectric performance to >150 °C.

The effect of nanoparticle size on the dielectric response of PEI nanocomposites is presented in Figure 3.21. Note that the peak position of the dielectric enhancement shifts to higher nanofiller volume content with nanoparticle size. For PEI/alumina (5 nm) nanocomposites, the peak is at ca. 0.24 vol.% with the dielectric constant $K = 5$, while for PEI/alumina (50 nm) nanocomposites the peak is at ca. 0.8 vol.% with the K near 4.9. For nanoparticle interfacial effects, large size nanofillers need higher volume content to reach a similar interfacial area surrounding the nanofillers.

3.4.1.2 Effect of nanoparticle types on dielectric response of PEI nanocomposites

PEI nanocomposites with 20 nm size SiO_2 ($K = 3.9$)/MgO ($K = 9.7$), and 70 nm size boron nitride (hexagonal BN, dielectric constant ~ 5–7 and dielectric loss <0.2%) were investigated. Dielectric constants measured at room temperature and 1 kHz are presented in Figure 3.22. For example, PEI/MgO films show a peak

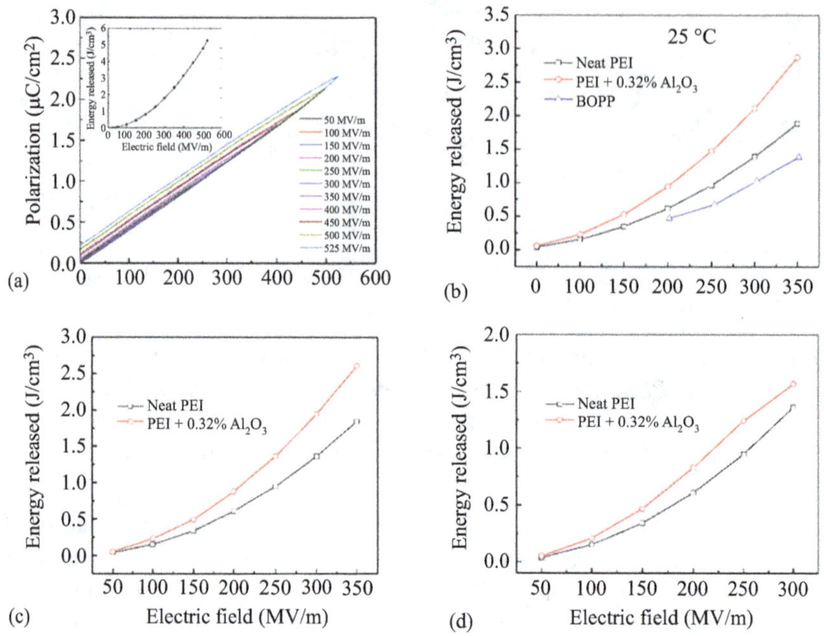

Figure 3.20 (a) Charge–discharge cycles of PEI/alumina (20 nm) nanocomposites with 0.32 vol.% alumina under different electrical fields at 10 Hz and room temperature. Inset: discharged energy density deduced from the charge–discharge cycle data. (b), (c), (d) Discharged energy density of the nanocomposites at different temperatures (room temperature, 100, and 150 °C), and their comparison with that of neat PEI and BOPP (room temperature), measured under 350 MV/m at 10 Hz. © 2017 The Royal Society of Chemistry. Reproduced, with permission, from [90]

dielectric constant of 4.95 at 0.35 vol.% MgO content, a trend very similar to that observed in PEI/alumina nanocomposites. It is interesting to note that the PEI/SiO_2 nanocomposite has a dielectric constant $K \sim 5$, which is higher than both the pure polymer matrix and the nanofiller. The PEI/BN films display an enhanced dielectric constant, peaking at ca. 0.83 vol.% with $K = 4.7$. These results indicate that the enhanced dielectric constant in the PEI films does not depend on the nanofiller types.

As a high dielectric nanofiller, $BaTiO_3$ (BTO) nanofiller-based PEI nanocomposites (50 nm particle size, $K > 500$) have been investigated earlier. To reach a dielectric constant of $K = 5$, required more than 12 vol.% BTO nanofiller [34,84]. As shown in Figure 3.22(b), the dielectric response of PEI/BTO (50 nm) is nearly the same as that of PEI/alumina (50 nm) in spite of the large difference in filler dielectric constant, i.e., the peak enhancement ($K = 4.9$) is at ca. 0.8 vol.%.

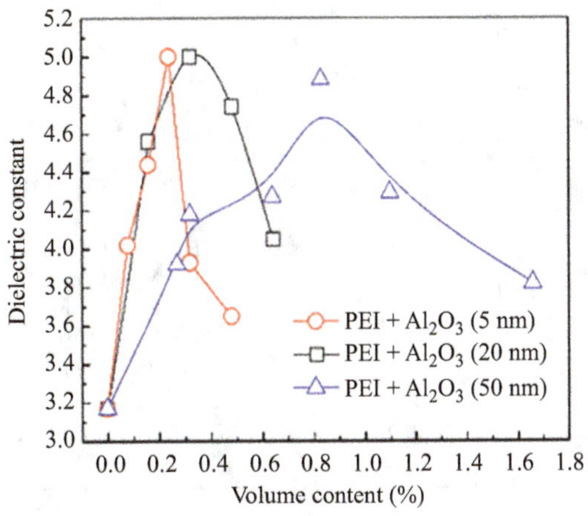

Figure 3.21 Effect of nanofiller size on the dielectric response (at 1 kHz) of PEI/alumina composite films vs. filler volume content. © 2017 The Royal Society of Chemistry. Reproduced, with permission, from [90]

3.4.1.3 Characterization and modeling of the PEI nanocomposites

The DSC data from both pure PEI and the PEI + 0.32 vol.% alumina nanocomposite film are shown in Figure 3.23, the nanocomposite film exhibits a reduced T_g (at ca. 210 °C) compared to that of PEI (T_g ~ 217 °C). In other words, average PEI segmental motion is somewhat faster in the presence of a low volume fraction of well-dispersed alumina nanoparticles. The effect of increment of dielectric constant is presumably dominated by changes in the dynamics of PEI segments' near-particle interfaces, leading to reduced constraints for dipole reorientations in interfacial regions in the presence of applied electric fields.

Yang *et al.* carried out a phenomenological modeling for the observed dielectric enhancement of PEI with very low volume loading of nanoparticles [90]. Here, a spatially varying local dielectric constant in the interfacial region in the polymer matrix around a nanoparticle is proposed:

$$K_{interface} = K_m + (K_2 - K_1)e^{-g}, g = (r/r_0)^2 \tag{3.4}$$

where r is the distance away from the surface of the nanofiller, K_m is the dielectric constant of the polymer matrix, K_1 and K_2 are two fitting parameters, and r_0 is a characteristic width of the interfacial region. For the PEI/Al_2O_3 nanocomposites, the fitting parameters are $r_0 = 50$ nm, $K_1 = 0.7$, and $K_2 = 8.5$ (see Figure 3.24(a)). These parameters are then fixed in modeling the dielectric response of PEI nanocomposites with different nanofiller sizes and volume contents, by the phase-field method [92,93].

As shown in Figure 3.24(b), the simulation results can reproduce the experimentally observed rapid increase in dielectric constant with a filler volume fraction

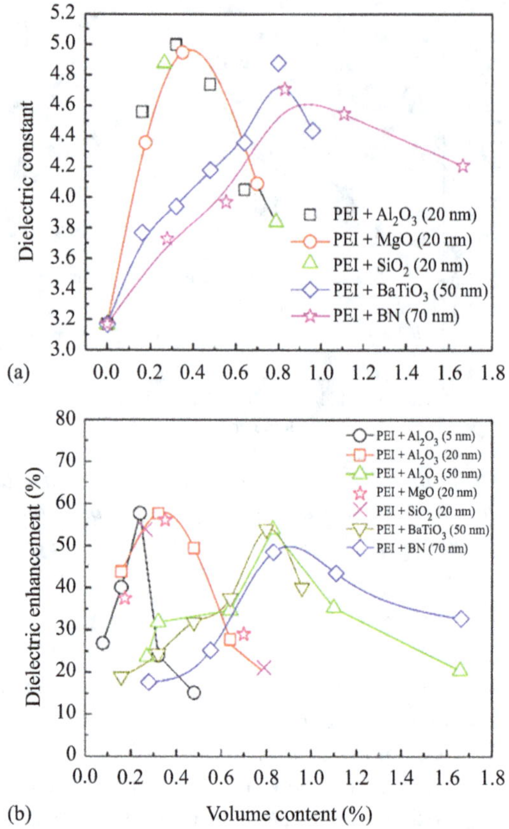

Figure 3.22 (a) Summary of the dielectric constants (at 1 kHz) of the various nanocomposites studied. (b) The percentage increase of the dielectric constant of the nanocomposites with respect to the original polymer matrix. © 2017 The Royal Society of Chemistry. Reproduced, with permission, from [90]

and the appearance of a peak at very small volume fractions of nanofillers, as well as the shift of the peak to a higher volume fraction as the filler size increases. The spatial distribution of the polarization response in the PEI nanocomposites to the applied field is presented in Figure 3.24(c) and (d). The model here, (3.4), does not take into consideration of nanofiller size.

3.4.1.4 Nanocomposites of polyimide and polystyrene

Analogous to PEI nanocomposites, the polyimide (PI) nanocomposites, as shown in Figure 3.25, also exhibit an enhanced dielectric constant occurring at very low filler volume loading. In contrast, nanocomposites of a non-polar polymer, polystyrene (PS), with alumina fillers of 20 and 50 nm, respectively, do not show

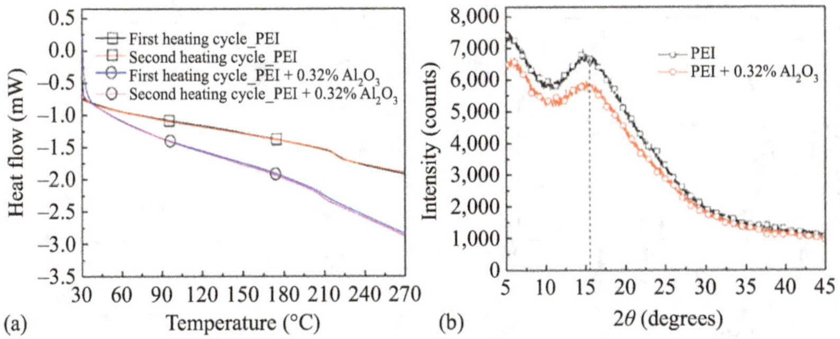

Figure 3.23 (a) DSC and (b) X-ray diffraction data of PEI and the PEI nanocomposite with 0.32 vol.% of alumina (20 nm). © 2017 The Royal Society of Chemistry. Reproduced, with permission, from [90]

dielectric enhancement, thus, confirming that presence of dipoles in high T_g polymers is important for the dielectric enhancement at low volume content of nanoparticles.

3.4.2 Nanocomposites with low content of nanofillers in high T_g semi-crystalline dipolar polymers

Our studies show that different from high T_g amorphous dipolar polymers, high T_g semi-crystalline polymers such as PEEU and PAEK (poly aromatic ether ketone) can display dielectric enhancement in both dielectric constant and breakdown strength over a broad temperature range [94]. PAEK is a commercial high temperature semi-crystalline dipolar polymer ($K = 3.6$, $T_g = 230$ °C, and $T_m = 350$ °C).

The dielectric constant K at 1 kHz of PEEU films at room temperature with various low nanofiller loading is presented in Figure 3.26(a), showing a dielectric enhancement peak $K = 7.4$ at ca. 0.21 vol.%. The films containing nanofillers display very similar dielectric loss and frequency dispersion as that of PEEU (see Figure 3.26(b)). Figure 3.26(c) presents the dielectric performance of the films with nanofillers to high temperature, showing thermal stability with low dielectric loss to 200 °C.

For dielectric materials such as polymers, the losses at high electric fields and high temperatures (such as at 150 °C) are caused mainly by high field conduction, which can be much higher than that at low fields (<10 MV/m, for example). High conduction loss at high electric field results in a low C/D efficiency, which lowers the energy density and may cause a low breakdown field [7]. Presented in Figure 3.27(a), (b), and (c) are the data for PEEU films with different nanofiller loadings at room temperature. Distinctly different from the amorphous polymer nanocomposites as discussed earlier, PEEU films with 0.21 vol.% filler exhibit 50% higher breakdown field (900 MV/m) than the base PEEU (600 MV/m). The C/D efficiency at high fields (conduction loss = 1 − efficiency) is deduced from

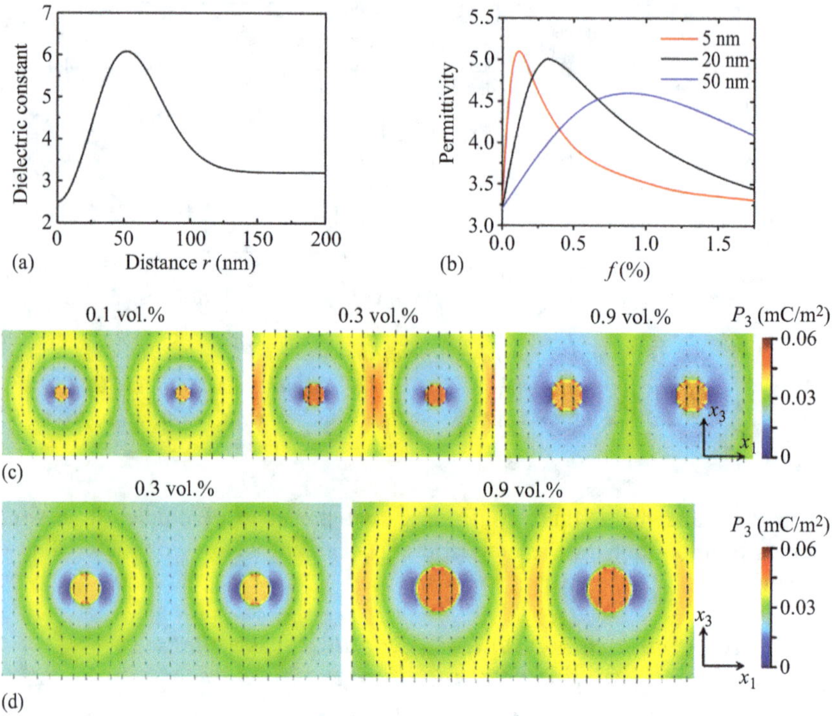

Figure 3.24 (a) Distribution of the dielectric constant of the polymer in the interfacial region as a function of the distance r from the surface of nanofiller. (b) Modeling results of the dielectric constant of PEI nanocomposites with 5, 20, and 50 nm nanofillers vs. the volume fraction of nanofillers. Polarization distribution in nanocomposites with (c) nanofillers of 20 nm at filler contents of 0.1, 0.3, and 0.9 vol.%; and (d) nanofillers of 50 nm at filler contents of 0.3 and 0.9 vol.%, on applying an electric field E3 = 1 MV/m, within the cross section passing through the centers of two nearest nanofillers. © 2017 The Royal Society of Chemistry. Reproduced, with permission, from [90]

the C/D data. Owing to enhancement in both the dielectric constant K and breakdown field E, the PEEU films with 0.21 vol.% nanofiller loading deliver a discharged energy density of 27 J/cm^3 with a C/D efficiency >90% under 900 MV/m, compared to 8.2 J/cm^3 of the base PEEU under 600 MV/m.

Wide-angle X-ray diffraction data for the base PEEU and nanocomposites are shown in Figure 3.28(a). PEEU with 0.21 vol.% displays a distinctly different X-ray pattern compared to the base PEEU and PEEU with higher alumina loadings. The data is analyzed to estimate the crystallinity and the peak position change of

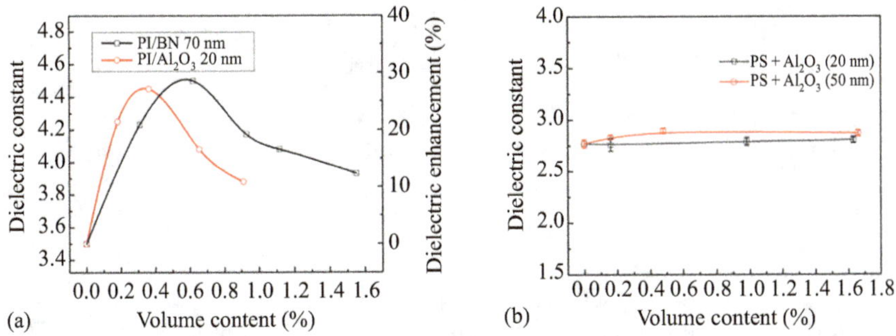

Figure 3.25 *(a) Dielectric constant (1 kHz) of nanocomposite films of PI/alumina (20 nm size) and PI/BN (70 nm size) vs. nanofiller volume load. (b) Dielectric constant (1 kHz) of nanocomposite films of PS/alumina (20 and 50 nm sizes) vs. nanofiller volume loading. © 2017 The Royal Society of Chemistry. Reproduced, with permission, from [90]*

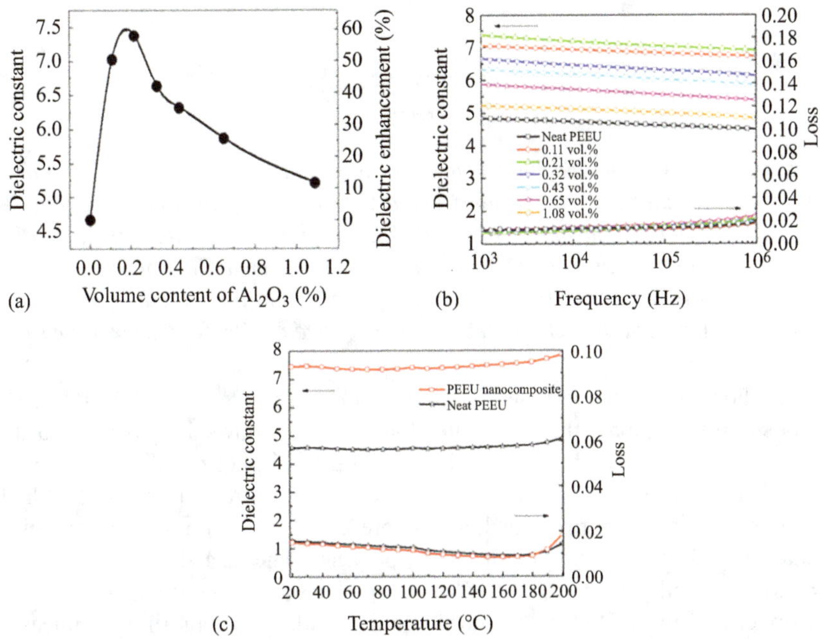

Figure 3.26 *(a) Dielectric constant at 1 kHz vs. the nanofiller loading and (b) the dielectric properties as functions of frequency of the PEEU nanocomposites with different nanofiller loadings measured at room temperature. (c) The dielectric properties at 1 kHz of the base PEEU and nanocomposite films with 0.21 vol.% filler loading vs. temperature. © 2020 AAAS. Reproduced, with permission, from [94]*

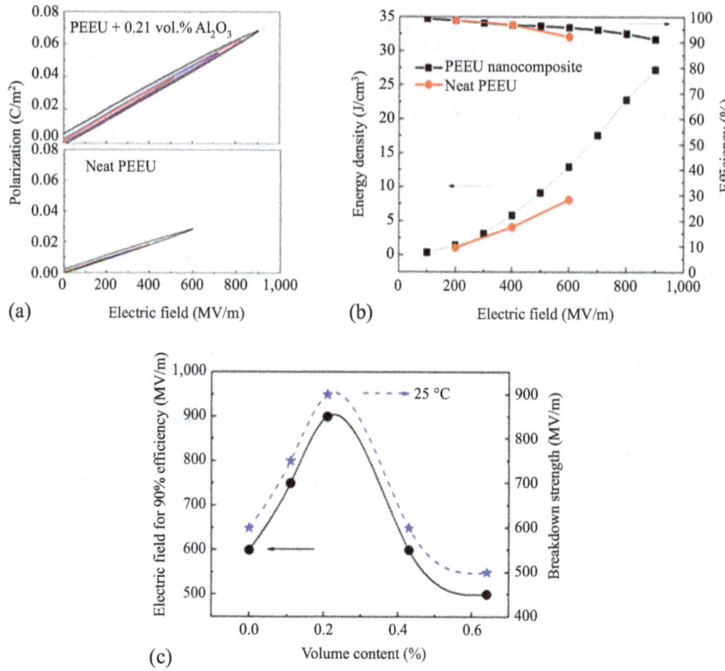

Figure 3.27 (a) Charge/discharge curves at different electric fields and (b) the discharged energy density and C/D efficiency as function of applied electric fields for the base PEEU and nanocomposite with 0.21 vol.% alumina at room temperature. (c) The electric field at 90% C/D efficiency and breakdown strength vs. nanofiller loading at room temperature. © 2020 AAAS. Reproduced, with permission, from [94]

the amorphous phase of the base PEEU and nanocomposite films [90]. Using the peak position to estimate the mean interchain spacing shows an expansion of the interchain spacing of about 5.8% for films with 0.21 vol.% of nanofillers compared to base PEEU and films with higher volume loading. Infrared spectra of the PEEU films with different nanofiller loadings, as presented in Figure 3.28(b), exhibited the changes of hydrogen bonding in PEEU polymer. The data show a softening of the hydrogen bonding in the PEEU films with 0.21 vol.% nanofiller loading compared to PEEU and PEEU with higher volume loading of nanofillers, indicating that the nanofillers of 0.21 vol.% partially disrupt the hydrogen bonding in the polymer, which reduces the constraints on the urea dipoles. Both the weakening of hydrogen bonding and expansion of the interchain spacing, which generates local free-space for dipoles, will enhance the dipolar response to the external field and increase the dielectric constant [37].

The data in Figure 3.28(a) also suggest that the films with 0.21 vol.% have a relatively smaller amorphous peak area compared to other films. By comparing the

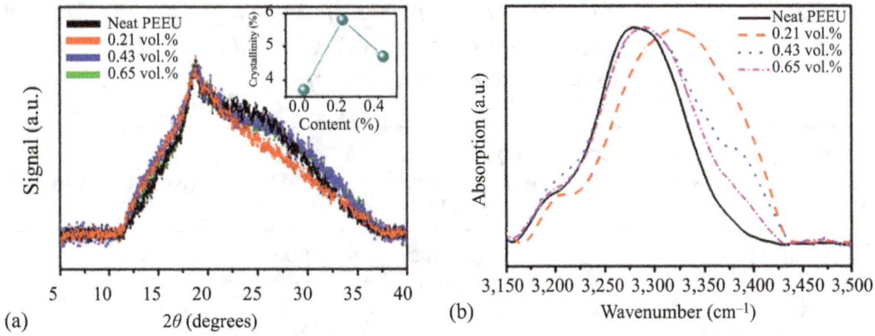

Figure 3.28 (a) Wide-angle XRD for PEEU with different alumina nanofiller loadings. The inset presents the crystallinity vs. the filler content, estimated from the XRD data. (b) FTIR spectra, showing the change of hydrogen bonding from the base PEEU to PEEU with different alumina nanofiller loadings. © 2020 AAAS. Reproduced, with permission, from [94]

peak areas of the broad amorphous peak and relatively sharp peak for the crystalline phase, we estimate the crystallinity of the PEEU films with different nanofiller loading, which is presented in the inset of Figure 3.28(a). The data show that there is a slight increase in crystallinity (and reduced crystallite size) of the PEEU films with 0.21 vol.% compared to the base PEEU. The results suggest that the nanofillers induced increase in crystallinity and reduced crystallite size in the PEEU 0.21 vol.% nanocomposites reduce the distance between crystallites, reducing the mean free path for the mobile charges. This may have a positive contribution to the reduced conductivity and enhanced breakdown strength. This mechanism is supported by the experimental results of Wang *et al.* on low-density polyethylene (LDPE)/alumina nanocomposites, in which 0.5 wt.% (0.12 vol.%) of alumina nanofiller (particles of 30-nm diameter) in LDPE enhances the breakdown field and reduces the conduction loss at high electric field, compared to the neat LDPE and composites with higher nanofiller loadings [95]. The nanofillers at 0.5 wt.% also increase the crystallinity of LDPE films.

In addition to PEEU, dielectric enhancement by very low volume loading of nanofillers in PAEK was also observed. It is presented in Table 3.3. Besides the high T_g dipolar polymers, we note that in several widely used low temperature (e.g., T_g near or below room temperature) semi-crystalline polymers such as LDPE, polypropylene (PP), early studies have also shown that nanofillers at very low volume loading (<0.5 vol.%) enhance the breakdown field and reduce the high field conduction loss at room temperature [95–97]. On the other hand, no enhancement in the dielectric constant was observed in these low-temperature dielectric polymers. The enhancements of capacitance performance from nanocomposites with low nanofiller volume loading of LDPE and PP from these earlier results are also included in Table 3.3.

Table 3.3 Summary of capacitance performance of nanocomposites of PEEU and PAEK at room temperature and at 150 °C (results from LDPE and PP nanocomposites are also included [95,96])

Polymers	Temperature (°C)	Polymer and nanocomposite (filler vol.%)	K	E_b (MV/m)	E_{CD} (MV/m)	U_e (J/cm^3)
PEEU (5 μm) Semi-crystal (T_g and T_m >250 °C)	25	Base polymer	4.7	600	600	8.2
		Nanocomposite (0.21 vol.%)	7.4	900	900	27
	150	Base polymer	4.7	400	200	0.83
		Nanocomposite (0.21 vol.%)	7.4	600	400	5
PAEK (10 μm) Semi-crystal (T_g = 230 °C, T_m = 350 °C)	25	Base polymer	3.6	400	400	2.5
		Nanocomposite (0.35 vol.%)	4.3	500	500	4.44
	150	Base polymer	3.6	350	150	0.36
		Nanocomposite (0.35 vol.%)	4.3	400	250	1.3
LDPE (0.1 mm) Semi-crystal (T_g < room temp.)	25	Base polymer	2.3	263	NA	NA
		Nanocomposite (0.12 vol.%)	2.3	355	NA	NA
PP (0.1 mm) Semi-crystal (T_g < room temp.)	25	Base polymer	2.2	305	NA	NA
		Nanocomposite (0.2 vol.%)	2.2	425	NA	NA

K, dielectric constant; E_b, breakdown field; E_{CD}, electric field at 90% C/D efficiency; U_e, discharged energy density at 90% C/D efficiency.

3.5 Conclusions

This chapter presents several recent works and strategies we take in enhancing the dielectric performance of polymer dielectric capacitors: dielectric constant, dielectric breakdown strength, high temperature operation (operation over a broad temperature range), and low conduction loss at high electric field and high temperatures (>150 °C). We note that dielectric metamaterials have been studied quite extensively in the past decades for high-frequency applications (from microwave to optical frequencies). In these dielectric metamaterials, local structures and interfaces, rather than averaged structures, play a significant role in enhancing the material performance and generating new material responses, which are not found in natural materials. The results of dielectric nanocomposites in which a very low volume loading of nanofillers generate remarkable changes in the dielectric performance of polymers presented in this chapter demonstrate that such a dielectric metamaterial strategy can also be explored at low frequencies for controlling and storing charges and electric energy in dielectric composites. Hence, we refer to this class of dielectric nanocomposites presented in this chapter as dielectric metamaterials. The low-cost and highly scalable approach demonstrated pave the way for the development of a totally new class of dielectric metamaterial with superior capacitance performance over a broad temperature range.

Compared to multilayer ceramic capacitors (MLCC) which require metal electrode in thickness >2 μm to ensure good electric conductivity, the electrode thickness of multilayer polymer capacitors (MLPC) is below 100 nm. Thus, the electrodes occupy very little capacitor volume in MLPC. This is significant since for many advanced energy applications, the dielectric layer thickness of multilayer capacitors is approaching to <1 μm. Thick electrode layers in MLCC, hence, will cause a large reduction of the capacitor performance at those thin dielectric layer thicknesses. MLPC with high-performance polymer dielectrics over a broad temperature (to temperatures >200 °C) may provide an attractive and low-cost alternative to MLCC in many electronic applications.

Acknowledgements

The Office of Naval Research (Grant No. N00024–14–1–0109, N00014–16–1–2454, and N00014–19–1–2028) supported the works presented in this chapter. We would like to acknowledge Minren Lin, J. Runt, J. Bernholc, B. Zhang, L. Q. Chen, and Tiannan Yang for their contributions to the works presented in this chapter, PolyK Technologies for providing PEI and PSU, and Polymics, Ltd., for providing PAEK used in these studies.

References

[1] Wu S., Shao M., Burlingame Q., *et al.* 'A high-K ferroelectric relaxor terpolymer as a gate dielectric for organic thin film transistors'. *Applied Physics Letters*. 2013;**102**(1):1.

[2] Sarjeant W. J., Zirnheld J., MacDougall F. W., *et al.* Capacitors—past, present, and future. In *Handbook of low and high dielectric constant materials and their applications*. Academic Press. 1999. pp. 423–91.

[3] Nalwa H. S. (ed.). *Handbook of low and high dielectric constant materials and their applications, two-volume set*. Elsevier, 1999.

[4] Chen Q., Shen Y., Zhang S., Zhang Q. M. 'Polymer-based dielectrics with high energy storage density'. *Annual Review of Materials Research*. 2015;**45**:433–58.

[5] Chu B., Zhou X., Re, K. *et al.* 'A dielectric polymer with high electric energy density and fast discharge speed'. *Science.* 2006;**313**(5785):334–6.

[6] Sharma V., Wang C., Lorenzini R. G., *et al.* 'Rational design of all organic polymer dielectrics'. *Nature Communications*. 2014;**5**:4845.

[7] Li Q., Chen L., Gadinski M. R., *et al.* 'Flexible high-temperature dielectric materials from polymer nanocomposites'. *Nature.* 2015;**523**(7562):576.

[8] Tan D., Zhang L., Chen Q., Irwin P. 'High-temperature capacitor polymer films'. *Journal of Electronic Materials*. 2014;**43**(12):4569–75.

[9] Cao Y., Irwin P. C., Younsi K. 'The future of nanodielectrics in the electrical power industry'. *IEEE Transactions on Dielectrics and Electrical Insulation*. 2004;**11**(5):797–807.

[10] Yoon M. H., Yan H., Facchetti A., Marks T. J. 'Low-voltage organic field-effect transistors and inverters enabled by ultrathin cross-linked polymers as gate dielectrics'. *Journal of the American Chemical Society*. 2005;**127**(29): 10388–95.

[11] Bertolazzi S., Wünsche J., Cicoira F., Santato C. 'Tetracene thin film transistors with polymer gate dielectrics'. *Applied Physics Letters*. 2011;**99**(1): 126.

[12] Roberts M. E., Queraltó N., Mannsfeld S. C., Reinecke B. N., Knoll W., Bao Z. 'Cross-linked polymer gate dielectric films for low-voltage organic transistors'. *Chemistry of Materials*. 2009;**21**(11):2292–9.

[13] Facchetti A. 'π-Conjugated polymers for organic electronics and photovoltaic cell applications'. *Chemistry of Materials*. 2010;**23**(3):733–58.

[14] Dang M. T., Hirsch L., Wantz G. 'P3HT: PCBM, best seller in polymer photovoltaic research'. *Advanced Materials*. 2011;**23**(31):3597–602.

[15] Huo L., Zhang S., Guo X., Xu F., Li Y., Hou J. 'Replacing alkoxy groups with alkylthienyl groups: a feasible approach to improve the properties of photovoltaic polymers'. *Angewandte Chemie International Edition*. 2011;**50**(41): 9697–702.

[16] Hanley T. L., Burford R. P., Fleming R. J., Barber K. W. 'A general review of polymeric insulation for use in HVDC cables'. *IEEE Electrical Insulation Magazine*. 2003;**19**(1):13–24.

[17] Ennis J. B., MacDougall F. W., Yang X. H., *et al*. 'Recent advances in high voltage, high energy capacitor technology'. *Proceedings of the 2007 16th IEEE International Pulsed Power Conference*. Albuquerque, NM, USA, June 2007 (IEEE, 2008), pp. 282–5.

[18] MacDougall F., Ennis J., Yang X. H., *et al*. 'High energy density capacitors for pulsed power applications'. *Proceedings of the 2009 IEEE Pulsed Power Conference*. Washington, DC, USA, June 2009 (IEEE, 2010), pp. 774–8.

[19] Bluhm H. *Power systems*. Berlin, Springer, 2006, p. 20.

[20] Osaka T., Datta M. *Energy storage systems in electronics*. CRC Press, 2000.

[21] Rabuffi M., Picci G. 'Status quo and future prospects for metallized polypropylene energy storage capacitors'. *IEEE Transactions on Plasma Science*. 2002;**30**(5):1939–42.

[22] Burress T. A., Coomer C. L., Campbell S. L., *et al*. *Evaluation of the 2007 Toyota Camry hybrid synergy drive system*. Report, ORNL/TM-2007/190, Revised. Oak Ridge National Laboratory (ORNL), Oak Ridge, TN, 2008.

[23] Burress T. A., Coomer C. L., Campbell S. L., *et al*. *Evaluation of the 2008 Lexus LS 600H hybrid synergy drive system*. Report ORNL/TM-2008/185. Oak Ridge National Laboratory (ORNL), Oak Ridge, TN, 2009.

[24] Chen Q., Wang Y., Zhou X., Zhang Q. M., Zhang S. 'High field tunneling as a limiting factor of maximum energy density in dielectric energy storage capacitors'. *Applied Physics Letters*. 2008;**92**(14):142909.

[25] Kao K. C. *Dielectric phenomena in solids*. Elsevier, 2004.
[26] Xu C., Boggs S. 'Measurement of resistive and absorption currents in capacitor films up to breakdown'. *Proceedings of the 2006 IEEE International Symposium on Electrical Insulation*. Toronto, Ont., Canada, June 2006 (IEEE, 2006), pp. 249–52.
[27] Ho J., Jow R. *Characterization of high temperature polymer thin films for power conditioning capacitors*. Report ARL-TR-4880. Army Research Lab Adelphi MD Sensors and Electron Devices Directorate, 2009.
[28] Smith N. J., Rangarajan B., Lanagan M. T., Pantano C. G. 'Alkali-free glass as a high energy density dielectric material'. *Materials Letters*. 2009;**63**(15): 1245–8.
[29] Zhu L., Wang Q. 'Novel ferroelectric polymers for high energy density and low loss dielectrics'. *Macromolecules*. 2012;**45**(7):2937–54.
[30] Zhou X., Zhao X., Suo Z., *et al.* 'Electrical breakdown and ultrahigh electrical energy density in poly (vinylidene fluoride-hexafluoropropylene) copolymer'. *Applied Physics Letters*. 2009;**94**(16):162901.
[31] Zhou X., Chu B., Neese B., *et al.* 'Electrical energy density and discharge characteristics of a poly (vinylidene fluoride-chlorotrifluoroethylene) copolymer'. *IEEE Transactions on Dielectrics and Electrical Insulation*. 2007;**14**(5): 1133–8.
[32] Zhang Q. M., Bharti V., Zhao X. 'Giant electrostriction and relaxor ferroelectric behavior in electron-irradiated poly (vinylidene fluoride-trifluoroethylene) copolymer'. *Science*. 1998;**280**(5372):2101–4.
[33] Thakur V. K., Gupta R. K. 'Recent progress on ferroelectric polymer-based nanocomposites for high energy density capacitors: Synthesis, dielectric properties, and future aspects'. *Chemical Reviews*. 2016;**116**(7):4260–317.
[34] Tan D., Tuncer E., Cao Y., Irwin P. 'Nanofiller dispersion in polymer dielectrics'. *Proceedings of the 2012 Annual Report Conference on Electrical Insulation and Dielectric Phenomena*. Montreal, QC, Canada, Oct 2012 (IEEE, 2012), pp. 916–18.
[35] Wang Y., Zhou X., Lin M., Zhang Q. M. 'High-energy density in aromatic polyurea thin films'. *Applied Physics Letters*. 2009;**94**(20):202905.
[36] Wu S., Li W., Lin M., *et al.* 'Aromatic polythiourea dielectrics with ultrahigh breakdown field strength, low dielectric loss, and high electric energy density'. *Advanced Materials*. 2013;**25**(12):1734–8.
[37] Thakur Y., Dong R., Lin M., *et al.* 'Optimizing nanostructure to achieve high dielectric response with low loss in strongly dipolar polymers'. *Nano Energy*. 2015;**16**:227–34.
[38] Cheng Z., Lin M., Wu S., *et al.* 'Aromatic poly (arylene ether urea) with high dipole moment for high thermal stability and high energy density capacitors'. *Applied Physics Letters*. 2015;**106**(20):202902.
[39] Frohlich H. *Theory of dielectrics: Dielectric constant and dielectric loss*. Clarendon Press, 1958.
[40] Thakur Y., Lin M., Wu S., Cheng Z., Jeong D. Y., Zhang Q. M. 'Tailoring the dipole properties in dielectric polymers to realize high energy density

with high breakdown strength and low dielectric loss'. *Journal of Applied Physics*. 2015;**117**(11):114104.

[41] Allen F. H., Kennard O., Watson D. G., Brammer L., Orpen A. G., Taylor R. 'Tables of bond lengths determined by X-ray and neutron diffraction. Part 1. Bond lengths in organic compounds'. *Journal of the Chemical Society, Perkin Transactions 2*. 1987;(12):S1–S19.

[42] Pan J., Li K., Li J., Hsu T., Wang Q. 'Dielectric characteristics of poly (ether ketone ketone) for high temperature capacitive energy storage'. *Applied Physics Letters*. 2009;**95**(2):022902.

[43] Damjanovic D., Demartin M. 'The Rayleigh law in piezoelectric ceramics'. *Journal of Physics D: Applied Physics*. 1996;**29**(7):2057.

[44] Wu S., Lin M., Lu S. G., Zhu L., Zhang Q. M. 'Polar-fluoropolymer blends with tailored nanostructures for high energy density low loss capacitor applications'. *Applied Physics Letters*. 2011;**99**(13):132901.

[45] Gaur U. 'The glass transition temperature of polyethylene'. *Macromolecules*. 1980;**13**:445–6.

[46] Carr S. H., Seanor D. A. (eds.). *Electrical properties of polymers*. New York, London, Academic Press, 1982, p. 215.

[47] Zhang Q. M., Su J., Kim C. H., Ting R., Capps R. 'An experimental investigation of electromechanical responses in a polyurethane elastomer'. *Journal of Applied Physics*. 1997;**81**(6):2770–6.

[48] McCrum N. G., Read B. E., Williams G. *Anelastic and dielectric effects in polymeric solids*. New York, Wiley, 1967, p. 617.

[49] Wu S., Lin M., Burlingame Q., Zhang Q. M. 'Meta-aromatic polyurea with high dipole moment and dipole density for energy storage capacitors'. *Applied Physics Letters*. 2014;**104**(7):072903.

[50] Thakur Y., Zhang B., Dong R., *et al*. 'Generating high dielectric constant blends from lower dielectric constant dipolar polymers using nanostructure engineering'. *Nano Energy*. 2017;**32**:73–9.

[51] Lee J. W., Takase Y., Newman B. A., Scheinbeim J. I. 'Effect of annealing on the ferroelectric behavior of nylon-11 and nylon-7'. *Journal of Polymer Science Part B: Polymer Physics*. 1991;**29**(3):279–86.

[52] Zhang Z., Litt M. H., Zhu L. 'Nature of ferroelectric behavior in main-chain dipolar glass nylons: Cooperative segmental motion induced by high poling electric field'. *Macromolecules*. 2018;**51**(5):1967–77.

[53] Zhang Q., Chen X., Zhang T., Zhang Q. M. 'Giant permittivity materials with low dielectric loss over a broad temperature range enabled by weakening intermolecular hydrogen bonds'. *Nano Energy*. 2019;**64**:103916.

[54] Fox T. G. 'Influence of diluent and of copolymer composition on the glass temperature of a polymer system'. *Bulletin of the American Physical Society*. 1956;**1**:123.

[55] Baker W. O., Yager W. A. 'The relation of dielectric properties to structure of crystalline polymers. II. Linear polyamides'. *Journal of the American Chemical Society*. 1942;**64**(9):2171–7.

[56] Ning N., Li S., Sun H., *et al.* 'Largely improved electromechanical properties of thermoplastic polyurethane dielectric elastomers by the synergistic effect of polyethylene glycol and partially reduced graphene oxide'. *Composites Science and Technology*. 2017;**142**:311–20.

[57] Baker W. O., Yager W. A. 'The relation of dielectric properties to structure of crystalline polymers. II. Linear polyamides'. *Journal of the American Chemical Society*. 1942;**64**(9):2171–7.

[58] Hosseini S. S., Teoh M. M., Chung T. S. 'Hydrogen separation and purification in membranes of miscible polymer blends with interpenetration networks'. *Polymer*. 2008;**49**(6):1594–603.

[59] Liu S., Tian M., Yan B., *et al.* 'High performance dielectric elastomers by partially reduced graphene oxide and disruption of hydrogen bonding of polyurethanes'. *Polymer*. 2015;**56**:375–84.

[60] Mattia J., Painter P. 'A comparison of hydrogen bonding and order in a polyurethane and poly (urethane–urea) and their blends with poly (ethylene glycol)'. *Macromolecules*. 2007;**40**(5):1546–54.

[61] Yoo S., Kim D. G., Ha T., Chan Won J., Jang K. S., Kim Y. H. 'Solution-processable, thin, and high-κ dielectric polyurea gate insulator with strong hydrogen bonding for low voltage organic thin-film transistors'. *ACS Applied Materials & Interfaces*. 2018;**10**(38);32462–70.

[62] Huang C., Zhang Q. M., Li J. Y., Rabeony M. 'Colossal dielectric and electromechanical responses in self-assembled polymeric nanocomposites'. *Applied Physics Letters*. 2005;**87**(18):182901.

[63] Li J., Seok S. I., Chu B., Dogan F., Zhang Q., Wang Q. 'Nanocomposites of ferroelectric polymers with TiO_2 nanoparticles exhibiting significantly enhanced electrical energy density'. *Advanced Materials*. 2009;**21**(2):217–21.

[64] Balasubramanian B., Kraemer K. L., Reding N. A., Skomski R., Ducharme S., Sellmyer D. J. 'Synthesis of monodisperse TiO_2–paraffin core–shell nanoparticles for improved dielectric properties'. *ACS Nano*. 2010;**4**(4):1893–900.

[65] Nelson J. K., Fothergill J. C. 'Internal charge behaviour of nanocomposites'. *Nanotechnology*. 2004;**15**(5):586.

[66] Nelson J. K. *Dielectric polymer nanocomposites*. New York, NY, USA, Springer, 2010, pp. 259–84.

[67] Kim P., Doss N. M., Tillotson J. P., *et al.* 'High energy density nanocomposites based on surface-modified $BaTiO_3$ and a ferroelectric polymer'. *ACS Nano*. 2009;**3**(9):2581–92.

[68] Li J., Claude J., Norena-Franco L. E., Seok S. I., Wang Q. 'Electrical energy storage in ferroelectric polymer nanocomposites containing surface-functionalized $BaTiO_3$ nanoparticles'. *Chemistry of Materials*. 2008;**20**(20):6304–6.

[69] Li Z., Fredin L. A., Tewari P., *et al.* 'In situ catalytic encapsulation of core-shell nanoparticles having variable shell thickness: Dielectric and energy

storage properties of high-permittivity metal oxide nanocomposites'. *Chemistry of Materials*. 2010;**22**(18):5154–64.

[70] Huang C., Zhang Q. M., Li J. Y., Rabeony M. 'Colossal dielectric and electromechanical responses in self-assembled polymeric nanocomposites'. *Applied Physics Letters*. 2005;**87**:182901–3.

[71] Guo N., DiBenedetto S. A., Tewari P., Lanagan M. T., Ratner M. A., Marks T. J. 'Nanoparticle, size, shape, and interfacial effects on leakage current density, permittivity, and breakdown strength of metal oxide–polyolefin nanocomposites: Experiment and theory'. *Chemistry of Materials*. 2010;**22**(4): 1567–78.

[72] Zou C., Kushner D., Zhang S. 'Wide temperature polyimide/ZrO_2 nanodielectric capacitor film with excellent electrical performance'. *Applied Physics Letters*. 2011;**98**(8):082905.

[73] Huang X., Zhi C. *Polymer nanocomposites*. Switzerland, Springer International Publishing, 2016.

[74] Dang Z. M., Zhou T., Yao S. H., *et al.* 'Advanced calcium copper titanate/polyimide functional hybrid films with high dielectric permittivity'. *Advanced Materials*. 2009;**21**(20):2077–82.

[75] Yang W., Yu S., Sun R., Du R. 'Nano-and microsize effect of CCTO fillers on the dielectric behavior of CCTO/PVDF composites'. *Acta Materialia*. 2011;**59**(14):5593–602.

[76] Prakash B. S., Varma K. B. R. 'Dielectric behavior of CCTO/epoxy and Al-CCTO/epoxy composites'. *Composites Science and Technology*. 2007; **67**(11–12):2363–8.

[77] Yuan J., Yao S., Poulin P. 'Dielectric constant of polymer composites and the routes to high-k or low-k nanocomposite materials'. *Polymer Nanocomposites*. Switzerland, Springer, 2016, pp. 3–28.

[78] Tang H., Lin Y., Andrews C., Sodano H. A. 'Nanocomposites with increased energy density through high aspect ratio PZT nanowires'. *Nanotechnology*. 2010;**22**(1):015702.

[79] Yao J., Xiong C., Dong L., *et al.* 'Enhancement of dielectric constant and piezoelectric coefficient of ceramic–polymer composites by interface chelation'. *Journal of Materials Chemistry*. 2009;**19**(18):2817–21.

[80] Banerjee S., Cook-Chennault K. A. 'Influence of Al particle size and lead zirconate titanate (PZT) volume fraction on the dielectric properties of PZT-epoxy-aluminum composites'. *Journal of Engineering Materials and Technology*. 2011;**133**(4):041016.

[81] Tang H., Lin Y., Sodano H. A. 'Enhanced energy storage in nanocomposite capacitors through aligned PZT nanowires by uniaxial strain assembly'. *Advanced Energy Materials*. 2012;**2**(4):469–76.

[82] Carr J. M., Mackey M., Flandin L., Schuele D., Zhu L., Baer E. 'Effect of biaxial orientation on dielectric and breakdown properties of poly (ethylene terephthalate)/poly (vinylidene fluoride-co-tetrafluoroethylene) multilayer films'. *Journal of Polymer Science Part B: Polymer Physics*. 2013;**51**(11): 882–96.

[83] Hanemann T., Szabó D. V. 'Polymer-nanoparticle composites: From synthesis to modern applications'. *Materials.* 2010;**3**(6):3468–517.
[84] Lomax J. F., Lomax E. A., Lu S. C., *et al.* 'Electrical properties of $BaTiO_3$ nanoparticles in poly (ether imide)'. *Smart Materials and Structures.* 2012;**21**(8):085017.
[85] Fredin L. A., Li Z., Lanagan M. T., Ratner M. A., Marks T. J. 'Sustainable high capacitance at high frequencies: Metallic aluminum–polypropylene nanocomposites'. *ACS Nano.* 2012;**7**(1):396–407.
[86] An L., Boggs S. A., Calame J. P. 'Energy storage in polymer films with high dielectric constant fillers'. *IEEE Electrical Insulation Magazine.* 2008;**24**(3): 5–10.
[87] Wang Q., Zhu L. 'Polymer nanocomposites for electrical energy storage'. *Journal of Polymer Science Part B: Polymer Physics.* 2011;**49**(20):1421–9.
[88] Lewis T. J. 'Interfaces are the dominant feature of dielectrics at the nanometric level'. *IEEE Transactions on Dielectrics and Electrical Insulation.* 2004;**11**(5);739–53.
[89] Dang Z. M., Yuan J. K., Yao S. H., Liao R. J. 'Flexible nanodielectric materials with high permittivity for power energy storage'. *Advanced Materials.* 2013;**25**(44):6334–65.
[90] Thakur Y., Zhang T., Iacob C., *et al.* 'Enhancement of the dielectric response in polymer nanocomposites with low dielectric constant fillers'. *Nanoscale.* 2017;**9**(31):10992–7.
[91] Teyssedre G., Laurent C. 'Advances in high-field insulating polymeric materials over the past 50 years'. *IEEE Electrical Insulation Magazine.* 2013;**29**(5):26–36.
[92] Chen L. Q. 'Phase-field models for microstructure evolution'. *Annual Review of Materials Research.* 2002;**32**(1):113–40.
[93] Yang T. N., Hu J. M., Nan C. W., Chen L. Q. 'Predicting effective magnetoelectric response in magnetic-ferroelectric composites via phase-field modeling'. *Applied Physics Letters.* 2014;**104**(5):052904.
[94] Zhang T., Chen X., Thakur Y., *et al.* 'A highly scalable dielectric metamaterial with superior capacitor performance over a broad temperature'. *Science Advances.* 2020;**6**(4):eaax6622.
[95] Wang W., Min D., Li S. 'Understanding the conduction and breakdown properties of polyethylene nanodielectrics: Effect of deep traps'. *IEEE Transactions on Dielectrics and Electrical Insulation.* 2016;**23**(1):564–72.
[96] Min D., Yan C., Mi R., *et al.* 'Carrier transport and molecular displacement modulated dc electrical breakdown of polypropylene nanocomposites'. *Polymers.* 2018;**10**(11):1207.
[97] Thakur Y., Lean M. H., Zhang Q. M. 'Reducing conduction losses in high energy density polymer using nanocomposites'. *Applied Physics Letters.* 2017;**110**(12):122905.

Chapter 4
Polymer/nanofiller composites

He Li[1] *and Qing Wang*[1]

4.1 Introduction

Polymers display intrinsic advantages of processability, flexibility, scalability, lightweight, and excellent dielectric strength in comparison to dielectric ceramics [1,2]. However, the organic nature endows relatively low dielectric constants (k) (e.g., <10) of polymers, which directly limits the energy density achieved in polymer film capacitors [3]. Over past decades, a composite approach has been developed via incorporating inorganic fillers with various functionalities into polymer matrices to improve multiple dielectric properties, including k and breakdown strength, and energy storage performance of polymers [4,5].

The commonly used nanofillers can be divided into three categories, i.e., conducting, semi-conducting, and insulating nanofillers, according to electrical conductivity [6]. One straightforward strategy to increase the k value of a polymer is the introduction of high-k nanofiller such as metal- and carbon-based conducting filler, semiconducting filler, and ceramic filler with relatively high k (e.g., ≥ 40). While the wide-band gap (e.g., ≥ 4) ceramic fillers typically exhibit a much lower k (e.g., ≤ 25), they are found to be effective in reducing leakage current and enhancing the charge-discharge efficiency [7,8]. The morphology of filler is also found to play a critical role in determining dielectric properties and capacitive performance of polymer/nanofiller composites. For example, one-dimensional (1-D) nanowires/nanofibers and two-dimensional (2-D) nanosheets/nanoplates are found to exhibit stronger dielectric properties in response to the applied electric field, attributing to their larger aspect ratio and specific surface area when compared with zero-dimensional (0-D) nanospheres [9,10]. Interestingly, it is verified that the parallelly arranged 2-D nanofillers are capable of dispersing the applied electric field throughout the polymer matrix to mitigate the local electric field distortions, resulting in largely improved breakdown strength [7,10].

Due to the ultrahigh surface energy, the nano-sized inorganic filler is prone to form agglomeration in polymer matrix. In addition, the sharp distinction between

[1]Department of Materials Science and Engineering, The Pennsylvania State University, University Park, PA, USA

organic and inorganic phases leads both physical and chemical mismatch, which accordingly gives rise to severe decreases in the breakdown strength and mechanical flexibility of the resulting polymer nanocomposites [11,12]. Thus, surface functionalization of inorganic nanofillers is of great significance and has been widely employed in the polymer composite approaches.

It is recognized well that the stored electrical energies are highly dependent on *k* and breakdown strength, namely, a high *k* along with a high breakdown strength will permit the achievement of a high energy density. However, these two dielectric parameters are typically in a negative correlation in conventional dielectric materials [13]. To improve the discharged energy density, e.g., the stored energy density subtracts the energy loss, dielectric loss, especially the conduction loss at high fields, has to be reduced. Thus, for the design of a high-performance dielectric capacitor, besides the improvement of *k*, the increase of breakdown strength and the decrease of energy loss are certainly of equal importance that needs to be prudently considered.

This chapter firstly summarized the recent utilization of different kinds of nanofillers with various functionalities and morphologies and their surface-modification methods in dielectric composites. Besides using fillers to increase *k* of polymer composites, a new design concept, in which the introduction of nanofiller with a moderate *k* and a wide bandgap to inhibit the leakage current and reduce energy loss, is highlighted. Second, the recent progress on the polymer/nanofiller composite films with heterogeneous and electrically topologic structures beyond simple 0–3 composites is described and discussed. The design paradigm aims to currently improve *k* value and breakdown strength of the resulting polymer/nanofiller composites through the elegant combination of different polymers and/or fillers with complementary functionalities. The control of the organic-inorganic interface properties allows tailoring of the physical and chemical compatibility of the filler and polymer phases. The architecture of topologically arranged nanofillers induces anisotropic dielectric responses and electric field distributions in the polymer/nanofiller composites.

4.2 Two-phase polymer composites with high-*k* nanofillers

Polymeric film capacitors with high dielectric breakdown strength and fast discharging speed are widely used in power systems and electronic devices [14]. Currently, the state-of-the-art, commercially available dielectric polymer for power capacitor is a biaxially oriented polypropylene (BOPP) film [15]. As a result of the nonpolar nature, the BOPP film capacitor exhibits negligible dielectric loss (e.g., <0.05% at 1 kHz) and ultrahigh charge-discharge efficiency (e.g., >95% at 500 MV/m). However, the stored energy density has been severely limited by its low *k* (e.g., ~2.2 at 1 kHz), which is merely around 2 J/cm^3 at 500 MV/m. In addition to polypropylene (PP), electrically linear polymers including polyethylene (PE), polystyrene (PS), polyurethane (PU), poly(methyl methacrylate) (PMMA), epoxy resin (EP), etc. were used as composite matrices in electrical insulation and capacitive energy storage applications [2–6]. More recently, with the emerging development of next-generation microelectronics and pulsed power systems that call for dielectric materials can operate efficiently at elevated temperatures, high-temperature polymer matrices with high glass transition temperature (T_g), including polyimide (PI), poly(ether imide) (PEI), and poly(ether ether ketone) (PEEK), have been systematically investigated [16] (see Chapter 5). Note that all these aforementioned polymers process low k (e.g., <5) that are greatly lower than most inorganic reinforcing phases. Therefore, they not only limit the further improvement in stored energy density but also lead to the formation of the electrically unmatched organic-inorganic interfaces in the resultant polymer composites.

To address this issue, nonlinear polymer matrices – ferroelectric poly(vinylidene fluoride) (PVDF) and its copolymers, terpolymers, and tetrapolymers with relatively high k (e.g., ≥ 10 at 1 kHz) have been developed and regarded as the greatest promising polymer candidates for high-energy polymeric film capacitors [3,17]. PVDF is one of the most widely used ferroelectric polymers in the development of advanced dielectric materials for energy storage devices and continues to receive extensive attention owing to its outstanding properties [18]. Due to the large electronegativity difference between fluorine atoms and hydrogen atoms, a large number of intrinsic dipole moments form the elementary unit that contributes to large electric polarization intensity, namely, high k value. Figure 4.1 demonstrates three major types of regular chain conformations of PVDF, including the TG^+TG^-, $TTTG^+TTTG^-$, and all-trans [17]. The T (trans) and $G^{\pm}$ (gauche$^{\pm}$) are associated with the torsional bond arrangements with substituents at 180° and ±60° to each other, respectively.

The variations of chain conformation result in four major crystalline phases of α-, δ-, γ-, and β-phase for PVDF as shown in Figure 4.2. Both the α- and δ-phase in PVDF adopt the TG^+TG^- chain conformation. α-phase in PVDF is

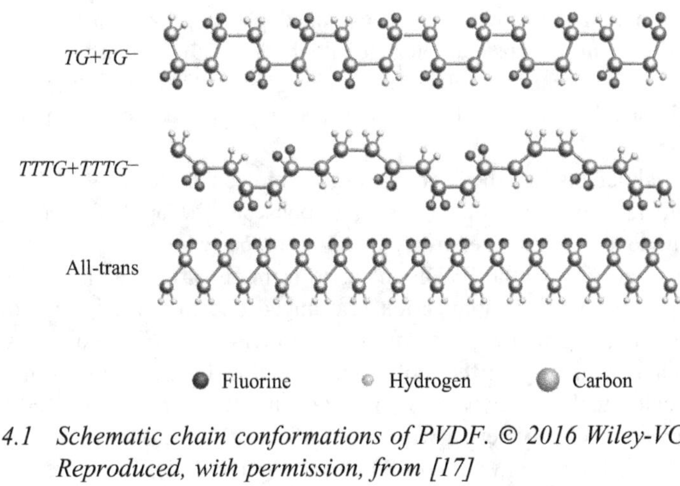

Figure 4.1 Schematic chain conformations of PVDF. © 2016 Wiley-VCH. Reproduced, with permission, from [17]

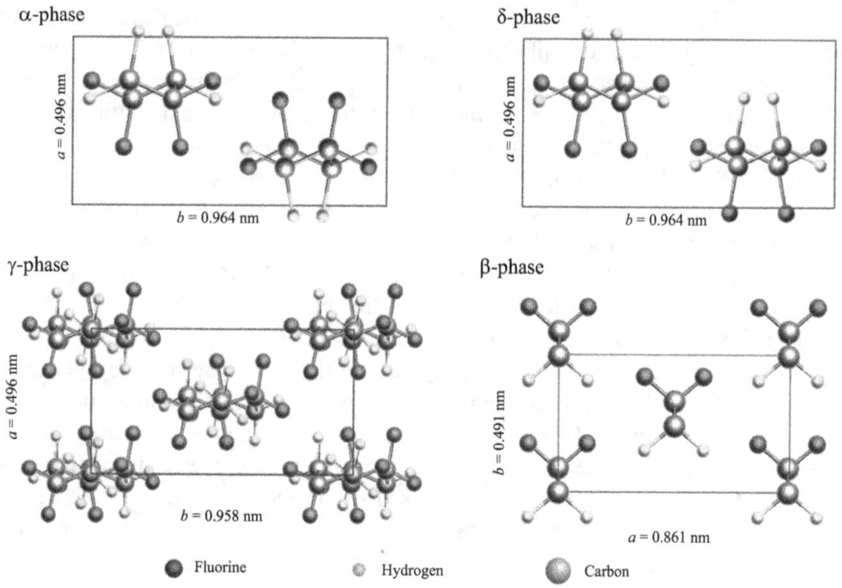

Figure 4.2 Schematic view of the unit cells of the four crystalline phases of PVDF. © 2016 Wiley-VCH. Reproduced, with permission, from [17]

macroscopically nonpolar due to the cancellation of dipole moments between the two antiparallel chains, which exhibits a kinetic steady state. On the contrary, δ-phase in PVDF is macroscopically polar because the two chains of polymorph are oriented in identical directions to exhibit a net electric polarization. δ-phase in PVDF is composed of a gauche bond after every three trans bonds of the all-trans conformation, in which three trans linked to a conformational sequence of $TTTG^{+}TTTG^{-}$. Particularly, β-phase is composed of molecular chains of the all-trans conformation that are the most extended planar zigzag fashion, and thus contributes to a unique pseudo hexagonal character of the lattice, which significantly facilitates the ferroelectric switching in PVDF, exhibiting a thermodynamically steady-state with largest ferroelectric, pyroelectric, and piezoelectric responses [17].

Structural defects have been adopted to the molecular chains of PVDF to yield fluoropolymers with novel behaviors in response to the applied electric fields. Trifluoroethylene (TrFE), a relatively bulky monomer when compared with VDF, has been introduced into the PVDF chains to form an earliest PVDF-based copolymer [19]. The interchain interaction is changed due to the existence of TrFE. Thus, the Curie temperature of the resulting P(VDF-*co*-TrFE) copolymers can be simply tailored with varying the molar ratio of TrFE. Later, a class of monomers including chlorotrifluoroethylene (CTFE) and hexafluoropropylene (HFP) have been employed to copolymerize with VDF [20,21]. The bulkier monomers facilitate dipole switching in P(VDF-*co*-CTFE) and P(VDF-*co*-HFP) by breaking apart the large crystals and decoupling the ferroelectric domains, resulting in slimmer electric displacement-electric field (*D-E*) loops, i.e., lower ferroelectric loss with respect to normal ferroelectric polymers of PVDF and P(VDF-*co*-TrFE). In addition, a third monomer such as CTFE and chlorofluoroethylene (CFE) have been incorporated into P(VDF-*co*-TrFE) in forming PVDF-based terpolymers with higher k [22,23]. The ferroelectric domain size and the energy barrier in phase transition are reduced in the resulting P(VDF-*ter*-TrFE-*ter*-CTFE) and P(VDF-*ter*-TrFE-*ter*-CFE) terpolymers, and thus leads the terpolymers exhibit relaxor ferroelectric behavior. Most recently, Chen *et al.* reported tetrapolymer compositions of P(VDF-*tetra*-TrFE-*tetra*-CFE-*tetra*-CTFE) containing four monomers [24]. The tetrapolymer with the VDF:TrFE:CFE:CTFE mole ratio of 56.1:33.2:5.9:4.8 exhibits relaxor ferroelectric behavior with a high k of ~50 at 25 °C and 100 Hz.

The utilization of high-k nanofillers (e.g., ≥ 40) is an effective approach that has been maturely developed to further improve the k of polymer composites. Commonly used high-k nanofillers can be divided into two main categories – conducting particles such as carbon- and metal-based fillers as well as high-k ceramic particles with a relatively narrow bandgap. The polymer/nanofiller composites consisting of conducting and high-k ceramic components are described and discussed in detail below.

4.2.1 Polymer/conductive filler nanocomposites

It is well known that the k of polymer composites can be significantly enhanced by incorporating conducting or semi-conducting nanofillers into the organic matrices. It is noteworthy that insulating polymer/conducting particle composites typically have an insulator-conductor transition with increasing conducting particle loading; the insulator-conductor transition is always observed with an abrupt change in conductivity and a divergence of the k near the percolation threshold of the resulting polymer composites [25]. Owing to the formation of inner-connected conducting paths, tens or even hundreds of times of the polymer matrices can improve the k of a composite. For instance, Dang *et al.* fabricated PVDF/Ni composites by using a simple blending and hot-modeling procedure [25]. Figure 4.3 clearly demonstrates both the conductivity (σ_{eff}) and k (ε_{eff}) of the PVDF/Ni composites have an insulator-conductor transition at the Ni volume fraction (f_{Ni}) of ~0.16–0.18.

Percolation theory allows one to present with power laws the electric conductivity of the composites near the insulator-conductor transition as follows,

$$\sigma_{\text{eff}} \propto (f_c - f_{\text{Ni}})^{-s'} \tag{4.1}$$

for $f_{\text{Ni}} < f_c$ and

$$\sigma_{\text{eff}} \propto (f_{\text{Ni}} - f_c)^{t} \tag{4.2}$$

for $f_{\text{Ni}} > f_c$, where f_c is the percolation threshold, t is the critical exponent in the conductive phase, and s' is the critical exponent in the insulation phase. Similarly, a

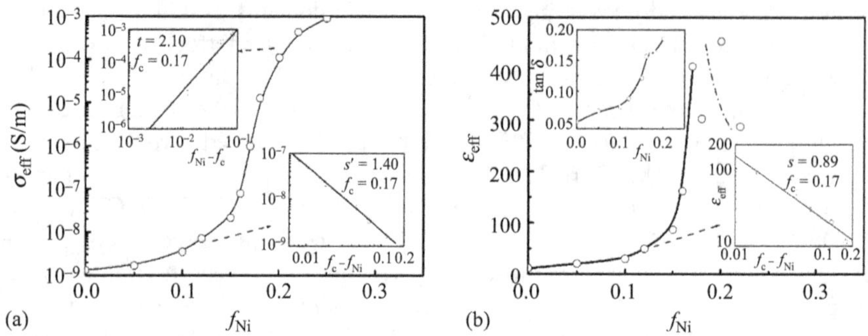

Figure 4.3 *(a) Dependence of the effective conductivity, σ_{eff}, of the Ni-PVDF composites on the Ni volume fraction, f_{Ni}, measured at room temperature and 100 Hz. The insets show the best fit of the σ_{eff} values to (4.1). (b) Variation of the effective dielectric constant, ε_{eff}, of the Ni-PVDF composites with Ni volume fraction, f_{Ni}, measured at room temperature and 100 Hz. The inset show the dependence of the loss factor, tan δ, on f_{Ni} and the best fit of the ε_{eff} data to (4.3). The broken lines are guides for the eye. © 2003 Wiley-VCH. Reproduced, with permission, from [25]*

power law also gives the variation of the *k* in the neighborhood of the percolation threshold,

$$\varepsilon \propto (f_c - f_{Ni})^{-s} \tag{4.3}$$

where *s* is the dielectric critical exponent. Greatly enhancement achieved in the *k* of the PVDF/Ni composites near the percolation threshold. The maximum *k* of ~400 was obtained in the percolation composite with ~0.17 vol.% Ni loading at room temperature and 100 Hz, which is ~40 times higher than that of the PVDF matrix.

Moreover, other metal particles such as Cu, Al, and Ag have been employed in varied polymer matrices to form high-*k* percolation composites. Zhu and coworkers studied the dielectric breakdown and high-field conduction behaviors of PP-based composites containing 18 and 100 nm-sized Al nanoparticles, respectively [26]. To achieve reasonable dispersion of nanofiller, phosphonic acid-terminated poly (ethylene-*co*-1-butene) was grafted to the alumina surface layer. It was found that compared to the 100 nm Al nanoparticles, smaller sized (e.g., 18 nm) Al nanoparticles needed a higher electric field to exhibit Fowler–Nordheim (FN) field electron emission. Although the *k* increased from 2.24 of neat PP to 5.7 of the PP nanocomposite with 25 vol.% Al (100 nm) loading at 25 °C and 1 kHz, the dielectric breakdown strength of this PP/Al nanocomposite sharply decreased to nearly 1/20 that of the neat PP. High-*k* flexible and transparent cellulose (CNF) composites, consisting of Ag filler with varied nanostructures, have been studied [27]. It was found that the addition of Ag nanofibers with high aspect ratio (i.e., ~700) showed the greatest enhancement in *k* of CNF-based nanocomposites with a relatively low filler loading of 1.2 wt.% ($k \sim 9.2$ at 120 kHz), compared to the cases of Ag nanowires with lower aspect ratios ($k \sim 5.5$ at 120 kHz) and Ag nanoparticles ($k \sim 4$ at 120 kHz). Later, Ag nanodots with a diameter of ~10 nm were in situ synthesized in a P(VDF-HFP) matrix [28]. Owing to the charge-transfer complexes formed between the nanodots and the polymer chains as well as the enhanced chain mobility with response to the applied electric field, concurrent enhanced k (~50%) and dielectric breakdown strength (~15%) were found in the P(VDF-HFP)/Ag nanocomposite filled with ultralow filler loading of 0.05 vol.%, resulting in a high discharged energy density of 27 J/cm^3.

Most recently, as newly developed high-k nanocomponents – liquid metal (LM) nano-droplets have been employed in polymer matrices. It is noteworthy that the addition of liquid metal nano-droplets could not only significantly improve k value but also maintain the mechanically flexible nature of the polymers [29–31]. Majidi *et al.* prepared a class of eutectic liquid gallium indium (EGaIn)-filled polymer composites utilizing poly(dimethylsiloxane) (PDMS) as a matrix (see Figure 4.4) [31]. The role of droplet size in determining the electrical and mechanical properties of the resulting composites has been studied. It was found that PDMS embedded with micro-size EGaIn droplets exhibits enhanced k values, e.g., ~11 of PDMS composite with 40 vol.% EGaIn (10μm) with respect to ~3 of neat elastomer. However, a dramatic reduction in breakdown strength was observed in this micro-composition (~1,000% drop from >100 to ~1 kV/mm). On the contrary, embedding PDMS with a nano-sized EGaIn inclusion (100 nm) could improve the k without severely decreasing the dielectric breakdown strength as well as elastic compliance and stretchability.

In particular, owing to a nomadic polarization mechanism, tetrameric Cu-phthalocyanine (CuPc) was shown to exhibit an extremely high k value of ~10^4–10^5. Zhang *et al.* investigated a class of P(VDF-TrFE) ferroelectric polymers consisting of or grafting with CuPc component, in which higher ks of ~35–50 were achieved with 25–40 wt.% loadings at room temperature and 10 kHz [32]. In addition to metal-based filler, the carbon-based nanofillers are well considered as an important example of conductive filler, also have great potential for achieving flexible high-k dielectric polymer/nanofiller composites. For example, high-k polymer dielectrics composed with thermoplastic poly(styrene-*co*-ethylene-*co*-butylene-*co*-styrene) (SEBS) and nano-sized carbon black (CB) was fabricated by Stoyanov *et al.* [33]. A percolation threshold value of 4.62 vol.% was observed in the SEBS/CB nanocomposites and further verified by the theoretical fitting. A significant enhancement in k of ~27, which is 13 times over the neat polymer matric, was achieved near the percolation threshold. However, due to the connected conductive CB networks, a rapid decrease in dielectric breakdown strength was inevitable. Compared to the 0-D nanofillers, 1-D carbon nanotubes (CNTs) and 2-D graphene nanosheets (GNSs) with larger aspect ratios could favor an easier formation of electrically conducting network under lower filler loadings, thereby lower percolation thresholds are able to be achieved [34]. The incorporation of CNTs into PVDF matrix led to the nanocomposites process a high k of 600 and a loss tangent (tan δ) of 2 at room temperature and 1 kHz [35]. Zhang and coworkers

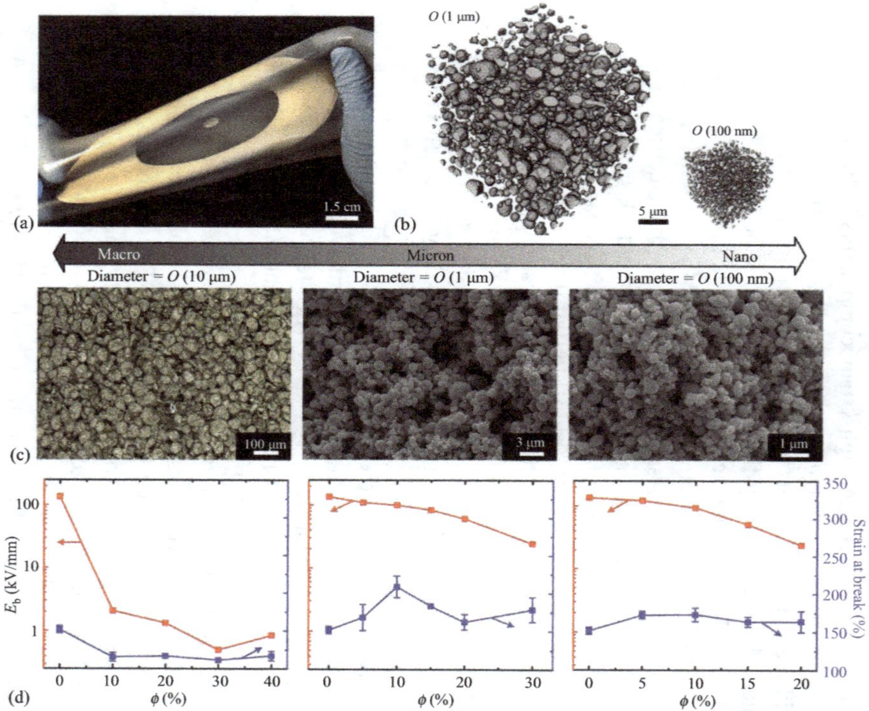

Figure 4.4 *LM–elastomer composites. (a) Image of stretched LM–elastomer nanocomposite (silver; annular shape) on a silicone elastomer carrier film (semitransparent). (b) Nanoscale X-ray computed tomography (Nano-CT) scan showing the 3-D microstructure indicating homogeneous dispersion of LM droplets in matrix for both composites with LM diameter of O (1 μm)—left and O (100 nm)—right on the same scale. (c) Microscopic images of LM droplets with diameter of O (10 μm)—left, O (1 μm)—middle, and O (100 nm)—right. The left image presents the top view of a composite with an elastomer matrix and O (10 μm) droplets made by shear mixing bulk LM with elastomer. (d) Weibull dielectric breakdown strength and mechanical strain at break for LM–elastomer composites with corresponding filler diameters of O (10 μm)—left, O (1 μm)—middle, and O (100 nm)—right. © 2019 Wiley-VCH. Reproduced, with permission, from [31]*

prepared P(VDF-TrFE-CFE)/CNTs nanocomposites with a k of ~102 along with a tan δ of 0.36 obtained at room temperature and 100 Hz and under a much lower filler loading (1.2 vol.%) [36]. Later, well-dispersed and highly aligned CNT fillers in PVDF matrix were reported in Bai group (Figure 4.5(a)) [37]. As can be seen in Figure 4.5(b), the activation energy (E_a) decreased with the increase of filler loading, which is owing to the ultrahigh surface areas of CNTs and enhanced interactions between filler and polymer matrix. Figure 4.5(c) shows that a giant k of 3,800 (at room temperature and 1 Hz) was obtained in the 12 vol.% CNTs-filled PVDF nanocomposite, which was ~380 times greater than that of neat polymer and even higher than the conducting filler (~2,000 of CNTs). This improvement could be explained clearly through a donor-acceptor system that boosts the interfacial polarization density. The AC conductivity increases constantly with f_{MWNT} and frequency when f_{MWNT} approaches f_c (0.104) (Figure 4.5(d)).

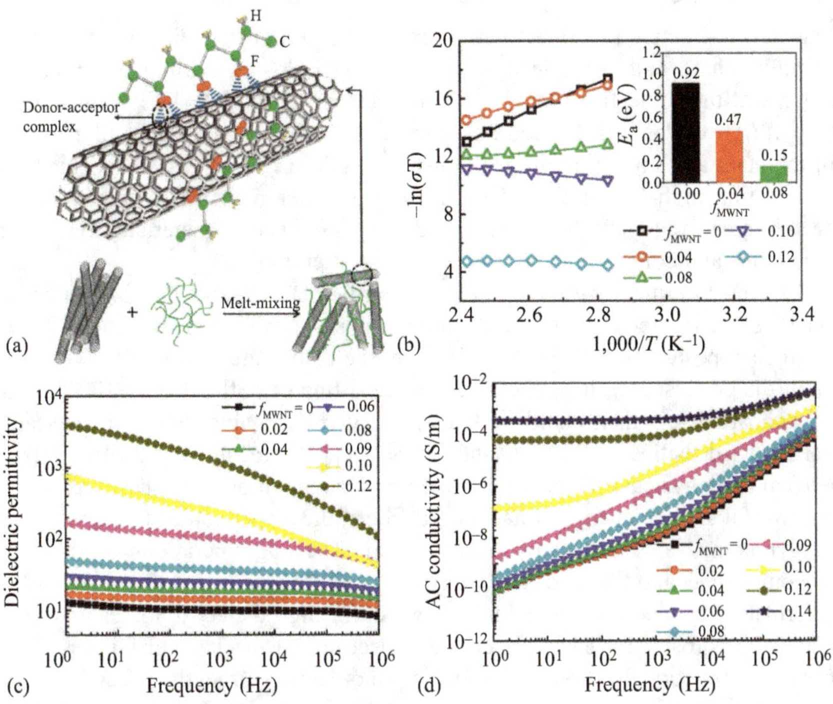

Figure 4.5 (a) Preparation scheme of melt-mixing PVDF/CNTs composites. (b)–ln(σT) versus 1,000/T for melt-mixing PVDF/CNTs composites with different loading of CNTs. The insert shows the activation energies at different volume fraction of CNTs. Frequency dependence of (c) permittivity and (d) AC conductivity on different loading of CNTs at room temperature. © 2011 American Chemistry Society. Reproduced, with permission, from [37]

As one of the most typical 2-D carbon filler, GNP is considered a better candidate for high-*k* polymer composites when compared with CNTs because of its distinctive layered structure and highly reactive properties [34]. Up to now, GNPs usually originate from graphene oxides (GOs). Zhang *et al.* firstly fabricated the high viscous PVDF-GOs solution in order to prevent agglomerations and then followed by a reduction procedure to obtain the well-dispersed PVDF/GNPs nanocomposites. A high *k* of ~63 (at 100 Hz) along with a low conductivity of $<10^{-6}$ S/m were achieved in the composite near the percolation threshold of 1.29 vol.% [38]. In order to address the dispersion issue of GNPs in polymers, Sodano and coworkers fabricated a class of PVDF/GNPs nanocomposites by using a directly in situ thermal process by pressing PVDF/GOs composites at 200 °C to reduction GOs into GNPs. In the thermal-treated nanocomposites, the GNPs fillers were isolated and dispersed well in PVDF matrix, resulting in a high *k* of >600 along with a low tan δ of <6 with 2.4 vol.% filler content [39].

4.2.2 Polymer/high-k ceramic filler nanocomposites

Compared to conducting fillers, the ferroelectric ceramics typically have higher threshold values, and hence lead to lower dielectric losses and higher breakdown strengths at the same filler loadings. For instance, the P(VDF-HFP) nanocomposites with 50 vol.% surface-modified barium titanate ($BaTiO_3$, BT) filler loading still maintain a relatively high breakdown strength of 210 MV/m [40]. Therefore, incorporating high-*k* ceramic fillers into the polymer matrix has always been a research area frequently covered in advanced dielectric materials for energy applications and continues to receive extensive attention.

As a most frequently used high-*k* ceramic filler, BT exhibits the size-dependent properties of ferroelectricity and dielectric responses. It was found that in the PVDF/BT nanocomposites with different filler size, the composites with ~50–80 nm BTs exhibit the greatest *k* with respect to those consisting of ~30–50 and ~100–150 nm BT nanoparticles, which could be led by the varied organic-inorganic interfacial polarization densities and BT crystal phases [41]. Seriously, due to the striking chemical and physical differences between organic and inorganic phases as well as the powerful surface energy of nano-sized fillers, the dispersion of inorganic fillers in polymer matrices is always problematical. The agglomerations of inorganic nanocomponents and the interface defects including air-pores and holes give rise to the formation of electron conductive networks for high dielectric losses as well as undesirable porosity for earlier occurred dielectric and mechanical failures. Thus, the surface functionalization of inorganic nanofillers is urgently needed and has been widely employed to yield homogeneously dispersed high-*k* polymer/ceramic nanocomposites. Dang and coworkers treated BT nanoparticles by using hydrogen peroxide, owing to the enhanced hydrogen bonds interaction between fluorine atoms and hydroxyl groups, the surface-hydroxylated BTs dispersed well in PVDF matrix [42]. Following this work, silane agents such as 3-aminopropyltriethoxysilane (KH550) were utilized, which can react with the hydroxyl groups to functionalize the surface of inorganic fillers with organic chains [43]. Perry *et al.* used phosphonic acids with different molecular structures for modification of BT nanoparticles and composited the modified fillers with PC and P(VDF-HFP) polymer matrices [44]. Particularly, 2-[2-(2-methoxyethoxy)ethoxy]ethyl phosphonic acid (PEGPA) with a tri(ethylene glycol) chain could enable BT nanoparticles compatible with polar, hydrophilic resins like PC, while pentafluorobenzyl phosphonic acid (PFBPA) contains a fluorinated aryl group could afford improved compatibility with fluoropolymers. The *k* was increased to ~20 and ~37 along with the calculation maximum energy density of ~3.9 and ~6.1 J/cm^3 of PC- and P(VDF-HFP)-based nanocomposites with the BT filler loading of 50 vol.%, respectively. Later, P(VDF-CTFE) and P(VDF-TrFE-CTFE)-based ferroelectric composites containing surface-modified BT nanoparticles were fabricated in Wang group [45]. The ethylene diamine moieties chains that grafted on the BT surface not only promote the dispersibility of the nanofillers in polar organic solvents but also favor the nanofillers homogeneous distributions in both copolymer and terpolymer matrices. As excepted, the *k* value of the resultant composites increases steadily with the increase of BT filler loading. For example, at a 20 vol.% filler loading of the BT nanoparticles, the *k* (at 1 kHz) of terpolymer-based nanocomposite increased from ~42 to ~50 whereas a *k* improved from ~12 to ~24 was found in the copolymer-based nanocomposite. Owing to the higher *k*, at an applied electric field of 150 MV/m, the greater discharged energy density of ~7.0 J/cm^3 was obtained in the P(VDF-TrFE-CTFE) nanocomposite filled with 30 vol.% BT particles with respect to that of ~3.7 J/cm^3 of the P(VDF-CTFE) nanocomposite filled with 23 vol.% BT particles. Recently, 1-D BT nanofibers surfaced modified by poly(vinyl pyrrolidone) (PVP) were utilized to fabricate P(VDF-HFP) nanocomposites [46]. It was believed that ascribing to the synergetic effects of organic surface functionalization and large aspect ratio of BT nanofibers, the P(VDF-HFP) nanocomposite containing 3 vol.% PVP-coated BT nanofibers delivers a largely enhanced discharged energy density of 8.55 J/cm^3 at a relatively low applied electric field of 300 MV/m, which processes >40% enhancement of that of the polymer matrix under the same field.

In addition to BT nanofiller, various high-*k* ceramic nano-additives such as strontium titanate ($SrTiO_3$, ST), barium strontium titanate ($BaSrTiO_3$, BST), barium zirconium titanate ($BaZrTiO_3$, BZT), and barium calcium zirconium titanate ($BaCaZrTiO_3$, BCZT) with tunable dielectric and ferroelectric properties have also been investigated for polymer/ceramic nanocomposite dielectrics. As shown in Figure 4.6(a)–(d), BST nanofillers with rationally designed nano-morphologies have been synthesized and form high-*k* polymer nanocomposites with the P(VDF-TrFE-CFE) polymer matrix in the Wang group [47]. The incorporation of BST nanofillers greatly enhanced the *k* (ε_r) of the nanocomposites. Owing to the larger aspect ratio of nanowires, the P(VDF-TrFE-CFE)/BST nanocomposites with the BST nanowires (NWs) exhibit a higher *k* than those of the polymer with the same loading of the BST nanoparticles (NPs), nanocubes (NCs), and nanorods (NRs). For example, as shown in Figure 4.6(e), for the ferroelectric polymer containing 15 vol.% BST nanofillers, the nanowires filled composite achieved a *k* of 72.5 (at 1 kHz), which is 20% higher than that of nanoparticles filled ones. Interestingly, dielectric breakdown strength of polymer nanocomposites is also strongly associated with the morphology of nanofillers, the polymer composites filled with the

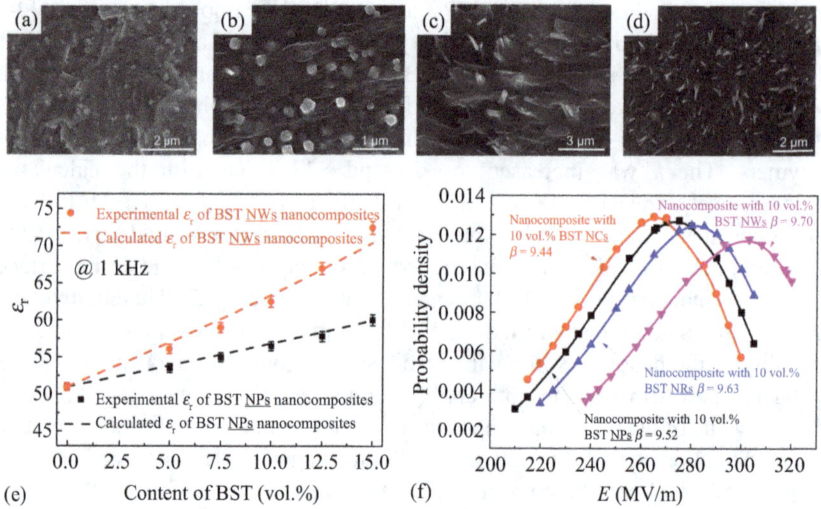

Figure 4.6 Cross-section SEM image of the P(VDF-TrFE-CFE)/10 vol.% (a) BST NPs, (b) BST NCs, (c) BST NRs, and (d) BST NWs nanocomposite, respectively. (e) Dependence of relative permittivity (ε_r) of the nanocomposites on the volume fraction of fillers measured at 1 kHz. The dash lines represent ε_r calculated by using the MG and PVS effective medium theories. (f) Weibull plots of the P(VDF-TrFE-CFE) nanocomposites with 10 vol.% BST NPs, NCs, NRs, and NWs. © 2015 American Chemistry Society. Reproduced, with permission, from [47]

BST nanowires exhibit the highest dielectric breakdown strength when compared with other nanostructured BSTs at the same loading. For instance, the breakdown strength of the P(VDF-TrFE-CFE) nanocomposite with 10 vol.% BST NWs is 303 MV/m, which is higher than 277, 267, and 282 MV/m of the nanocomposites containing 10 vol.% BST NPs, NCs, and NRs, respectively (Figure 4.6(f)). This difference could be attributed to the largest Young's modulus obtained in the nanowires-filled composites than the polymers filled with nanoparticles, nanocubes, and nanorods.

A similar study regarding the effect of filler morphology on dielectric properties of PVDF/BST nanocomposites has been conducted by Sonado and coworkers [48]. It was found that the nanocomposites with larger aspect ratio BST nanowires process higher k values than the composites filled with BST nanorods at the same filler loading. Owing to the significant increase in k while retaining the desired dielectric strength, an improved discharged energy density of 14.86 J/cm^3 was achieved in the composites with 7.5% BST nanowires at 450 MV/m, which represents a 42.9% increase with respect to the neat PVDF at the same electric field. Huang *et al.* selected four kinds of dopamine (dopa)-functionalized high-k nanowires with different inherent characteristics to fabricate P(VDF-HFP) polymer

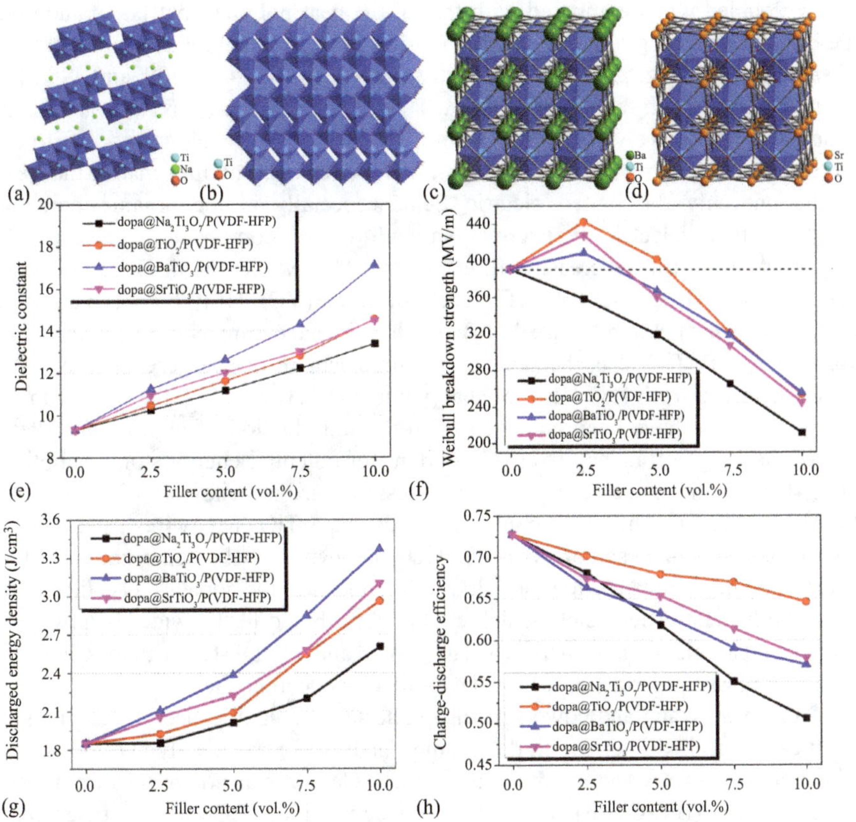

Figure 4.7 Crystal structures of (a) $Na_2Ti_3O_7$, (b) anatase TiO_2, (c) $BaTiO_3$, and (d) $SrTiO_3$ with TiO_6 polyhedron. (e) Dielectric constant (k) (at 1 kHz), (f) Weibull breakdown strength, (g) discharged energy density, and (h) charge-discharge efficiency of P(VDF-HFP)-based nanocomposites with different volume fractions of nanowires. © 2015 American Chemistry Society. Reproduced, with permission, from [49]

nanocomposites [49]. Among the nanowires, sodium titanate ($Na_2Ti_3O_7$) and titanium dioxide (TiO_2) are non-ferroelectric materials, while $BaTiO_3$ and $SrTiO_3$ exhibits ferroelectric and paraelectric nature, respectively. The crystal structures of the $Na_2Ti_3O_7$, anatase-phased TiO_2, $BaTiO_3$, and $SrTiO_3$ are depicted in Figure 4.7(a)–(d). Figure 4.7(e) and (f) represents the k (at 1 kHz) and breakdown strength of P(VDF-HFP) nanocomposite consisting of high-k nanowires as a function of filler loading. Among those four kinds of nanowires, the introduction of dopa@$Na_2Ti_3O_7$ brings the faintest improvement of k at the same filler loading, while the dopa@$BaTiO_3$-filled nanocomposites deliver the greatest k value. For instance, the k of the P(VDF-HFP) composite with 10 vol.% dopa@$BaTiO_3$ is nearly doubled when compared with that of the neat polymer matrix. In addition, the breakdown strength of the nanocomposites is improved upon the incorporation of small loading (e.g., 2.5 vol.%) of high-*k* nanowires, e.g., 442, 408, and 428 MV/m of the composites containing dopa@TiO_2, dopa@$BaTiO_3$, and dopa@$SrTiO_3$ nanowires, respectively, vs. 391 MV/m of the neat polymer. However, due to the large electrical mismatch between high-*k* nanowires and polymer matrix that lead to the unevenly distributed electric fields especially at the organic-inorganic interface, the dielectric breakdown strength of all composites monotonously decreases with further increasing filler loading. Meanwhile, the discharged energy density and charge-discharge efficiency of the P(VDF-HFP)/high-*k* nanowires composites are shown in Figure 4.7(g) and (h). The nanocomposite consisting of 10 vol.% dopa@$BaTiO_3$ has the greatest improvement in discharge energy density, which increased from 1.85 J/cm^3 of neat polymer to 3.37 J/cm^3. Although enhanced energy densities have been achieved upon the introduction of high-*k* nanofillers, charge-discharge efficiency is also important for potential energy storage applications. Due to the paraelectric phase and less hysteresis behavior of $SrTiO_3$, the P(VDF-HFP) nanocomposites consisting of dopa@$SrTiO_3$ nanowires usually have high efficiency than that of dopa@$BaTiO_3$ nanowires at the same content. It is noteworthy that compared to other kinds of composites, the dopa@TiO_2-filled ones show the highest efficiencies, which could be attributed to the wider band gap of TiO_2, comparable *k* value between TiO_2 filler and P(VDF-HFP) matrix and the smaller remnant polarization of the TiO_2-filled nanocomposites.

Moreover, Wang and coworkers also indicated the significance of designing an electrically matched interface in determining high-performance polymer composites with enhanced energy storage properties [50]. A class of nanocomposites from combining P(VDF-TrFE-CTFE) matrix and surface-modified TiO_2 nanoparticles were fabricated. The higher degree of crystallinities and smaller crystalline domains were found in P(VDF-TrFE-CTFE)/TiO_2 nanocomposites in comparison with pristine terpolymer. Besides, the existence of the organic-inorganic interfaces of the nanocomposites is evidenced in the temperature-dependent dielectric spectra as shown in Figure 4.8(a) and (b). A shift of the dielectric relaxation peak toward a lower temperature and a reduced loss tangent have been evidence of increased trap density and enhanced interface polarization intensity of the resultant nanocomposites. Due to the close *k* value of the terpolymer (i.e., ~42 at 1 kHz) and TiO_2 filler (~47), the *k* of the nanocomposites remained broadly stable while tan δ slightly decreases with increasing of filler loading (Figure 4.8(c)). However, the comparable *k* values mitigate the formation of the electric field distortion as a result of the components with sharply varied dielectric properties, and hence contributes the improvements of the discharged energy density especially at high fields in the P(VDF-TrFE-CTFE)/TiO_2 nanocomposites. For example, as shown in Figure 4.8(d), the polymer nanocomposite containing 10 vol.% TiO_2 nanoparticles exhibits a stored energy density of ~6.9 J/cm^3 at 200 MV/m, which displays an improvement of ~45% with respect to the neat P(VDF-TrFE-CTFE) matrix with that of ~4.7 J/cm^3 under the same electric field.

In brief, numerous efforts have been made to enhance the *k* value of dielectric materials by adding high-*k* components into organic phases. The incorporation of

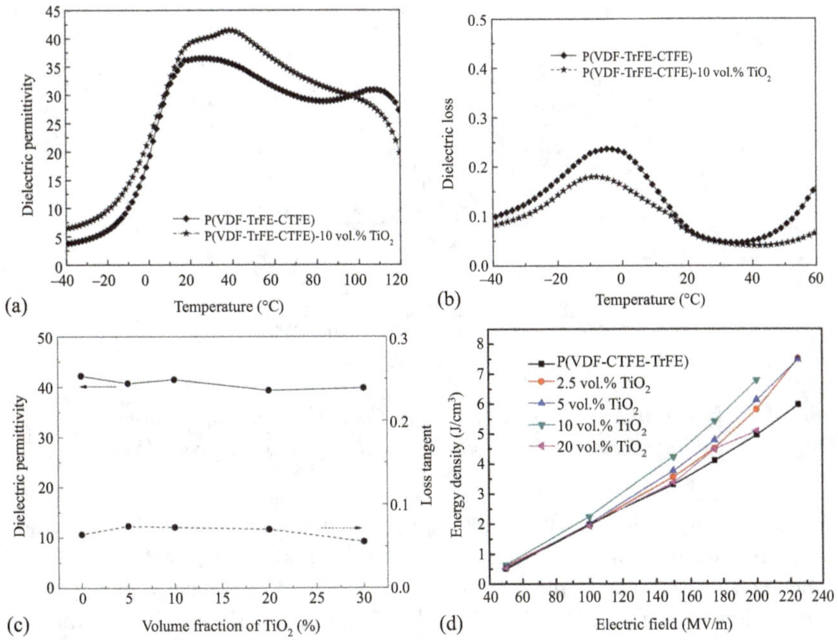

Figure 4.8 Temperature dependence of the (a) dielectric permittivity (k) and (b) dielectric loss (tan δ) of the polymer and nanocomposite measured at 1 kHz. (c) Dielectric permittivity (k) and dielectric loss (tan δ) of the P(VDF-TrFE-CTFE)/TiO_2 nanocomposites measured at 1 kHz and room temperature with a 1 V bias. (d) The stored energy density of the polymer and nanocomposites as a function of the applied field. © 2009 Wiley-VCH. Reproduced, with permission, from [50]

metal- and carbon-based filler, as well as ceramic fillers with ultrahigh k value, could improve the k of the fabricated composites by hundreds of times of the polymer matrices. In spite of greatly increased k values; however, the resulting polymer nanocomposites typically suffer from the electrical unmatched interfaces, e.g., significantly increased dielectric loss and severely decreased dielectric breakdown strength, which as a result preclude the substantial gain in discharged energy density. How to resolve the contradiction is still considered as a major challenge that critically needs to be addressed.

4.3 Two-phase polymer composites with wide-band gap nanofillers

In Section 4.2, the currently existed issues regarding polymer composites containing high-k nanofillers have been proposed and discussed. In order to further improve dielectric breakdown strength along with reducing the dielectric loss under

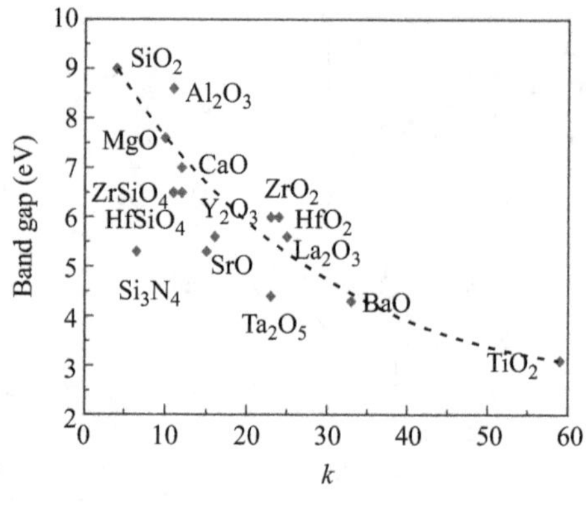

Figure 4.9 Variation of dielectric constant (k) with band gap of nanofillers. © 2005 Elsevier. Reproduced, with permission, from [51]

both low and high electric fields, the composite approaches by utilizing ceramic nanofillers with wide bandgap (e.g., ≥ 4) have been attacked great attentions. As can be seen in Figure 4.9, the ceramic filler with wide bandgap typically shows relatively low k (e.g., ≤ 25) [51], which has been revealed effective in impeding electron hopping and enhancing breakdown strength of polymer nanocomposites in recent years [7].

Metal oxides are most commonly used reinforcing filler for polymer composites, which are not only widely employed in advanced dielectrics but also thermal managements and structural engineering applications. Surfaced modified silicon dioxide (SiO_2) nanoparticles with wide bandgap (~9 eV) were incorporated into the epoxy resin matrix by Li *et al.* [52]. The results show that the small-loading doping of SiO_2 could not only enhance the breakdown strength and decrease dielectric loss tangent but also synergistically improve the mechanical properties including the tensile strength and elongation at break of the epoxy/SiO_2 nanocomposites. For PP nanocomposites consisting of aluminum oxide (Al_2O_3) nanoparticles (bandgap of ~8.8 eV) [53], it was also found that small loading of Al_2O_3 (0.5 wt.%) could largely improve the dielectric strength by 30%, which is attributed to the increase of deep traps, resulting in the reduction in mobility of charge carriers, the enhanced height of barrier in PP/Al_2O_3 nanocomposites. Beyond improved breakdown strength, electrical conduction behavior is also a critical issue. Especially, for polymer dielectrics used at high temperatures and under high electric fields, the incorporation of wide bandgap Al_2O_3 has also proved to be effectively reducing the leakage currents at high fields by Zhang and coworkers [8], e.g., a more than two orders of magnitude reduction in high-field conduction current density was obtained in a semi-crystalline polymer – poly(tetrafluoroethylene-*ter*-hexafluoropropylene-*ter*-vinylidene fluoride) (THV) filled with less than 1 wt.% of Al_2O_3 nanoparticles. Upon the addition of wide bandgap nanofillers, the carrier hopping in the THV polymer is significantly reduced owing to a large decrease in the mobile carrier concentrations and increased trap depth.

Furthermore, zirconium dioxide (ZrO_2)- and tantalum oxide (Ta_2O_5)-contained polymer nanocomposites were prepared in Wang group [54,55]. The uniform nanoparticle dispersion and the increase of the *k* induced by the ZrO_2 nanoparticles (bandgap of ~5.8 eV, *k* of ~22) boosted a largely enhanced energy density of ~11.2 J/cm^3 (at 270 MV/m) of the P(VDF-CTFE) composite filled with 9.1 wt.% phosphonic-acid functionalized nano-ZrO_2, which represents a 60% improvement with respect to the neat polymer. The organic-inorganic hybrids of P(VDF-CTFE)/ Ta_2O_5 composites were prepared from the covalent incorporation of tantalum species into ferroelectric polymers via in situ sol-gel condensation (Figure 4.10(a)) [55]. The composites containing 5, 9, 13, 17, and 23 wt.% $Ta(OCH_2CH_3)_5$ are referred to as H1–5, respectively. As can be seen in Figure 4.10(b) and (c), the composites with the optimal composition exhibit largely reduced leakage current densities and Weibull breakdown strength of 505 MV/m, which are more than an order of magnitude lower and 40% higher than the polymer matrix, respectively. The enhanced dielectric properties could be ascribed to the deep traps generated in the composites at the molecular level, which is further evidenced by thermally stimulated discharge current (TSDC) results of Figure 4.10(d). Compared to neat P(VDF-CTFE), the H3 hybrid composite exhibits a maximum energy density of ~18 J/cm^3, which is more than 180% greater of the neat polymer (Figure 4.10(e)).

Most recently, hafnium oxide (HfO_2) nanorods, which exhibit very similar band structure and physical properties with ZrO_2, were synthesized by Loos *et al.* [56]. A P2VP-*b*-P(VDF-TrFE)-*b*-P2VP block copolymer exhibits characteristics of ferroelectric relaxor was used as a matrix template to guide the dispersion of inorganic phases. The formation of strong hydrogen bonds between the surface of nanofillers and P2VP domains drives the selective homogeneous dispersion of HfO_2 nanorods inside lamellar domains made via a self-assembly process. Owing to the moderate *k* value (~22) of the HfO_2 filler that prevents local electric field distortions, and the homogeneous dispersion via self-assembly, the polymer nanocomposite with 12.5 wt.% HfO_2 display enhanced discharged energy densities along with a ~34% improvement in efficiency compared to the neat block copolymer.

In addition to the metal oxide, nitride ceramics, e.g., boron nitride (BN) with a wide bandgap (~6 eV) also exhibits superb insulating properties. Hexagonal boron nitride (*h*-BN), the most stable crystalline form of BN, is a layered van der Waals crystal with *k* ranging from ~3–4 and dielectric strength as higher as ~800 MV/m. Recently, a library of *h*-BN nanosheets (BNNSs)-filled polymer nanocomposites has been fabricated by using different polymer matrices including P(VDF-TrFE-CFE), PMMA, PEI, and cross-linked divinyltetramethyldisiloxane-bis(benzocyclobutene) (*c*-BCB) [1,57–59]. Multiple excellent dielectric performances including dielectric breakdown, resistivity, and dielectric loss were achieved in these polymer nanocomposites. As shown in Figure 4.11(a)–(c), the BNNSs are intimately and homogeneously dispersed in the P(VDF-TrFE-CFE) polymer matrix. The polar surface of B-N bonds favors the dispersion of BNNSs in both organic solvents and fluoropolymer matrices [59]. The BNNS acts as an insulating barrier against high-field leakage current, and thus permits to prevent dielectric failure, resulting in

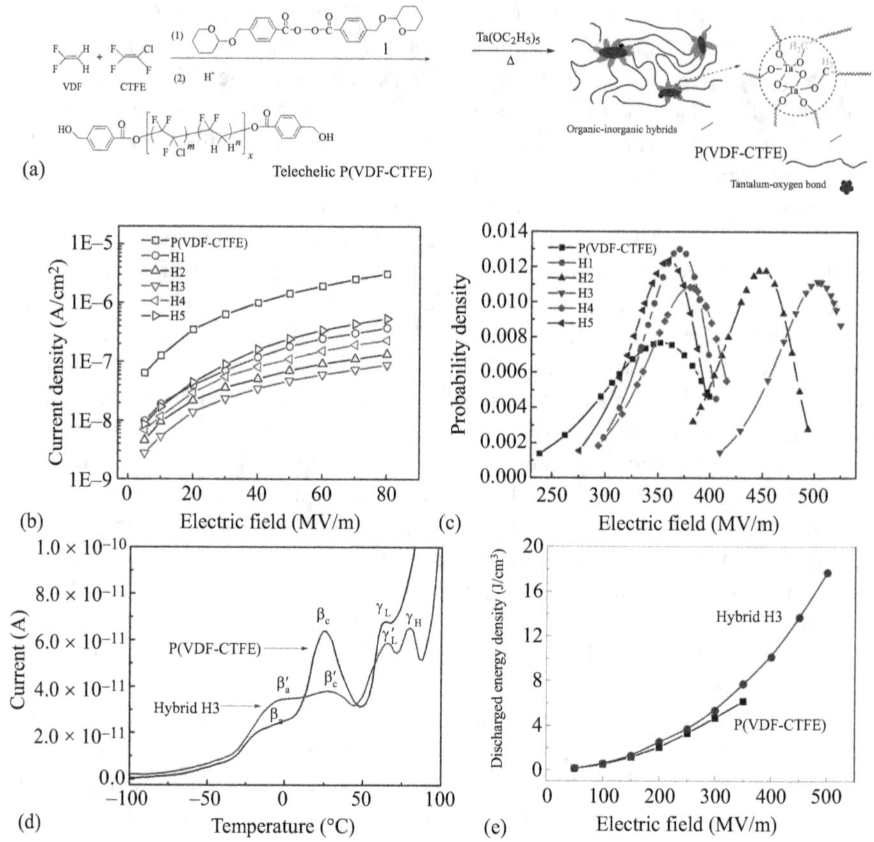

Figure 4.10 (a) Synthesis of the organic-inorganic hybrids from sol-gel condensation of chain-end functionalized P(VDF-CTFE). (b) The dependence of DC current density on the applied electric field of the P(VDF-CTFE) and the hybrids. (c) The Weibull distribution of breakdown fields of the P(VDF-CTFE) and the hybrids. (d) TSDC curves of the pristine P(VDF-CTFE) and the hybrid H3. (e) The discharged energy density of solution-processed P(VDF-CTFE) and the hybrid H3 as a function of electric field. © 2015 Wiley-VCH. Reproduced, with permission, from [55]

significantly enhanced breakdown strength, charge-discharge efficiency, and discharged energy density of the P(VDF-TrFE-CFE)/BNNSs nanocomposites. As can be seen in Figure 4.11(d) and (e), the nanocomposite with 12 wt.% BNNSs shows the lowest remnant polarization of *D-E* loops and extremely high charge-discharge efficiencies even under very high electric fields, e.g., ~83% at 300 MV/m and ~80% at 600 MV/m. Compared to neat P(VDF-TrFE-CFE) terpolymer, the discharged energy density of 12 wt.% BNNSs-filled nanocomposite improved by 121%, thus achieving an ultrahigh value of 20.3 J/cm^3 at 650 MV/m. Interestingly, it was

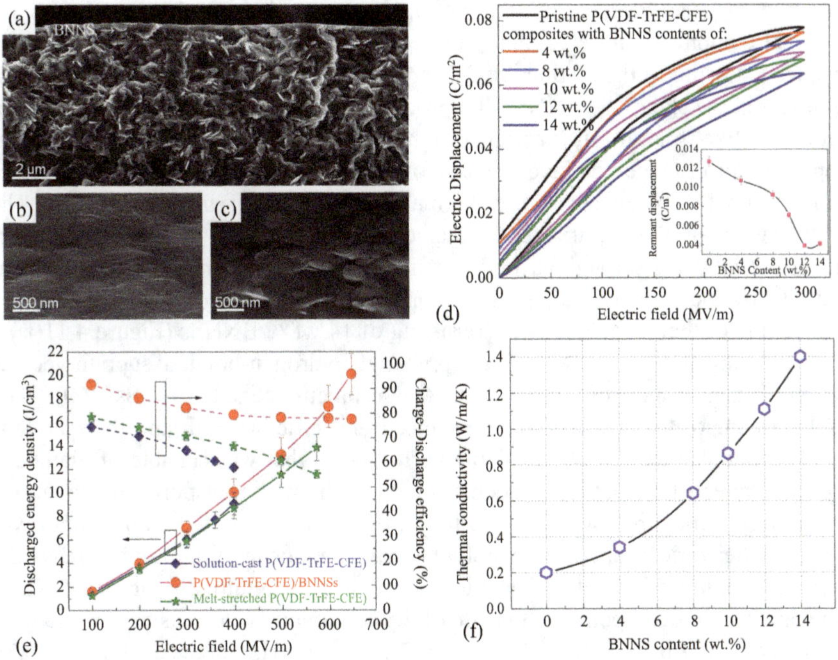

Figure 4.11 *(a) Large-scale cross-sectional SEM image of the P(VDF-TrFE-CFE)/ BNNS composite film with 12 wt.% BNNSs. (b) Zoom-in top-view and (c) cross-sectional SEM images of the P(VDF-TrFE-CFE)/BNNS composite film with 12 wt.% BNNSs. (d) Comparison of D-E loops of pristine P(VDF-TrFE-CFE) and a series of P(VDF-TrFE-CFE)/BNNS nanocomposites at an electric field of 300 MV/m. Inset shows remnant displacement of P(VDF-TrFE-CFE)/BNNS nanocomposites as a function of filler content. (e) Comparison of discharged energy density and charge-discharge efficiency of pristine P(VDF-TrFE-CFE) and P (VDF-TrFE-CFE)/BNNS nanocomposites with 12 wt.% of BNNSs at different electric fields. (f) Thermal conductivity of P(VDF-TrFE-CFE)/ BNNS nanocomposites with different filler contents. © 2015 The Royal Society of Chemistry. Reproduced, with permission, from [59]*

reported that the dielectric strength of *h*-BN-filled polymer composites could be further enhanced by reducing the thickness of the *h*-BN sheet to a few layers. The terpolymer filled with two types of BNNSs with different average thicknesses of 8 nm (8t-BNNSs) and 15 nm (15t-BNNSs) exhibit different Weibull breakdown strength and dielectric reliability, e.g., ~475 vs. ~440 MV/m of the nanocomposites filled with 10 wt.% 8t-BNNSs and 15t-BNNSs, respectively. Beyond dielectric strength and energy density, the thermal conductivity and thermal stability are comparably important for dielectric materials especially those for high-temperature energy storage applications. The great significance releasing the heat issues for electrical equipment and electronic devices is reviewed and discussed in detail in Chapter 5. Owing to the distinctive 2-D crystal structure with strong B-N covalent sp^2 bonds in the plane, the BNNS filler processes the advantages of the (002) face of a graphitic-like structure such as super-high thermal conductivity equally as good as graphene. For instance, the incorporation of BNNSs in P(VDF-TrFE-CFE) polymer matrix dramatically enhances the thermal conductivity of nanocomposites. The BNNSs with ultrahigh intrinsic thermal conductivity (~300 Wm/K) form interconnected networks throughout polymer matrix, which increases the thermal conductivity by more than 6 times, from ~0.2 Wm/K of neat P(VDF-TrFE-CFE) to ~1.4 Wm/K of the nanocomposite consisting of 14 wt.% BNNSs (Figure 4.11(f)).

Following this work, Xin *et al.* prepared hydrogen-bonded supramolecular polymer nanocomposites filled with amide-functionalized BNNSs [60]. The resultant composites not only have enhanced dielectric strength and thermal conductivity but also exhibit self-healable performance that was capable of maintaining simultaneously electrical, mechanical, and thermal transport functionalities after multiple fractures. Xiong *et al.* [61,62] have also reported similar improvements of dielectric properties in polymer nanocomposites upon adding BNNSs. Their group utilized biomaterials including cellulose and chitin as matrices, BNNSs as enhancing nanocomponents. In spite of the continuously decreasing k value with the increase of filler content, e.g., from 7.73 (at 1 kHz) of neat chitin to 6.78 (at 1 kHz) of the composites with 10 wt.% BNNSs, a desirable high k of ~7.1 (at 1 kHz) and a maximum Weibull breakdown value of 451 MV/m were still obtained in chitin nanocomposite with optimal filler loading (6 wt.% BNNSs), resulting in a marvelous discharge energy density of 8.67 J/cm^3 at 450 MV/m [62].

To sum up, while it is found that the addition of nano-components with wide bandgap is very helpful for decreasing dielectric loss and promoting breakdown strength, the relatively low k of these fillers (e.g., ~3–4 of BN) still limit the further enhancements of stored energy density in the resulting polymer nanocomposites.

4.4 Polymer composites with hetero-phase composition

In this section, “hetero-phase composition” refers to the inclusion of two or more inorganic nanocomponents in a polymer/nanofiller composite. This approach has great potential in combining the distinct functionalities of different inorganic phases to enhance the multiple physical performances of multi-phased polymer nanocomposites.

Manias *et al.* described a ternary composition approach by adding layered organically modified montmorillonites (o-MMTs) with large aspect ratio and spherical $BaTiO_3$ nanoparticles with high k value into epoxy resin matrix [63]. Based on the dielectric relaxation behaviors, it was found that the organic-inorganic interface plays an important role in determining the local dynamics and physical properties of the polymer composites. The epoxy/o-MMTs composites present approximately three orders of magnitude slower interfacial dynamics than those at the epoxy-BT interfaces, indicating more robust interfaces in the o-MMT composites than that of in the BT-based composites. Later, it was proved that the o-MMT filler surface-grafted with organic functional groups processes the ability to react with the resin matrix, which is beneficial for reducing interfacial defects and better dispersing of the nanofillers [64]. Compared to the binary epoxy/o-MMTs composite systems, the presence of the third-phase $BaTiO_3$ nanoparticles improved the maximum electric displacement by 40% while maintaining high breakdown strength and low conduction loss. As a high-*k* filler, calcium carbonate ($CaCO_3$) also added into PP/o-MMTs compositions as the third reinforcing nanocomponents [65]. It was postulated that the pseudo-2-D nanoclays could preferentially physically absorb on the surfaces of $CaCO_3$ nanoparticles to tailor the nature of the organic-inorganic interfaces of ternary PP/$CaCO_3$/o-MMT nanocomposites. As a result, the dielectric strength and reliability are substantially improved upon the elegant combination of two kinds of nanofillers, far exceeding the property of the respective binary PP/$CaCO_3$ and PP/o-MMT composites (e.g., ~425 MV/m of the ternary composites with 6 wt.% o-MMT and 12 wt.% $CaCO_3$ vs. ~376 MV/m of the composites with 6 wt.% o-MMT and ~295 MV/m of the composites with 12 wt.% $CaCO_3$).

Another important heterogeneous composition example of the polymer nanocomposites consisting of two inorganic fillers with complementary functionalities, including BNNSs with high dielectric strength and $BaTiO_3$ nanoparticles with large electric polarization intensity, was proposed by Wang and coworkers [66]. As can be seen in Figure 4.12(a)–(d), considerable aggregations of BT nanoparticle were observed in the binary P(VDF-CTFE)/BT nanocomposites. Interestingly, the addition of 2-D-structured BNNSs with high specific surface area favors the dispersion of the second phase by physically dividing the matrix into numerous sub-blocks and placing impervious barriers to prevent BT nanospheres from aggregating in P(VDF-CTFE)/BT/BNNSs terpolymer nanocomposites. The unipolar electric *D-E* field loops presented in Figure 4.12(e) clearly demonstrate a synergistically improvement in both electric displacement and breakdown strength of ternary nanocomposites with respect to pristine P(VDF-CTFE). An elegant composition of 12 wt.% BNNSs and 15 wt.% BT gives rise to a 20% improvement in *k* along with a ~40% increase in breakdown strength in comparison with the copolymer matrix, resulting in an ultrahigh discharged energy density of 21.2 J/m^3 (Figure 4.12(f)). Moreover, the slimmer hysteresis loop with lower remnant polarization in Figure 4.12(e) is evident in the ternary nanocomposites, denoting the lower energy loss, e.g., a significantly improved charge-discharge efficiency of ~78% is achieved at 500 MV/m (Figure 4.12(g)).

Following this work, very similar inorganic compositions were employed in the P(VDF-CTFE)-based polymer matrix with a cross-linked structure, which further shreds of evidence that the incorporation of the third phase filler (i.e., BNNSs) could largely increase breakdown strength and decrease leakage current [67]. As a result, the ternary composite with the optimal filler loadings of 6 wt.% BNNSs and 5 wt.% BT displays ~40% and ~30% improvement in breakdown strength and discharged energy density, respectively, compared to those of the binary composite

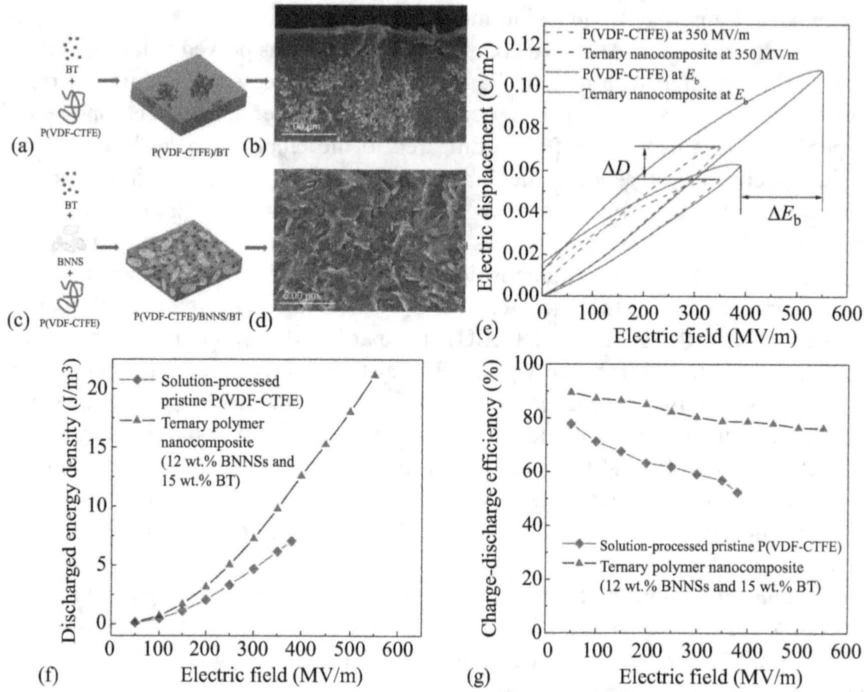

Figure 4.12 (a)–(d) Schematic and cross-sectional SEM images of P(VDF-CTFE)/BT composite film with 15 wt.% BT, and of P(VDF-CTFE)/BNNS/BT ternary nanocomposite film with 12 wt.% BNNSs and 15 wt.% of BT. (e) D-E loops measured under unipolar electric fields of 10 Hz for pristine P(VDF-CTFE) and P(VDF-CTFE)/BNNS/BT ternary nanocomposite with 12 wt.% BNNSs and 15 wt.% BT. (f) Discharged energy density and (g) charge-discharge efficiency of solution-processed pristine P(VDF-CTFE) and P(VDF-CTFE)/BNNS/BT ternary nanocomposite with 12 wt.% of BNNSs and 15 wt.% of BT as a function of electric field. © 2014 Wiley-VCH. Reproduced, with permission, from [66]

with 5 wt.% BT. BST nanowires with larger aspect ratio and as a direct consequence of higher *k* and electric displacement than its nanoparticles were also utilized as a third component for preparing ternary BNNSs-based polymer nanocomposites [68]. Owing to the concurrently improved breakdown strength and charge-discharge efficiency, the P(VDF-TrFE-CFE)/BST/BNNSs ternary nanocomposite with the optimal composition of nanofillers delivers a discharged energy density of 24.4 J/cm^3, which is 295% that of neat terpolymer. A novel hybrid nanoparticle approach – embedding $BaTiO_3$ nanoparticles in 2-D BNNSs – has been proposed by Luo *et al.* as shown in Figure 4.13(a) [69]. At a high calcination temperature of ~700 °C, the randomly dispersed $BaTiO_3$ were tightly enwrapped by

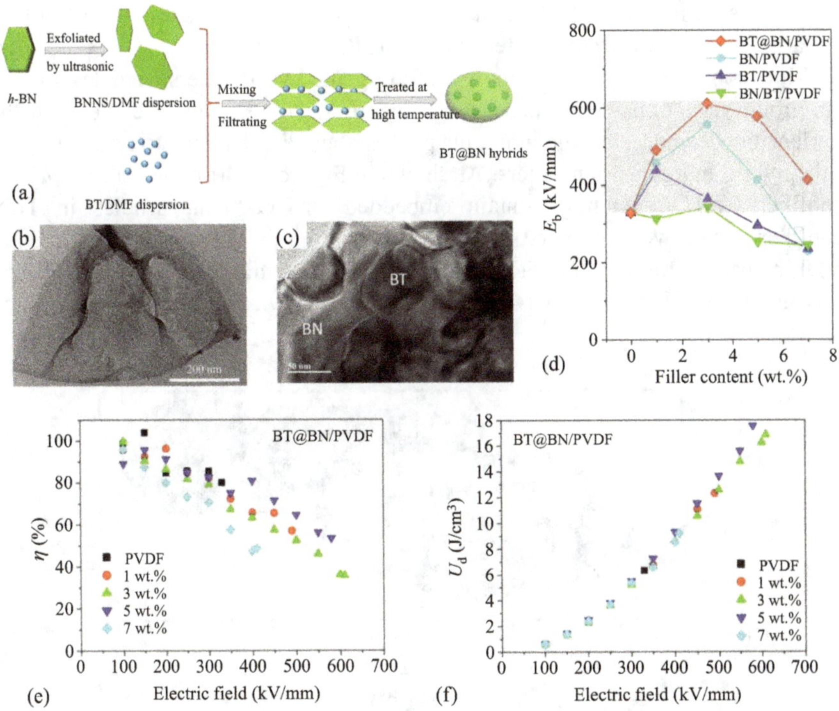

Figure 4.13 *(a) Schematic preparation process of the BT@BN hybrids. (b) and (c) TEM images of the exfoliated BNNS and the prepared BT@BN hybrids, respectively. (d) Variation of Weibull breakdown strength (E_b) with the filler content of the four types of films. (e) Variation of the calculated charge-discharge efficiency (η) from the D-E loops of the BT@BN/PVDF films with the external applied electric field. (f) The calculated discharged energy density (U_d) of the BT@BN/PVDF films as a function of the external applied electric field. © 2019 Wiley-VCH. Reproduced, with permission, from [69]*

the adjacent BNNSs to form BT@BN hybrids during a slow melting process of the BNNSs. Figure 4.13(b) shows the exfoliated BNNSs with the large planar size of ~0.6 × ~0.5 μm, while the BT@BN hybrids demonstrate that the BT nanoparticles with an average diameter of ~100 nm were incorporated in the BNNSs, which act as insulating hosts (Figure 4.13(c)). As can be seen in Figure 4.13(d), the PVDF/BT@BN nanocomposites with 3 wt.% hybrid filler loading delivers the highest breakdown strength of 610 kV/mm with respect to the polymer composites containing single-phase BT or BN filler, which is 1.85 times of pristine PVDF (e.g., ~329 kV/mm). A largely improved discharged energy density of 17.6 J/cm^3 and the highest charge-discharge efficiency were achieved in the nanocomposites with an optimal BT@BN hybrid filler loading (i.e., 3 wt.%), as shown in Figure 4.13(e) and (f), which is attributed to the combining of superior insulating strength of BN host and large electric polarization intensity of $BaTiO_3$ dopant.

More complex structures of inorganic-inorganic hybrid fillers were designed in Nan group via atomic-scale interface engineering [70]. Beyond the focus on the interface between organic and inorganic phases, they directed attention to the inside of the hybrid inorganic nanofillers. As shown in Figure 4.14(a), the $BaTiO_3$@TiO_2 nanofibers (BTO@TO_nfs) contain embedded $BaTiO_3$ nanoparticles in TiO_2 nanofiber, which were prepared by an electrospinning process and incorporated to PVDF matrix to improve interfacial polarization. Structural models of $BaTiO_3$ and TiO_2 are shown in Figure 4.14(b). It was revealed via resolution transmission

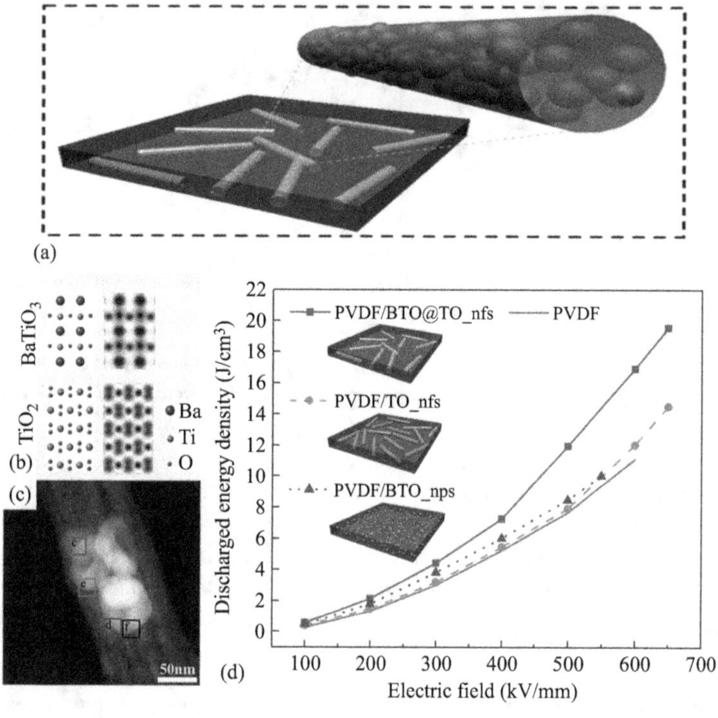

Figure 4.14 (a) Schematic illustration of BTO@TO_nfs and PVDF/BTO@TO_nfs nanocomposites. (b) Structural models of $BaTiO_3$ and TiO_2 and corresponding simulated ABF images along the [110] direction. (c) Low-magnification HAADF image of $BaTiO_3$ particles coated by a TiO_2 nanowire. (d) Discharged energy density of PVDF nanocomposites embedded with BTO@TO_nfs, TO_nfs, BTO_nps, and pure PVDF films as a function of electric field, the volume fraction of the three nano-inclusions were fixed at 3% in all composites. © 2015 Wiley-VCH. Reproduced, with permission, from [70]

electron microscopy (HRTEM) study (Figure 4.14(c)) that in the TiO_2-$BaTiO_3$ interfacial regions, mutual occupation of Ba^{2+} and Ti^{4+} cations, as well as the expansion of $BaTiO_3$ lattice, could induce additional dipoles or larger dipolar moment, which contributes to enhanced interfacial polarization. As compared in Figure 4.14(d), as a result of the concomitantly improved breakdown strength and *k*, the maximum discharged energy density of ~20 J/cm^3 was obtained at ~650 kV/mm in the PVDF nanocomposite filled with 3 vol.% of BTO@TO_nfs, which is far exceeding the composites containing TO_nfs and BTO_nps with the same filler loading.

Later, owing to the higher breakdown strength and lower remnant polarization of P(VDF-HFP) than PVDF matrix, the P(VDF-HFP) nanocomposites filled with 3 vol.% BTO@TO_nfs exhibit a much higher discharged energy density of ~31.2 J/cm^3 along with a great charge-discharge efficiency of ~78% at ~800 kV/mm [71]. Moreover, another heterogeneous composition example of the polymer nanocomposites simultaneously filled with two kinds of high-*k* inorganic fillers with different morphologies was proposed by Gerhardt and coworkers [72]. The solution-casted ternary P(VDF-HFP) composites with an elegant filler combination of 37.1 vol.% 0-D BT nanoparticles and 3 vol.% 1-D multi-wall CNTs displays a *k* of ~71.7 and a tan δ of ~0.045 at 1 kHz along with a calculated energy density of ~19.82 J/cm^3 at 250 MV/m.

Numerous efforts centering the hetero-phase composition strategies have been successfully performed through introducing two or more inorganic nanocomponents into polymer matrices. More complex electrically anisotropic design paradigms, including interfacial engineering and topological arrangement of nanofillers, are described and discussed in detail in Sections 4.5 and 4.6, respectively.

4.5 Polymer composites with core@shell structure design of nanofillers

While numerous organic modifiers, including surfactants, silane coupling agents, dopamines, and phosphonic acids, have been intensively employed in polymer/nanofiller composite strategies with relatively simple treatment processes to reduce the agglomerations of fillers and improve the chemical compatibility of polymer-filler interfaces. Still, there exist some limitations in realizing the full potential of polymer nanocomposites, e.g., the organic modifiers themselves typically do not play a significant role in improving dielectric properties of the resulting polymer nanocomposites. To further address the superfluous surface energy of the inorganic nanofillers, as well as the dissimilar physical and chemical surface properties between the organic and inorganic phases in polymer nanocomposites. Numerous design of core@shell structure of inorganic nanofillers have been employed in polymer nanocomposites and are considered as an effective route for improving the chemical compatibility and alleviating electrical mismatches between nanocomponents and polymer matrices [73].

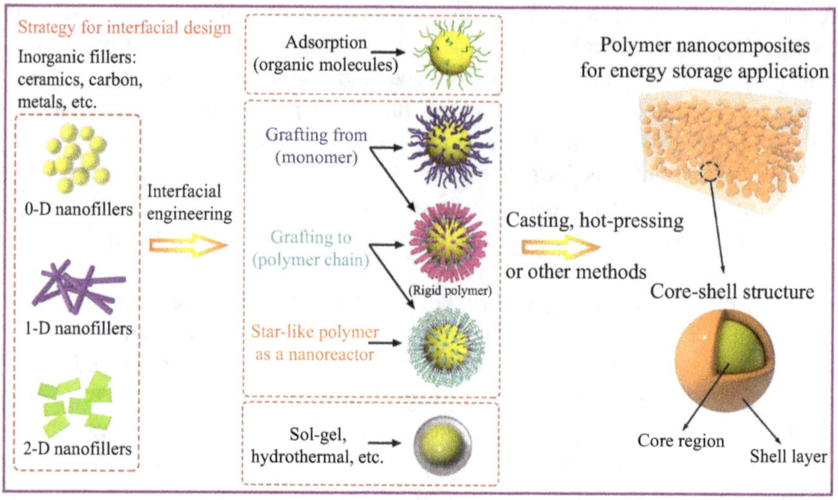

Figure 4.15 General methods associated with the design and control the interface of core-shell structured fillers for dielectric capacitor application. The left column shows the range of filler types, the central column indicates the core-shell interfacial control methods, and the right column shows the final nanocomposite structure. © 2019 The Royal Society of Chemistry. Reproduced, with permission, from [74]

Figure 4.15 presents some general methods of the fabrication of polymer nanocomposites for energy storage applications using core@shell strategies [74]. First, organic molecules could be physically adsorbed onto the inorganic filler's surface by hydrogen bonding or through electrostatic interactions. In addition, the organic shell could be directly, chemically grafted onto the inorganic filler surfaces via "grafting to" or "grafting from" methods. Particularly, the "grafting from" approach is typically based on the creation of a core@shell structure via in situ controlled/living radical polymerization. For example, atom transfer radical polymerization (ATRP) and reversible addition-fragmentation chain transfer (RAFT) polymerization of monomers on the inorganic filler's surface. While the "grafting to" approach favors the formation of a core@shell structure by grafting the pre-prepared polymer chains onto the inorganic filler's surface by triggering a chemical reaction between the end-groups of polymer molecular and the functional groups on the surface of inorganic fillers. Moreover, an unusual strategy for the formation of core@shell-structured nanofillers has been developed in Lin group by utilizing star-like polymers as nanoreactors [75]. Last, it is noteworthy that the sol-gel and hydrothermal approaches have been widely used for the design of core@shell nanofillers with inorganic shells, which will be covered and discussed in detail below.

Dang and coworkers synthesized a class of core@shell $BaTiO_3$ nanoparticles with ethylene propylene diene monomer (EPDM) and PMMA shells in varied thickness to prepare PP-based nanocomposite films [76,77]. For example, the "grafting from" strategy, e.g., the ATRP approach was used to prepare high-*k* PMMA@$BaTiO_3$ nanoparticles, as shown in Figure 4.16(a) [77]. The core@shell structured nanoparticles with different average thicknesses of shell (e.g., 4, 6, 8, and 10 nm) are shown in Figure 16(b)–(e). The influence of the thickness of PMMA shell on the dielectric and energy storage performances has been studied. The incorporation of 10 wt.% nanocomponents largely increases the *k* of PP matrix, e.g., from 2.2 of neat PP to ~3 of BT/PP and to ~3.8 of PMMA@BT-10/PP at 1 kHz. Interestingly, the loss tangent of PMMA@BT/PP nanocomposites was significantly reduced when compared with PP nanocomposite containing raw BT. As compared in Figure 4.16(f), upon the addition of 10 wt.% unmodified BT, the breakdown strength severely decreased from 361 MV/m of neat PP to 256 MV/m of BT/PP nanocomposites. In a sharp contrast, the PP nanocomposites consisting of core@shell BT with thin PMMA shell (8 nm) exhibit greatly improved breakdown strength of 448 MV/m. The electrical breakdown strength of polymer nanocomposites could be mainly affected by three factors enabled by the core@shell strategies. First, the organic PMMA shell improves the interface compatibility between the inorganic BT cores and PP matrix, which thereby reduces the interfacial defect density. Second, the PMMA shell with a high intrinsic breakdown strength acts as a robust phase that can block the development of breakdown paths. More notably, the material composition in PMMA@BT/PP nanocomposites shows an excellent example of electrically gradient materials. The PMMA shell with an intermediate *k* between $BaTiO_3$ core and PP matrix acts as a transition layer to homogenize the local electric field of organic-inorganic interfaces. At 250 kV/mm as shown in Figure 4.16(g), the discharged energy density enhances from 0.72 J/cm^3 of neat PP to 1.51 J/cm^3 of the PP nanocomposite filled with 10 wt.% PMMA@BT (8-nm shell thickness), the charge-discharge efficiency of the resulting composites also reaches a maximum value of ~95%.

Huang and Jiang group proposed a class of core@shell-structured $BaTiO_3$ nanoparticles by employing the ATRP approach [78]. For instance, the polymers of poly(2-hydroxylethyle methacrylate), poly(hydroxyethyl methacrylate), and sodium polyacrylate were respectively coated on the surfaces of $BaTiO_3$ nanoparticles. Moreover, the $BaTiO_3$ nanoparticles with core@double-shell structures were also fabricated by grafting block copolymers by using the ATRP method. Note that a core@shell structured "matrix-free" composite system has been designed and prepared based on this technology, in which the polymer matrix is directly pre-grafted to the nanofillers and the polymer nanocomposites are obtained by pressing the core@shell structured fillers together at an appropriate temperature. Compared to the commonly used fabrication methods, i.e., mixing nanofillers with polymer matrices, this approach could overcome the filler agglomeration issues of polymer nanocomposites with relatively high loadings. In this way, inorganic nanofillers were separated forcibly by the thick grafted polymer shells with nearly uniform spacing, resulting in much reduced dielectric loss in the resultant

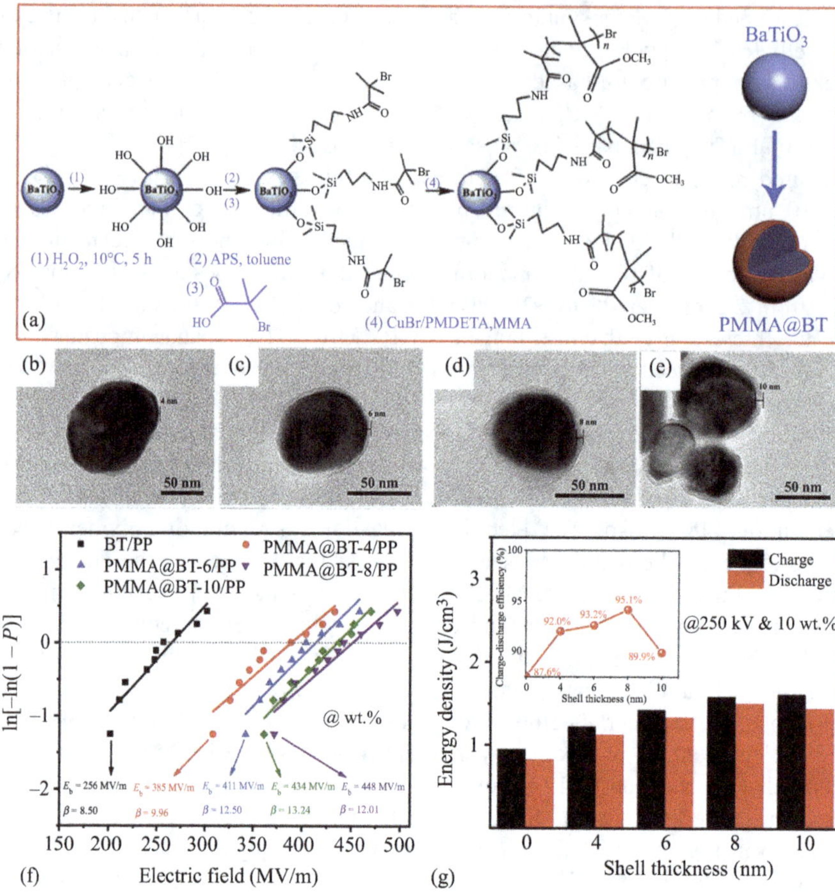

Figure 4.16 *(a) Schematic diagram illustrating the preparation of core-shell structured PMMA@BT nanoparticles by ATRP method. TEM images of PMMA@BT nanoparticles with different thicknesses of PMMA shells: (b) PMMA@BT-4, (c) PMMA@BT-6, (d) PMMA@BT-8, and (e) PMMA@BT-10 nanoparticles. (f) Weibull distributions for the electrical breakdown strengths of PP nanocomposites with 10 wt.% loading of raw BT nanoparticles and PMMA@BT nanoparticles with different thicknesses of PMMA shells. (g) The dependences of charged and discharged energy densities at 250 MV/m of PP nanocomposites with 10 wt.% loading of fillers on the thickness of PMMA shell. The inset in (d) shows the dependence of charge-discharge efficiency at 250 MV/m of PP nanocomposites with 10 wt.% loading of fillers on the thickness of PMMA shell.*

"matrix-free" polymer nanocomposites [78]. Compared to the "grafting from" approach, the "grafting to" method paves a way to tailoring the molecular composition and the molecular weight of the polymer chains of the inorganic nanofillers to achieve the expected homogeneous structures and multiple physical properties of the resulting nanocomposites. Based on the "grafting to" approach, Huang and Jiang group prepared core@shell $BaTiO_3$ nanoparticles with linear polymer shells by utilizing a highly efficient click-chemistry reaction without any usage of transition metal catalysis; the fabrication processes are illustrated in Figure 4.17(a). Thiol-terminated PS or PMMA molecular chains with different molecular weights were first synthesized by RAFT polymerization, and the core@shell $BaTiO_3$ nanoparticles were prepared by grafting the polymer chains on the surface of vinyl-functionalized inorganic nanofillers via a thiol-ene click reaction. Figure 4.17(b) and (c) demonstrates the TEM images of core@shell structured PS@$BaTiO_3$ and PMMA@$BaTiO_3$ nanoparticles. As shown in Figure 4.17(d) and

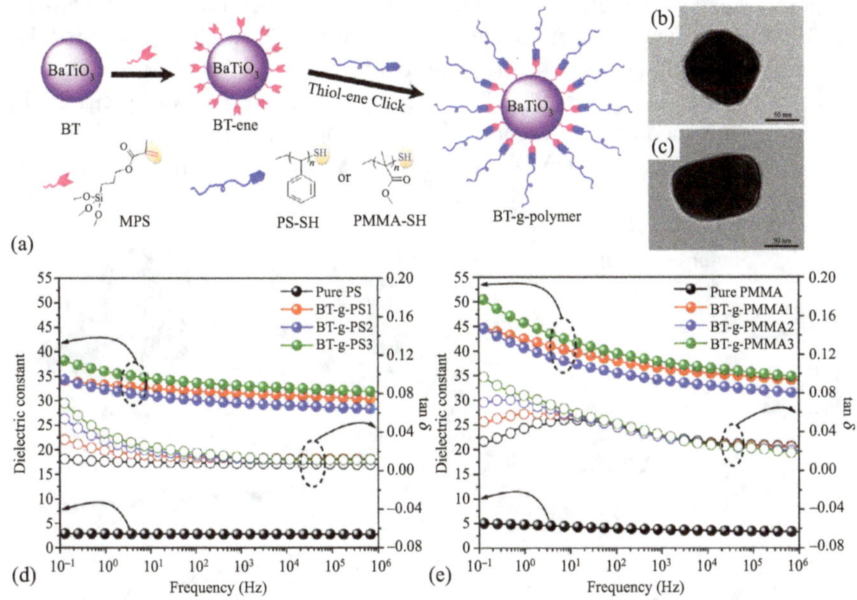

Figure 4.17 *(a) Illustrations of the synthesis process for PS@$BaTiO_3$ and PMMA@$BaTiO_3$ nanocomposites by thiol-ene click reactions. TEM images of core-shell (b) PS@$BaTiO_3$ and (c) PMMA@$BaTiO_3$. Frequency-dependent dielectric constant and tan δ for (d) PS@$BaTiO_3$ and (e) PMMA@$BaTiO_3$ nanocomposites. The calculated molecular weights of PS1, PS2, and PS3 were 10.3, 41.6, and 80.5 kg/mol, respectively. The calculated molecular weights of PMMA1, PMMA2, and PMMA3 were 10.7, 42.6, and 81.3 kg/mol, respectively. © 2014 American Chemical Society. Reproduced, with permission, from [79]*

(e), it was found that the *k* and the tan *δ* of the core@shell polymer@$BaTiO_3$ nanocomposites were highly dependent on the molecular weight of the grafted chains and the grafting density of the core@shell-structured nanofillers, especially at low-frequency ranges. In general, the grafted molecular chains with higher molecular weights results in lower grafting density and thereby contribute to higher loss tangent in the core@shell nanocomposites [79].

Furthermore, ultrasmall-sized nanodots have been carved on the polymeric shells to fabricate core@shell nanofillers with satellite-like structured shells [80–83]. For example, Ag nanodots were embedded onto the poly(vinyl pyrrolidone) (PVP) shell of $BaTiO_3$; the fabrication processes are shown in Figure 4.18(a), and the obtained nano-assemblies were marked as Ag@BT (Figure 4.18(b)) [80]. It was postulated that the quantum confinement effect leads the ultra-small Ag nanodots no more electrical conductors. Moreover, the ultra-small Ag nanoparticles could act as Coulomb islands to hinder the movement of the electrons, resulting in higher charge-discharge efficiency and lower leakage current density in PVDF nanocomposite containing 20 vol.% satellite-like core@shell $BaTiO_3$ nanoparticles (Ag1%@BT-20) with respect to the one containing 20 vol.% unmodified BT nanoparticles (BT-20) (Figure 4.18(c) and (d)).

Later, a strawberry-like BT@PDA-Ag hybrid nanofiller was prepared by decorating Ag nanodots onto a core@shell polydopamine (PDA)-coated $BaTiO_3$ nanoparticle to fabricate the P(VDF-HFP)-based nanocomposites [81]. It was found

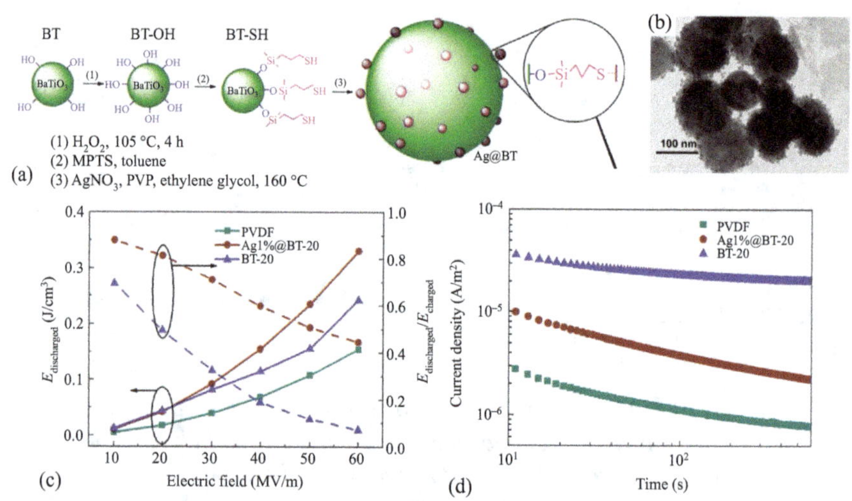

Figure 4.18 (a) Scheme of the preparation process of the Ag@BT core-satellite nano-assemblies. (b) TEM image of Ag1%@BT core-satellite nano-assemblies. (c) Discharged energy density ($E_{discharged}$) and (d) current density of neat PVDF, PVDF/BT-20, and PVDF/Ag1%@BT-20 nanocomposites. © 2013 The Royal Society of Chemistry. Reproduced, with permission, from [80]

that the Ag nanodots on the PDA shell inhibit electron migration in the P(VDF-HFP) composites and effectively suppress the space charge accumulation at the interface between the polymer matrix and inorganic $BaTiO_3$ nanoparticles, resulting in higher discharged energy densities in the P(VDF-HFP)/BT@PDA-Ag nanocomposites in comparison with the ferroelectric polymer matrix consisting of unmodified BT and core@shell BT@PDA nanoparticles.

Besides an organic shell layer for nanofillers, inorganic shells have been employed to prepare all-inorganic core@shell high-*k* nanofillers. Note that the inorganic shells using relatively low-*k* oxide (e.g., Al_2O_3) [84] or moderate-*k* material (e.g., TiO_2) [85] have also been designed as a gradient buffer between high-*k* ceramic nanofillers and low-*k* polymer matrices, which helps to alleviate the distortion of local field in the polymer/nanofiller composites resulting from a severe interfacial mismatch in *k* values.

As an excellent insulator with the wide bandgap (~9), SiO_2 has been intensively utilized as shell material for all-inorganic nanofillers [86–89]. For instance, $BaTiO_3$ nanoparticles with an inorganic@inorganic core@shell structure containing SiO_2 shells were synthesized by Wang and coworkers [87]. The PVDF nanocomposites containing $BaTiO_3$@SiO_2 nanofillers exhibit lower remnant polarization and much reduced dielectric loss in comparison with the composites filled with unmodified $BaTiO_3$ nanoparticles, which are attributed to the SiO_2 shell with wide bandgap acting as an insulating barrier. A similar core@shell approach was reported by Huang *et al.*, in which a thin shell layer of SiO_2 was coated onto the surface of smaller-sized $BaTiO_3$ nanoparticles (i.e., diameter <10 nm) [88]. In particular, the smaller size of the nanofiller gives rise to the larger interfacial area between inorganic and organic components, e.g., the 10 nm $BaTiO_3$ nanoparticle has almost tenfold larger interface areas than the conventional ones of 100 nm, which boosts greater interfacial polarization and dielectric response. The PVDF nanocomposite with the optimal filler loading (i.e., 5 vol.%) core@shell structured nanocomposites process enhanced breakdown strength and electric displacement, thus delivering a high discharged energy density of 11.5 J/cm^3 at 420 kV/mm. SiO_2 shell was also employed in a 1-D core@shell structure. The calcium copper titanate (CCTO) nanowires with giant *k* (i.e., $>10^5$) and high aspect ratio were surface coated by SiO_2 shell to prepare PVDF-based nanocomposites [89]. Compared to uncoated CCTO nanowires, the addition of CCTO@SiO_2 nanowires significantly decreases the dielectric loss while maintaining the high *k* of PVDF composites. For example, a desirable *k* of ~68 (at 1 kHz) and relatively low loss tangent of 0.081 were obtained in the composites containing 40 vol.% CCTO@SiO_2 nanowires, which is almost seven times greater than the PVDF matrix and 430% lower than the composites with 40 vol.% uncoated CCTO nanowires. As a wide bandgap insulator with higher *k* than SiO_2 (~4 of SiO_2 vs. ~9.5 of Al_2O_3), Al_2O_3 was utilized to form buffer layer for CCTO nanofibers by Chi *et al.* [90]. It was found the PVDF composites filled with core@shell-structured CCTO@Al_2O_3 nanofibers not only have desirable *k* value as high as the composites containing uncoated CCTO nanofibers but also exhibit comparable electric breakdown strength as Al_2O_3 nanofiber-filled composites. The PVDF/CCTO@Al_2O_3 nanofiber composites with the optimal filler loading of 4 vol.% deliver more than a doubling of improvement in high energy density with respect to polymer matrix, e.g., 8.46 J/cm^3 at 340 MV/m of composites and 3.68 J/cm^3 at 330 MV/m of PVDF.

Zhai and coworkers prepared a class of core@shell-structured high-k nanofiller with inorganic shells via electrospinning or hydrothermal processes [84,85]. Owing to the matched k value between Al_2O_3 shell and PVDF matrix, the PVDF nanocomposites containing 5 vol.% core@shell structure $BaTiO_3$@Al_2O_3 nanofibers (BT@AO NFs) exhibit a maximum discharge energy density of 12.18 J/cm^3 at 400 MV/m, which shows more than 250% enhancement of neat PVDF (e.g., 4.8 J/cm^3 at 350 MV/m) [84]. The improved capacitive performance is attributed to the Al_2O_3 shell layer that could effectively confine the charge mobility and reduce the Maxwell–Wagner–Sillars (MWS) interfacial polarization as well as space charge polarization between the polymer/nanofiller interfaces, thus simultaneously improving the breakdown strength and charge-discharge efficiency. Following this work, all-inorganic $BaTiO_3$-based nanofibers with a promising double-shell gradient structure were prepared by using a simple one-step coaxial electrospinning approach, as shown in Figure 4.19(a) [85]. As can be clearly seen in Figure 4.19(b) and (c), the material

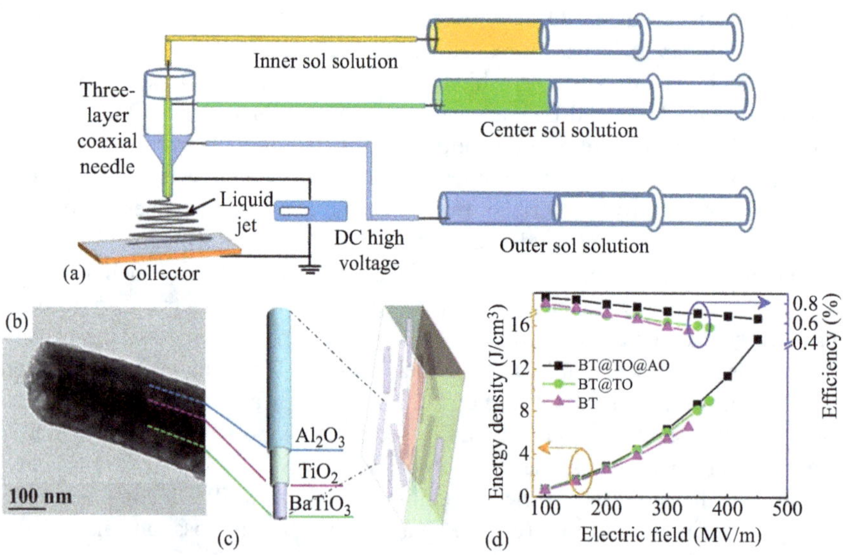

Figure 4.19 (a) The schematic of the three-layer coaxial electrospinning of the core-double-shell structured BT@TO@AO NFs. (b) TEM image of the core@double-shell structured BT@TO@AO NFs. (c) Schematic of the nanocomposite embedded core@double-shell structured BT@TO@AO NFs. (d) Energy density and efficiency of the PVDF composites filled with BT, BT@TO, and BT@TO@AO NFs.

composition of the core@double shell-structured nanofiber from inside to outside is $BaTiO_3$, TiO_2, and Al_2O_3 with gradually decreased *k* value. The novel design of hierarchical $BaTiO_3$@TiO_2@Al_2O_3 nanofibers (BT@TO@AO NFs) is not only advantageous to reduce the leakage current density but also greatly helps to alleviate the local field distortion in the PVDF/nanofiller composites. In comparison with the PVDF nanocomposites consisting of uncoated BT NFs and single shell BT@TO NFs, the composites with double shell BT@TO@AO NFs exhibit suppressed low-field loss tangents and high-field current densities, much-improved breakdown strengths, and charge-discharge efficiencies at the same filler loading. As a consequence, the PVDF-based nanocomposite containing 3.6 vol.% BT@TO@AO NFs exhibits the highest energy storage density of 14.84 J/cm^3 at 450 MV/m (Figure 4.19(d)).

Most recently, Zhang *et al.* combined hydrothermal process and precipitation method to in situ prepared SnO_2 nanodots on the surface of TiO_2 nanowires [91]. Different from the method that decorating small nanodots on polymer shells, this approach describes a class of satellite-like nanofiller without an organic middle layer. In the fabricated P(VDF-TrFE-CTFE)/TiO_2 NWs@SnO_2 nanocomposites, the synergetic effects of quantum size and Coulomb blockade of embedded SnO_2 nanodots effectively hinder the carrier transport and inhibit the electric conductivity and dielectric loss. As a result, a high breakdown strength of >250 MV/m was achieved in the high-*k* terpolymer-based nanocomposites with 5 vol.% TiO_2 NWs@SnO_2 loading, which is comparable with neat polymer.

In summary, significant signs of progress have been made by designing and preparing core@shell structures of nanofillers, which provide opportunities to engineering the organic/inorganic interfaces to favor the dispersion of nanofillers as well as to customize the multiple physical properties of the polymer nanocomposites via tailoring the structure and composition of shell materials. Considering that the core@shell structures typically require precise control over the nano-architectures, the challenges still lie ahead, e.g., the large-scale production of polymer/nanofiller composite films.

4.6 Polymer composites with topologically arranged nanofillers

The dielectric anisotropy in polymer nanocomposites enabled by the core@shell structure of nanofillers, e.g., the formation of double-shell-structured nanofibers with gradient-varying *k* value, could effectively enhance energy storage properties, which are summarized in Section 4.5. Moreover, numerous approaches centered on inducing anisotropy in polymer dielectrics by arranging nanofillers in topological architectures, which is considered as a predictable approach and beneficial to tailor-specific dielectric factors including dielectric constant, dielectric loss, and/or dielectric breakdown strength in different dimensions [10,92].

Up to now, numerous fabrication methods to control the nanofillers' arrangement have been performed by using electrostatic spinning, freeze-drying, nano-confined alignment, and mechanical pressing and stretching, etc. [93–100]. For polymer/nanofiller composites, dielectric anisotropy could be achieved by two commonly used means. (1) pre-attaching nanofillers onto an anisotropic skeleton, e.g., cellulose, followed by mixing with a polymer matrix to form topological nanostructures. In addition, compositing high-aspect-ratio 1-D and/or 2-D nanofillers with anisotropic nature and polymer matrices through external assistance including electrical field induction and mechanical stretching. In order to efficiently enhance the k, e.g., electric displacement of polymer composites through dielectric orientation effect, Wong and coworkers designed a high-k 3-D network of $BaTiO_3$ nanoparticles and filler epoxy resin to prepare high-k polymer nanocomposites [93]. As illustrated in Figure 4.20(a), by using lignocellulose as a skeleton, the BT nanocomposites are connected tightly and form a porous scaffold through the step-by-step processes of ball milling, freeze-drying, and high-temperature sintering. Figure 4.20(b) shows the surface morphology SEM image of the BT-lignocellulose skeleton filled with epoxy before sintering. It was clearly found that the BT nanoparticles were enwrapped and supported upon the existence of the lignocelluloses (Figure 4.20(c)). After being sintered at high temperature, the lignocellulose skeletons were burned off, and the adjacent nanoparticles were connected and formed a 3-D-BT network (Figure 4.20(d) and (e)). Notably, greater k values were achieved in the epoxy composites containing 3-D-BT networks in comparison with the ones with simply mixed epoxy/$BaTiO_3$ composites. For instance, a high k of 34.5 and a low tan δ of 0.07 were achieved at 1 kHz of 16 vol.% 3-D BT-network filled epoxy resin, while the traditional 0–3 structured epoxy composites with 17 vol.% $BaTiO_3$ loading exhibits lower k (~11 at 1 kHz) and higher tan δ of 0.14. The obtained high k is also helpful with the transmission

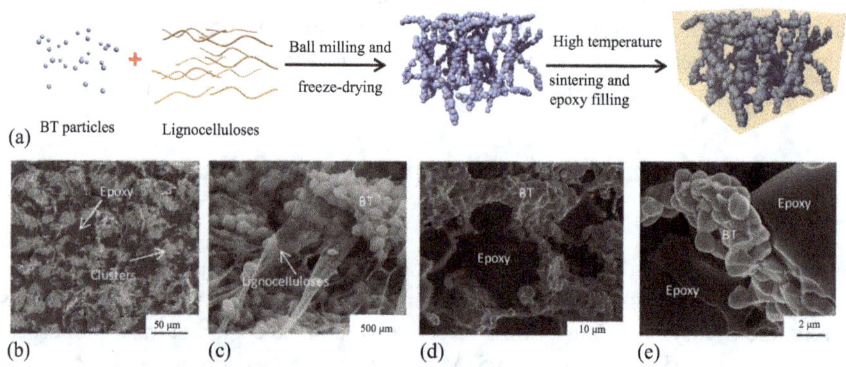

Figure 4.20 (a) Schematic of the preparation of the 3-D BT/epoxy composites. (b) and (c) Cross-sectional SEM images of the freeze-dried BT-lignocellulose injected with epoxy resin. (d) Fractured-surface SEM image of the 3-D BT/epoxy composites under low magnification. (e) Fractured-surface SEM image of the 3-D BT/epoxy composites under high magnification. © 2017 The Royal Society of Chemistry. Reproduced, with permission, from [93]

of electric polarization through high-*k* internetworks under an applied electric field in the resulting epoxy composites. As a result, a high discharged energy density of 8.3×10^3 J/cm^3 along with a high charge-discharge efficiency of 90.3% were achieved at 100 kV/cm in the epoxy composite with 16 vol.% 3-D-BT network.

Based on the dielectric anisotropy effect, an enhanced *k*, hence greater electrical polarization could be achieved along the direction in which the inorganic phases (e.g., 1-D, 2-D, and/or connected 0-D fillers) are parallelly aligned or oriented in organic matrices with the external applied electric field. Zhang *et al.* combined titanium dioxide/lead zirconate titanate (TiO_2@PZT) nanowire arrays and high-*k* P(VDF-TrFE-CTFE) terpolymer matrix to prepare polymer/nanowire array composites [94]. The cross-sectional SEM images of TiO_2 nanowire array-3, TiO_2@PZT nanowire array-3, and P(VDF-TrFE-CTFE)/TiO_2@PZT nanowire array-3 composites are shown in Figure 4.21(a), (b), and (c), respectively. Note that the nanowire array-1 to -3 corresponds to an intensified density of the nanowire array in polymer composites. As shown in Figure 4.21(d), the polymer composite containing nanowire array-3 delivers the highest *k* (i.e., 218.9 at 1 kHz), hence giving rise to the highest electric polarization (Figure 4.21(e)) when compared with nanowire array-1 and -2 filled polymer composites. Accordingly, as shown in Figure 4.21(f), a high discharge energy density of 6.9 J/cm^3 was obtained at a relatively low electric field (i.e., ~140 kV/mm), which is owing to the high *k* value of both PZT ceramic array and terpolymer matrix as well as the enhanced interface polarization between polymer/PZT and PZT/TiO_2 interfaces.

Yao *et al.* reported a lead-free 1-D array-based polymer composite with a similar structure. PVDF polymer is embedded into the TiO_2 nanorod array by the spin-coating method followed by annealing and quenching processes [95]. The TiO_2 nanorod array gives rise to greater electrical polarization along the direction of the external applied electric field. The pure polymer layer maintains high breakdown strength and low leakage current of the polymer nanocomposites, resulting in a high energy density of 10.62 J/cm^3 at 340 MV/m achieved in the PVDF composite with 0.18 height ratio of nanorod array. In addition to the pre-grown 1-D nano-arrays, the alignment of 1-D nanowires in different orientations with applied electric field also plays a significance role in determining *k* and breakdown strength of the polymer/nanowire composites. For example, the high-*k* nanowires aligned in the polymer matrix in the axis of the applied electric field can typically induce greater electric polarization, while the nanowires perpendicularly arranged to the applied electric field are able to maintain or even benefit the breakdown strength of polymer matrix. Xie *et al.* utilized modified scraping- and nanoconfining-assisted methods to induce the parallel and vertical alignment of BT nanowires in P(VDF-CTFE) matrix (Figure 4.22(a) and (b)) [96]. The cross-sectional SEM images of X-Y-aligned and Y-aligned P(VDF-CTFE)/BT nanowire composites are shown in Figure 4.22(c) and (d), respectively. The introduction of high-*k* BT nanowires largely improved the dielectric constant and electric displacement of both X-Y- and Z-aligned polymer nanocomposites, but the dielectric responses show different improvement ratios between these two different filler-oriented nanocomposites. As excepted, both *k* and electric conductivity of the

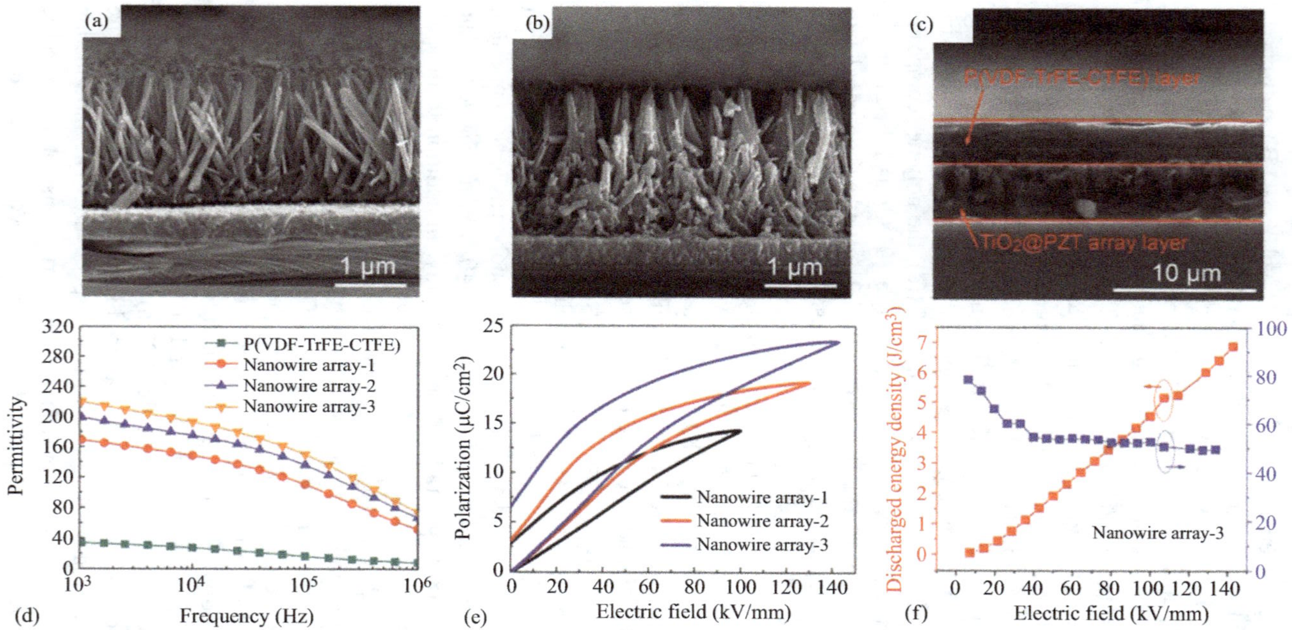

Figure 4.21 *Cross-sectional SEM images of (a) TiO_2 nanowire array-3 grown on FTO, (b) TiO_2@PZT nanowire array-3 and (c) TiO_2@PZT nanowire array-3/P(VDF-TrFE-CTFE) nanocomposite. (d) Relative permittivity and (e) D-E loops of the nanocomposites with different nanowire arrays. (f) The discharged energy density and efficiency of TiO_2@PZT nanowire array-3/P(VDF-TrFE-CTFE) nanocomposite. © 2018 Wiley-VCH. Reproduced, with permission, from [94]*

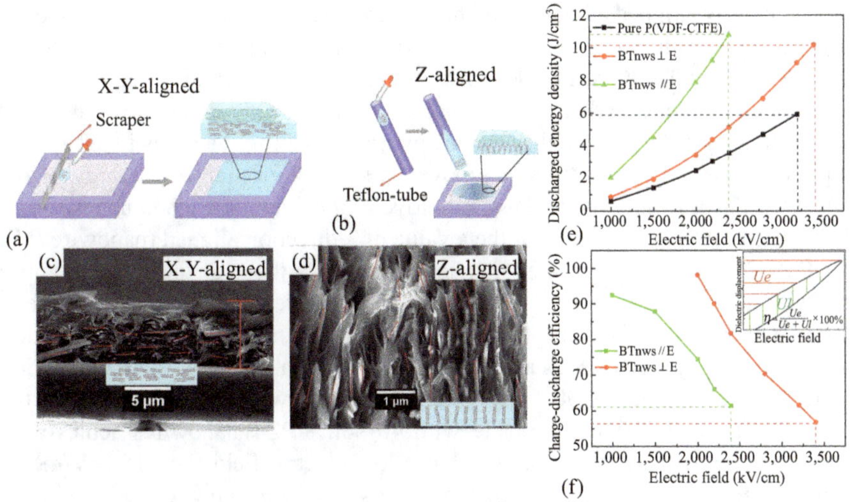

Figure 4.22 Schematic illustration of the preparation process of the (a) X-Y-aligned nanocomposite and (b) Z-aligned nanocomposite. Representative cross-sectional SEM image of the (c) X-Y-aligned nanocomposite film and (d) Z-aligned nanocomposite film. (e) Variations of discharge energy density and (f) charge-discharge efficiency of pure P(VDF-CTFE), X-Y-aligned, and Z-aligned nanocomposites. © 2017 The Royal Society of Chemistry. Reproduced, with permission, from [96]

Z-aligned nanocomposites are larger than those of the X-Y-aligned nanocomposites, which is attributed to the enhanced polarization intensity along with the Z-aligned directions of the composites. As shown in Figure 4.22(e) and (f), the Z-aligned (nanowire // E) nanocomposite with 3 vol.% BT nanowires exhibits higher discharged energy density buy lower charge-discharge efficiency than those of X-Y-aligned (nanowire ⊥ E) nanocomposite with the same filler content and under the same electric fields. However, the breakdown strength of the Z-aligned nanocomposite is lower than the neat P(VDF-CTFE) matrix and X-Y-aligned nanocomposite, because the Z-aligned nanowires are parallel to the direction in which the tree is growing. These act as vulnerable phases of breakdown. As a result, the Z-aligned nanocomposite exhibited a maximal discharged energy density of 10.8 J/cm^3 at 240 MV/m while the maximal discharged energy density of 10.1/cm^3 was obtained in the X-Y-aligned nanocomposite under a much higher electric field (i.e., 340 MV/m).

An electrical field-assisted approach was reported to achieve the orientation of epoxy resin/sodium titanate ($Na_2Ti_6O_{13}$) nanowires composites [97]. DC electric field was applied to the composite mixtures during the curing process of thermosetting epoxy materials. It was found that with a small loading of $Na_2Ti_6O_{13}$ nanowires, the electric breakdown strength of the X-Y-aligned (nanowires perpendicularly aligned along with the electric field directions) nanocomposite improved by 87% in comparison with that of the no-alignment sample. Moreover, further studies of TSDC results demonstrated that more electronic deep traps were perpendicularly generated along the nanowire alignment direction while more shallow traps parallelly existed in the nanowire orientation, which is in agreement with the breakdown test results. Moreover, a simpler mechanical strategy of the uniaxial-strain assembly method was employed for achieving aligned nanowires in PVDF matrix [98]. It is found that the *k* value of 3-direction aligned (nanowire // E) PVDF/PZT nanowire composites is much higher than those of the PVDF composites with randomly dispersed PZT nanowires and 1-direction aligned (nanowire $\perp$ E) PZT nanowires at the same filler content. For instance, ~34 of the 3-direction aligned composite versus ~25 and ~18 of the random dispersed and 1-direction aligned nanocomposite at 30 vol.% filler loading. Owing to the highest *k*, the 3-direction aligned PVDF composite with 40 vol.% PZT nanowires achieve the greatest energy density of ~1.28 J/cm^3 at a low electric field (i.e., 15 MV/m). A similar mechanical force-assisted method by using a twin-screw extruder to yield PE/o-MMT nanocomposites with 2-D oriented nanostructures was proposed by Manias and coworkers [99]. It was found that the highly oriented o-MMT nanoplates show a measurable positive effect on enhancing the breakdown strength of the polymer nanocomposites. For example, a more than 20% improvement in breakdown strength was achieved in the nanocomposite with 6 wt.% oriented o-MMT with respect to neat PE matrix. Combining the distinct advantages of the inducing of 1-D nanofiller orientation and the construction of core@shell structure, Chi and coworkers prepared aligned (nanowire $\perp$ E) PVDF-based composites consisting of SiO_2-coated $0.5Ba(Zr_{0.2}Ti_{0.8})O_3$–$0.5(Ba_{0.7}Ca_{0.3})TiO_3$ nanofibers (BZCT NFs) [100]. The aligned BCZT NFs give rise to larger electric displacement while maintaining high breakdown strength of the polymer matrix. The SiO_2 shell acts as a buffer to reduce the leakage current. As a result, a high discharged energy density of ~18.9 J/cm^3 was achieved in the PVDF composite with 3 vol.% aligned SiO_2-coated BZCT NFs. More importantly, benefiting from the highly aligned nanofibers that head-to-tail connected with each other in the in-plane direction, the in-plane thermal conductivity of the resulting composites was greatly improved as well.

Most recently, it was found that the existence of topologically arranged nanofillers in multi-layer structured polymer composites could not only regulate the electric field distribution in dielectrics at a macroscopic level but also combine the intrinsic advantages of anisotropic dielectric properties from each layer [101–104]. Thus, the layered structure in dielectric films – combining rationally designed nanostructures and nanocomponent configurations – has been considered well to be a very efficient approach to achieve a simultaneous improvement in both breakdown strength and energy density, along with a reduction in conduction loss [105,106], as highlighted and further computationally verified by finite element simulations [107] and phase-field simulations [108], which will be discussed in detail in Chapter 6.

To conclude, the importance of dielectric anisotropy induced by topologically arranged nanofillers in a polymer matrix is as equal as the rational choice of reinforcing nanocomponents and the interfacial modulation between organic and inorganic phases in determining energy storage properties of polymer nanocomposites. However, similar to the core@shell approach, the challenges we are facing on this strategy are also overwhelming, as several crucial issues such as the controlling methods of nanofiller alignment and the processability of large-scale polymer composite films remain to be addressed.

4.7 Conclusions

Polymer/nanofiller composites are emerging as an important class of dielectric materials for potential applications in advanced electronic and electrical power systems. There have been many exciting developments in the field of dielectric polymer composites over the past decade, and the pace of progress has continued to accelerate. However, currently available polymer composites and fabrication approaches, which are most in laboratory scales, still fall far short of the requirement specifications in practical capacitor applications. There is plenty of room for further improvement as several key issues remain to be addressed. There is a striking lack of fundamental understanding of the inorganic filler-polymer matrix interfacial coupling in the composites and the composite structure/high-field dielectric property relationship. Multiscale simulations are needed to understand the electrical polarization, electric field distribution, conduction loss, and breakdown mechanisms of the polymer composites with different structure architectures. Self-healing properties, which is critical for practical capacitor applications, have yet been demonstrated in polymer composites [60,109]. Large-scale fabrication of dielectric polymer composites into capacitor-grade thin films (thickness $\leq$ 10 μm), by using industrially friendly and low-cost methods, has not been demonstrated. Prototype capacitors based on polymer composites need to be developed in order to truly demonstrate the advantages of the composite approach. Future development of dielectric polymer composites shall be an effort combining improved fundamental understanding, rational design of materials, scalable synthesis, processing approaches, etc. Rapid advances are anticipated through highly collaborative efforts from synthetic chemists, materials scientists and processing engineers, and electrical engineers.

References

[1] Li Q, Chen L, Gadinski MR, *et al.* 'Flexible high-temperature dielectric materials from polymer nanocomposites'. *Nature*, 2015;523(7562):576–579.

[2] Wang Q, Zhu L. 'Polymer nanocomposites for electrical energy storage'. *Journal of Polymer Science Part B: Polymer Physics*, 2011;49(20):1421–1429.

[3] Li H, Liu F, Fan B, Ai D, Peng Z, Wang Q. 'Nanostructured ferroelectric-polymer composites for capacitive energy storage'. *Small Methods*, 2018; 2(6):1700399.

[4] Dang ZM, Yuan JK, Yao SH, Liao RJ. 'Flexible nanodielectric materials with high permittivity for power energy storage'. *Advanced Materials*, 2013;25(44):6334–6365.

[5] Dang ZM, Yuan JK, Zha JW, Zhou T, Li ST, Hu GH. 'Fundamentals, processes and applications of high-permittivity polymer–matrix composites'. *Progress in Materials Science*, 2012;57(4):660–723.

[6] Thakur VK, Gupta RK. 'Recent progress on ferroelectric polymer-based nanocomposites for high energy density capacitors: synthesis, dielectric properties, and future aspects'. *Chemical Reviews*, 2016;116(7):4260–4317.

[7] Li H, Ai D, Ren L, *et al.* 'Scalable polymer nanocomposites with record high-temperature capacitive performance enabled by rationally designed nanostructured inorganic fillers'. *Advanced Materials*, 2019;31(23): 1900875.

[8] Thakur Y, Lean MH, Zhang QM. 'Reducing conduction losses in high energy density polymer using nanocomposites'. *Applied Physics Letters*, 2017;110(12):122905.

[9] Guo N, DiBenedetto SA, Tewari P, Lanagan MT, Ratner MA, Marks TJ. 'Nanoparticle, size, shape, and interfacial effects on leakage current density, permittivity, and breakdown strength of metal oxide-polyolefin nanocomposites: experiment and theory'. *Chemistry of Materials*, 2010; 22(4):1567–1578.

[10] Shen ZH, Wang JJ, Lin Y, Nan CW, Chen LQ, Shen Y. 'High-throughput phase-field design of high-energy-density polymer nanocomposites'. *Advanced Materials*, 2018;30(2):1704380.

[11] Zhang X, Li BW, Dong L, *et al.* 'Superior energy storage performances of polymer nanocomposites via modification of filler/polymer interfaces'. *Advanced Materials Interfaces*, 2018;5(11):1800096.

[12] Pan Z, Yao L, Zhai J, Yao X, Chen H. 'Interfacial coupling effect in organic/inorganic nanocomposites with high energy density'. *Advanced Materials*, 2018;30(17):1705662.

[13] Li H, Ren L, Ai D, *et al.* 'Ternary polymer nanocomposites with concurrently enhanced dielectric constant and breakdown strength for high-temperature electrostatic capacitors'. *InfoMat*, 2020;2(2):389–400.

[14] Chu B, Zhou X, Ren K, *et al.* 'A dielectric polymer with high electric energy density and fast discharge speed'. *Science*, 2006;313(5785):334–336.

[15] Zhou Y, Li Q, Dang B, *et al.* 'A scalable, high-throughput, and environmentally benign approach to polymer dielectrics exhibiting significantly improved capacitive performance at high temperatures'. *Advanced Materials*, 2018;30(49):1805672.

[16] Li Q, Yao FZ, Liu Y, Zhang G, Wang H, Wang Q. 'High-temperature dielectric materials for electrical energy storage'. *Annual Review of Materials Research*, 2018;48:219–243.

[17] Li Q, Wang Q. 'Ferroelectric polymers and their energy-related applications'. *Macromolecular Chemistry and Physics*, 2016;217(11):1228–1244.

[18] Zhu L, Wang Q. 'Novel ferroelectric polymers for high energy density and low loss dielectrics'. *Macromolecules*, 2012;45(7):2937–2954.

[19] Zhang QM, Bharti V, Zhao X. 'Giant electrostriction and relaxor ferroelectric behavior in electron-irradiated poly (vinylidene fluoride-trifluoroethylene) copolymer'. *Science*, 1998;280(5372):2101–2104.

[20] Zhou X, Chu B, Neese B, Lin M, Zhang QM. 'Electrical energy density and discharge characteristics of a poly (vinylidene fluoride-chlorotrifluoroethylene) copolymer'. *IEEE Transactions on Dielectrics and Electrical Insulation*, 2007;14(5):1133–1138.

[21] Guan F, Pan J, Wang J, Wang QM, Zhu L. 'Crystal orientation effect on electric energy storage in poly (vinylidene fluoride-co-hexafluoropropylene) copolymers'. *Macromolecules*, 2009;43(1):384–392.

[22] Zhang Z, Chung TM. 'Study of VDF/TrFE/CTFE terpolymers for high pulsed capacitor with high energy density and low energy loss'. *Macromolecules*, 2007;40(4):783–785.

[23] Xia F, Cheng ZY, Xu H, *et al.* 'High electromechanical responses in a poly (vinylidene fluoride–trifluoroethylene–chlorofluoroethylene) terpolymer'. *Advanced Materials*, 2002;14(21):1574–1577.

[24] Chen X, Xu W, Lu B, Zhang T, Wang Q, Zhang QM 'Towards electrocaloric heat pump—A relaxor ferroelectric polymer exhibiting large electrocaloric response at low electric field'. *Applied Physics Letters*, 2018;113(11): 113902.

[25] Dang ZM, Lin YH, Nan CW. 'Novel ferroelectric polymer composites with high dielectric constants'. *Advanced Materials*, 2003;15(19):1625–1629.

[26] Zhang G, Li Y, Tang S, Thompson RD, Zhu L. 'The role of field electron emission in polypropylene/aluminum nanodielectrics under high electric fields'. *ACS Applied Materials & Interfaces*, 2017;9(11):10106–10119.

[27] Ji S, Jang J, Cho E, *et al.* 'High dielectric performances of flexible and transparent cellulose hybrid films controlled by multidimensional metal nanostructures'. *Advanced Materials*, 2017;29(24):1700538.

[28] Huang X, Zhang X, Ren G, *et al.* 'Non-intuitive concomitant enhancement of dielectric permittivity, breakdown strength and energy density in percolative polymer nanocomposites by trace Ag nanodots'. *Journal of Materials Chemistry A*, 2019;7(25):15198–15206.

[29] Bartlett MD, Fassler A, Kazem N, Markvicka EJ, Mandal P, Majidi C. 'Stretchable, high-k dielectric elastomers through liquid-metal inclusions'. *Advanced Materials*, 2016;28(19):3726–3731.

[30] Yan J, Malakooti MH, Lu Z, *et al.* 'Solution processable liquid metal nanodroplets by surface-initiated atom transfer radical polymerization'. *Nature Nanotechnology*, 2019;14:684–690.

[31] Pan C, Markvicka EJ, Malakooti MH, *et al.* 'A liquid-metal–elastomer nanocomposite for stretchable dielectric materials'. *Advanced Materials.*, 2019;31(23):1900663.

[32] Zhang QM, Li H, Poh M, *et al.* 'An all-organic composite actuator material with a high dielectric constant'. *Nature*, 2002;419(6904):284–287.

[33] Stoyanov H, Mc Carthy D, Kollosche M, Kofod G. 'Dielectric properties and electric breakdown strength of a subpercolative composite of carbon black in thermoplastic copolymer'. *Applied Physics Letters*, 2009;94(23):232905.

[34] Dang ZM, Zheng MS, Zha JW. '1D/2D carbon nanomaterial-polymer dielectric composites with high permittivity for power energy storage applications'. *Small*, 2016;12(13):1688–1701.

[35] Dang ZM, Wang L, Yin Y, Zhang Q, Lei QQ. 'Giant dielectric permittivities in functionalized carbon-nanotube/electroactive-polymer nanocomposites'. *Advanced Materials*, 2007;19(6):852–857.

[36] Zhang S, Zhang N, Huang C, Ren K, Zhang QM 'Microstructure and electromechanical properties of carbon nanotube/poly (vinylidene fluoride—trifluoroethylene—chlorofluoroethylene) composites'. *Advanced Materials*, 2005;17(15):1897–1901.

[37] Yuan JK, Yao SH, Dang ZM, Sylvestre A, Genestoux M, Bai J. 'Giant dielectric permittivity nanocomposites: realizing true potential of pristine carbon nanotubes in polyvinylidene fluoride matrix through an enhanced interfacial interaction'. *The Journal of Physical Chemistry C*, 2011; 115(13):5515–5521.

[38] Shang J, Zhang Y, Yu L, Shen B, Lv F, Chu PK. 'Fabrication and dielectric properties of oriented polyvinylidene fluoride nanocomposites incorporated with graphene nanosheets'. *Materials Chemistry and Physics*, 2012; 134(23):867–874.

[39] Tang H, Ehlert GJ, Lin Y, Sodano HA. 'Highly efficient synthesis of graphene nanocomposites'. *Nano Letters*, 2011;12(1):84–90.

[40] Kim P, Doss NM, Tillotson JP, *et al.* 'High energy density nanocomposites based on surface-modified $BaTiO_3$ and a ferroelectric polymer'. *ACS Nano*, 2009;3(9):2581–2592.

[41] Fan BH, Zha JW, Wang D, Zhao J, Dang ZM. 'Size-dependent low-frequency dielectric properties in the $BaTiO_3$/poly (vinylidene fluoride) nanocomposite films'. *Applied Physics Letters*, 2012;100(1):012903.

[42] Zhou T, Zha JW, Cui RY, Fan BH, Yuan JK, Dang ZM. 'Improving dielectric properties of $BaTiO_3$/ferroelectric polymer composites by employing surface hydroxylated $BaTiO_3$ nanoparticles'. *ACS Applied Materials & Interfaces*, 2011;3(7):2184–2188.

[43] Dang ZM, Wang HY, Xu HP. 'Influence of silane coupling agent on morphology and dielectric property in $BaTiO_3$/polyvinylidene fluoride composites'. *Applied Physics Letters*, 2006;89(11):112902.

[44] Kim P, Jones SC, Hotchkiss PJ, *et al.* 'Phosphonic acid-modified barium titanate polymer nanocomposites with high permittivity and dielectric strength'. *Advanced Materials*, 2007;19(7):1001–1005.

[45] Li J, Claude J, Norena-Franco LE, Seok SI, Wang Q. 'Electrical energy storage in ferroelectric polymer nanocomposites containing surface-functionalized $BaTiO_3$ nanoparticles'. *Chemistry of Materials*, 2008;20(20):6304–6306.

[46] Li Z, Liu F, Yang G, *et al.* 'Enhanced energy storage performance of ferroelectric polymer nanocomposites at relatively low electric fields induced by surface modified $BaTiO_3$ nanofibers'. *Composites Science and Technology*, 2018;164:214–221.

[47] Zhang G, Zhang X, Yang T, *et al.* 'Colossal room-temperature electrocaloric effect in ferroelectric polymer nanocomposites using nanostructured barium strontium titanates'. *ACS Nano*, 2015;9(7):7164–7174.

[48] Tang H, Sodano HA. 'Ultra high energy density nanocomposite capacitors with fast discharge using $Ba_{0.2}Sr_{0.8}TiO_3$ nanowires'. *Nano Letters*, 2013; 13(4):1373–1379.

[49] Wang G, Huang X, Jiang P. 'Tailoring dielectric properties and energy density of ferroelectric polymer nanocomposites by high-k nanowires'. *ACS Applied Materials & Interfaces*, 2015;7(32):18017–18027.

[50] Li J, Seok SI, Chu B, Dogan F, Zhang QM, Wang Q. 'Nanocomposites of ferroelectric polymers with TiO_2 nanoparticles exhibiting significantly enhanced electrical energy density'. *Advanced Materials*, 2009;21(2): 217–221.

[51] Robertson J. 'Band offsets of high dielectric constant gate oxides on silicon'. *Journal of Non-Crystalline Solids*, 2002;303(1):94–100.

[52] Li H, Liu F, Tian H, *et al.* 'Synergetic enhancement of mechanical and electrical strength in epoxy/silica nanocomposites via chemically-bonded interface'. *Composites Science and Technology*, 2018;167:539–546.

[53] Wang W, Min D, Li S. 'Understanding the conduction and breakdown properties of polyethylene nanodielectrics: effect of deep traps'. *IEEE Transactions on Dielectrics and Electrical Insulation*, 2016;23(1):564–572.

[54] Li J, Khanchaitit P, Han K, Wang Q. 'New route toward high-energy-density nanocomposites based on chain-end functionalized ferroelectric polymers'. *Chemistry of Materials*, 2010;22(18):5350–5357.

[55] Han K, Li Q, Chanthad C, Gadinski MR, Zhang G, Wang Q. 'A hybrid material approach toward solution-processable dielectrics exhibiting enhanced breakdown strength and high energy density'. *Advanced Functional Materials*, 2015;25(23):3505–3513.

[56] Meereboer NL, Terzić I, Portale G, Loos K. 'Improved energy density and charge-discharge efficiency in solution processed highly defined ferroelectric block copolymer-based dielectric nanocomposites'. *Nano Energy*, 2019;64:103939.

[57] Liu F, Li Q, Li Z, *et al.* 'Poly (methyl methacrylate)/boron nitride nanocomposites with enhanced energy density as high temperature dielectrics'. *Composites Science and Technology*, 2017;142:139–144.

[58] Li H, Xie Z, Liu L, *et al.* 'High-performance insulation materials from poly (ether imide)/boron nitride nanosheets with enhanced DC breakdown strength and thermal stability'. *IEEE Transactions on Dielectrics and Electrical Insulation*, 2019;26(3):722–729.

[59] Li Q, Zhang G, Liu F, *et al.* 'Solution-processed ferroelectric terpolymer nanocomposites with high breakdown strength and energy density utilizing

boron nitride nanosheets'. *Energy & Environmental Science*, 2015;8(3):922–931.

[60] Xing L, Li Q, Zhang G, *et al.* 'Self-healable polymer nanocomposites capable of simultaneously recovering multiple functionalities'. *Advanced Functional Materials*, 2016;26(20):3524–3531.

[61] Yang J, Xie H, Chen H, *et al.* 'Cellulose nanofibril/boron nitride nanosheet composites with enhanced energy density and thermal stability by interfibrillar cross-linking through Ca^{2+}'. *Journal of Materials Chemistry A*, 2018;6(4):1403–1411.

[62] Wang J, Chen H, Li X, *et al.* 'Flexible dielectric film with high energy density based on chitin/boron nitride nanosheets'. *Chemical Engineering Journal*, 2019:123147.

[63] Tomer V, Polizos G, Manias E, Randall CA. 'Epoxy-based nanocomposites for electrical energy storage. I: Effects of montmorillonite and barium titanate nanofillers'. *Journal of Applied Physics*, 2010;108(7):074116.

[64] Polizos G, Tomer V, Manias E, Randall CA. 'Epoxy-based nanocomposites for electrical energy storage. II: Nanocomposites with nanofillers of reactive montmorillonite covalently-bonded with barium titanate'. *Journal of Applied Physics*, 2010;108(7):074117.

[65] Li B, Salcedo-Galan F, Xidas PI, Manias E. 'Improving electrical breakdown strength of polymer nanocomposites by tailoring hybrid-filler structure for high-voltage dielectric applications'. *ACS Applied Nano Materials*, 2018;1(9):4401–4407.

[66] Li Q, Han K, Gadinski MR, Zhang G, Wang Q. 'High energy and power density capacitors from solution-processed ternary ferroelectric polymer nanocomposites'. *Advanced Materials*, 2014;26(36):6244–6249.

[67] Xie Y, Wang J, Yu Y, Jiang W, Zhang Z. 'Enhancing breakdown strength and energy storage performance of PVDF-based nanocomposites by adding exfoliated boron nitride'. *Applied Surface Science*, 2018;440:1150–1158.

[68] Liu F, Li Q, Li Z, Dong L, Xiong C, Wang Q. 'Ternary PVDF-based terpolymer nanocomposites with enhanced energy density and high power density'. *Composites Part A: Applied Science and Manufacturing*, 2018;109:597–603.

[69] Luo S, Yu J, Yu S, *et al.* 'Significantly enhanced electrostatic energy storage performance of flexible polymer composites by introducing highly insulating-ferroelectric microhybrids as fillers'. *Advanced Energy Materials*, 2019;9(5):1803204.

[70] Zhang X, Shen Y, Zhang Q, *et al.* 'Ultrahigh energy density of polymer nanocomposites containing $BaTiO_3$@TiO_2 nanofibers by atomic-scale interface engineering'. *Advanced Materials*, 2015;27(5):819–824.

[71] Zhang X, Shen Y, Xu B, *et al.* 'Giant energy density and improved discharge efficiency of solution-processed polymer nanocomposites for dielectric energy storage'. *Advanced Materials*, 2016;28(10):2055–2061.

[72] Jin Y, Xia N, Gerhardt RA. 'Enhanced dielectric properties of polymer matrix composites with $BaTiO_3$ and MWCNT hybrid fillers using simple phase separation'. *Nano Energy*, 2016;30:407–416.

[73] Huang X, Jiang P. 'Core–shell structured high-k polymer nanocomposites for energy storage and dielectric applications'. *Advanced Materials*, 2015;27(3):546–554.

[74] Luo H, Zhou X, Ellingford C, *et al*. 'Interface design for high energy density polymer nanocomposites'. *Chemical Society Reviews*, 2019;48(16):4424–4465.

[75] Jiang B, Pang X, Li B, Lin Z. 'Organic-inorganic nanocomposites via placing monodisperse ferroelectric nanocrystals in direct and permanent contact with ferroelectric polymers'. *Journal of the American Chemical Society*, 2015;137(36):11760–11767.

[76] Zheng MS, Zheng YT, Zha JW, *et al*. 'Improved dielectric, tensile and energy storage properties of surface rubberized $BaTiO_3$/polypropylene nanocomposites'. *Nano Energy*, 2018;48:144–151.

[77] Liu B, Yang M, Zhou WY, *et al*. 'High energy density and discharge efficiency polypropylene nanocomposites for potential high-power capacitor'. *Energy Storage Materials*, 2020;27:443–452.

[78] Huang Y, Huang X, Schadler LS, He J, Jiang P. 'Core@double-shell structured nanocomposites: a route to high dielectric constant and low loss material'. *ACS Applied Materials & Interfaces*, 2016;8(38):25496–25507.

[79] Yang K, Huang X, Zhu M, Xie L, Tanaka T, Jiang P. 'Combining RAFT polymerization and thiol–ene click reaction for core–shell structured polymer@$BaTiO_3$ nanodielectrics with high dielectric constant, low dielectric loss, and high energy storage capability'. *ACS Applied Materials & Interfaces*, 2014;6(3):1812–1822.

[80] Xie L, Huang X, Li BW, Zhi C, Tanaka T, Jiang P. 'Core–satellite Ag@$BaTiO_3$ nanoassemblies for fabrication of polymer nanocomposites with high discharged energy density, high breakdown strength and low dielectric loss'. *Physical Chemistry Chemical Physics*, 2013;15(40):17560–17569.

[81] Yang K, Huang X, He J, Jiang P. 'Strawberry-like Core–Shell Ag@Polydopamine@ $BaTiO_3$ hybrid nanoparticles for high-k polymer nanocomposites with high energy density and low dielectric loss'. *Advanced Materials Interfaces*, 2015;2(17):1500361.

[82] Luo S, Yu S, Sun R, Wong CP. 'Nano Ag-deposited $BaTiO_3$ hybrid particles as fillers for polymeric dielectric composites: toward high dielectric constant and suppressed loss'. *ACS Applied Materials & Interfaces*, 2013;6(1):176–182.

[83] Ren L, Meng X, Zha JW, Dang ZM. 'Coulomb block effect inducing distinctive dielectric properties in electroless plated barium titanate@silver/poly(vinylidene fluoride) nanocomposites'. *RSC Advances*, 2015;5(80):65167–65174.

[84] Pan Z, Yao L, Zhai J, *et al*. 'Excellent energy density of polymer nanocomposites containing $BaTiO_3$@Al_2O_3 nanofibers induced by moderate

interfacial area'. *Journal of Materials Chemistry A*, 2016;4(34):13259–13264.

[85] Pan Z, Zhai J, Shen B. 'Multilayer hierarchical interfaces with high energy density in polymer nanocomposites composed of $BaTiO_3$@TiO_2@Al_2O_3 nanofibers'. *Journal of Materials Chemistry A*, 2017;5(29):15217–15226.

[86] Han K, Li Q, Chen Z, *et al.* 'Suppression of energy dissipation and enhancement of breakdown strength in ferroelectric polymer–graphene percolative composites'. *Journal of Materials Chemistry C*, 2013;1(42): 7034–7042.

[87] Yu K, Niu Y, Bai Y, Zhou Y, Wang H. 'Poly(vinylidene fluoride) polymer based nanocomposites with significantly reduced energy loss by filling with core-shell structured $BaTiO_3/SiO_2$ nanoparticles'. *Applied Physics Letters*, 2013;102(10):102903.

[88] Bi K, Bi M, Hao Y, *et al.* 'Ultrafine core-shell $BaTiO_3$@SiO_2 structures for nanocomposite capacitors with high energy density'. *Nano Energy*, 2018;51:513–523.

[89] Tang H, Zhou Z, Bowland CC, Sodano HA. 'Synthesis of calcium copper titanate ($CaCu_3Ti_4O_{12}$) nanowires with insulating SiO_2 barrier for low loss high dielectric constant nanocomposites'. *Nano Energy*, 2015;17:302–307.

[90] Chi Q, Wang X, Zhang C, *et al.* 'High energy storage density for poly (vinylidene fluoride) composites by introduced core–shell $CaCu_3Ti_4O_{12}$@Al_2O_3 nanofibers'. *ACS Sustainable Chemistry & Engineering*, 2018;6(7):8641–8649.

[91] Zhang Q, Zhang Z, Xu N, Yang H. 'Dielectric properties of P(VDF-TrFE-CTFE) composites filled with surface-coated TiO_2 nanowires by SnO_2 nanoparticles'. *Polymers*, 2020;12(1):85.

[92] Zhang X, Jiang J, Shen Z, *et al.* 'Polymer nanocomposites with ultrahigh energy density and high discharge efficiency by modulating their nanostructures in three dimensions'. *Advanced Materials*, 2018;30(16):1707269.

[93] Luo S, Shen Y, Yu S, *et al.* 'Construction of a 3D-$BaTiO_3$ network leading to significantly enhanced dielectric permittivity and energy storage density of polymer composites'. *Energy & Environmental Science*, 2017;10(1): 137–144.

[94] Zhang D, Liu W, Guo R, Zhou K, Luo H. 'High discharge energy density at low electric field using an aligned titanium dioxide/lead zirconate titanate nanowire array'. *Advanced Science*, 2018;5(2):1700512.

[95] Yao L, Pan Z, Liu S, Zhai J, Chen HH. 'Significantly enhanced energy density in nanocomposite capacitors combining the TiO_2 nanorod array with poly (vinylidene fluoride)'. *ACS Applied Materials & Interfaces*, 2016; 8(39):26343–26351.

[96] Xie B, Zhang H, Zhang Q, *et al.* 'Enhanced energy density of polymer nanocomposites at a low electric field through aligned $BaTiO_3$ nanowires'. *Journal of Materials Chemistry A*, 2017;5(13):6070–6078.

[97] Chen S, Cheng Y, Xie Q, *et al.* 'Enhanced breakdown strength of aligned-sodium-titanate-nanowire/epoxy nanocomposites and their anisotropic

dielectric properties'. *Composites Part A: Applied Science and Manufacturing*, 2019;120:84–94.

[98] Tang H, Lin Y, Sodano HA. 'Enhanced energy storage in nanocomposite capacitors through aligned PZT nanowires by uniaxial strain assembly'. *Advanced Energy Materials*, 2012;2(4):469–476.

[99] Tomer V, Polizos G, Randall CA, Manias E. 'Polyethylene nanocomposite dielectrics: implications of nanofiller orientation on high field properties and energy storage'. *Journal of Applied Physics*, 2011;109(7):074113.

[100] Zhang Y, Zhang C, Feng Y, *et al.* 'Excellent energy storage performance and thermal property of polymer-based composite induced by multi-functional one-dimensional nanofibers oriented in-plane direction'. *Nano Energy*, 2019;56:138–150.

[101] Wang Y, Cui J, Yuan Q, Niu Y, Bai Y, Wang H. 'Significantly enhanced breakdown strength and energy density in sandwich-structured barium titanate/poly(vinylidene fluoride) nanocomposites'. *Advanced Materials*, 2015;27(42):6658–6663.

[102] Li Z, Liu F, Li H, *et al.* 'Largely enhanced energy storage performance of sandwich-structured polymer nanocomposites with synergistic inorganic nanowires'. *Ceramics International*, 2019;45(7):8216–8221.

[103] Wang Y, Wang L, Yuan Q, *et al.* 'Ultrahigh energy density and greatly enhanced discharged efficiency of sandwich-structured polymer nanocomposites with optimized spatial organization'. *Nano Energy*, 2018;44:364–370.

[104] Wang Y, Wang L, Yuan Q, *et al.* 'Ultrahigh electric displacement and energy density in gradient layer-structured $BaTiO_3$/PVDF nanocomposites with an interfacial barrier effect'. *Journal of Materials Chemistry A*, 2017;5(22):10849–10855.

[105] Liu F, Li Q, Cui J, *et al.* 'High-energy-density dielectric polymer nanocomposites with trilayered architecture'. *Advanced Functional Materials*, 2017;27(20):1606292.

[106] Jiang J, Shen Z, Cai X, *et al.* 'Polymer nanocomposites with interpenetrating gradient structure exhibiting ultrahigh discharge efficiency and energy density'. *Advanced Energy Materials*, 2019;9(15):1803411.

[107] Wang Y, Chen J, Li Y, Niu Y, Wang Q, Wang H. 'Multilayered hierarchical polymer composites for high energy density capacitors'. *Journal of Materials Chemistry A*, 2019;7(7):2965–2980.

[108] Shen ZH, Wang JJ, Jiang JY, *et al.* 'Phase-field modeling and machine learning of electric-thermal-mechanical breakdown of polymer-based dielectrics'. *Nature Communications*, 2019;10(1):1843.

[109] Yang Y, He J, Li Q, *et al.* 'Self-healing of electrical damage in polymers using superparamagnetic nanoparticles'. *Nature Nanotechnology*, 2019;14(2):151–155.

Chapter 5

High-temperature polymer-based dielectrics for electrostatic energy storage

Sang Cheng[1] *and Qi Li*[1]

5.1 Introduction

Dielectric materials are the core elements of dielectric capacitors that are one of the most important passive components in advanced electrical and electronic systems [1–4]. Among the currently available dielectric materials, polymers have various advantages over their counterparts, e.g., ceramics, such as the high breakdown strength, low loss, great processability, as well as self-clearing behavior [5–8]. The most practiced polymer dielectrics in mainstream capacitors are biaxially oriented polypropylenes (BOPPs), which have high breakdown strength (>700 MV/m) and low dielectric loss (<0.02%) at room temperature [9,10]. However, their dielectric performance deteriorates significantly under elevated temperature and the maximum operating temperature of BOPP is only 105 °C [11], which cannot meet the urgent demands for film capacitors operating under harsh environment. For example, the ambient temperature of power inverters near the engines in hybrid electric vehicles (HEVs) exceeds 140 °C. To maintain the regular work of DC/AC inverters, a secondary cooling system is introduced into present HEVs, bringing extra volume and weight [12–14]. High-temperature conditions also exist in deep oil and gas exploration, where the electronics experience temperature as high as 200 °C [13]. In aerospace industry, the control systems including sensors and signal conditioning electronics operate at temperature of 300–400 °C [15,16]. The development of wide-bandgap semiconductors, such as silicon carbide (SiC) and gallium nitride (GaN), is expected to bring about the revolution of electronic devices, enabling higher operating temperature of over 150 °C. These electronic devices cannot be employed practically unless the assorted film capacitors are able to tolerate such high temperature [13].

Although many heat-resistant polymers have been developed for other applications, e.g., electrical and electronic insulation, their applications to capacitor films are less mentioned due to the small markets [17,18]. Notable characteristics

[1]State Key Lab of Power Systems, Department of Electrical Engineering, Tsinghua University, Beijing, China

of these polymers for capacitor dielectrics need to be further discussed. Moreover, some nanocomposites capitalizing on the advantages of both polymer dielectrics and ceramic dielectrics are also developed to meet the demanding requirements [19,20]. In this chapter, we provide an overview of polymer-based high-temperature dielectric materials, highlighting their superior features as well as major challenges and limitations for capacitor applications. General considerations for design of high-performance polymer-based dielectric materials for high-temperature applications will be discussed firstly.

5.2 Basic principles of developing high-temperature dielectric materials

While relevant characteristics of polymer dielectrics have been discussed in the previous chapters, there are some special problems calling to be tackled under elevated temperature. In this section, we will summarize the basic principles and some critical impediments of developing polymer-based dielectric materials for high-temperature applications.

5.2.1 Thermal stability and processability

Excellent thermal stability, i.e., the capability of maintaining mechanical integrity and great dielectric performance under elevated temperature, is one of the prerequisite properties for dielectric materials operating at high temperature. Thermal stability of polymers is usually characterized by the glass-transition temperature (T_g), where materials transit from relatively hard glassy states into viscous or rubbery states [21]. As the temperature passes through T_g, polymer dielectrics lose their mechanical stiffness and the thermal expansion, free volume, shear modulus, and many other properties of them show an abrupt change [22], decreasing the performance and reliability of capacitors. For instance, loose of polymer chains generates numerous free volumes and eases the propagation of space charges, thus increasing the electronic and ionic conduction in polymers. Melting temperature (T_m) of crystalline or semi-crystalline polymers, usually higher than T_g, is another parameter associated with the thermal stability, where the crystalline-to-amorphous phase transition occurs and polymers may experience major property change.

Polymers, in principle, suffer from poor thermal stability and low T_g as a result of the weak interactions between the molecule chains [23]. For example, T_g of polyethylene terephthalates (PETs), a type of polar polymers with relatively high permittivity, is only 70 °C, and their dielectric constant as well as dissipation factor increase remarkably as the temperature increases above T_g due to increased mobility of polar groups [24]. A common approach to raising the thermal stability of polymers is to introduce rigid structures, e.g., π conjunction (imide ring, imidazole ring), aromatic/heteroaromatic molecular skeletons and high-strength chemical bonds (C–F etc.), into polymer main chains, restricting the rotation of main chains [17]. Besides, strengthening the intermolecular interaction by incorporating hydrogen bond or cross-linking is another effective method.

Considering the capacitance of film capacitors is inversely proportional to the thickness of dielectrics, polymer films cannot be practically used for capacitor applications unless they can be processed thin enough (usually <12 μm) (Figure 5.1) [25]. High thermal stability and mechanical stiffness, however, yield poor processability of high-temperature polymers, which is one of the major impediments to employ currently high-temperature resistant polymer dielectrics in capacitors [18].

5.2.2 Dielectric permittivity and dissipation factor

The dielectric permittivity (ε_r) of dielectric materials is positively correlated with the energy storage density, indicating that employing high-permittivity polymers as dielectric materials can reduce the physical size of capacitors to acquire a specific energy, which is of great significance to the miniaturization of electrical and electronic devices. The energy loss during polarization and depolarization of dielectric materials is reflected by dissipation factor (tan δ), which contributes to continuous heating inside the dielectric materials and deteriorates the energy storage capability [26]. While a series of ferroelectric polymers with high ε_r (usually >10) have been developed for room-temperature applications [27–32], they are not reliable for high-temperature applications owing to the competing mechanism between ε_r and tan δ, especially at elevated temperature, e.g., the dissipation factor of PVDF increases from ~2.9% at 30 °C to ~17.5% at 120 °C at 1 kHz [33]. Almost all high-temperature polymer dielectric materials, therefore, are linear dielectrics with relatively low ε_r and tan δ and, in turn, need further modification to achieve higher energy density.

Both dielectric permittivity and dissipation factor vary with temperature and electrical-field frequency since the polarization of dipoles and space charges is dependent on their mobility and polarization time (Figure 5.2). Polar polymers with polar groups in their main chains, e.g., PETs, are usually more susceptible to temperature and frequency, whereas dielectric performance of nonpolar polymers, e.g., BOPPs, is less dependent on these two conditions owing to lack of polar

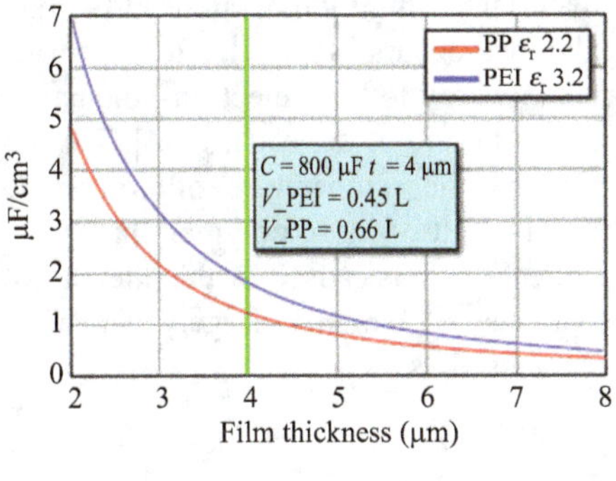

Figure 5.1 Film thickness dependence of capacitance. © 2014 Springer Nature. Reprinted, with permission, from [16]

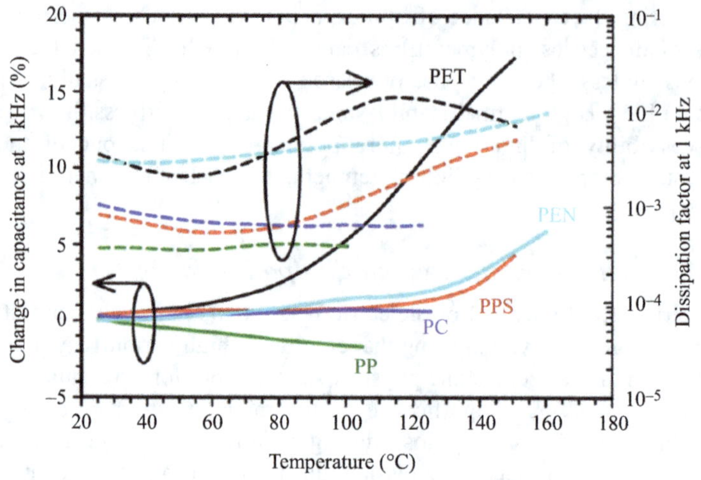

Figure 5.2 Typical change in capacitance and dissipation factor of various polymer film capacitors as a function of temperature at 1 kHz. © 2018 American Chemical Society. Reprinted, with permission, from [18]

groups [18]. According to the standard of Electronic Industries Alliance (Arlington, VA), capacitive variation of qualified X7R capacitors should be less than ±15% from room temperature over the temperature range of –55 to 125 °C [19].

5.2.3 Breakdown strength

Breakdown strength, defined by the maximum electric field the dielectric can tolerate, is one of dominant parameters for the energy storage density of dielectric materials, i.e., quadratically correlated with the maximum energy density. Among various breakdown mechanisms which have been proposed, free volume breakdown, electromechanical breakdown, and thermal breakdown most probably occur in polymer dielectric materials at high temperature [34–36].

Free volume breakdown occurs when electrons are accelerated by electric fields and obtain sufficient energy to form electron-hole pairs in the free volume of polymers, finally leading to failure of dielectrics [34]. This breakdown mechanism is dominant in polymers at the temperature of around T_g since free volume increases significantly as a result of loose polymer chains.

Electromechanical breakdown is caused by the internal Maxwell stress induced by electric fields, the process of which is largely dependent on the mechanical strength of polymers, expressed as

$$E_b = 0.606\left(\frac{Y}{\varepsilon_0\varepsilon_r}\right)^{0.5}$$

where E_b is the breakdown strength, Y is the Young's modulus, ε_0 is the permittivity of vacuum, and ε_r is the relative permittivity [35]. Young's modulus, i.e., the mechanical stiffness, of polymers commonly decreases with elevated temperature and therefore electromechanical breakdown occurs in many high-temperature conditions. It is noteworthy that dielectric failures caused by the two mechanisms mentioned above can be relieved by improving T_g of polymers; yet, this strategy may be inapplicable to thermal breakdown.

Thermal breakdown describes the process that the internal heat generation of polymers exceeds the heat dissipation to the external environment, consequently leading to the temperature rising and thermal runaway within polymers [36]. Specifically, since polymers possess relatively low thermal conductivity (0.2–0.5 W/(m·K)), Joule heat generated by power dissipation cannot dissipate fast enough to retain the thermal equilibrium within polymers and cause the temperature rising. Also, the dissipation factor and conductivity of polymers increase with elevated temperature and more Joule heat is produced. Once this positive cycle is out of control during continuous operation of capacitors, catastrophic thermal runaway may occur (Figure 5.3). Mechanism of thermal breakdown indicates that high thermal stability of high-temperature polymers does not necessarily result in high breakdown strength under high temperature unless thermal runaway, which can be averted by suppressing the power dissipation or improving the thermal conductivities of polymers is unlikely to happen.

5.2.4 Leakage current

While dielectric loss of polymers from leakage current is negligible compared to loss caused by polarization at room temperature, significant conduction loss

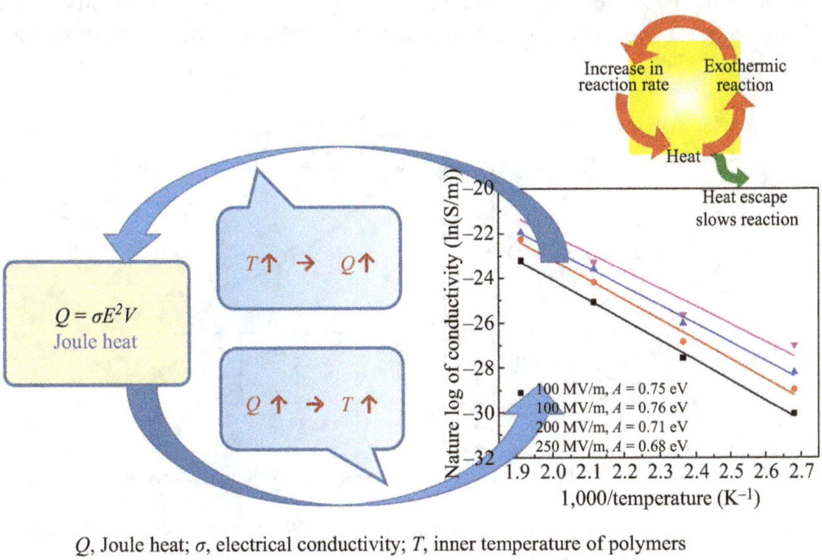

Q, Joule heat; σ, electrical conductivity; T, inner temperature of polymers

Figure 5.3 Illustration of thermal runaway

induced by many temperature-dependent conduction mechanisms [37–40], e.g., Schottky emission, thermionic-field emission, Poole–Frenkel emission, hopping conduction, and ohmic conduction, occurs in the high-temperature region. Among various conduction mechanisms in high-temperature conditions, Schottky emission and Poole–Frenkel emission are dominant since they are exponentially scaled with temperature [37].

Schottky emission (sometimes called thermionic emission), one of electrode-limited conduction mechanism, is the process that thermally activated charge carriers obtain enough energy to pass through the energy barrier at the electrode-dielectric interface and form current (Figure 5.4) [37]. Such a conduction mechanism is largely dependent on the barrier height at the electrode-dielectric interface, which is determined by not only band structure at the interface but also the applied electric field, i.e., the energy barrier height at the interface may be lowered by the image force induced by the applied electric field [41]. Poole–Frenkel emission is a bulk-limited conduction mechanism that charge carriers obtain enough energy by thermal excitation and escape from traps into conduction band of polymers to form current, which is associated with the trap energy level in polymer dielectrics.

Conduction loss is the main source of heat generation in polymers under high temperature and high electric fields, and thus restricting leakage current in a low level is critical to avoid the thermal runaway of capacitors.

5.2.5 Self-clearing behavior

Self-clearing behavior, which is essential to secure the operation of metalized film capacitors, denotes a phenomenon in which the energy released in the breakdown process melts the breakdown spot to form an open circuit rather than a short circuit, preventing the capacitors from complete breakdown [42,43]. Previous studies confirm that poor self-clearing behaviors are associated with free carbon produced

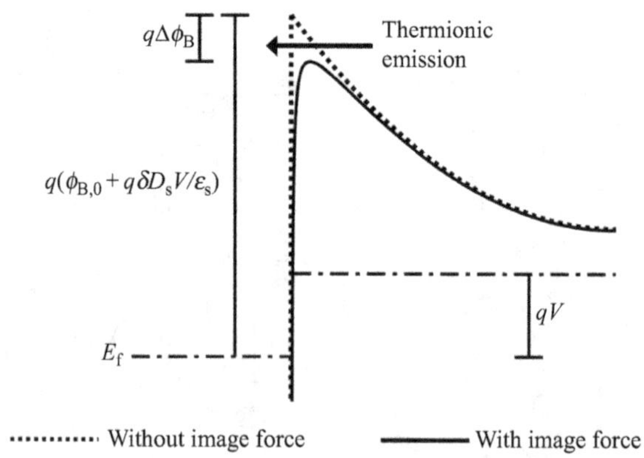

Figure 5.4 Illustration of Schottky emission. © 2019 Springer Nature. Reprinted, with permission, from [161]

during the breakdown process and polymers with high ratios of (hydrogen + oxygen) to carbon tend to show better self-clearing behaviors [44]. However, high-temperature polymers usually contain numerous high-carbon-content structures, e.g., imide ring, imidazole ring, aromatic/heteroaromatic molecular skeletons, etc., to improve their thermal stability, which consequently leads to their poor self-clearing capability. For example, BOPPs have a high ratio of (hydrogen + oxygen) to carbon of 2 and show good self-clearing capability, while the ratio of (hydrogen + oxygen) to carbon for Kapton, one of commercial high-temperature polymers, is only 0.625 and its self-clearing capability of is poor [44,45]. It is necessary to tackle the poor self-clearing behavior of high-temperature polymers before they can be practically used in film capacitors.

5.3 High-temperature polymer dielectrics

Driven by the rapid development of aerospace industry and advanced electrical and electronic systems, multiple high-temperature polymer dielectrics have been developed and achieved commercialization since their originating time of 1960s [46,47]. Commercial high-temperature polymer dielectrics usually possess high thermal stability, high breakdown strength, and good processability; yet, high dielectric constant, low dissipation factors, and high volumetric resistivity are also needed for their capacitor applications [16–20]. In this section, the properties of current high-temperature polymer dielectrics excavated for capacitor applications (Table 5.1), along with the relationship between their capacitive performance and molecular structures, are briefly discussed. Figure 5.5 illustrates the chemical structure these polymers. Most of them contain aromatic or heteroaromatic structures to secure their high thermal stability, while a few of non-aromatic polymers that rely on the high crystallinity or high-strength chemical bonds to realize high T_g are also presented.

5.3.1 Polyimide, poly(ether imide), and poly(amide imide)

These polymers are characterized by the imide ring in the backbone, which is in conjunction with the aromatic ring and yields high thermal stability, high mechanical strength, as well as excellent insulating properties [48].

Polyimides (PIs), synthesized from the condensation of dianhydride and diamine (or diisocyanate) monomers and the following imidization reaction, are one of the most widely used commercial high-temperature polymers [49,50]. Commercial PI families include general-purpose PI films (Kapton by Dupont, USA) and various functional PI films such as high-temperature resistant PI films (Upilex-S by Ube, Japan), corona-resistant PI films (Kapton CRC by Dupont, USA), and thermal conductive PI films (Kapton MT by Dupont, USA) [17].

Kapton with a high T_g of over 350 °C shows high breakdown strength (>300 MV/m), low dissipation factor (<0.1% at 1 kHz), and relatively high dielectric constant (3.2–3.4 at 1 kHz) at the temperature ranging from room temperature to 200 °C [51,52]. The dielectric performance of Kapton, however,

Table 5.1 Overview of some high-temperature polymer dielectric films

Material	T_g (°C)	ε_r (1 kHz)	tan δ (1 kHz)	E_b (MV/m)	Ref.
PI (Kapton)	360–410	2.7–3.5	0.0013–0.0026	154–303 (7.6–127 μm)	[19]
PI (Uplex-S)	285–500	3.2–3.5	0.0013–0.007	147–320 (12.5–125 μm)	[19]
PEI (Ultem)	217–247	3.15	0.0012	200 (25 μm)	[19]
PAI	>270	4.5	<0.01	>85 (6.5 μm)	[60]
PBI	427	3.3	0.001	402 (55 μm)	[19]
PBO	N/A	2.9 (1 MHz)	0.009 (1 MHz)	470 (3 μm)	[17]
PBT	N/A	2.9	0.002	N/A	[73]
FPE	330	3.2–3.5	0.0031–0.007	220–320 (100–120 μm)	[19]
BCB	>350	2.75	0.0012	300 (10 μm)	[19]
PC	145	2.8	0.001	600 (3 μm)	[160]
PET	75	3.3	0.003	580 (0.7 μm)	[160]
PPS	90	2.8	0.0005	500 (1.2 μm)	[160]
PEN	120	3.0	0.003	600 (1 μm)	[160]
PTFE	327 (T_m)	2.1	0.0005	443 (7 μm)	[122]
PFA	305 (T_m)	2.1	0.0002	350 (20 μm)	[93]
PEEK	150	3.1	0.003	150 (50 μm)	[19]
PEKK	162	2.6	0.007	84 (100 μm)	[19]
PPEK	250	3.5	0.006	470 (20–40 μm)	[99]
ArPU	N/A	4.1	0.005	680 (2.5 μm)	[100]
PEEK	150	3.1	0.003	150 (50 μm)	[139]

Note: T_g, glass-transition temperature; ε_r, relative dielectric constant; tan δ, dissipation factor; E_b, dielectric strength.

deteriorates with further increase in temperature. For example, the dissipation factor of Kapton increases from 0.1% at 25 °C and 1 kHz to 6% at 300 °C and 1 kHz and the dielectric constant decreases from 3.2 to 2.8 [53]. Moreover, Kapton has a high conduction loss of 24% under the electric field of 200 MV/m at the temperature of 150 °C, leading to a charge-discharge efficiency as low as 76% and a low energy density of only 0.44 J/cm^3 (Figure 5.6) [54]. The conduction loss even reaches 100% at 250 °C, making thermal breakdown easy to occur [54]. Kapton exhibits moisture sensitivity and relatively poor thermal-oxidative resistance due to the existence of imide ring structure and diphenylether moiety, while Upilex-S deriving from biphenyl-type monomers possesses excellent thermo-oxidative stability and lower water absorption [55]. Upilex-S shows stable electrical properties over a wide range of temperatures and frequencies, e.g., the dielectric strength of Upilex-S maintains relatively high (>200 MV/m) up to 400 °C. In addition, the dielectric constant and dissipation factor are 3.3 and 0.1% at 25 °C and 1 kHz, respectively, which remain constant at 300 °C [53,55].

One major impediment to employing PIs in film capacitors is their poor processability deriving from their insoluble and infusible nature, e.g., the manufacture of Kapton films below 12 μm thickness is difficult [56]. Poly(ether imides) (PEIs), produced by incorporating flexible ether linkages into the backbone of PIs, sacrifice some thermal stability with respect to PIs but obtain enhanced melt processability and solubility in solvents, i.e., PEI films can be processed to 5 μm thickness

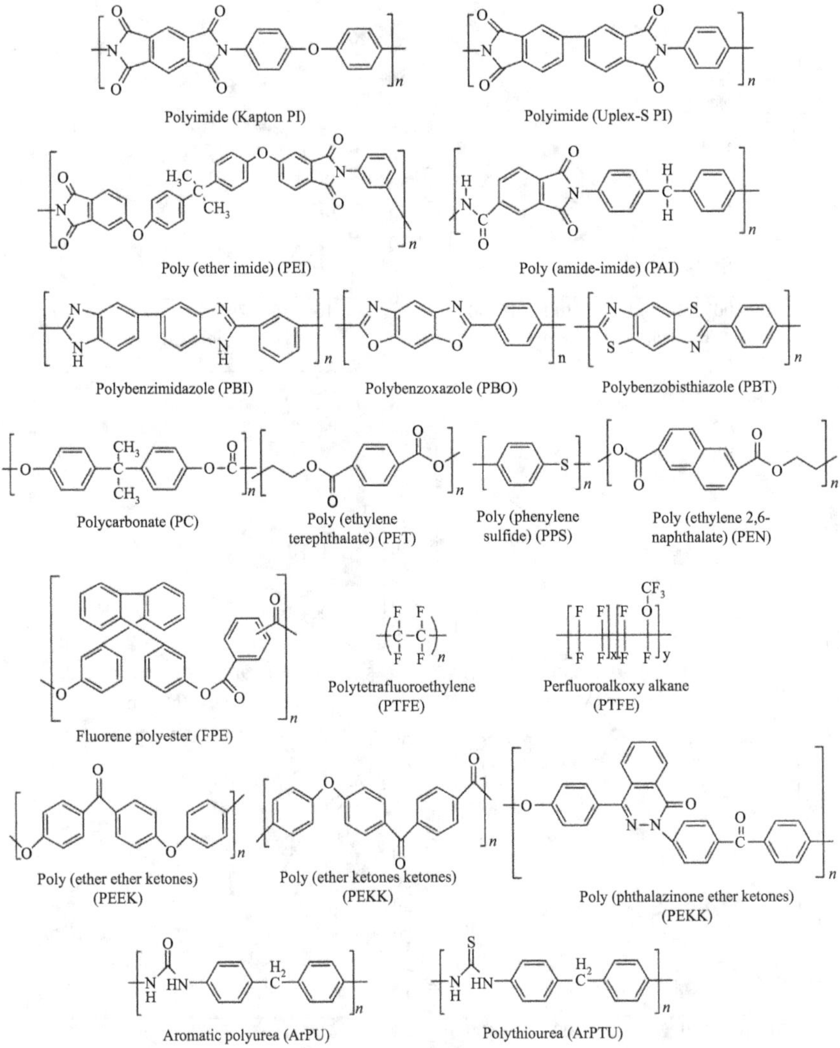

Figure 5.5 Chemical structures of high-temperature dielectric materials

by the melt-extrusion method [16,57]. One of commercial PEI trademarked as ULTEM by SABIC, which is synthesized from the disodium salt of bisphenol A and 1,3-bis(4-nitrophthalimido) benzene has a compromised T_g of 217 °C. ULTEM exhibits a dielectric constant of 3.1 and dissipation of 0.3% at the temperature from 25 to 200 °C and the frequency from 100 Hz to 10 kHz, similar to that of PIs (Figure 5.7) [11]. Notably, PEIs surpass all other commercial polymers in the energy-storage performance at 150 °C, e.g., they discharge the energy density of 0.5 J/cm^3 with a charge-discharge efficiency of 90% under the electric field of 200 MV/m and the temperature of 150 °C (Figure 5.6) [54]. Poly(amide imides)

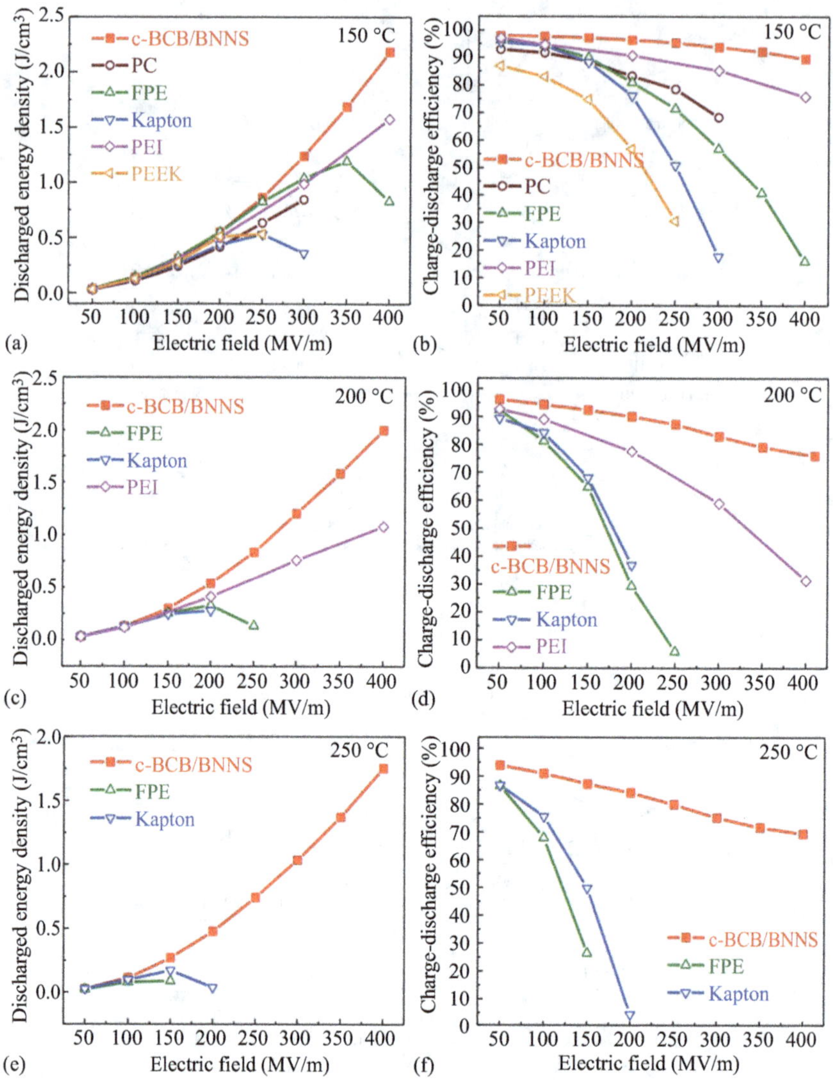

Figure 5.6 Electrical energy storage performance of various high-T_g polymer dielectrics under different temperature. © 2015 Springer Nature. Reprinted, with permission, from [54]

(PAIs) are another class of soluble imide polymer modified from PIs, which are synthesized from the condensation of 4,42-diphenylmethane diisocyanate (MDI) and trimellitic anhydride (TMA) in a polar solvent [58]. PAIs inherit the good thermal stability of PIs and can retain the mechanical stiffness up to 260 °C [59]. As for the dielectric performance, PAIs have a relatively high dielectric constant of

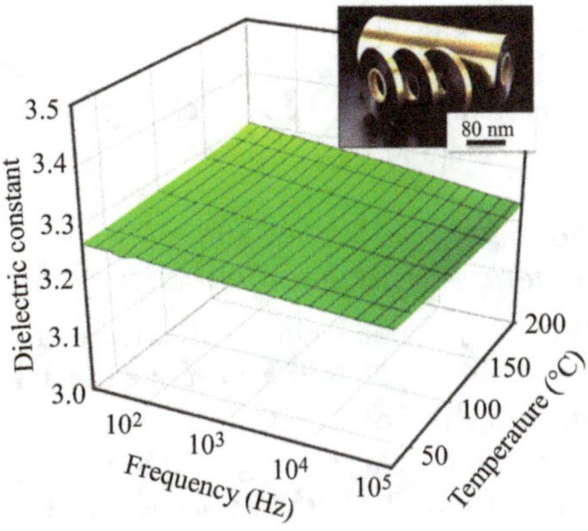

Figure 5.7 Dielectric properties of PEI. © 2014 Springer Nature. Reprinted, with permission, from [16]

4.5 at 25 °C and 1 kHz, while their relatively high dissipation factor (~1%) may limit their applications in capacitors [60].

5.3.2 Polybenzimidazole, polybenzoxazole, and polybenzobisthiazole

These polymers contain para-linked aromatic and heteroaromatic rings in their main chains, which are in conjunction with each other and induce a high mechanical strength as well as a high thermal stability.

Polybenzimidazoles (PBIs) are a type of thermoplastic with an ultrahigh T_g of over 400 °C, which are universally used in textile applications [61]. Among PBI families, poly(2,2′-(*m*-phenylene)-5,5′-bisbenzimidazole) (mPBI) synthesized from the high-temperature polycondensation of monomers of tetraaminobiphenyl and either diphenyl-isophthalate or isophthalic acid is most commercialized [62–64]. While mPBI have great chemical stability, excellent thermal resistance, and good insulating properties, it is not suitable for capacitor applications owing to its relatively high dissipation factor. The dissipation factor of mPBI is ~2% at 22 °C from 60 Hz to 100 kHz and rises sharply to ~20% as the temperature reaches 200 °C [61,65]. Moreover, the dielectric constant of mPBI is also largely dependent on the temperature, i.e., 3.4 at 25 °C and 10.4 at 300 °C [65].

Polybenzoxazoles (PBOs), which are developed by a unique spinning technology, have been commercialized since 1998, trademarked as Zylon [66–69]. Zylon is attractive for its high electrical resistivity, thermo-oxidative resistance, and breakdown strength. Polybenzobisthiazoles (PBTs) are modified from PBOs by replacing the O atoms by S atoms [68,70]. PBO and PBT have no T_g or T_m since they decompose before the state transition occurs, whereas they are able to continuously operate at 350 °C [71,72]. Both PBO and PBT have the advantages of low dielectric loss and low water absorption resulting from the lack of polar groups in their molecular structures [17]. The dissipation factor of PBO and PBT are 0.67 and

0.02% at 1 kHz and 25 °C, respectively, which increase to 4.98 and 0.25% as the temperature exceeds 250 °C [73]. Additionally, their capacitance is quite stable with respect to the temperature, i.e., the change is less than 10% over the temperature range from –55 to 300 °C [73].

5.3.3 Fluorene polyester and cross-linked divinyltetramethyldisiloxane-bis(benzocyclobutene)

Fluorene polyesters (FPEs), prepared from the reaction of fluorene bisphenol and phthaloyl chloride, are promising dielectric materials for high-temperature capacitors accounting for their high T_g of 335 °C and low dissipation factor (<0.3% at 1 kHz) at the high temperature (250 °C) (Figure 5.8) [18,73]. The dielectric constant of FPEs is similar to that of PIs but more stable under variable temperatures, e.g., the dielectric constant decreases from 3.5 at 25 °C and 1 kHz to 3.3 at 300 °C and 1 kHz [73]. At 150 °C, FPEs discharge a maximum energy density of over 1.2 J/cm^3, twice as high as PIs (Figure 5.6) [54]. Considering that thin (<5 μm) FPE films have been manufactured at the industrial scale, FPE capacitors have been available and investigated since 2011 [74,75]. Metalized FPE films exhibit a relatively high breakdown strength and a high insulation resistance, and pass the lifetime tests of 250 h at 200 °C, proving their potential for high-temperature capacitor applications [75,76].

c-BCBs are obtained from the thermally cross-linked or photo-cross-linked from the BCB monomer, which have a decomposing temperature of over 350 °C [77]. However, they begin to lose weight during continuous operation as the temperature exceeds 300 °C since they are easily oxidized under the elevated temperature [78]. The dielectric performance of c-BCB is quite stable over a large range of temperatures and frequencies, i.e., the dielectric constant of c-BCB is ~2.7 at 25 °C and 1 kHz and varies by only –2.4% as the temperature increases to 350 °C. The dissipation also remains as low as <0.2% over the temperature from 25 to 300 °C and the frequency from 1 kHz to 2 MHz [54].

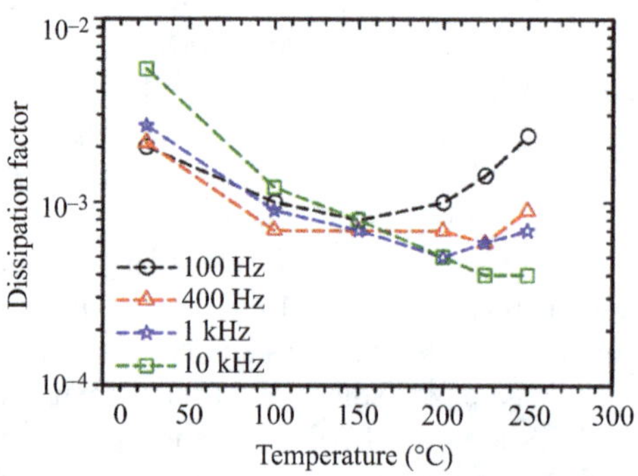

Figure 5.8 Dissipation factor of FPE as a function of temperature at various frequencies. © 2018 American Chemical Society. Reprinted, with permission, from [18]

5.3.4 Polycarbonate, poly(ethylene terephthalate), poly(phenylene sulfide), and poly(ethylene 2,6-naphthalate)

Compared to polymers containing both heterocyclic rings and aromatic rings in the skeletons mentioned above, these polymers only contain aromatic rings in their molecular chains and exhibit relatively low T_g. However, such aromatic polymer films are usually cheaper and cost-effective for general applications. Figure 5.2 shows their dielectric constant and dissipation factor as a function of temperature.

Polycarbonates (PCs), which have been employed as dielectric materials in capacitors since 1950s, are synthesized from carbonic acids and dihydric alcohols [79]. PC capacitors stand out for the low temperature-dependence of their capacitance, i.e., variation of ~3% over the temperature from 25 to 125 °C. The dissipation of PC capacitors is also as low as <0.1% from 25 to 125 °C [80,81]. Moreover, the breakdown strength of solvent-cast PC films is very high, even similar to that of BOPPs, leading to a high energy density of 9.2 J/cm^3 [16]. The operating temperature of PC films is limited to <125 °C accounting for their relatively low T_g of 150 °C [57].

Poly(phenylene sulfides) (PPSs) consisting of aromatic rings linked with sulfides are considered good substitutes for PCs as high-temperature dielectric materials [56,81]. PPSs have the dielectric constant and dissipation factor similar to PCs, while the temperature coefficient of capacitance (TCC) of PPSs is only 1.5% over the temperature from −55 to 125 °C [82]. Although PPSs have a relatively low T_g of 85 °C, they can be used at the temperature up to 150–170 °C due to the high degree of crystallinity (60%–65%) [57,82].

Poly(ethylene terephthalates) (PETs), synthesized from terephthalic acid and ethylene glycol, show similar thermal stability and dielectric constant to PCs [82]. However, PETs exhibit better self-clearing behavior than PCs attributed to the higher content of O atoms in their molecular structures. Very thin (<0.5 μm) PET films with a high breakdown strength of over 500 MV/m can be manufactured by a biaxially oriented method, which is beneficial for their capacitor applications [45,83]. The dielectric constant and dissipation factor of PETs, however, are strongly dependent on the temperature as well as the frequency due to the presence of polar groups in the backbone [24,83].

Poly(ethylene 2,6-naphthalates) (PENs) are a type of polyester synthesized from naphthalene-2,6-dicarboxylic acid and ethylene glycol, which have similar chemical structures to PETs [84]. The dielectric performance of PENs, consequently, is similar to that of PETs; yet, PEN films can tolerate higher temperature benefiting from the higher T_g [84,85].

5.3.5 Polytetrafluoroethylene and perfluoroalkoxy alkane

Such fluoropolymers are characterized by the high-strength carbon-fluorine bonds, which result in a high resistance to solvents, acids, and bases as well as an excellent thermal stability.

Polytetrafluoroethylenes (PTFEs) consist wholly of carbon and fluorine and possess a high T_g of over 260 °C. As nonpolar polymers with no molecular dipoles, PTFEs exhibit a low but stable dielectric constant of 2.1 over a wide range of temperatures and frequencies [86,87]. The dissipation factor of PTFEs remains as low as 0.05% even at the temperature of 200 °C and the frequency of 1 kHz [88]. The poor processability can be one of the major drawbacks for PTFE films employed in high-temperature film capacitors. To be specific, traditional methods to manufacture polymer films such as melt-extrusion or solvent-casting are not practical to the production of thin PTFE films, owing to the high melt viscosity deriving from the highly crystalline structures of PTFEs [89,90]. Current strategies to manufacture PTFE films, e.g., sintering after solution dispersion of PTFE powder, will introduce pinholes and faults into the films, resulting in a large reduction of breakdown strength and a low energy density [88,90].

Perfluoroalkoxy alkanes (PFAs) are copolymers of tetrafluoroethylene and perfluoroethers, which have a similar chemical structure to PTFEs but contain an oxygen atom at the branches. As a result, PFAs are able to be melt-processed into thin films at the expense of poorer thermal stability compared to PTFEs [91], i.e., the T_m of PFA is 305 °C, slightly lower than that of PTFE (327 °C). In addition, PFAs possess similar dielectric properties to PTFEs. Namely, PFAs present a low dielectric of ~2.1 and a low dissipation factor of <0.1% at 60 Hz and 22 °C, which vary slightly over a large temperature range (22–200 °C) [92]. The breakdown strength of PFAs is also temperature independent and remains high at the temperature up to 150 °C [92].

5.3.6 Polyketone

Polyketones represent a class of polymers containing ketone groups and ether groups in their backbones, whose chemical, mechanical, and electrical properties are dependent on the ratio of ketone to ether.

Poly(ether ether ketones) (PEEKs) are one of the most commercialized members in polyketone families, which are typically synthesized from the polymerization of 4,4′-difluorobenzophenone and bisphenolate salt [93,94]. While PEEKs possess high thermal resistance, good mechanical properties, and great processability, their high-temperature dielectric applications are limited by their low T_g (150 °C). Both the dissipation factor and conduction loss of PEEKs increase significantly as the temperature approaches T_g [95]. As a result, the maximum dischargeable energy density of PEEK at 150 °C is only ~0.5 J/cm^3 (with a low efficiency of 55%), which is similar to that of Kapton (Figure 5.6) [54].

Poly(ether ketone ketones) (PEKKs) are a kind of semi-crystalline thermoplastic developed by replacing some ether groups in PEEKs with ketone groups (i.e., the ratio of ketone to ether is 1:2 for PEEK while 2:1 for PEKK) [96]. Compared to PEEKs, PEKKs have a higher T_g varying from 160 to 165 °C depending on the ratio of terephthalic moieties to isophthalic moieties and the dielectric performance of PEKKs is stable over the temperature from 25 to 150 °C and a wide frequency range. PEKK films deliver a maximum energy density of over 2.4 J/cm^3 under an electric field of over 400 MV/m at 150 °C [97].

By incorporating the rigid asymmetric phenyl phthalazinone moiety into the polyketone backbone, poly(phthalazinone ether ketones) (PPEKs) have a high T_g of 250 °C [98]. PPEKs have a considerably stable dielectric constant of ~3.5 and dissipation factor of ~0.6% at 1 kHz over the temperature from 25 to 225 °C. The breakdown strength of PPEKs also decreases slightly from room temperature to 190 °C (470 to 441 MV/m). PPEKs are able to discharge an energy density of >2.5 J/cm^3 with an efficiency of >90% under the electrical field of 400 MV/m at 130 °C (Figure 5.9) [98].

5.3.7 Polyurea and polythiourea

Polyureas are a class of polymers containing two amine groups (–NH_2) joined by a carbonyl functional group (C=O) (also called polyurea linkage) in the main chains. Aromatic polyurea (ArPU) contains both polyurea linkage and aromatic rings, which is fabricated through vapor deposition polymerization of monomers of 4,4′-diaminodiphenylmethane (MDA) and 4,4′-diphenylmethane diisocyanate (MDI) [99–101]. Due to the high dipole moment (4.56 D) in the urea unit, ArPU possesses a high dielectric constant of 4.1 at 1 kHz and room temperature, which shows no variation over the temperature from –25 to 150 °C. The dissipation factor of ArPU is as low as 0.5% at 1 kHz and 25 °C and increases to 1% as the temperature reaches 150 °C. Notably, ArPU delivers a pretty high and stable energy density over a large temperature range from 25 to 180 °C (e.g., 4.6 J/cm^3 under the electric field of 491 MV/m) [99]. By substituting the CH_2 group in ArPU for the polar ether group, poly(arylene ether urea) (PEEU) with higher dielectric constant (4.7 at 1 kHz) and changeless dielectric loss (1.1% at 1 kHz) is synthesized from the thermal polycondensation from (*m*-phenylenedioxy) dianiline and diphenyle carbonate (Figure 5.10) [102]. Both the dielectric constant and the dissipation factor are relatively stable over the temperature range from 25 to 200 °C. As a result,

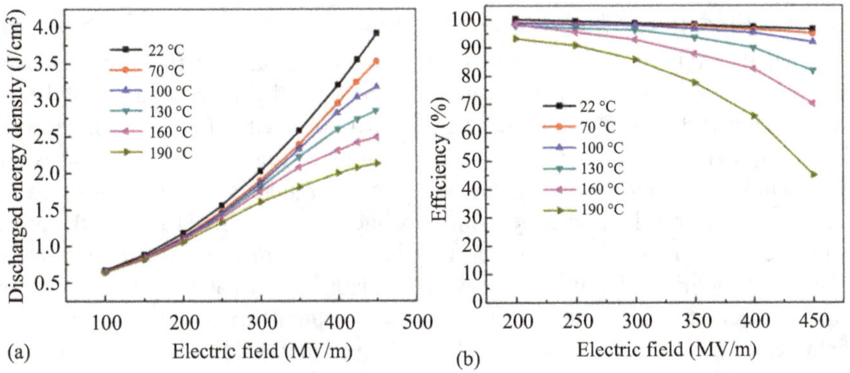

Figure 5.9 (a) Discharged energy density of PPEK as a function of temperature and the applied field. (b) Efficiency versus temperature under different electric fields. © 2010 American Chemical Society. Reprinted, with permission, from [98]

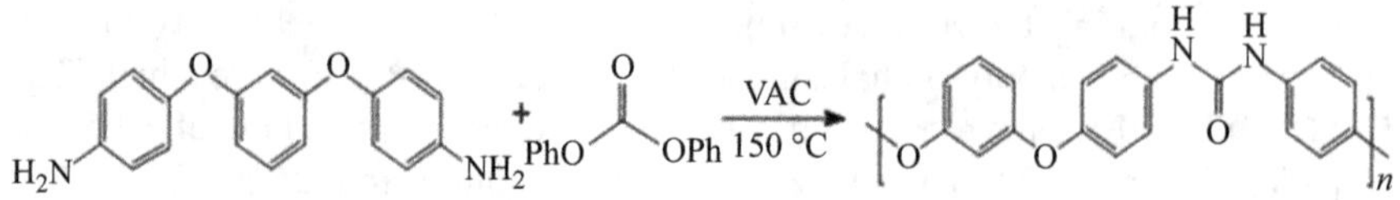

Figure 5.10 Schematic of synthesis and chemical structures of poly (arylene ether urea). © 2015 AIP Publishing. Reprinted, with permission, from [102]

PEEU films discharge a high energy density of 9 J/cm^3 and maintain a high breakdown strength of >600 MV/m at 120 °C.

The aromatic polythiourea (ArPTU) has a similar structure to ArPU and the replacement of oxygen atom by sulfur atom increases the dielectric constant to 4.4 at 1 kHz and 25 °C at the expense of slightly enhanced dielectric loss [100]. High-temperature energy storage performance of ArPTU has not been investigated before.

5.4 Modified high-temperature polymers

Although there are many high-T_g dielectric materials from either commercial or non-commercial sources, few of them meets all the requirements needed for high-temperature capacitor applications. This is the main reason why few of high-temperature capacitors have been commercialized currently. To address this issue, great efforts have been paid to modify the molecular structures of these polymers to make them more suitable for capacitor applications.

5.4.1 Modified polyimides

5.4.1.1 Fluorinated polyimides

Fluorinated polyimides are a class of polyimides containing fluorine atoms in their repeating units, which are usually synthesized by incorporation of hexafluoroisopropylidene (–C[CF_3]$_2$–) or pendant trifluoromethyl (–CF_3) into the main chains of polyimides [103–106]. The incorporation of fluorine atoms increases the free volume in polymers due to the larger size of fluorine atoms compared to hydrogen atoms, resulting in better solubility and processability [107]. On the other hand, the high-energy C–F bonds ensure the high thermal and thermo-oxidative stability of fluorinated polyimides [106]. Notably, fluorinated polyimides with fluorine symmetrically located in the backbone exhibit lower dielectric constant compared to their non-fluorinated analogues, since increased free volume in polymers would decrease the polar groups in a unit volume and the electronic polarizability of the C–F bond is smaller compared to the C–H bond [107,108]. When the fluorine is positioned non-symmetrically; however, the low-frequency dielectric constant may increase owing to the high electronegativity of fluorine atoms [108].

For example, Kim *et al.* [109] prepared fluorinated polyimides with various F/C ratios under different fluorine pressure between 0 and 50 kPa, finding that the dielectric constant decreased from ~3.1 to ~2.7 with increasing F/C ratios in the polyimides. Chisca *et al.* [110] synthesized a serious of aromatic polyimide films by a two steps polycondensation reaction, revealing that methyl groups or isopropylidene or hexafluoroisopropylidene units all induced a great solubility of polyimides while groups with ether linkages reduced the T_g to 200–228 °C and hexafluoroisopropylidene units kept the T_g as high as 278–287 °C. Moreover, the substitution of hydrogen atoms by fluorine atoms is confirmed to decrease the dissipation factor of polyimides [110,111]. SIXEF-44 synthesized from 2,2-bis(3,4dicarboxyphenyl) hexafluoropropane dianhydride (6FDA) and 2,2-bis(4-aminophenyl) hexafluoropropane diamine (4,4′-6F diamine) has a high T_g of 323 °C and exhibits a low dissipation factor of ~0.1% over the temperature from –25 to 250 °C at 1 kHz, which is much lower than Kapton [73]. The perfluoro polyimide (PFPI) synthesized from 2,2-bis(4,4-aminophenoxy)-phenyl-hexafluoropropane (4-BDAF) and pyromellitic dianhydride (PMDA) also has a dissipation factor as low as ~0.1% from 25 to 300 °C [112].

5.4.1.2 High-permittivity polyimides

Many researches have focused on improving the polarization of polyimides because the energy density stored in polymer films increases linearly with the dielectric constant. Incorporation of polar groups, e.g., nitrile groups (–CN), bipyridine units and sulfonyl groups (SO_2), into the backbones or branches of polyimides is a common approach to achieve higher dielectric constant; yet in turn, this method usually leads to a large increase in the dissipation factor due to the enhanced polarization. More importantly, the dielectric performance of polymers with a large number of polar groups in their main chains are commonly largely dependent on temperatures and frequencies, which is undesirable in capacitor applications.

Wang *et al.* [113–115] developed a series of polyimides containing varied amount of nitrile groups, i.e., 0, 1, and 3, respectively, in each repeating unit and compared their thermal, mechanical as well as dielectric properties. The synthetic process was based on the condensation polymerization of diamines with nitrile moieties and different kinds of dianhydrides including 2,2′-bis(3,4-dicarboxyphenyl)-1,1,1,3,3,3-hexafluoropropane dianhydride (6FDA), 4,4′-oxydiphthalic dianhydride (OPDA), 4,4′-benzophenonetetracarboxylic dianhydride (BTDA), and pyromellitic dianhydride (PMDA) followed by the imidization reaction. It was found that incorporating nitrile groups into the backbone promoted the T_g and the PI with one nitrile group exhibited higher thermal stability than the PI with three nitrile groups. The dielectric constant of the PI with polar –CN groups increased significantly, which was 3.1 for 0 CN samples, 4 for 1 CN samples, and 4.6 for 3 CN samples at 190 °C and 1 kHz, respectively. But, it varied a lot by ~13% for 1 CN samples and 21.6% for 3 CN samples at 1 kHz over the temperature from 25 to 190 °C (Figure 5.11) [115]. The dissipation factor of polyimides containing 1 CN group also increased sharply from 0.2% at 25 °C and 1 kHz to 3% at 190 °C and 1 kHz. The energy storage performance of nitrile-containing PIs was in comparison

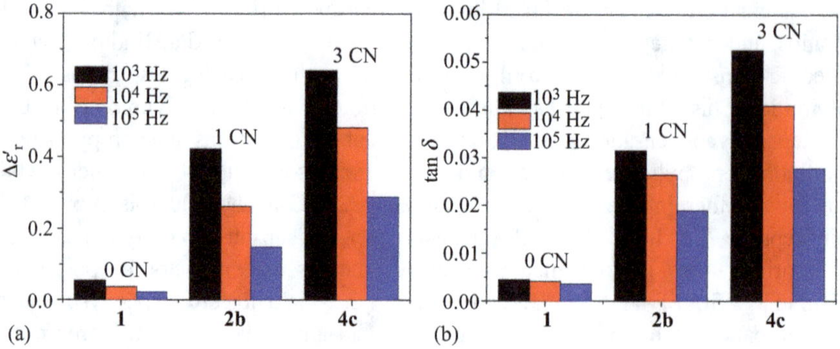

Figure 5.11 (a) Increase of the real part of permittivity and (b) tan δ at 190 °C at 10^3, 10^4, and 10^5 Hz for samples. © 2014 The Royal Society of Chemistry. Reprinted, with permission, from [115]

with that of commercial high-temperature polymer films including Kapton and Ultem under the temperature of 190 °C and the electric field of 100 MV/m, where nitrile-containing PIs discharged more energy density with a slightly lower charge-discharge efficiency.

While they brought some problems in the meantime, researchers also utilized many other polar groups to improve the dielectric responses of PIs. For example, the addition of bipyridine units into the backbone of PIs improved the dielectric constant to 5.5–7.2 but induced a high dissipation factor of 4% owing to the difficulty of frictionless dipole switching [116,117]. Ma *et al.* introduced polar groups of –O– and –C(=O)– into the skeleton of PI by the polymerization of commercial polyether aliphatic diamines and dianhydrides monomers, which enhanced the dielectric constant and energy density of PI films at the expense of much compromised thermal stability (T_g = 53–174 °C) [118]. More recently, Tong *et al.* developed a new PI-possessing high dielectric constant, low dissipation factor, and excellent heat resistance simultaneously by introducing strong polar sulfonyl groups (–SO_2–) together with flexible linkages between benzene groups into the main chains of PIs [119]. One of the samples with a T_g as high as 290 °C exhibited a high dielectric constant of 5.98 and a low dissipation factor of 0.37% at 1 kHz and room temperature attributed to the high dipole moment (4.30 D) of sulfonyl groups and the effective reduction of the long-range dipolar coupling among strong polar structures. Both the dielectric constant and dissipation factor were almost constant over the temperature from 25 to 150 °C, where the breakdown strength reduced slightly from 536 to 468 MV/m. Consequently, the dischargeable energy density and charge-discharge efficiency of the PI film were as high as 7.04 J/cm^3 and 91.3% under an electric field of 500 MV/m, 2.3 times that of BOPP films.

5.4.2 Modified fluorine resin

The potential high thermal stability induced by the high-energy C–F bonds and the potential high dielectric constant derived from the strong electronegativity of fluorine atoms make fluorine resin attractive for high-temperature capacitor applications.

To address the poor quality of PTFE films produced by traditional method, Gore developed thin PTFE films (<10 μm) with better processability presumably by melt-extrusion [120,121]. The 6-μm Gore PTFE film exhibited a large increase in the breakdown strength, i.e., 756.2 MV/m of Gore PTFEs versus 443 MV/m of the cast ones at 40 °C. However, the breakdown strength declined sharply to 270 MV/m as the temperature reached 250 °C. The temperature and frequency dependence of capacitance and dissipation factor were also studied at the capacitor level, showing that the capacitance of the capacitor with 6-μm Gore PTFE film decreased by ~7% and the dissipation retained relatively stable (~0.1%) at 1 kHz from 25 to 250 °C.

Introducing large monomers such as HFP into the PTFE, which effectively declines the melting temperature and crystallinity, is another potential approach to enable the manufacture of PTFE films by melt-extrusion methods [123]. Zhang *et al.* prepared copolymer P(TFE-HFP) with a molar ratio of HFP/TFE of 13/87 mol%, finding that the decomposition temperature of P(TFE-HFP) was reduced from 500 °C of PTFE to about 450 °C and the crystallinity decreased to 61.6% [124]. In terms of the dielectric performance, P(TFE-HFP) films exhibited a slightly higher (10%) dielectric constant than BOPP and a rather low dissipation factor (<0.05%), which remained stable over a wide range of temperatures and frequencies. Moreover, the maximum dischargeable energy density of P(TFE-VDF) reached 6 J/cm^3 and the charge-discharge efficiency retained >80% at the temperature of 110 °C.

Capitalized on the idea of combining the high dielectric constant of PVDF and the high thermal stability of PTFE, Zhang *et al.* [124] developed P(TFE-HFP-VDF) (THV) terpolymers with different ratios of TFE/HFP/VDF (THV-A: 59/22/19, THV-B: 67.5/17.5/15, THV-C: 76/13/11), where HFP was added as minor component to improve the processability of the films (Figure 5.12). Direct copolymerization of monomers was chosen instead of blending PVDF with PTFE, which put the monomers in the same polymer chains without causing the phase separation. It was found that T_m of THV increased from 174 to 228 °C as TFE content increased from 59 to 76 wt.%, all of which were clearly higher than that of PVDF. The research conducted by Lovinger proved that P(TFE-VDF) copolymers with TFE content higher than 40% generally exhibited higher T_m than PVDF [125]. As for the dielectric constant, THV-A and THV-B had relatively high dielectric constant of 6–8, while the dielectric constant of THV-C reduced significantly to 4.4 at 25 °C and 1 kHz. Notably, the dielectric constant of THV-C was still approximately two times that of PTFE homopolymers and BOPP. As a result, THV-C delivered a much higher energy (4.6 J/cm^3) than BOPP (2.8 J/cm^3) with a similar charge-discharge efficiency of 98% under the electric field of 500 MV/m, indicating that it

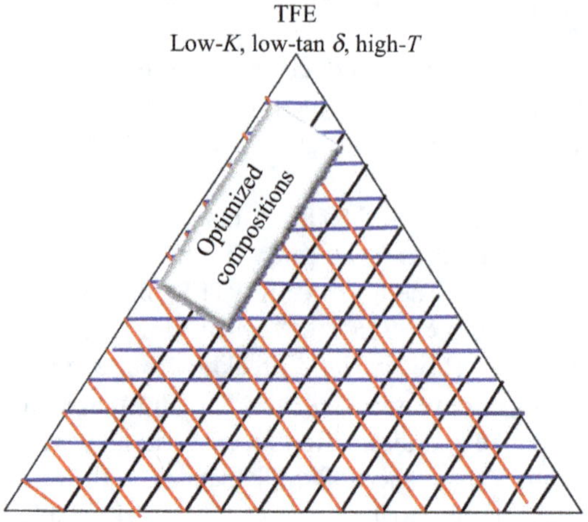

Figure 5.12 Molecular design of the high-temperature high energy density terpolymer dielectric materials. © 2012 IEEE. Reprinted, with permission, from [124]

is a good alternative to BOPP as dielectric materials in film capacitors. On the other hand, THV-C had a stable dissipation factor of $<1\%$ over the temperature from 0 to 150 °C at 1 kHz and a reasonably high breakdown strength of 603 MV/m at room temperature.

5.4.3 Modified heteroaromatic polymers

Heteroaromatic dielectric polymers are attractive for ultrahigh-temperature applications since they usually have ultrahigh T_g induced by the strong π conjunction. To further improve their thermal or dielectric properties, Venkat *et al.* have prepared a series of polymers based on PBO, PI, and FPE, and evaluated their dielectric performance at variable temperatures (Figure 5.13) [125].

The fluorinated PBO (OH-6F-PBO/12F-PBO copolymer) was synthesized by the copolymerization of hydroxyphenyl-6F-PBO (OH-6F-PBO) and 12F-PBO with the ratio of 1:1 [126]. OH-6F-PBO/12F-PBO copolymer had an extremely high degradation temperature of over 500 °C in both nitrogen and air atmosphere. The capacitance of metalized OH-6F-PBO/12F-PBO copolymer films increased by 2–3% as the temperature rose from room temperature to 250 °C, which showed a hysteresis effect during a complete heating and cooling process, i.e., the capacitance of this film cannot return to its original value over a heating and cooling cycle. A dielectric constant (2.95) slightly higher than that of commercial PBO was derived from the measured capacitance. Additionally, OH-6F-PBO/12F-PBO copolymer exhibited a stable dissipation factor of 1.4%–1.7% in the 25–250 °C range.

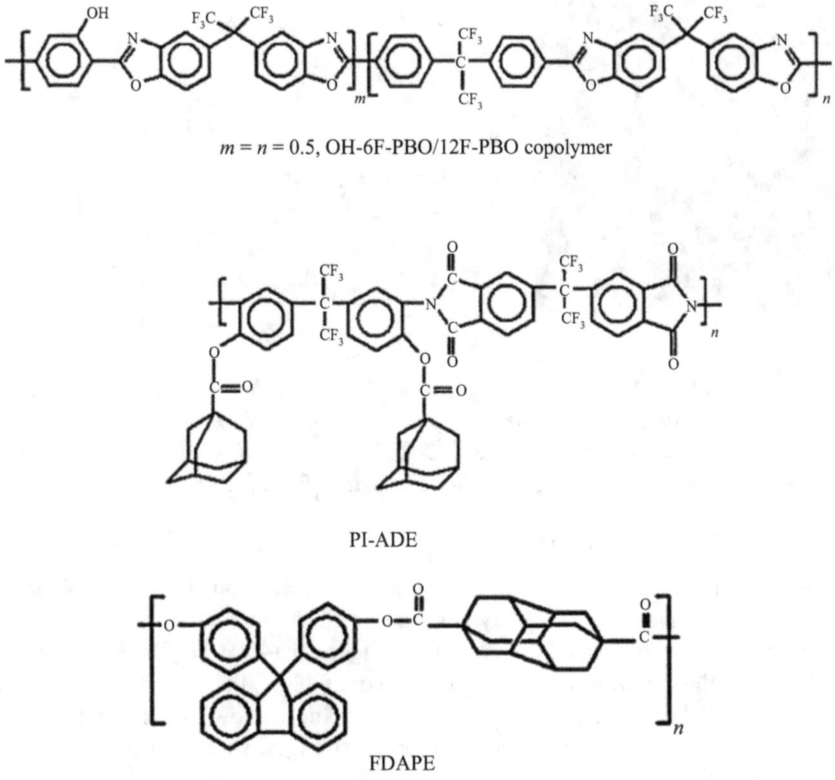

Figure 5.13 Molecular structures of modified heteroaromatic polymers. © 2009 Elsevier B.V. Reprinted, with permission, from [125]

The fluorinated PI (PI-ADE) synthesized from the post-polymer reaction of a hydroxy polyimide with adamantane-1-carboxylic acid chloride had a T_g of 305 °C and degraded at 350 °C in the air. The capacitance of PI-ADE films only varied by ~1.8%–2.5% from 25 to 250 °C, which was much more stable than that of Kapton. The capacitance, however, decreased more sharply as the temperature exceeded 250 °C due to the relatively low T_g of PI-ADE. Moreover, the dissipation factor of PI-ADE retained low (0.5–0.8%) at 1 kHz from 25 to 250 °C.

To further improve the T_g of FPE, Venkat synthesized FDAPE, including more aromaticity and heterocyclic rings in their molecular structures. FDAPE consists of a FPE group and a 4,9-diamantyl group in its repeating unit, resulting in a high T_g of over 450 °C. FDAPEs exhibited a rather low dissipation factor (<0.4%) and an extremely stable dielectric constant at 10 kHz over the temperature from 25 to 350 °C, which made them attractive candidates for high-temperature dielectric materials.

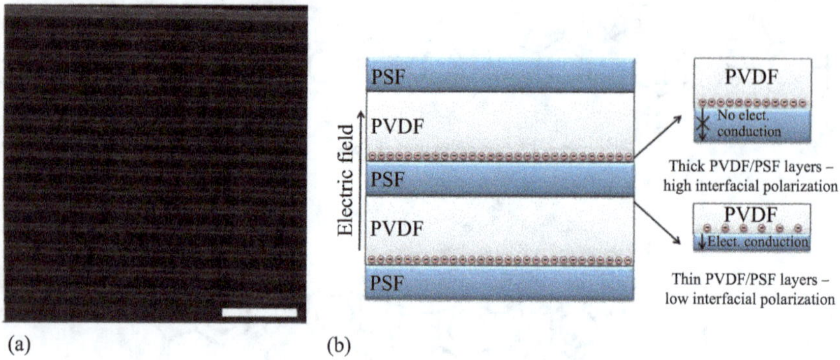

Figure 5.14 (a) SEM images of multilayer films. (b) Schematic of the interfacial polarization from both impurity ions and free electrons for the PSF/PVDF multilayer films with either thick or thin PSF layers. © 2013 Elsevier. Reprinted, with permission, from [129]

5.4.4 Multilayer polymers

By utilizing the state-of-the-art coextrusion system developed by Case Western Reserve University, multilayer polymer films with the individual layer as thin as 10 nm have been fabricated successfully (Figure 5.14(a)) [127]. Compared to single-layer films, multilayer films composed of high dielectric constant, high conductivity PVDF-based polymers and high breakdown, low-loss linear dielectric polymers tend to exhibit enhanced breakdown strength, superior energy density, and suppressed hysteresis loss [128].

According to the research conducted by Zhu *et al.* [128], there exist some unique phenomena in multilayer films. For example, the large contrast of dielectric constant between two types of films can induce the Maxwell–Wagner–Sillars (MWS) interfacial polarization, which results in the accumulation of charges and ions at the interface (Figure 5.14(b)). These interfacial charges can reduce the local electric fields and trap electrons injected from the metal electrodes, decreasing the charge conduction and improving the breakdown strength. In addition, it is found that the hysteresis loss of multilayer films decreases with the increase in layer numbers as the thickness of individual layer declines to nanoscale. This effect, usually called nanoconfinement effect, ascribes to the suppressed ion migration in thinner high conductivity layers, which can be understood by both experiment and theoretical calculation. However, ultrathin (<100 nm) individual layer may not provide sufficient ions to form interfacial polarization, in turn, leads to increased conductivity and reduced breakdown strength [129]. Notably, the electrical properties of multilayer films are closely related to the interfacial adhesion. Namely, local breakdown of multilayer films may generate layer lamination adjacent breakdown spots, where further breakdown process is likely to occur owing to the low breakdown strength of air [128].

Although various multilayer systems have been established [129–132], only the multilayer film consisting of polysulfone (PSF) and poly(vinylidene fluoride)

(PVDF) is feasible to be employed in high-temperature capacitors owing to the high T_g of PSF. The PSF/PVDF multilayer film investigated was about 12-μm thick with 33 alternating layers [132]. In terms of the components, PVDF ($T_m = 175$ °C) had a high dielectric constant ($\varepsilon_r = 12$) and high hysteresis loss due to the presence of ferroelectric domains, whereas leaner dielectric PSF ($T_g = 185$ °C) had a relatively low dielectric constant ($\varepsilon_r = 3$) and high breakdown strength. Based on the bilayer model [133], the electric field applied in PSF layers is four times that in PSF layers in multilayer films. Consequently, the hysteresis loss of PVDF can be suppressed because of the low electric field and the high breakdown strength of PSF ensures the overall dielectric strength of multilayer films. Driven by multiple phenomena mentioned above, the PSF/PVDF multilayer film retained a high Weibull breakdown strength of over 500 MV/m from 25 to 150 °C. While hysteresis loss of the multilayer film was relatively low at room temperature, it increased significantly with elevated temperature. It has been reported that manipulate the crystal orientation by introducing confined crystallization is an effective method to further reduce the hysteresis loss [134]. Remarkably, current studies on multilayer polymer films focus mainly on room-temperature applications, the synthesis as well as dielectric performance of high-temperature multilayer polymer films need to be further investigated.

5.5 High-temperature composites

Some common limitations of polymers such as low dielectric permittivity, low Young's modulus, and low thermal conductivity have hindered the further development of high-performance polymer dielectric materials by just modifying the molecular structures of polymers. In contrary, while ceramic materials suffer from high loss, low breakdown strength and poor processability, they have superior thermal stability and mechanical strength, along with high dielectric constant. To combine the advantages of polymers and ceramics as well as overcome their limitations, multiple nano-sized ceramic fillers have been introduced into polymer matrix to form nanocomposites [135–137]. Apart from improving the dielectric constant, suppressing the dielectric loss is another major issue for high-temperature nanocomposites since both polarization loss and conduction loss increase sharply at elevated temperature. More recently, it is found that a new strategy by coating a nanoscale layer onto the surface of polymers may effectively suppress the charge injection and conduction loss. In this section, we will make a summary of current high-temperature nanocomposites and put forward some potential routes to high-performance nanocomposites under high-temperature conditions.

5.5.1 Nano-doped composites

5.5.1.1 High-permittivity nanocomposites

High permittivity is essential for achieving high energy density of dielectric materials. Numerous researches have confirmed that introducing high permittivity nanoparticles (e.g., $BaTiO_3$, $Ba_xSr_{1-x}TiO_3$, $CaCu_3Ti_4O_{12}$ (CCTO), $Pb(Zr,Ti)O_3$

(PZT), and TiO_2) into polymers is able to promote the dielectric constant of nanocomposites [136]. Meanwhile, this method can produce some problems, which may be not significant at room temperature but need to be thoughtfully considered at elevated temperature. For example, polymers and nanoparticles cannot closely attached in nanocomposites due to the large difference of physical and chemical properties. Such a phase separation may bring about a large amount of free volume into the nanocomposites and provide migration path for charge carriers, leading to increased leakage current. Moreover, the introduction of high-conductivity nanoparticles will naturally generate more space charges to form current in nanocomposites. The enhanced leakage current may be negligible at room temperature owing to the low mobility of charge carriers but it becomes the main source of dielectric loss under high temperature. Therefore, the doping content of nanofillers for promoting high-temperature dielectric responses need to be carefully controlled to avoid unexpected increase in dielectric loss.

By incorporation of $BaTiO_3$ nanoparticles into PI, the dielectric constant at 25 °C and 1 kHz was promoted from 3.1 of PI to 12.5, 18.5, 23.5, and 33 of nanocomposites with the $BaTiO_3$ content of 5, 10, 15, and 20 vol.%, respectively, which led to a large enhanced energy density [138]. The dielectric constant and energy density were relatively stable from 20 to 120 °C. However, the breakdown strength reduced sharply due to the addition of numerous nanoparticles, especially at elevated temperature. The dielectric loss of nanocomposites under high temperature and high electric fields was not presented in this study. An effective method to realize high-permittivity nanocomposites with relatively low content of nanofillers is to introduce 1-D materials such as nanorods, nanofibers, and nanowires into polymer matrix [139–141]. Compared to nanoparticles, 1-D nanofillers possess a larger lattice volume and a higher aspect ratio in lattice constant, which can induce a higher ion polarization and a higher dielectric permittivity [142]. Moreover, the relatively low surface energy derived from the small specific surface of 1-D nanofillers mitigates their agglomeration in polymer matrix, significantly reducing the unexpected dielectric loss. Wang *et al.* [143] prepared PI-based nanocomposites with $BaTiO_3$ nanoparticles and $BaTiO_3$ nanowires using an in situ dispersion polymerization method, respectively, proving that the addition of nanowires improved the dielectric constant more significantly than the addition of nanoparticle, e.g., 2 vol.% of nanowires and 10 vol.% of nanoparticles led to a similar improvement in the dielectric constant of PI. However, the nanocomposite with nanowires exhibited higher conductivity than the one with the same content of nanoparticles since nanowires were easier to reach the percolation threshold than nanoparticles. Notably, the dielectric constant and dissipation factor of PI-$BaTiO_3$ nanowires were stable over the temperature from 25 to 150 °C [144]. Improving the compatibility between nanofillers and polymer matrix by modifying the surface of nanofillers is supposed to be a candidate for further decreasing the dielectric loss stemming from the introduction of nanofillers. Although many surface-modification technologies have been developed for room-temperature nanocomposites [145–148], their contributions to high-temperature dielectric materials need to be further discussed.

More recently, it was reported that considerable enhancement in dielectric constant with minor influence on dielectric loss may be achieved by the introduction of a small amount of ceramic nanofillers, regardless of their dielectric constant (Figure 5.15(a)) [149]. For example, both the PEI nanocomposite with 0.82 vol.% of Al_2O_3 (50 nm in diameter) and the nanocomposite with 0.8 vol.% of $BaTiO_3$ (50 nm in diameter) nanoparticles exhibited ~55% higher dielectric constant than the pristine PEI. Owing to the low doping content of nanofillers, the dielectric loss retained low as temperature exceeded 150 °C (Figure 5.15(b)). This phenomenon may be attributed to the multi-layered interfacial region formed between nanoparticles and polymer matrix, where the inner layer has a low dielectric constant since the orientation of dipoles is restricted by chemical bonding, and the outer layer has a high dielectric constant induced by more active dipoles. In this regard, the nanocomposites with optimal content of nanoparticles contain most polarized regions between each pair of nanoparticles, yet the regions disappear with higher or lower content of the nanofillers.

5.5.1.2 Low-loss nanocomposites

Due to various temperature-dependent conduction mechanisms, largely enhanced conduction loss is considered the major challenge polymer-based dielectric materials are faced with under high-temperature conditions. Incorporation of excellent insulating materials into polymers may significantly suppress the conduction loss and improve the insulating performance.

The introduction of a low fraction of Al_2O_3 (~0.5 vol.%) nanoparticles into THV reduced the conduction (>40 MV/m) at high temperature (>80 °C) by over two orders of magnitude, which may result from reduced mobile carrier concentration and enhanced trap depth [150]. Specifically, the structure changes in numerous interface regions introduced by nanofillers could increase the density and

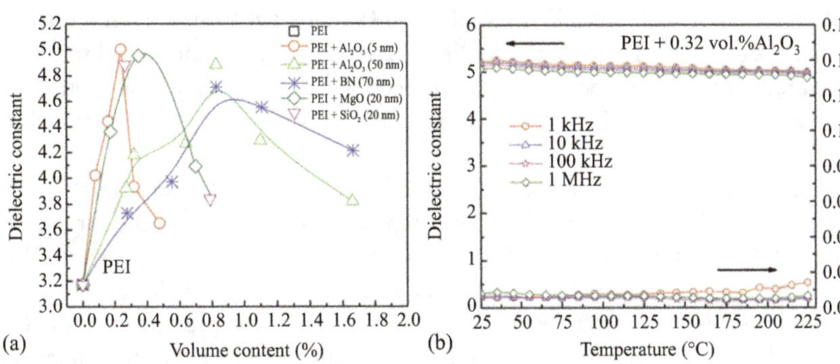

Figure 5.15 (a) Summary of dielectric constants (at 1 kHz) of PEI nanocomposites with different nanofillers. (b) Dielectric properties versus temperature of the PEI/alumina (20 nm size) nanocomposites with 0.32 vol.% alumina loading at different frequencies. © 2017 The Royal Society of Chemistry. Reprinted, with permission, from [149]

depth of the charge carrier trapping sites. The simulation results identified the effectiveness of the introduction of Al_2O_3 nanoparticles and the hopping conduction mechanism occurring in THV. Notably, the conductivity of THV/Al_2O_3 nanocomposites retained constant from room temperature to 90 °C, while that of pristine THV rose remarkably with elevated temperature.

Compared to nanoparticles, high-aspect-ratio (2-D) insulating dopants such as nanosheets are more attractive in suppressing the leakage current and improving the breakdown strength of nanocomposites for two reasons. First, 2-D nanofillers can prolong the path length as well as provide more traps and scattering centers for charge carriers [151]. Second, they work as physical barriers in polymer matrix, preventing the development of electrical trees. The c-BCB-based nanocomposite with 10 vol.% of boron nitride nanosheets (BNNSs) prepared from liquid-phase exfoliation of hexagonal boron nitride (*h*-BN) powders showed a >50% improvement in breakdown strength compared to neat c-BCB, i.e., 403 MV/m for c-BCB/BNNS and 262 MV/m for c-BCB [54]. Meanwhile, the conductivity of the c-BCB/BNNS nanocomposite was as low as 9.2×10^{-14} S/m, which was more than an order of magnitude lower than that of c-BCB. Under 200 MV/m and 150 °C, consequently, the charge-discharge efficiency of the nanocomposite retained a high value of ~97%, similar to that of BOPP film capacitors used in HEVs under the same electric field. Yet, the nanocomposite delivered a 40% higher energy density than BOPP because of the higher dielectric constant. Furthermore, the c-BCB/BNNS nanocomposite discharged a maximum energy density of 2.4 J/cm^3 with an efficiency of >90% at 150 °C and 400 MV/m, surpassed all the commercial polymer films (Figure 5.6). In terms of the dielectric performance, the variation in dielectric constant was only 1.3% and the dissipation factor increased slightly from 0.09 to 0.13% at 10 kHz over the temperature from room temperature to 300 °C. Intriguingly, BNNSs are not only great insulating materials but also excellent thermal conductive materials, and in turn, the introduction of BNNSs promoted the thermal conductivity from 0.2 W/m·K of the c-BCB to 1.8 W/m·K of the c-BCB/BNNS nanocomposite. The enhanced thermal conductivity has alleviated the inner heat accumulation and temperature rise, protecting the nanocomposite from thermal runaway. The effect of BNNSs is also identified by the phase field modeling [152].

To improve the properties of the interfacial region between BNNSs and polymers, Xu *et al.* [153] developed nanocomposites by anchoring poly (aryl ether sulfone) (DPAES) on the surface of functionalized boron nitride nanosheets (BN-BCB) (Figure 5.16). The nanocomposite delivered an energy density of 2.7 J/cm^3 with the charge-discharge efficiency above 90%, higher than previous works. The suppressed dielectric loss may be attributed to the nanoconfinement effect between BNNSs and DPAES, which restricts the molecular segment and decreases the charge hopping. On the other hand, the covalent bonds between BNNSs and polymers provide deep traps for charge carriers, reducing the leakage current. Recently, the dielectric performance of nanocomposites was further promoted by employing 2-D Al_2O_3 nanoplates (Al_2O_3 NPLs), which have higher dielectric constant and wider bandgap than BNNSs as nanofillers [154]. The addition of

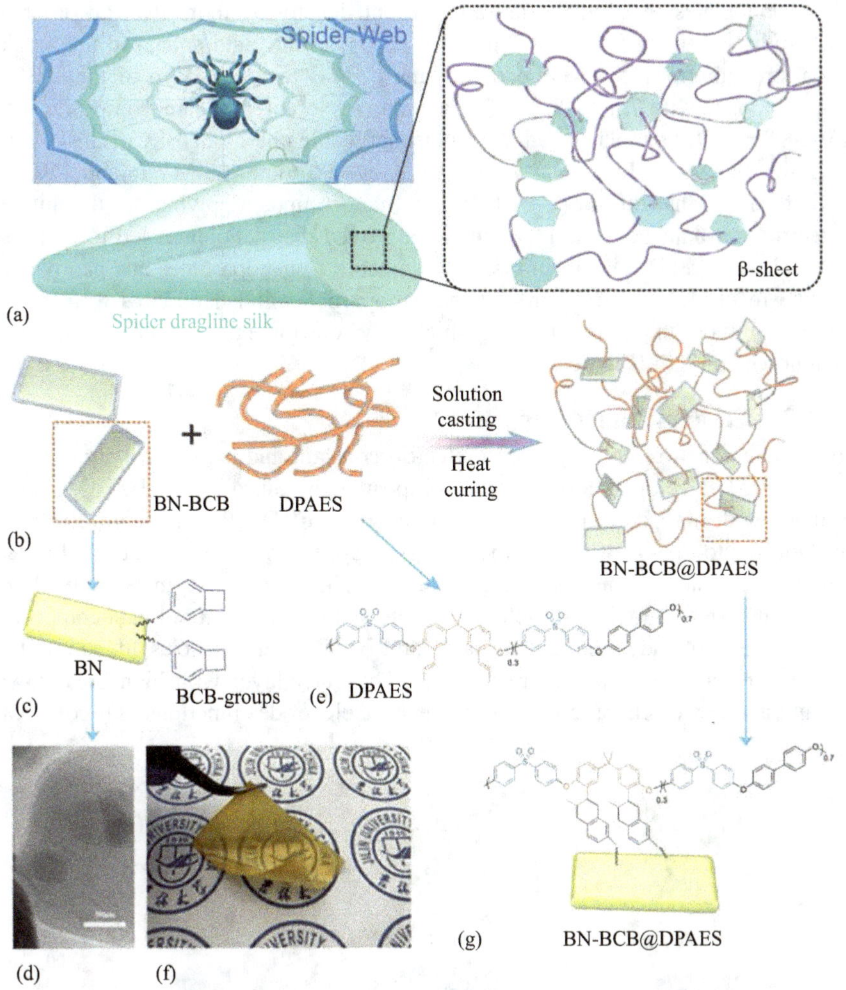

Figure 5.16 Bioinspired design of the polymer composite films for high-temperature electric energy storage. (a) Spider silk and the hierarchical microscopic structure of the spider silk. The nanoconfinement effect of the multiple chemical bonds between the crystal protein nanosheets and the polymer matrix makes the spider silk exhibits excellent mechanical properties. (b) Schematic of preparation process of BN-BCB@DPAES films. (c) Schematic of BCB group functionalized BN. (d) TEM images of BN-BCB fillers. (e) Chemical structure of polymer matrix (DPAES). (f) A digital photo of BN-BCB@DPAES film. (g) The chemical structure of BN-BCB@DPAES. © 2019 Wiley-VCH. Reprinted, with permission, from [153]

Al_2O_3 nanoplates reduced the leakage current density by more than an order of magnitude, i.e., from 5.5×10^{-8} to 1.12×10^{-9} A/cm^2 and enhanced the breakdown strength from 284 MV/m of pristine c-BCB to 489 MV/m of the c-BCB nanocomposite with 7.5 vol.% Al_2O_3 nanoplates. Notably, the nanocomposite with 7.5 vol.% Al_2O_3 nanoplates had a high dischargeable energy density of 3.31 J/cm^3 (with the charge-discharge energy density above 90%) at 400 MV/m and 150 °C, outperforming that of the c-BCB/BNNS nanocomposite owing to the higher dielectric constant. Phase-field simulations proved that nanoplates led to less distorted electric fields in nanocomposites compared to nanoparticles and nanowires, i.e., the local electric fields around the nanoparticles and nanowires were much higher than those in polymer matrix, which may contribute to more enhancement in breakdown strength (Figure 5.17).

5.5.1.3 Ternary nanocomposites

To achieve concurrently improved dielectric constant and breakdown strength, Li *et al.* [155] developed ternary nanocomposites by introducing $BaTiO_3$ nanoparticles with high dielectric constant together with BNNSs with high electrical insulation into the c-BCB polymer matrix. Considering that electrode-limited Schottky emission is the main conduction mechanism in polymers under high temperature and high electric fields, a unique sandwich structured nanocomposite with BNNSs embedded in the outer layer and $BaTiO_3$ nanoparticles added into the inner layer was prepared (Figure 5.18(a)). The outer layer with high breakdown strength and low electrical conductivity near the electrodes functions as topological

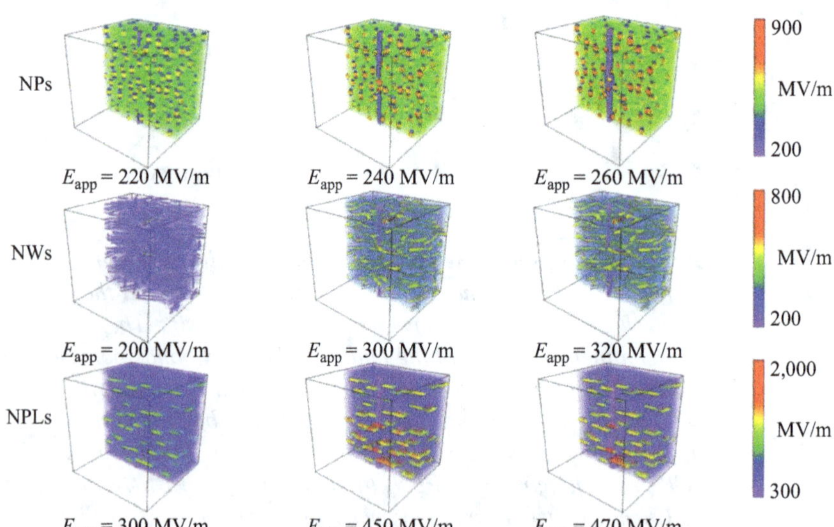

Figure 5.17 Polarization distribution in nanocomposites with nanofillers of 20 nm at filler contents of 0.1, 0.3, and 0.9 vol.%. © 2019 Wiley-VCH. Reprinted, with permission, from [154]

barriers against the injection of thermal-activated charge carriers, whereas the inner layer with high dielectric constant ensures high energy density of the nanocomposite. Moreover, it was reported that the low electric field regions formed at the interfaces between high-permittivity layers and low-permittivity layers might prevent the sandwich-structured nanocomposite from total breakdown [156]. As a result, the nanocomposite with 10 vol.% BNNSs and 25 vol.% $BaTiO_3$ nanoparticles exhibited a high dielectric constant (~5.5) and a relatively high breakdown strength (366 MV/m) at 150 °C. The energy density of such nanocomposite reached 1.1 J/cm^3 (with the charge-discharge efficiency above 90%) at 200 MV/m and 150 °C, around two times that of c-BCB/BNNS nanocomposite (Figure 5.18(b)). The energy density maintained constant from room temperature to 150 °C and even exceeded 4 J/cm^3 at an applied field of 400 MV/m.

Apart from sandwich-structured ternary nanocomposites, the electrical performance of single-layer ternary nanocomposites is also supposed to be better than that of single-doped nanocomposites [157,158]. Li *et al.* [157] investigated the electrical performance of ternary nanocomposites containing BNNSs and $BaTiO_3$ nanoparticles at room temperature, finding that the 2-D dimension of the BNNSs was beneficial to the homogeneous dispersion of $BaTiO_3$ nanoparticles and the nanocomposite showed a large increase in the dischargeable energy density. Furthermore, Luo *et al.* [158] employed BNNSs with $BaTiO_3$ nanoparticles tightly encapsulated in them as nanofillers, which lead to the much lower space charge density and leakage current in the ternary nanocomposites compared with randomly dispersed BNNSs and $BaTiO_3$ nanoparticles.

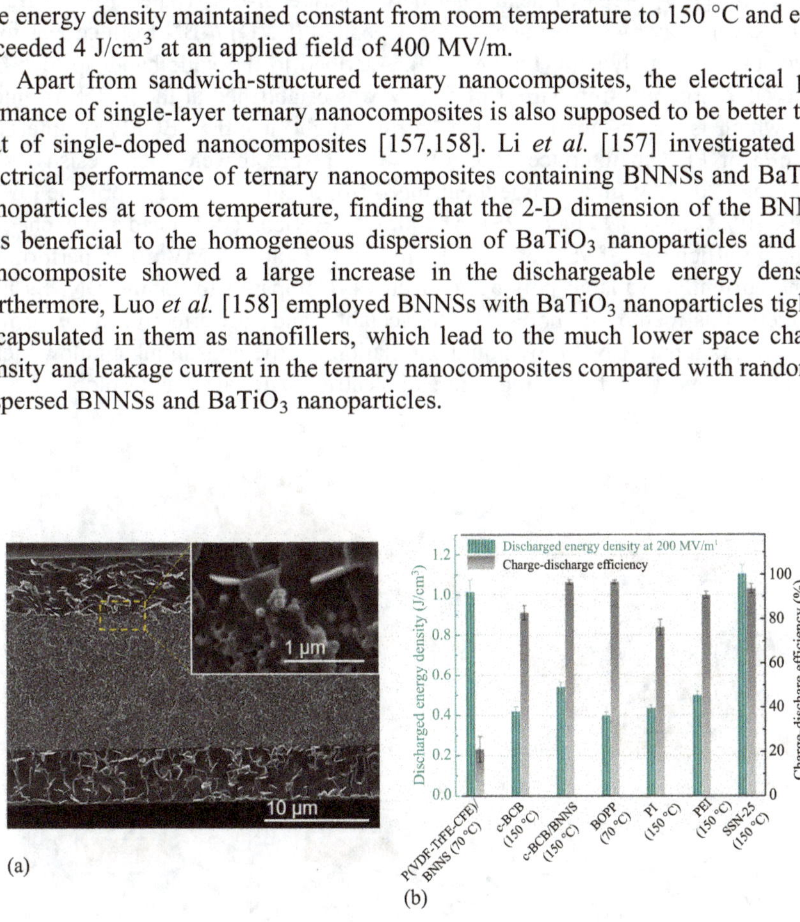

Figure 5.18 (a) Cross-sectional SEM images of sandwich-structured nanocomposite. (b) Comparison of discharge energy density and charge-discharge efficiency under 200 MV/m and elevated temperatures.

5.5.2 *Surface-coated polymer films*

As mentioned above, the energy barrier at interfaces between dielectrics and electrodes plays a dominant role in determining the injection of thermal-activated charge carriers and the electrical conduction is the main source of energy loss for polymers at high temperature, indicating that the dielectric loss of high-temperature polymer dielectrics can be suppressed by promoting the barrier height at interfaces.

Following this idea, a new strategy of coating a dense layer of wide-bandgap inorganic materials onto the surface of polymers is adopted to obtain high-performance dielectric materials for high-temperature applications. For example, Azizi *et al.* [159] prepared hexagonal boron nitride (*h*-BN)-coated PEI by transferring several layers of chemical-vapor-deposited (CVD) *h*-BN from copper foils to the surface of polymers (Figure 5.19). Ascribed to the wide bandgap of *h*-BN (5.9 eV), a large electron barrier of 5.1 eV was established at the *h*-BN/Au interface, which is much larger than the barrier height at either BOPP/Au interface (3.2 eV) or PEI/Au interface (2.93 eV). As a result, the electrical resistivity at 150 °C and 200 MV/m was largely enhanced from 2.34×10^{12} Ω of pristine PEI to 3.69×10^{13} Ω of *h*-BN-coated PEI. At 200 °C, *h*-BN-coated PEI had a high charge-discharge efficiency of 96% at 200 MV/m and 91% at 200 MV/m, outperforming all the currently available polymer dielectrics. Compared to doping BNNSs into polymers, this method is more feasible, scalable, and cost-effective since current approach to prepare BNNSs by liquid exfoliation is time-consuming and low yield. In addition, most of high-temperature dielectric materials are insoluble in the

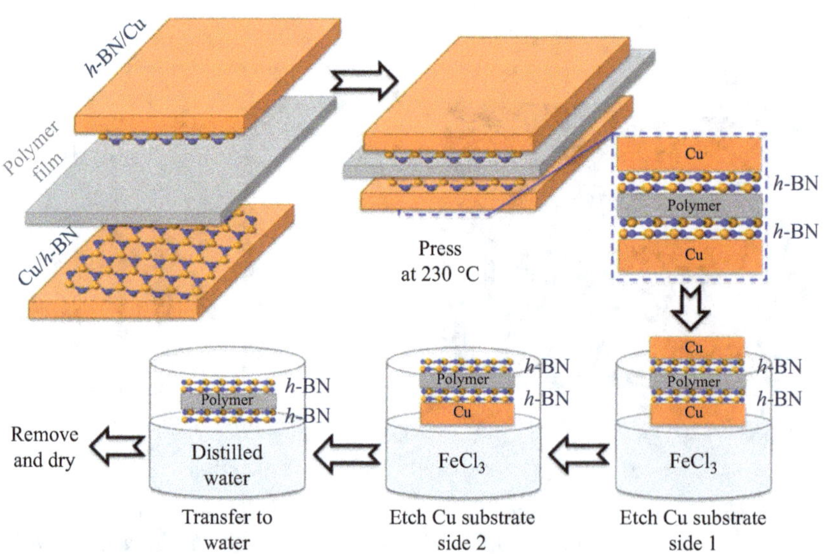

Figure 5.19 Schematic illustration showing the transfer process of the CVD-grown h-BN films on the surface of polymer films. © 2017 Wiley-VCH. Reprinted, with permission, from [159]

solvents, making it unpractical to prepare nanocomposites using the solution mixing and casting method.

However, such CVD-based approach still has its drawbacks. First, CVD process is usually conducted with a low deposition rate under ultrahigh vacuum conditions, which is incompatible with current techniques employed to fabricate polymer films. Second, the complicated transferring procedure included in the preparation of *h*-BN-coated PEI yields a low throughput. To address these issues, Zhou *et al.* [41] proposed a new approach to improve the barrier height at the interface between the electrode and the polymer dielectric, i.e., depositing a nanoscale layer of SiO_2 onto the surface of polymer dielectrics via dielectric-barrier-discharge plasma-enhanced chemical vapor deposition (PECVD). It was experimentally identified that the barrier height for electrons and holes were promoted from 3.9 and 1.3 eV of Au/BOPP interface to 4.3 and 4.7 eV of Au/SiO_2 interface. The coating layers suppressed the charge injection as well as the conduction loss of various commercial polymer dielectrics and in turn improved their energy storage performance under elevated temperature. For example, the SiO_2-coated BOPP had a high dischargeable energy density of 1.33 J/cm^3 with the charge-discharge efficiency above 90% at 120 °C, which was almost 18 times that of pristine BOPP. Remarkably, the technique was implemented with a fast processing rate at room temperature and atmospheric pressure with a fast manufacturing rate, which may meet the requirements for the manufacture of polymer capacitor films in industry (Figure 5.20).

5.6 Conclusions and prospects

With the development of HEVs, deep oil & gas exploration, aerospace industry, and next-generation switching technology, the demands for high-temperature film capacitors have become more urgent. Various aromatic polymers, heteroaromatic polymers, and fluoropolymers, which possess relatively high thermal stability, have been excavated as high-temperature dielectric materials. However, few of them meet all the requirements for capacitor applications including high dielectric constant, low dissipation factor, high breakdown strength, high volume resistivity, along with great processability and self-clearing capability. The dielectric performance of these polymers can be further improved by modifying the molecular structures. In addition, combination of polymer and inorganic nanomaterials to form the nanocomposites may overcome some common limitations of polymer dielectrics and obtain favorable features for high-temperature applications.

There are still some critical issues needing to be tackled for the further development of high-temperature polymer dielectrics. First, the fundamental theories on the relationship between the chemical structures and thermal stability or dielectric performance of polymers, which are essential to the rational design of high-T_g polymer dielectrics, are largely unexplored. Second, almost all high-temperature polymer dielectrics are linear dielectrics with low dielectric constant due to the competing mechanism between dielectric constant and dissipation factor,

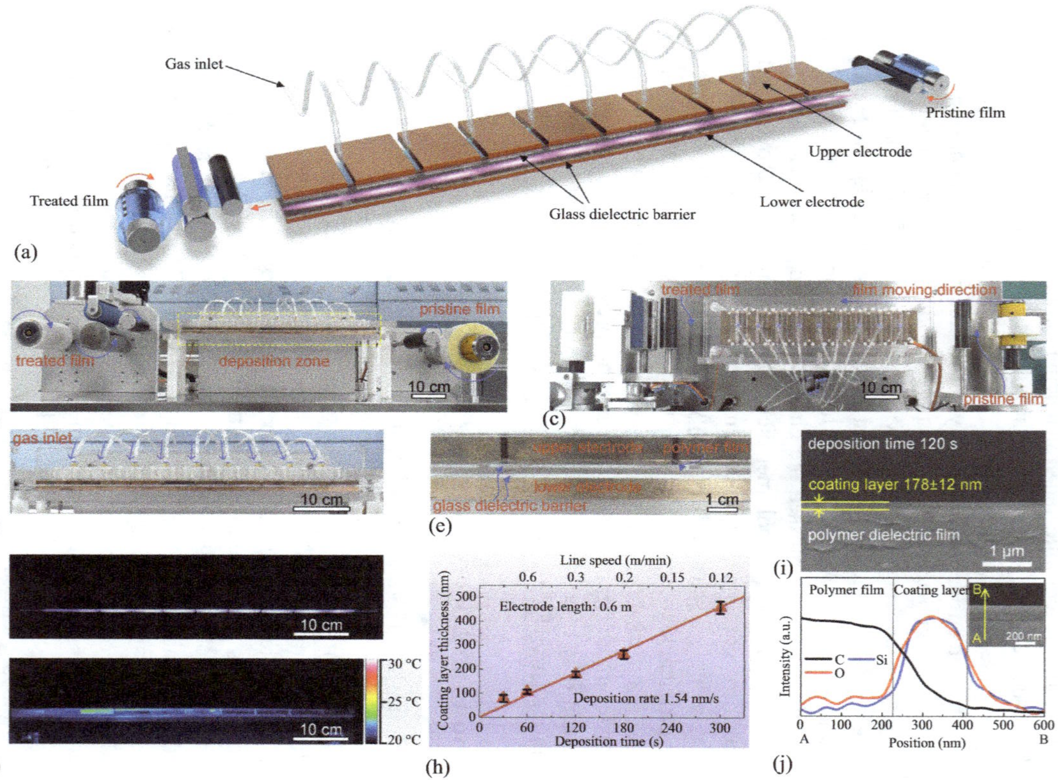

Figure 5.20 *(a) Schematic of the roll-to-roll PECVD. (b) Front view and (c) top view of the homemade prototype for continuous coating of polymer dielectrics. (d), (e) Photographs of the deposition region. (f) Photograph of the glowing discharge corresponding to the region shown in (d). (g) Infrared thermal image of the operating discharge corresponding to the region shown in (d). (h) Thickness of the coating layer and prototype line speed versus deposition time. (i) Cross-sectional SEM image of the coating layer on polymer film. (j) Element concentration from EDX scanned across the coating layer deposited on polymer film. © 2018 Wiley-VCH. Reprinted, with permission, from [41]*

especially at elevated temperature, i.e., the improvement in dielectric constant of polymers is usually accompanied by increased dissipation factor. Third, the abruptly increased conduction loss induced by the various temperature-dependent conduction mechanisms at high temperature may lead to the thermal runaway. This phenomenon implies that high-T_g polymer dielectrics do not necessarily possess high breakdown strength at high temperature unless the leakage current is effectively suppressed. Fourth, although the nanocomposite method can improve some unfavorable properties inherent to polymer dielectrics such as low dielectric constant and low thermal conductivity, it brings about deterioration in other properties at the same time. For example, incorporation of high-permittivity nanofillers into polymers indeed is effective in acquiring high dielectric constant; yet, the largely increased conduction loss induced by the introduction of the various types of material defects is not acceptable under high temperature conditions. Moreover, the lack of a mature manufacturing process hinders scalable production of high-quality nanocomposite dielectric films, which results in a high cost of the material.

References

[1] Irvine J. T., Sinclair D. C., West A. R. 'Electroceramics: Characterization by impedance spectroscopy'. *Advanced Materials.* 1990; 2(3): 132–138.

[2] Sarjeant W. J., Clelland I. W., Price R. A. 'Capacitive components for power electronics'. *Proceedings of the IEEE*. 2001; 89(6): 846–855.

[3] Tan Q., Irwin P., Cao Y. 'Advanced dielectrics for capacitors'. *IEEJ Transactions on Fundamentals and Materials*. 2006; 126(11): 1153–1159.

[4] Reaney I. M., Iddles D. 'Microwave dielectric ceramics for resonators and filters in mobile phone networks'. *Journal of the American Ceramic Society*. 2006; 89(7): 2063–2072.

[5] Pan J., Li K., Li J., Hsu T., Wang Q. 'Dielectric characteristics of poly (ether ketone ketone) for high temperature capacitive energy storage'. *Applied Physics Letters*. 2009; 95(2): 022902.

[6] Islam M. S., Qiao Y., Tang C., Ploehn H. J. 'Terthiophene-containing copolymers and homopolymer blends as high-performance dielectric materials'. *ACS Applied Materials & Interfaces*. 2015; 7(3): 1967–1977.

[7] Chu B., Zhou X., Ren K., *et al.* 'A dielectric polymer with high electric energy density and fast discharge speed'. *Science.* 2006; 313(5785): 334–336.

[8] Khanchaitit P., Han K., Gadinski M. R., Li Q., Wang Q. 'Ferroelectric polymer networks with high energy density and improved discharged efficiency for dielectric energy storage'. *Nature Communications.* 2013; 4: 2845.

[9] Laihonen S. J., Gafvert U., Schutte T., Gedde U. W. 'DC breakdown strength of polypropylene films: Area dependence and statistical behavior'. *IEEE Transactions on Dielectrics and Electrical Insulation*. 2007; 14(2): 275–286.

[10] Laihonen S. J. Polypropylene: Morphology, defects and electrical breakdown. KTH, Doctoral dissertation; 2005.

[11] Ho J., Jow R. *Characterization of high temperature polymer thin films for power conditioning capacitors.* No. ARL-TR-4880. Army Research Lab Adelphi MD Sensors and Electron Devices Directorate, 2009.

[12] Buttay C., Planson D., Allard B., *et al.* 'State of the art of high temperature power electronics'. *Materials Science and Engineering: B.* 2011; 176(4): 283–288.

[13] Watson J., Castro G. 'A review of high-temperature electronics technology and applications'. *Journal of Materials Science: Materials in Electronics.* 2015; 26(12): 9226–9235.

[14] Caliari L., Bettacchi P., Boni E., *et al.* 'KEMET film capacitors for high temperature, high voltage and high current'. *CARTS International Proceedings–ECA*, 2013.

[15] Weimer J. A. 'Electrical power technology for the more electric aircraft'. *[1993 Proceedings] AIAA/IEEE Digital Avionics Systems Conference*; IEEE, October 1933, 445–450.

[16] Tan D., Zhang L., Chen Q., Irwin P. 'High-temperature capacitor polymer films'. *Journal of Electronic Materials.* 2014; 43(12): 4569–4575.

[17] Zhang X. M., Liu J. G., Yang S. Y. 'A review on recent progress of R&D for high-temperature resistant polymer dielectrics and their applications in electrical and electronic insulation'. *Reviews on Advanced Material Science.* 2016; 46(1): 22–38.

[18] Ho J. S., Greenbaum S. G. 'Polymer capacitor dielectrics for high temperature applications'. *ACS Applied Materials & Iterfaces.* 2018; 10(35): 29189–29218.

[19] Li Q., Yao F. Z., Liu Y., Zhang G., Wang H., Wang Q. 'High-temperature dielectric materials for electrical energy storage'. *Annual Review of Materials Research.* 2018; 48: 219–243.

[20] Fan B., Liu F., Yang G., *et al.* 'Dielectric materials for high-temperature capacitors'. *IET Nanodielectrics.* 2018; 1(1): 32–40.

[21] Androsch R., Wunderlich B. 'A study of annealing of poly (ethylene-co-octene) by temperature-modulated and standard differential scanning calorimetry'. *Macromolecules.* 1999; 32(21): 7238–7247.

[22] Zarzycki J. *Glasses and the vitreous state.* Cambridge University Press, no. 9; 1991.

[23] Young R. J., Lovell P. A. *Introduction to polymers.* CRC Press; 2011.

[24] Williams G., Thomas D. K. 'Phenomenological and molecular theories of dielectric and electrical relaxation of materials'. *Novocontrol Application Note Dielectrics.* 1998; 3: 1–29.

[25] Cozens J. 'Development of plastic dielectric capacitors'. *IRE Transactions on Component Parts.* 1959; 6(2): 114–118.

[26] Zhang S., Zou C., Kushner D. I., *et al.* 'Semicrystalline polymers with high dielectric constant, melting temperature, and charge-discharge efficiency'.

IEEE Transactions on Dielectrics and Electrical Insulation. 2012; 19(4): 1158–1166.

[27] Zhu L. 'Exploring strategies for high dielectric constant and low loss polymer dielectrics'. *The Journal of Physical Chemistry Letters*. 2014; 5(21): 3677–3687.

[28] Gregorio R., Ueno E. M. 'Effect of crystalline phase, orientation and temperature on the dielectric properties of poly (vinylidene fluoride)(PVDF)'. *Journal of Materials Science*. 1999; 34(18): 4489–4500.

[29] Zhu Y., Jiang P., Zhang Z., Huang X. 'Dielectric phenomena and electrical energy storage of poly (vinylidene fluoride) based high-k polymers'. *Chinese Chemical Letters*. 2017; 28(11): 2027–2035.

[30] Lu T., Studer A. J., Cortie D., *et al.* 'Susceptible ferroelectric/antiferroelectric phase transition near the surface of Nb-doped lead zirconate stannate titanate from surface processing'. *ACS Applied Materials & Interfaces*. 2016; 8(23): 14313–14317.

[31] Xia W., Zhang Z. 'PVDF-based dielectric polymers and their applications in electronic materials'. *IET Nanodielectrics*. 2018; 1(1): 17–31.

[32] Wang S., Li Q. 'Design, synthesis and processing of PVDF-based dielectric polymers'. *IET Nanodielectrics*. 2018; 1(2): 80–91.

[33] Gregorio R., Ueno E. M. 'Effect of crystalline phase, orientation and temperature on the dielectric properties of poly (vinylidene fluoride)(PVDF)'. *Journal of Materials Science*. 1999; 34(18): 4489–4500.

[34] Ieda M. 'Dielectric breakdown process of polymers'. *IEEE Transactions on Electrical Insulation*. 1980; (3): 206–224.

[35] Stark K. H., Garton C. G. 'Electric strength of irradiated polythene'. *Nature*. 1955; 176(4495): 1225–1226.

[36] Zebouchi N., Bendaoud M., Essolbi R., Malec D., Ai B., Giam H. T. 'Electrical breakdown theories applied to polyethylene terephthalate films under the combined effects of pressure and temperature'. *Journal of Applied Physics*. 1996; 79(5): 2497–2501.

[37] Chiu F. C. 'A review on conduction mechanisms in dielectric films'. *Advances in Materials Science and Engineering*, 2014.

[38] Ho J., Jow T. R. 'High field conduction in biaxially oriented polypropylene at elevated temperature'. *IEEE Transactions on Dielectrics and Electrical Insulation*. 2012; 19(3): 990–995.

[39] Ieda M. 'Electrical conduction and carrier traps in polymeric materials'. *IEEE Transactions on Electrical Insulation*. 1984; (3): 162–178.

[40] Ambegaokar V., Halperin B. I., Langer J. S. 'Hopping conductivity in disordered systems'. *Physical Review B*. 1971; 4(8): 2612.

[41] Zhou Y., Li Q., Dang B., *et al.* 'A scalable, high-throughput, and environmentally benign approach to polymer dielectrics exhibiting significantly improved capacitive performance at high temperatures'. *Advanced Materials*. 2018; 30(49): 1805672.

[42] McLean D. A. 'Metallized paper for capacitors'. *Proceedings of the IRE*. 1950; 38(9): 1010–1014.

[43] Qin S., Ma S., Boggs S. A. 'The mechanism of clearing in metalized film capacitors'. *2012 IEEE International Symposium on Electrical Insulation*; IEEE, June, 2012, 592–595.
[44] Reed C. W., Cichanowskil S. W. 'The fundamentals of aging in HV polymer-film capacitors'. *IEEE Transactions on Dielectrics and Electrical Insulation*. 1994; 1(5): 904–922.
[45] Caliari L., Bettacchi P., Boni E., *et al*. 'KEMET SMD film capacitors for high temperature applications'. *Additional Papers and Presentations*. 2013; 2013(HITEN): 000013–000024.
[46] Mitsui H. 'Progress in Japan in electrical insulation at high temperatures'. *IEEE Electrical Insulation Magazine*. 1996; 12(3): 16–27.
[47] Everett J., Garrou P., Dalman D. 'New polymers for emerging interconnection applications'. *Plastics for Electronics*; Springer, Dordrecht, 1999, pp. 333–353.
[48] Liaw D. J., Wang K. L., Huang Y. C., Lee K. R., Lai J. Y., Ha C. S. 'Advanced polyimide materials: Syntheses, physical properties and applications'. *Progress in Polymer Science*. 2012; 37(7): 907–974.
[49] Kirby A. J. *Polyimides: Materials, processing and applications*. iSmithers Rapra Publishing; 1992, vol. 59.
[50] Stenzenberger H. D., Hergenrother P. M. *Polyimides*. D. Wilson (ed.). Glasgow: Blackie; 1990. p. 34.
[51] Hanley T. L., Burford R. P., Fleming R. J., Barber K. W. 'A general review of polymeric insulation for use in HVDC cables'. *IEEE Electrical Insulation Magazine*. 2003; 19(1): 13–24.
[52] Diaham S., Zelmat S., Locatelli M. L., Dinculescu S., Decup M., Lebey T. 'Dielectric breakdown of polyimide films: Area, thickness and temperature dependence'. *IEEE Transactions on Dielectrics and Electrical Insulation*. 2010; 17(1): 18–27.
[53] Jones R. J., Wright W. F. *High temperature polymer dielectric film insulation*. TRW Space and Defense Sector Redondo Beach, CA, 1992.
[54] Li Q., Chen L., Gadinski M. R., Zhang S., Zhang G., Li H. U. 'Flexible high-temperature dielectric materials from polymer nanocomposites'. *Nature*. 2015; 523(7562): 576.
[55] Tsukiji M., Bitoh W., Enomoto J. 'Thermal degradation and endurance of polyimide films'. *IEEE International Symposium on Electrical Insulation*; IEEE, June 1990, pp. 88–91.
[56] Carter M. A. *U.S. Patent No. 6,687,115*. Washington, DC: U.S. Patent and Trademark Office, 2004.
[57] Odian G. *Principles of polymerization*. John Wiley & Sons; New York, 2004.
[58] Chung I. S., Park C. E., Ree M., Kim S. Y. 'Soluble polyimides containing benzimidazole rings for interlevel dielectrics'. *Chemistry of Materials*. 2001; 13(9): 2801–2806.
[59] Park J. J., Shin S. S., Lee J. Y., Han S. W., Kang D. P. 'Effect of conductor radius of polyesterimide-polyamideimide enameled round wire on insulation

breakdown voltage and insulation lifetime'. *Transactions on Electrical and Electronic Materials*. 2015; 16(3): 146–150.

[60] Diaham S., Locatelli M. L., Lebey T., Dinculescu S. 'Dielectric and thermal properties of polyamide-imide (PAI) films'. *2009 IEEE Conference on Electrical Insulation and Dielectric Phenomena*; IEEE, October, 2009, pp. 482–485.

[61] Bourbigot S., Flambard X. 'Heat resistance and flammability of high performance fibres: A review'. *Fire and Materials*. 2002; 26(4–5): 155–168.

[62] Vogel H., Marvel C. S. 'Polybenzimidazoles, new thermally stable polymers'. *Journal of Polymer Science*. 1961; 50(154): 511–539.

[63] Seymour R. B., Kirshenbaum G. S. *High performance polymers, their origin and development*. New York: Elsevier; 1986.

[64] Li Q., Jensen J. O., Savinell R. F., Bjerrum. 'High temperature proton exchange membranes based on polybenzimidazoles for fuel cells'. *Progress in Polymer Science*. 2009; 34(5): 449–477.

[65] Hammoud A. N., Suthar J. L. 'Characterization of polybenzimidazole (PBI) film at high temperatures'. *IEEE International Symposium on Electrical Insulation*; IEEE, June 1990, 449–451.

[66] Wolfe J. F., Arnold F. E. 'Rigid-rod polymers. 1. Synthesis and thermal properties of para-aromatic polymers with 2,6-benzobisoxazole units in the main chain'. *Macromolecules*. 1981; 14(4): 909–915.

[67] Choe E. W., Kim S. N. 'Synthesis, spinning, and fiber mechanical properties of poly (p-phenylenebenzobisoxazole)'. *Macromolecules*. 1981; 14(4): 920–924.

[68] Wolfe J. F., Sybert P. D., Sybert J. R. *U.S. Patent No. 4,533,693*. Washington, DC: U.S. Patent and Trademark Office, 1985.

[69] Wang J., Liang G., Zhao W., Zhang Z. 'Enzymatic surface modification of PBO fibres'. *Surface and Coatings Technology*. 2007; 201(8): 4800–4804.

[70] Wolfe J. F., Loo B. H., Arnold F. E. 'Rigid-rod polymers. 2. Synthesis and thermal properties of para-aromatic polymers with 2,6-benzobisthiazole units in the main chain'. *Macromolecules*. 1981; 14(4): 915–920.

[71] Sperling L. H. *Introduction to physical polymer science*. John Wiley & Sons; 2005.

[72] Yang H. H. *Kevlar aramid fiber*. Wiley; 1993.

[73] Mandelcorn L., Miller R. L. 'High temperature, > 200 degrees C, polymer film capacitors'. *IEEE 35th International Power Sources Symposium*; IEEE, June 1992, 369–372.

[74] Maercklein E., Maki E. *Propulsion and Power Rapid Response R&D Support: Task Order 14: Research and Analysis of 5 μm Fluorene Polyester (FPE) Thin Films for High Voltage, Pulse Power Applications*. AFRL-RZWP-TR-2009–2057, 2009.

[75] Stricker J., Scofield J., Brar N., *et al.* 'Evaluation of fluorene polyester film capacitors'. *CARTS USA-2010*. 2010; 15–18.

[76] Brown T. 'High-temperature polymer film capacitor development'. *Retrieved*. 2009; 1(28): 2010.

[77] Sadana A. K., Saini R. K., Billups W. E. 'Cyclobutarenes and related compounds'. *Chemical Reviews.* 2003; 103(4): 1539–1602.

[78] Schwödiauer R., Neugschwandtner G. S., Bauer-Gogonea S., Bauer S., Wirges W. 'Low-dielectric-constant cross-linking polymers: Film electrets with excellent charge stability'. *Applied Physics Letters.* 1999; 75(25): 3998–4000.

[79] Cozens J. 'Development of plastic dielectric capacitors'. *IRE Transactions on Component Parts.* 1959; 6(2): 114–118.

[80] *Metallized Polycarbonate Film Capacitors* [online]. Available from http://www.ecicaps.com/products/5mc-series/ [accessed December 7, 2016]

[81] *Film & Mica Capacitors Dearborn Electronics* [online]. Available from http://www.exxeliausa.com/wp-content/themes/dearborn/catalogs/FILM_AND_MICA_2012.pdf [accessed November 27, 2016]

[82] Caliari L., Bettacchi P., Boni E., *et al.* 'KEMET film capacitors for high temperature, high voltage and high current'. *CARTS International Proceedings–ECA*, 2013.

[83] Reddish W. 'The dielectric properties of polyethylene terephthalate (terylene)'. *Transactions of the Faraday Society.* 1950; 46: 459–475.

[84] Yang P., Tian F., Ohki Y. 'Dielectric properties of poly (ethylene terephthalate) and poly (ethylene 2,6-naphthalate)'. *IEEE Transactions on Dielectrics and Electrical Insulation.* 2014; 21(5): 2310–2317.

[85] Yoon K. H., Lee S. C., Park O. O. 'Thermal properties of poly (ethylene 2,6-naphthalate) and poly (butylene 2,6-naphthalate) blends'. *Polymer Journal.* 1994; 26(7): 816.

[86] Ehrlich P., Amborski L. E., Burton R. L. 'Dielectric properties of teflon from room temperature to 314 C and from frequencies of 102 to 105 c/s'. *Journal of Research of the National Bureau of Standards.* 1953; 51(4): 185–188.

[87] Critchley J. P., Knight G. J., Wright W. W. *Heat-resistant polymers: Technologically useful materials.* Springer Science & Business Media, Berlin, 2013.

[88] Tan Q., Irwin P., Cao Y. 'Advanced dielectrics for capacitors'. *IEEJ Transactions on Fundamentals and Materials.* 2006; 126(11): 1153–1159.

[89] Wunderlich B. *Macromolecular physics: Crystal melting.* Academic Press; 2013.

[90] Sperati C. A. 'Polytetrafluoroethylene: History of its development and some recent advances'. *High Performance Polymers: Their Origin and Development*; Springer, Dordrecht, 1986, 267–278.

[91] Putnam R. E. 'Development of thermoplastic fluoropolymers'. *High Performance Polymers: Their Origin and Development*; Springer, Dordrecht, 1986, 279–286.

[92] Suthar J. L., Laghari J. R. 'Dielectric breakdown studies of teflon perfluoroalkoxy at high temperature'. *Journal of Materials Science.* 1992; 27(7): 1795–1800.

[93] Attwood T. E., Dawson P. C., Freeman J. L., Hoy L. R. J., Rose J. B., Staniland P. A. 'Synthesis and properties of polyaryletherketones'. *Polymer.* 1981; 22(8): 1096–1103.

[94] Kemmish D. *Update on the technology and applications of polyaryletherketones*. Canada: ISmithers, 2010.

[95] Ho J., Jow T. R. 'Effect of crystallinity and morphology on dielectric properties of PEEK at elevated temperature'. *2013 IEEE International Conference on Solid Dielectrics (ICSD)*; IEEE, June 2013, 385–388.

[96] Cheng S. Z., Ho R. M., Hsiao B. S., Gardner K. H. 'Polymorphism and crystal structure identification in poly (aryl ether ketone ketone)'. *Macromolecular Chemistry and Physics*. 1996; 197(1): 185–213.

[97] Pan J., Li K., Li J., Hsu T., Wang Q. 'Dielectric characteristics of poly (ether ketone ketone) for high temperature capacitive energy storage'. *Applied Physics Letters*. 2009; 95(2): 022902.

[98] Pan J., Li K., Chuayprakong S., Hsu T., Wang Q. 'High-temperature poly (phthalazinone ether ketone) thin films for dielectric energy storage'. *ACS Applied Materials & Interfaces*. 2010; 2(5): 1286–1289.

[99] Wang Y., Zhou X., Lin M., Zhang Q. M. 'High-energy density in aromatic polyurea thin films'. *Applied Physics Letters*. 2009; 94(20): 202905.

[100] Wang Y., Zhou X., Lin M., *et al.* 'Nonlinear conduction in aromatic polyurea thin films and its influence on dielectric applications over a broad temperature range'. *IEEE Transactions on Dielectrics and Electrical Insulation*. 2010; 17(1): 28–33.

[101] Thakur Y., Lin M., Wu S., Cheng Z., Jeong D. Y., Zhang Q. M. 'Tailoring the dipole properties in dielectric polymers to realize high energy density with high breakdown strength and low dielectric loss'. *Journal of Applied Physics*. 2015; 117(11): 114104.

[102] Cheng Z., Lin M., Wu S., *et al.* 'Aromatic poly (arylene ether urea) with high dipole moment for high thermal stability and high energy density capacitors'. *Applied Physics Letters*. 2015; 106(20): 202902.

[103] Coleman M. R., Koros W. J. 'Isomeric polyimides based on fluorinated dianhydrides and diamines for gas separation applications'. *Journal of Membrane Science*. 1990; 50(3): 285–297.

[104] Tanaka K., Kita H., Okano M., Okamoto K. I. 'Permeability and permselectivity of gases in fluorinated and non-fluorinated polyimides'. *Polymer.* 1992; 33(3): 585–592.

[105] Tanaka K., Kita H., Okamoto K. I. 'Sorption of carbon dioxide in fluorinated poyimides'. *Journal of Polymer Science Part B: Polymer Physics.* 1993; 31(9): 1127–1133.

[106] Ghosh A., Mistri E. A., Banerjee S. 'Fluorinated polyimides: Synthesis, properties, and applications'. *Handbook of Specialty Fluorinated Polymers.* William Andrew Publishing; 2015, 97–185.

[107] Ando S., Matsuura T., Sasaki S., Hougham G. 'Fluoropolymers 2: properties'. Kluwer Academic/Plenum Publishers, New York. 1999.

[108] Verbicky J. W., Mark H. F., Bikales N. M., Overberger C. G., Menges G., Kroschwitz J. I. 'Encyclopedia of polymer science and engineering'. H. F. Mark, N. M. Bikales, C. G. Overberger, and G. Menges (eds.). 1988; 12: 364.

[109] Park S. J., Cho K. S., Kim S. H. 'A study on dielectric characteristics of fluorinated polyimide thin film'. *Journal of Colloid and Interface Science*. 2004; 272(2): 384–390.

[110] Chisca S., Musteata V. E., Sava I., Bruma M. 'Dielectric behavior of some aromatic polyimide films'. *European Polymer Journal*. 2011; 47(5): 1186–1197.

[111] Guan Y., Wang D., Song G., *et al.* 'Novel soluble polyimides derived from 2,2′-bis[4-(5-amino-2-pyridinoxy)phenyl]hexafluoropropane: Preparation, characterization, and optical, dielectric properties'. *Polymer*. 2014; 55(16): 3634–3641.

[112] Jones R. J., Wright W. F. *High temperature polymer dielectric film insulation*. TRW Space and Defense Sector Redondo Beach, CA, 1992.

[113] Wang D. H., Riley J. K., Fillery S. P., Durstock M. F., Vaia R. A., Tan L. S. 'Synthesis and characterization of unsymmetrical benzonitrile-containing polyimides: Viscosity-lowering effect and dielectric properties'. *Journal of Polymer Science Part A: Polymer Chemistry*. 2013; 51(23): 4998–5011.

[114] Wang D. H., Kurish B. A., Treufeld I., Zhu L., Tan L. S. 'Synthesis and characterization of high nitrile content polyimides as dielectric films for electrical energy storage'. *Journal of Polymer Science Part A: Polymer Chemistry*. 2015; 53(3): 422–436.

[115] Treufeld I., Wang D. H., Kurish B. A., Tan L. S., Zhu L. 'Enhancing electrical energy storage using polar polyimides with nitrile groups directly attached to the main chain'. *Journal of Materials Chemistry A*. 2014; 2(48): 20683–20696.

[116] Peng X., Wu Q., Jiang S., Hanif M., Chen S., Hou H. 'High dielectric constant polyimide derived from 5,5′-bis[(4-amino)phenoxy]-2,2′-bipyrimidine'. *Journal of Applied Polymer Science*. 2014; 131(24).

[117] Peng X., Xu W., Chen L., Ding Y., Xiong T., Chen S., Hou H. 'Development of high dielectric polyimides containing bipyridine units for polymer film capacitor'. *Reactive and Functional Polymers*. 2016; 106: 93–98.

[118] Ma R., Baldwin A. F., Wang C., *et al.* 'Rationally designed polyimides for high-energy density capacitor applications'. *ACS Applied Materials & Interfaces*. 2014; 6(13): 10445–10451.

[119] Tong H., Fu J., Ahmad A., Fan T., Hou Y., Xu J. 'Sulfonyl-containing polyimide dielectrics with advanced heat resistance and dielectric properties for high-temperature capacitor applications'. *Macromolecular Materials and Engineering*. 2019; 304(4): 1800709.

[120] Kerwien C. M., Malandro D. L., Broomall J. R. 'Large area DC dielectric breakdown voltage measurement of BOPP and PTFE thin films'. *2016 IEEE Conference on Electrical Insulation and Dielectric Phenomena (CEIDP)*; IEEE, October 2016, 486–489.

[121] Donhowe M., Lawler J., Souffie S., Stein Jr E. L. '250 °C operating temperature dielectric film capacitors'. *Additional Papers and Presentations*. 2011; 2011(HITEN): 000201–000206.

[122] Zhang X., Zhao Y., Wu Y., Zhang Z. 'Poly (tetrafluoroethylene-hexafluoropropylene) films with high energy density and low loss for high-temperature pulse capacitors'. *Polymer*. 2017; 114: 311–318.

[123] Lovinger A. J., Davis D. D., Cais R. E., Kometani J. M. 'On the Curie temperature of poly (vinylidene fluoride)'. *Macromolecules*. 1986; 19(5): 1491–1494.

[124] Zhang S., Zou C., Kushner D. I., *et al*. 'Semicrystalline polymers with high dielectric constant, melting temperature, and charge-discharge efficiency'. *IEEE Transactions on Dielectrics and Electrical Insulation*. 2012; 19(4): 1158–1166.

[125] Venkat N., Dang T. D., Bai Z., *et al*. 'High temperature polymer film dielectrics for aerospace power conditioning capacitor applications'. *Materials Science and Engineering: B*. 2010; 168(1–3): 16–21.

[126] Dang T. D., Mather P. T., Alexander Jr M. D., *et al*. 'Synthesis and characterization of fluorinated benzoxazole polymers with high *Tg* and low dielectric constant'. *Journal of Polymer Science Part A: Polymer Chemistry*. 2000; 38(11): 1991–2003.

[127] Ponting M., Hiltner A., Baer E. 'Polymer nanostructures by forced assembly: Process, structure, and properties'. *Macromolecular Symposia*. Weinheim: Wiley-VCH Verlag; August 2010. 294(1): 19–32.

[128] Baer E., Zhu L. '50th anniversary perspective: Dielectric phenomena in polymers and multilayered dielectric films'. *Macromolecules*. 2017; 50(6): 2239–2256.

[129] Tseng J. K., Tang S., Zhou Z., Mackey M., Carr J. M., Mu R. 'Interfacial polarization and layer thickness effect on electrical insulation in multilayered polysulfone/poly (vinylidene fluoride) films'. *Polymer*. 2014; 55(1): 8–14.

[130] Mackey M., Hiltner A., Baer E., Flandin L., Wolak M. A., Shirk J. S. 'Enhanced breakdown strength of multilayered films fabricated by forced assembly microlayer coextrusion'. *Journal of Physics D: Applied Physics*. 2009; 42(17): 175304.

[131] Chen X., Tseng J. K., Treufeld I., Mackey M., Schuele D. E., Li R. 'Enhanced dielectric properties due to space charge-induced interfacial polarization in multilayer polymer films'. *Journal of Materials Chemistry C*. 2017; 5(39): 10417–10426.

[132] Yin K., Zhang J., Li Z., Feng J., Zhang C., Chen X. 'Polymer multilayer films for high temperature capacitor application'. *Journal of Applied Polymer Science*. 2019; 136(20): 47535.

[133] Mackey M., Schuele D. E., Zhu L., Flandin L., Wolak M. A., Shirk J. S. 'Reduction of dielectric hysteresis in multilayered films via nanoconfinement'. *Macromolecules*. 2012; 45(4): 1954–1962.

[134] Wang H., Keum J. K., Hiltner A., Baer E., Freeman B., Rozanski A., Galeski A. 'Confined crystallization of polyethylene oxide in nanolayer assemblies'. *Science.* 2009; 323(5915): 757–760.

[135] Wang Q., Zhu L. 'Polymer nanocomposites for electrical energy storage'. *Journal of Polymer Science Part B: Polymer Physics*. 2011; 49(20): 1421–1429.

[136] Dang Z. M., Yuan J. K., Yao S. H., Liao R. J. 'Flexible nanodielectric materials with high permittivity for power energy storage'. *Advanced Materials.* 2013; 25(44): 6334–6365.

[137] Chen Q., Shen Y., Zhang S., Zhang Q. M. 'Polymer-based dielectrics with high energy storage density'. *Annual Review of Materials Research*. 2015; 45; 433–458.

[138] Rajib M., Martinez R., Shuvo M., Karim H., Delfin D., Afrin S. 'Enhanced energy storage of dielectric nanocomposites at elevated temperatures'. *International Journal of Applied Ceramic Technology*. 2016; 13(1): 125–132.

[139] Zhang G., Zhang X., Yang T., *et al.* 'Colossal room-temperature electrocaloric effect in ferroelectric polymer nanocomposites using nanostructured barium strontium titanates'. *ACS Nano.* 2015; 9(7): 7164–7174.

[140] Pan Z. B., Yao L. M., Zhai J. W., *et al.* 'Fast discharge and high energy density of nanocomposite capacitors using $Ba_{0.6}S_{r0.4}TiO_3$ nanofibers'. *Ceramics International.* 2016; 42(13): 14667–14674.

[141] Pan Z., Yao L., Zhai J., Wang H., Shen B. 'Ultrafast discharge and enhanced energy density of polymer nanocomposites loaded with 0.5 $(Ba_{0.7}Ca_{0.3})$ TiO_3–0.5 Ba $(Zr_{0.2}Ti_{0.8})$ O_3 one-dimensional nanofibers'. *ACS Applied Materials & Interfaces*. 2017; 9(16): 14337–14346.

[142] Wiesmann H. J., Zeller H. R. 'A fractal model of dielectric breakdown and prebreakdown in solid dielectrics'. *Journal of Applied Physics*. 1986; 60(5): 1770–1773.

[143] Wang M., Li W. L., Feng Y., *et al.* 'Effect of $BaTiO_3$ nanowires on dielectric properties and energy storage density of polyimide composite films'. *Ceramics International.* 2015; 41(10): 13582–13588.

[144] Wu Y. H., Zha J. W., Yao Z. Q., Sun F., Li R. K., Dang Z. M. 'Thermally stable polyimide nanocomposite films from electrospun $BaTiO_3$ fibers for high-density energy storage capacitors'. *RSC Advances.* 2015; 5(56): 44749–44755.

[145] Rao Y., Wong C. P. 'Material characterization of a high-dielectric-constant polymer–ceramic composite for embedded capacitor for RF applications'. *Journal of Applied Polymer Science*. 2004; 92(4): 2228–2231.

[146] Li J., Khanchaitit P., Han K., Wang Q. 'New route toward high-energy-density nanocomposites based on chain-end functionalized ferroelectric polymers'. *Chemistry of Materials*. 2010; 22(18): 5350–5357.

[147] Tchoul M. N., Fillery S. P., Koerner H., *et al.* 'Assemblies of titanium dioxide-polystyrene hybrid nanoparticles for dielectric applications'. *Chemistry of Materials*. 2010; 22(5): 1749–1759.

[148] Paniagua S. A., Kim Y., Henry K., Kumar R., Perry J. W., Marder S. R. 'Surface-initiated polymerization from barium titanate nanoparticles for hybrid dielectric capacitors'. *ACS Applied Materials & Interfaces*. 2014; 6(5): 3477–3482.

[149] Thakur Y., Zhang T., Iacob C., *et al.* 'Enhancement of the dielectric response in polymer nanocomposites with low dielectric constant fillers'. *Nanoscale.* 2017; 9(31): 10992–10997.

[150] Thakur Y., Lean M. H., Zhang Q. M. 'Reducing conduction losses in high energy density polymer using nanocomposites'. *Applied Physics Letters*. 2017; 110(12): 122905.

[151] Zhong S. L., Dang Z. M., Zhou W. Y., Cai H. W. 'Past and future on nanodielectrics'. *IET Nanodielectrics.* 2018; 1(1): 41–47.

[152] Shen Z. H., Wan J. J., Jiang J. Y., *et al.* 'Phase-field model of electrothermal breakdown in flexible high-temperature nanocomposites under extreme conditions'. *Advanced Energy Materials.* 2018; 8(20): 1800509.

[153] Xu W., Liu J., Chen T., *et al.* 'Bioinspired polymer nanocomposites exhibit giant energy density and high efficiency at high temperature'. *Small.* 2019; 1901582.

[154] Li H., Ai D., Ren L., *et al.* 'Scalable polymer nanocomposites with record high-temperature capacitive performance enabled by rationally designed nanostructured inorganic fillers'. *Advanced Materials.* 2019; 1900875.

[155] Li Q., Liu F., Yang T., *et al.* 'Sandwich-structured polymer nanocomposites with high energy density and great charge–discharge efficiency at elevated temperatures'. *Proceedings of the National Academy of Sciences*. 2016; 113(36): 9995–10000.

[156] Wang Y., Cui J., Yuan Q., Niu Y., Bai Y., Wang H. 'Significantly enhanced breakdown strength and energy density in sandwich-structured barium titanate/poly (vinylidene fluoride) nanocomposites'. *Advanced Materials.* 2015; 27(42): 6658–6663.

[157] Li Q., Han K., Gadinski M. R., Zhang G., Wang Q. 'High energy and power density capacitors from solution-processed ternary ferroelectric polymer nanocomposites'. *Advanced Materials.* 2014; 26(36): 6244–6249.

[158] Luo S., Yu J., Yu S., *et al.* 'Significantly enhanced electrostatic energy storage performance of flexible polymer composites by introducing highly insulating-ferroelectric microhybrids as fillers'. *Advanced Energy Materials.* 2019; 9(5): 1803204.

[159] Azizi A., Gadinski M. R., Li Q., *et al.* 'High-performance polymers sandwiched with chemical vapor deposited hexagonal boron nitrides as scalable high-temperature dielectric materials'. *Advanced Materials.* 2017; 29(35): 1701864.

[160] Li Z., Chen X., Zhang C., *et al.* 'High dielectric constant polycarbonate/nylon multilayer films capacitors with self-healing capability'. *ACS Applied Polymer Materials*. 2019; 1(4): 867–875.

[161] Nicholls J., Dimitrijev S., Tanner P., Han J. 'Description and verification of the fundamental current mechanisms in silicon carbide Schottky barrier diodes'. *Scientific Reports.* 2019; 9(1): 3754.

Chapter 6

Design and simulations of capacitor dielectrics by phase-field computations

Zhonghui Shen[1], Mengfan Guo[2], Long-Qing Chen[3] and Yang Shen[2]

6.1 Introduction

With the advent of advanced preparation methods and characterization techniques, it is now possible to design diverse dielectric materials for intended applications related to modern electronic and electrical systems [1–4]. By changing the component, microstructure, and the corresponding properties in dielectrics, the complex correlations within the multiple structural parameters and dielectric performance have been intensively studied by experimental explorations [5–7]. Nowadays, precise interpretation with high time and space resolution on dielectric capacitor at external stimuli could be achieved with the aids of powerful characterization methods [8–11], including broadband dielectric spectroscopy, thermally stimulated current spectroscopy, and scanning tip microscopy. Thus, a deeper understanding on dielectrics with regard to the influence of fabrications, compositions, and configurations has been gained [12–14]. However, it is still a big challenge to detect the local dynamic behaviors of dielectrics in a more realistic situation, where high electric field or extreme temperature is applied. On one hand, non-equilibrium process related to fast energy dissipation is dominant, which is difficult to be captured. On the other hand, the physical or chemical degradations caused by synergistic effects between electrical, thermal, and mechanical factors are buried in the bulk dielectrics, hindering the direct observation of their local evolution. Therefore, theoretical simulation has become an alternative and effective method to help us investigate the underlying mechanisms and design of advanced dielectrics for electrostatic capacitors.

For the past decades, a class of powerful theoretical methods have been developed [15], including effective-medium approximation, density functional

[1]State Key Laboratory of Advanced Technology for Materials Synthesis and Processing, Wuhan University of Technology, Wuhan, China
[2]School of Materials Science and Engineering, State Key Lab of New Ceramics and Fine Processing, Tsinghua University, Beijing, China
[3]Department of Materials Science and Engineering, The Pennsylvania State University, University Park, USA

theory, and molecular dynamics theory. With the help of these computational progresses, we now have the ability to interpret the experimental results and to predict the properties of dielectric materials at different space and time scales. Besides the improvements on simulation theory, the computational power has been drastically increased and massive parallel computational strategies have been proposed. Hence, it is possible to model various dielectric materials at multiscale, from atomic or molecular interaction, to nano-regions of polar domain structure, to mesoscopic composition distribution and up to the performance of macroscopic devices. Furthermore, since 2011 has witnessed the launch of the materials genome innovative (MGI), in which the materials fabrication and computational simulation are combined to screen, to optimize, and to predict novel materials in an unparalleled scale and rate [15–17], the typical trial-and-error approach, which is common in the current material researches, could be replaced by a more time-saving and cost-effective methodology from the laboratory to the marketplace. Dielectrics materials, on which element, polymorph, chemical composition, and microstructure have great impacts, have also been benefited from this materials development, as high-throughput calculation and machine learning help to screen and design the candidates for capacitor materials [18,19].

Generally, there are three key parameters in the dielectrics for assessment of dielectric performance: the dielectric constant, breakdown strength, and dielectric loss. The first two parameters are related to the capacity of the electrostatically induced charges, i.e., the capacitance, whereas the latter two parameters account for the reliability of the capacitor. Moreover, the size, shape, and spatial arrangement of the local structures, and their interactions with defects, dipoles, domains, and space charges in a dielectric play critical roles in determining the dielectric properties. Therefore, a number of theoretical calculations on different scales have been employed to study those influencing factors and then improve the dielectric properties. Among them, phase-field method, a mesoscale computation approach, is one the most effective and powerful tools for understanding the microstructure-property correlations and for guiding the design principals in dielectric materials [20,21], including ceramic, polymer, polymer/polymer composite, and ceramic/polymer composite. Hereby, in this chapter, we will give a comprehensive introduction on the theory of phase-field simulation, and summarize its recent applications on interpreting dielectric behaviors observed in experiment, and instructing experimental efforts toward advanced dielectrics.

6.2 Phase-field theory

The simulation method based on phase-field theory has emerged as a powerful computational approach to modeling and predicting mesoscale morphological and microstructure evolution in the past decade, giving coherent evolutionary simulations on not only grain growth and coarsening, crack propagation, dislocation, and phase transformations [21,22], but also functional materials including ferroic thin films, lithium-ion batteries, and dielectric capacitors. The phase-field model describes a microstructure by using a set of order parameters that are continuous across the interfacial regions. The order parameters are divided into two types: the conserved and the nonconserved. The latter one has to satisfy the local conservation condition. Then, the free energy of an inhomogeneous microstructure system could be described by a set of conserved ($c_1, c_2, \ldots$) and nonconserved ($\eta_1, \eta_2, \ldots$) order parameters as follows:

$$F = \int \left[f\left(c_1, c_2, \ldots, c_n, \eta_1, \eta_2, \ldots \eta_p\right) + \sum_{i=1}^{n} \alpha_i (\nabla c_i)^2 + \sum_{i=1}^{3} \sum_{j=1}^{3} \sum_{k=1}^{p} \beta_{ij} \nabla_i \eta_k \nabla_j \eta_k \right] d^3 r + \int\int G(r - r') d^3 r d^3 r', \tag{6.1}$$

where f is the local free-energy density and α_i and β_{ij} are the gradient energy coefficients, respectively. The first volume integral denotes the contribution to the total free energy from short-range chemical interactions, and the second one represents the long-range interactions, such as electrostatic interactions, electric dipole interactions. Therefore, all free energy is dependent on the order parameters.

As mentioned above, one of the key terms in the free energy is the local free-energy density function. In most cases of phase-field simulations, the free-energy density function is described by a double-well model as:

$$f(\phi) = 4\Delta f \left(-\frac{1}{2}\phi^2 + \frac{1}{4}\phi^4 \right), \tag{6.2}$$

where ϕ is an order parameter and Δf is the potential barrier between the two stable states. The free-energy function has a pair of doubly degenerate minima when $\phi = -1$ and $\phi = +1$. For example, in the case of solidification, $\phi = -1$ and $\phi = +1$ represent the liquid and solid states, respectively. Furthermore, if ϕ stands for a conserved composition field variable, the two minima will represent the two equilibrium phases with different compositions. For this case, Δf can be regarded as the driving force of the transformation between different phases. Otherwise, if ϕ represents a nonconserved field variable, the two minima will describe the degenerate antiphase domain states. In certain cases, the two minima of the free energy density are set at $\phi = 0$ and $\phi = 1$, where the formula (6.2) can be described as

$$f(\phi) = 4\Delta f \phi^2 \left(1 - \phi^2\right). \tag{6.3}$$

There are also other types of free energy function for different applications, such as double-obstacle potential model and crystalline energy. Usually, to simulate evolutionary process by phase-field model, introduction of multiple order parameters instead of merely one is necessary to describe the free energy due to the coupling effects among different physical fields. For example, even in the phase-field model for solidification of a pure liquid, the order parameter is coupled to the temperature field.

Interface is quite common in non-homogeneous materials, located between different phases, and usually endowed with properties different from the bulk state.

Moreover, the interface has play an important role in the microstructure evolution process, as the boundary, nuclei, and meeting site of neighboring phases. The phase-field model is also capable of taking interface into computational consideration, by including an extra energy term related to the interfacial regions. For a simple system with a single order parameter, the total free energy F of a microstructure with interface could be expressed by

$$F = F_{\text{bulk}} + F_{\text{int}} = \int_{v} \left[F(\phi) + \frac{1}{2}\kappa_{\phi}(\nabla\phi)^2 \right] dV, \tag{6.4}$$

where F_{bulk} and F_{int} are the bulk and interfacial energy terms respectively, and κ_{Φ} is the gradient energy coefficient. If a double-well model is applied to describe the free-energy function, the interfacial energy γ (interfacial energy per unit area) can be obtained, and written as

$$\gamma = \frac{4\sqrt{2}}{3}\sqrt{\kappa_{\phi}\Delta f}. \tag{6.5}$$

For the evolution of a microstructure, the free energy may include the bulk chemical free energy, interfacial energy, elastic strain energy, magnetic energy, electrostatic energy, and/or other external physical fields such as stress/strain, electric field, and temperature.

With the total free energy discussed above, the dynamic process of a microstructure evolution obeys the rules depicted by Cahn-Hilliard and Allen-Cahn equations,

$$\frac{\partial c_i(\mathbf{r},t)}{\partial t} = \nabla M_{ij}\nabla \frac{\delta F}{\delta c_j(\mathbf{r},t)}, \tag{6.6}$$

$$\frac{\partial \eta_p(\mathbf{r},t)}{\partial t} = -L_{pq}\frac{\delta F}{\delta \eta_q(\mathbf{r},t)}, \tag{6.7}$$

where M_{ij} and L_{pq} are the related to atom or interface mobility. After constructing the free energy and evolution equations, we can solve them to model the microstructure evolution by several numerical methods, for example, finite-difference method and fast Fourier transform method.

Initially, the phase-field model is applied to simulate the structural evolutions in grain growth and coarsening, solidification, and solid-state phase transformation. Nevertheless, with the supplement of energy terms related to magnetic, electrostatic, and chemical potential, the phase-field simulations have also been implemented in modeling the field-induced crack propagation, ferroelectric domain switching, and lithium dendrite growth. In the following sections, we mainly focus on the applications in dielectric materials, where energy terms may related to compositional changes, electrostatic force, Joule heating, and electrostatic force have been taken into considerations.

6.3 Phase-field models for capacitor dielectrics

In this section, we summarize recent progresses of phase-field methods on modeling the dielectrics categorized by four computational targets: dielectric constant, dielectric breakdown, charge behavior, and domain switching. These methods are favored by dielectrics with inhomogeneous structures such as polymer nanocomposites, ceramics, and multi-domain ferroelectrics, as phase-field model is a mesoscale computational tool matching the scale of the microstructures within these systems.

6.3.1 Calculation of dielectric constants

Computational calculations of dielectric constant have been intensively conducted to simulate the dielectric responses in polymer nanocomposites, which are a class of competitive candidate for next-generation high-energy-density electrostatic capacitors, due to their combination of high E_b (300–900 MV/mm) from polymeric matrix and high ε_r (10^2–10^4) from ceramic filler. Since experimental studies have confirmed that the dielectric properties of polymer nanocomposites depend on both the properties of constituent materials and the mesoscale structure related to volume fraction, size, shape, and orientation of the fillers, phase-field model has become a perfect method to investigate these effects on the dielectric response, accompanied by a large range of experimental assistance. Therefore, a number of phase-field simulations have been employed to elucidate the influence of microstructure on the effective dielectric properties of polymer nanocomposites, and to shed light on how better capacitors could be designed by using composite materials.

When taking the polarization vector field as the order parameter, the dielectric property of a composite under applied electric field could be predicted by considering the electrostatic energy. Wang *et al.* developed a phase-field model of dielectric composites containing circular fillers to simulate heterogeneous distributions of polarization, charge density, local electric field, and effective dielectric constant [23]. It solves electrostatic equations using the fast Fourier transform technique and parallel computing algorithm. The simulations demonstrate the strong dependence of overall dielectric response on the filler distribution, as shown in Figure 6.1(a). It is evident that the dielectric susceptibility $\chi^{\mathrm{eff}}/\chi_{\mathrm{M}}$ could reach 2.4 when fillers are arranged into chains connecting the electrodes. Since continuous path with high dielectric constant is formed parallel to the external electric field, the filler interaction largely reduces the depolarization effect, resulting in significantly improved susceptibility under applied electric field. In contrast, the susceptibilities moderately decrease to 1.6 when the fillers are uniformly dispersed into hexagonal or square patterns, due to the restricted inter-filler interaction. The value is further reduced to 1.4 with fillers arranged into translational chains. Moreover, the matrix also contributes to the dielectric response due to the inhomogeneous electric field in the composites. The calculation also reveals that the electric field will be highly concentrated in regions between closely located fillers

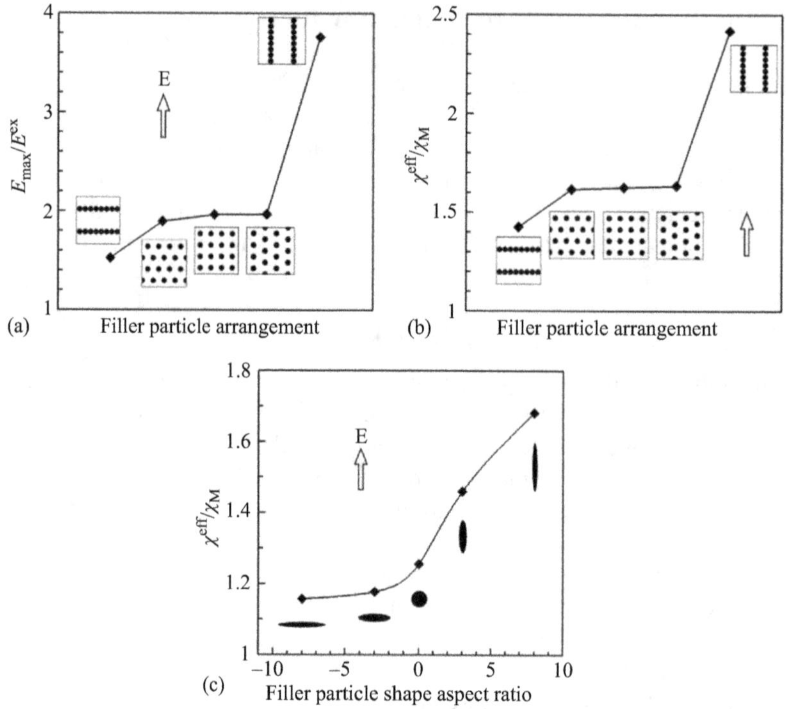

Figure 6.1 (a) The effect of filler arrangement on effective dielectric susceptibility with 20% volume fraction. (b) The effect of filler arrangement on maximum electric field concentration with 20% volume fraction and dielectric susceptibility $\chi_F/\chi_M = 10$ $\chi_F/\chi_M = 10$. E_{max}/E^{ex} is the ratio of maximal local electric field to externally applied electric field. (c) Effect of filler shape and orientation on effective dielectric susceptibility with 8.7% volume fraction of fillers. © 2011 AIP Publishing. Reprinted, with permission, from [23]

when chains are formed parallel to the external field, rendering more fraction of matrix actively responding to the electric field. The ratios between highest local electric field and external field E_{max}/E^{ex} with different arrangements of fillers are also plotted in Figure 6.1(b). While E_{max}/E^{ex} reaches 3.8 with fillers arranged into vertical chains, the electric field is more moderate in composites with uniform dispersion along y-axis, with E_{max}/E^{ex} around 2.0. However, concentrated electric field may be unsatisfying when the breakdown strength is considered. As regions with severely concentrated electric field may become the hot spot where breakdown will start, a much lower breakdown strength would be obtained when fillers are too close in the direction parallel to the external field. These results may suggest an adversely coupled correlation between dielectric constant and breakdown strength in terms of the microstructure effect. Moreover, the shape effect of the fillers is also investigated by this work, as displayed in Figure 6.1(c). It is found that both the orientation and the shape of fillers affect the effective dielectric susceptibility of composites due to different anisotropic depolarization field. Elongated particles or short fibers with their long axes oriented in the same direction of applied electric field could lead to a higher effective dielectric susceptibility, consistent with the same effect observed in vertical chains. Thus, the effective dielectric properties of composites are closely related to filler microstructures, when their distribution and shape are taken into consideration.

Besides, the dielectric response of fillers with complex structure can also be resolved by this phase-field model [24]. As mentioned above, the dielectric breakdown could begin at the interface of polymeric matrix and ceramic filler, attributed to the large chemical and dielectric mismatch between the two phases. Therefore, fillers with core-shell structure have been widely applied to mitigate the concentrated local electric field, and meanwhile, to preserve the dielectric response. As illustrated in Figure 6.2(a) and (b), core-shell circular fillers with different size parameter and dielectric susceptibility can be modelled by phase-field method.

Figure 6.2(c) gives the variation of the maximal local electric field in the matrix $E_{\max}^{\text{matrix}}$ and shell $E_{\max}^{\text{shell}}$ as a function of shell dielectric susceptibility. With the increase of shell dielectric susceptibility, $E_{\max}^{\text{shell}}$ decreases, and $E_{\max}^{\text{matrix}}$ increases gradually. When $\chi_{\text{S}} : \chi_{\text{M}} = 1.81 : 1$, a quite similar levels of susceptibility between shell and matrix, the local electric field becomes more homogeneous with $E_{\max}^{\text{shell}} = E_{\max}^{\text{matrix}}$ as a result of depolarization competition between different phases. Moreover, the effect of shell thickness on local electric field concentration and optimal dielectric susceptibility is illustrated in Figure 6.2(d). With the increase of shell thickness, the maximum local electric field magnitude decreases gradually, and nearly saturates when $t/r = 1$. However, the thicker the shell is, the higher the optimal dielectric susceptibility goes. When the t/r is close to the value of 1, the optimal dielectric susceptibility is ~2, implying a rather low dielectric susceptibility of coating layer considering the value from matrix.

6.3.2 Dielectric breakdown process

Understanding the dielectric breakdown behavior is crucial to the design of high-performance and reliable dielectric materials. However, it is a challenge to detect and predict the breakdown behavior due to various complex factors in this highly non-equilibrium process, including blocked vision from electrodes, high electric field applied, and fast propagation of microscale cracks. Especially in inhomogeneous dielectric system where the local electric field may become more distorted at the interfaces, the breakdown path will evolve in more complicated patterns. Therefore, phase-field method has provided supports to study the breakdown behavior under physical stimuli, as a complement to the characterization techniques. Recently, Shen *et al.* developed a comprehensive phase-field model to investigate the electrostatic breakdown process of polymer nanocomposites [25]. As displayed in Figure 6.3, a two-dimensional simulation is performed to predict the evolution of breakdown process in a nanocomposite system consisting of poly

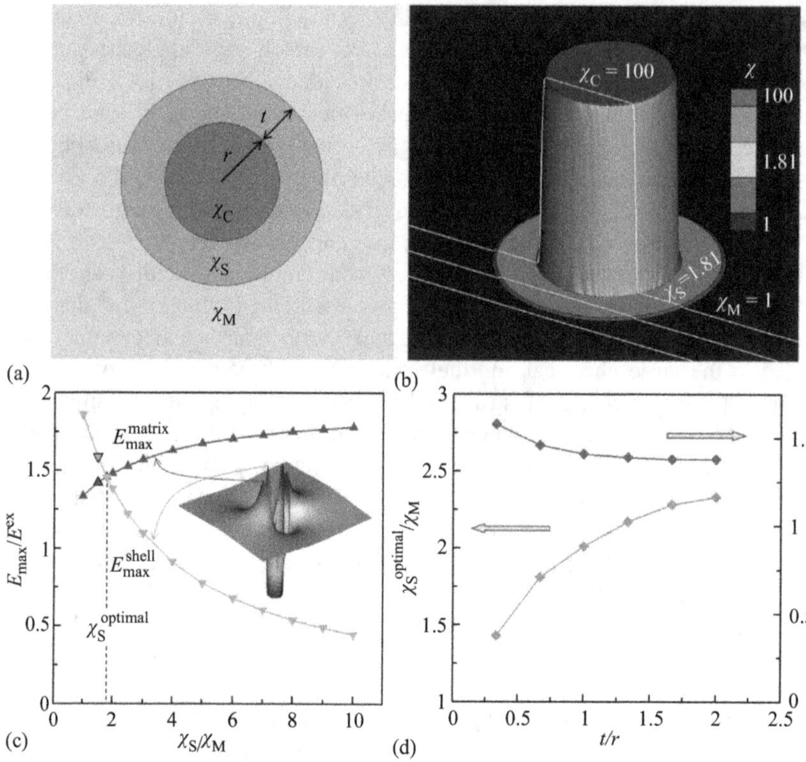

Figure 6.2 (a) Schematic of a particle with core-shell structure embedded in the matrix, where t and r are the thickness of shell thickness and radius of core, χ_C, χ_S, and χ_M are the dielectric susceptibility of core, shell, and matrix, respectively. (b) Three-dimensional visualization of dielectric susceptibility distribution with $\chi_C : \chi_S : \chi_M = 100 : 1.81 : 1$ and t/r = 2:3. (c) The effect of shell dielectric susceptibility on local electric field with $\chi_C : \chi_M = 100 : 1$. (d) The effect of shell thickness on maximal local electric field and dielectric susceptibility. © 2011 AIP Publishing. Reproduced, with permission, from [24]

(vinyldene fluoride) (PVDF) matrix and $BaTiO_3$ (BTO) fillers. Two specific shapes of the fillers, i.e., nanoparticles and nanofibers, are considered to study the microstructure effect on the breakdown process. In composite containing nanoparticles, the breakdown will first start at the interfaces between the matrix and the nanoparticles at low electric field (~165 kV/mm), and proceed to grow into breakdown path, during which nanoparticles near the breakdown path are gotten round (Figure 6.3(a) and (b)). However, the breakdown path in nanofiber-containing composites prefers to penetrate through the nanofibers, rather than get

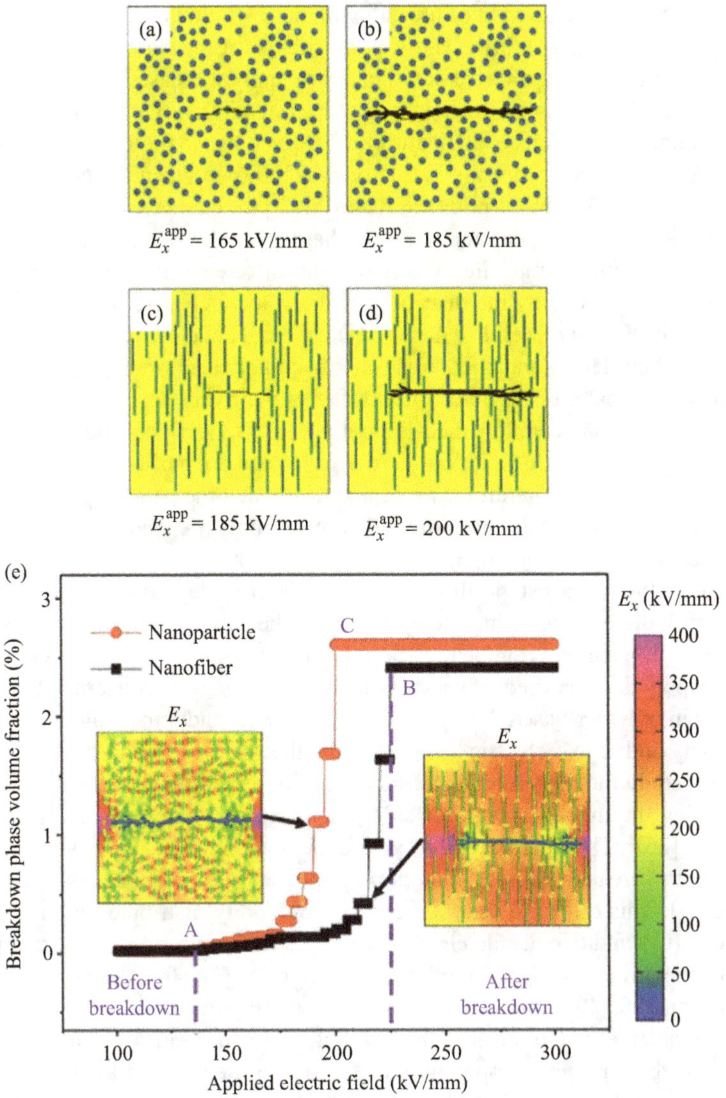

Figure 6.3 The breakdown path evolution of composites with (a), (b) nanoparticles and (c), and (d) nanofibers at volume fraction of 10%. (e) Variations of corresponding breakdown phase volume fraction as a function of applied electric field E_x^{app}. Insets are the map of local electric field distribution. © 2018 Wiley-VCH. Reproduced, with permission, from [25]

round the fillers, leading to a much higher breakdown field (Figure 6.3(c) and (d)). More quantitative evaluations on breakdown behavior can be presented by plotting the changes in volume fraction of breakdown phase with increasing applied electric field, as shown in Figure 6.3(e). The A point stands for the start of breakdown, and then the volume fraction of breakdown path gradually increases. For both systems, the volume fraction of breakdown path increases steadily at first, then abruptly, and finally saturates at a plateau, indicating the happening of the overall breakdown. More specifically, the composite is totally broken down when the applied electric field reaches 225 kV/mm (B point) in nanofiber system and 200 kV/mm (C point) in nanoparticle system, consistent with the harder breakdown process observed in nanofiber-containing composite. Moreover, the difference in breakdown strength could also be explained by the distribution of the local electric field, as illustrated in the insets of Figure 6.3(e). For nanoparticles, the local electric field concentrates at the two shoulders along the direction of applied electric field; while in the nanofibers, the local electric field concentrates at their vertices instead. Thus, when the breakdown path encounters the fillers, it is easier to grow around a nanoparticle at a lower applied electric field, but harder to bypass a nanofiber until the electric field gets high enough. Therefore, we can conclude that the path propagation speed, path tortuosity, and breakdown strength are strongly dependent on the microstructure of polymer nanocomposites.

Besides the electrostatic breakdown, dielectric degradation may also be induced by Joule heating or high temperature in the surroundings, especially under applications in harsh environments where high electric field and high temperature are both applied. In consequence, thermal breakdown is also a crucial problem, especially in polymer-based dielectrics where bad heat endurance, ultralow thermal conductivity and high dielectric loss makes the thermal deterioration more favored. In order to study the thermal effect, Shen *et al.* improved the phase-field model by incorporating thermal energy from Joule heating in a polyimide-$SrTiO_3$ (PI-STO) composite [26]. The thermal effect on breakdown process is investigated by simulating the breakdown path evolution at different temperatures, as displayed in Figure 6.4. In this case, the nanoparticles are randomly distributed with a volume fraction of 10% and two needle electrodes are placed at two sides of the composite, taken as the initial nucleation of breakdown path. The breakdown process is simulated at 300, 400, and 500 K, as shown in Figure 6.4(b)–(d), (e)–(g), and (h)–(j), respectively. Obviously, the starting electric field for the breakdown decreases with increasing temperature, as the staring field is 225 kV/mm at 300 K, 125 kV/mm at 400 K, and 75 kV/mm at 500 K. Also, the breakdown strength shows similar trend, decreasing from 250 kV/mm at 300 K to 84 kV/mm at 500 K. The deterioration in breakdown strength is attributed to the increasing Joule heating energy arising from increasing electrical conductivity. For example, with an applied electric field of 100 kV/mm, the average electrostatic energy is almost constant with the value of 10^6 J/cm^3, while the average heating energy density exponentially rises from 45 to 3.5×10^9 J/cm^3 with the temperature increasing from 300 to 500 K. Therefore, the rise in the Joule heating energy accelerates the growth

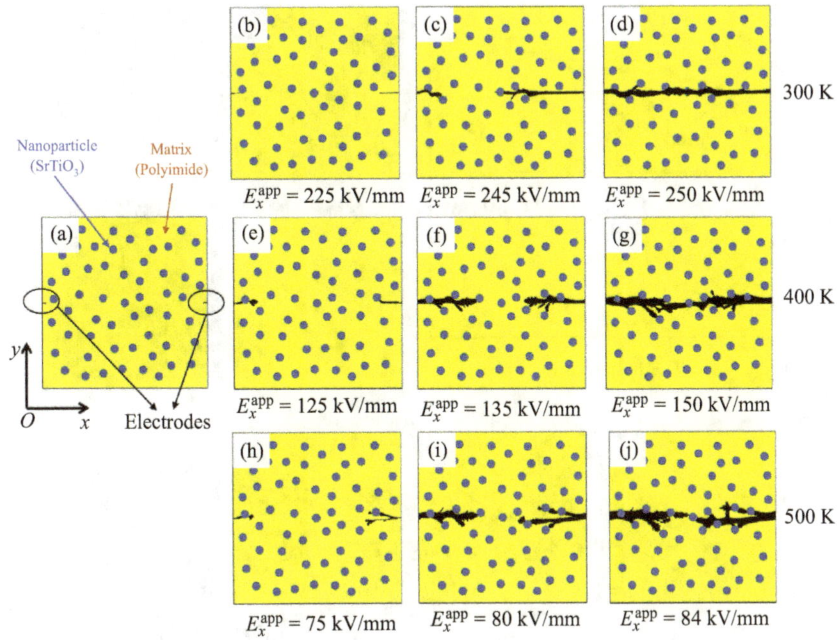

Figure 6.4 The breakdown path evolutions of (a) PI-$SrTiO_3$ nanocomposites at (b)–(d) 300 K, (e)–(g) 400 K, and (h)–(j) 500 K. © 2018 Wiley-VCH. Reproduced, with permission, from [26]

of the breakdown path, resulting in more reduced breakdown strength at higher temperatures.

6.3.3 Polarization response and domain structure

Dielectric capacitors adopting ferroelectric materials are promising for the field-induced switching of ferroelectric domains, which gives rise to the high polarization and hence the enhanced energy density. Usually, the energy density of capacitor dielectrics can be calculated from the polarization-electric (*P-E*) field loop, which is closely related to the domain switching behavior. By modulating the domain structure, parameters of the *P-E* loop, e.g., the remnant polarization, the nonlinearity, and the maximal polarization, could be changed, altering the corresponding energy density. Thereby, elucidation of domain behavior under different applied electric field is helpful to the design of ferroelectric materials for dielectric capacitors. Phase-field method is famous for its success in precise prediction of domain evolution and switching behavior under external stimuli based on Landau theory. Here, some typical cases are taken as examples to introduce the application of phase-field method on understanding the dielectric properties in ferroelectrics. Ma *et al.* used the phase-field model to study the depolarization field effect on the dielectric and piezoelectric properties [27]. Figure 6.5 shows the simulated *P-E*

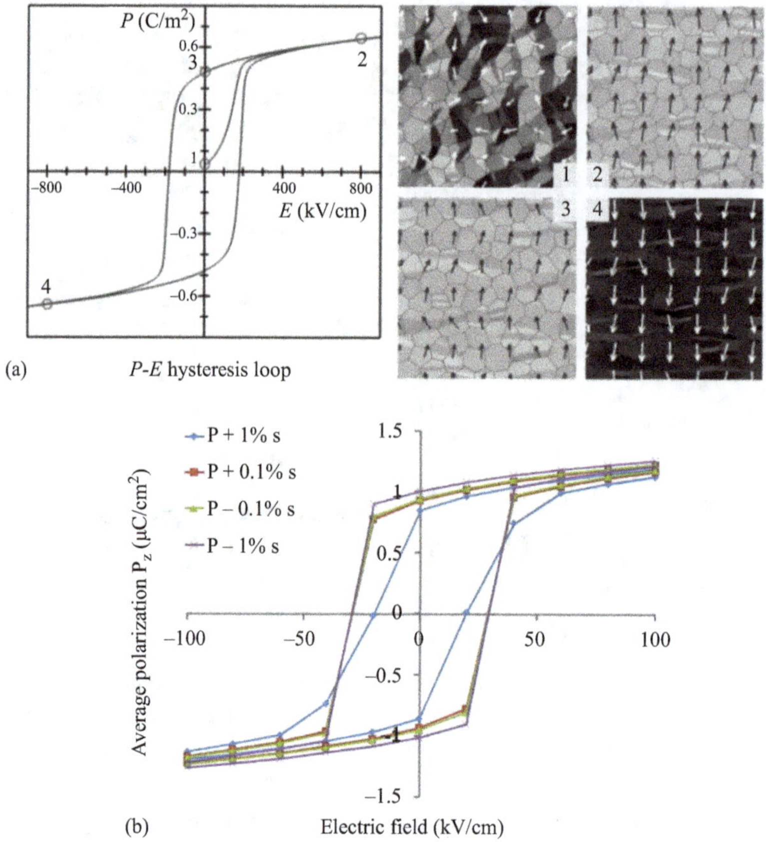

Figure 6.5 (a) Simulated P-E field hysteresis loop. (1–4) Simulated domain states of four typical stages. © AIP Publishing. Reproduced, with permission, from [27]. (b) P-E loop pf PZT(20/80) under different strains. © 2018 American Ceramic Society. Reproduced, with permission, from [28]

loop and corresponding domain structures. The state 1 corresponds to initially random polarization without any applied electric field. The states 2 and 4 show the fully poled domain structures where the polarization dipoles are all aligned along the direction of applied electric field. The state 3 corresponds to the domain structures when the applied electric field is withdrawn and the remnant polarization is reserved. In order to obtain high energy density in ferroelectrics, high-saturated polarization (states 2 and 4), and low remnant polarization (state 3) are desirable. Recently, Lin *et al.* found that the *P-E* loop can be modulated by the strain effect and lower remnant polarization can be induced under +1% strain in PZT (20/80) thin films by phase-field simulations [28], as shown in Figure 6.5(b). Four different mismatch strains, including strong tensile strain (+1%), weak tensile strain

(+0.1%), weak compressive strain (–0.1%), and strong compressive strain (–1%), are considered to investigate the strain effect on domain switching. It is found that a strong tensile strain of +1% could result in a much decreased remnant polarization than other strains, which can be explained by the stable middle states of in-plane domain configurations under strong tensile strain. Those metastable domains could provide more switching path and more nucleation points of anisotropic domains during their switching, and contribute to a more diffused collective domain distribution.

Phase field method is also applied to explain the variation of polarization response in polymer nanocomposites. Thakur *et al.* reported a significant increase of the dielectric constant in polyetherimide nanocomposites containing low dielectric constant fillers with ultralow volume fraction [29]. The phase-field simulation helps to provide explanation to the novel phenomenon, while the experimental investigation could be extremely difficult. Here, by simulating the local distribution of polarization around the nanoparticles, the evident interfacial polarization can be confirmed. In this model, a spatially varying local dielectric constant around the particle is proposed to describe the features of interface region. As shown in Figure 6.6(a), the dielectric constant in the interfacial region first increases and then decreases gradually, with the distance from the filler surface getting larger. Based on the above assumption, the polarization response to an external field could be simulated by solving the electrostatic equilibrium equation in this system, and then the effective dielectric constant of the nanocomposites can be calculated by considering the distribution of local polarizations. The simulation results could reproduce the experimentally observed increase in dielectric constant at a low volume fraction of fillers. Moreover, with the particle size reduced to smaller scale, the peak of dielectric constant becomes sharper and maximal dielectric response can be obtained under lower volume fraction. These dielectric phenomena could be understood from the local polarization response around the particles (Figure 6.6(c) and (d)). For particular fillers with radius around 20 nm, the local polarization could be enhanced between two nearest fillers at 0.3% volume fraction, corresponding the maximal dielectric constant in Figure 6.6(b). When the distance between two fillers decreases or increases, both the polarization response will drop and the dielectric constant reduces. Due to the reduced surface-area-to-volume ratio, the dielectric peak shifts to a higher volume fraction of 0.9 vol.% when the radius of fillers increases to 50 nm, as the polarization distribution displayed in Figure 6.6(d). Therefore, a good match between experimental and simulated performance is achieved by the phase-field method, giving a good explanation of the dielectric response in nanocomposites.

Recently, a phase-field model relying on the polarization of both ceramic and polymeric ferroelectrics has been employed to study the mechanism of interfacial coupling in polymer nanocomposites [30]. In this model, a simplified BZT-P(VDF-TrFE-CFE) bilayer system with an interface layer with 5-nm thickness is considered, as displayed in Figure 6.7. Due to the different ferroelectric properties between BZT and the polymer, the polarization response may affect each other upon the application of electric field. It can be observed that only the BZT layer

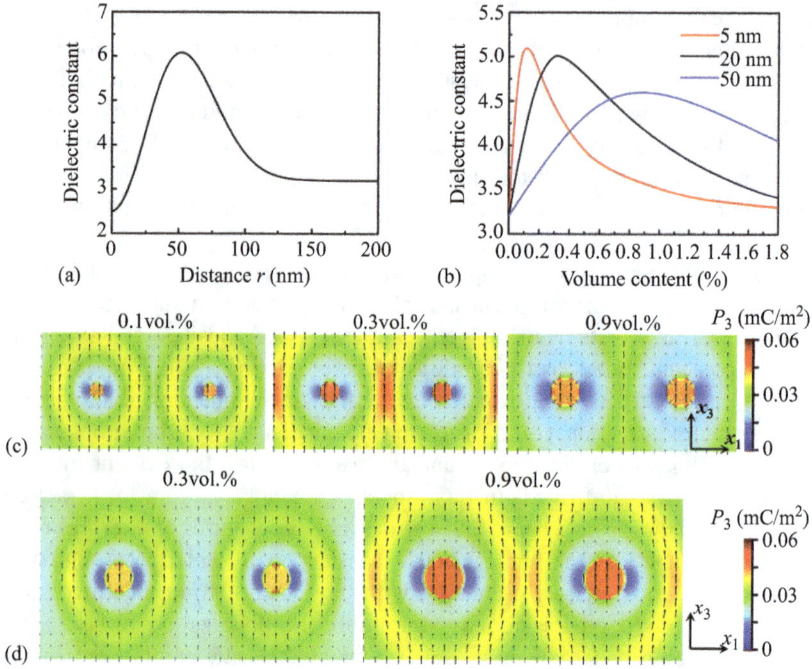

Figure 6.6 *(a) Distribution of local electric constant around the particle. (b) Predicted dielectric constant of nanocomposites with different volume fractions with 5, 20, and 50 nm particle nanofillers. Polarization distribution in nanocomposites with (c) 20 nm nanofillers at 0.1, 0.3, and 0.9 vol.% and (d) 50 nm nanofillers at 0.3 and 0.9 vol.%. © 2017 The Royal Society of Chemistry. Reproduced, with permission, from [29]*

exhibits polar regions at the initial state, corresponding to polarization of nano-region domains in relaxor ferroelectrics. After applying an electric voltage of 0.15 V along z direction, the dipoles in BZT response faster than P(VDF-TrFE-CFE), leading to a substantial change in P_z. However, at the interface region, the dipoles in polymer change remarkably as much as those in BZT. As the P_z in BZT increase, a higher interfacial polarization can be induced as shown in different states, attributed to the dipole rotation to Pz in P(VDF-TrFE-CFE) at the interface where BZT nanofiber induces preferential crystallization and easy ferroelectric switching of P(VDF-TrFE-CFE).

6.4 Applications of phase-field computation on experimental design

As MGI launched in 2011, the paradigm of materials discovery is changed from trial-and-error method to the parallel model of experiments and calculations.

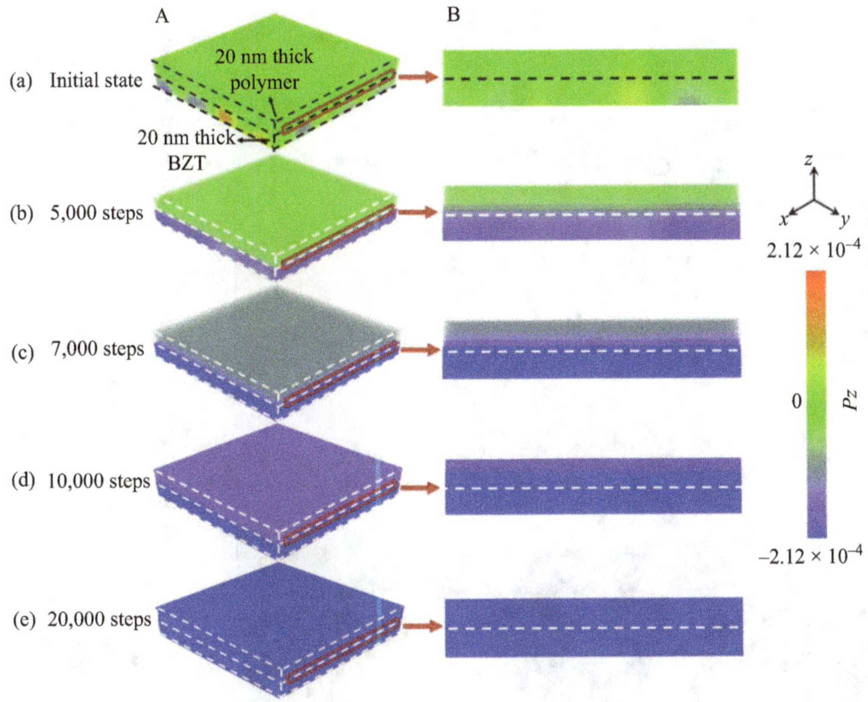

Figure 6.7 The evolution of polarization exchange coupling around the interface region under an applied voltage of 0.15 V along z-axis. © 2018 Wiley-VCH. Reproduced, with permission, from [30]

Today, the computational simulation has become one of the most important components in the development of material research to assist the discovery and design of new materials [31]. A large number of computing paradigms have been successfully proposed to study the underlying mechanisms and to design new materials. In this section, applications of phase-field simulations are taken as examples to introduce how the computational simulation assists the experimental design in dielectric materials for capacitors.

6.4.1 *Inorganic dielectric*

Inorganic dielectric has been a commercialized candidate for energy storage in electrical and electronic systems and extensive efforts have been made to improve the energy density to meeting the increasing demand of advanced systems. Recently, Pan *et al.* synthesized lead-free $BiFeO_3$-$BaTiO_3$-$SrTiO_3$ (BFO-BTO-STO) solid-solution film with ultrahigh energy density by polymorphic nanodomain design, guided by the predicted results of phase-field simulations [32]. According to Landau theory, it is easier for polarization to switch between <111>R and <001>T in polymorphic nanodomain structure, as the energy profile

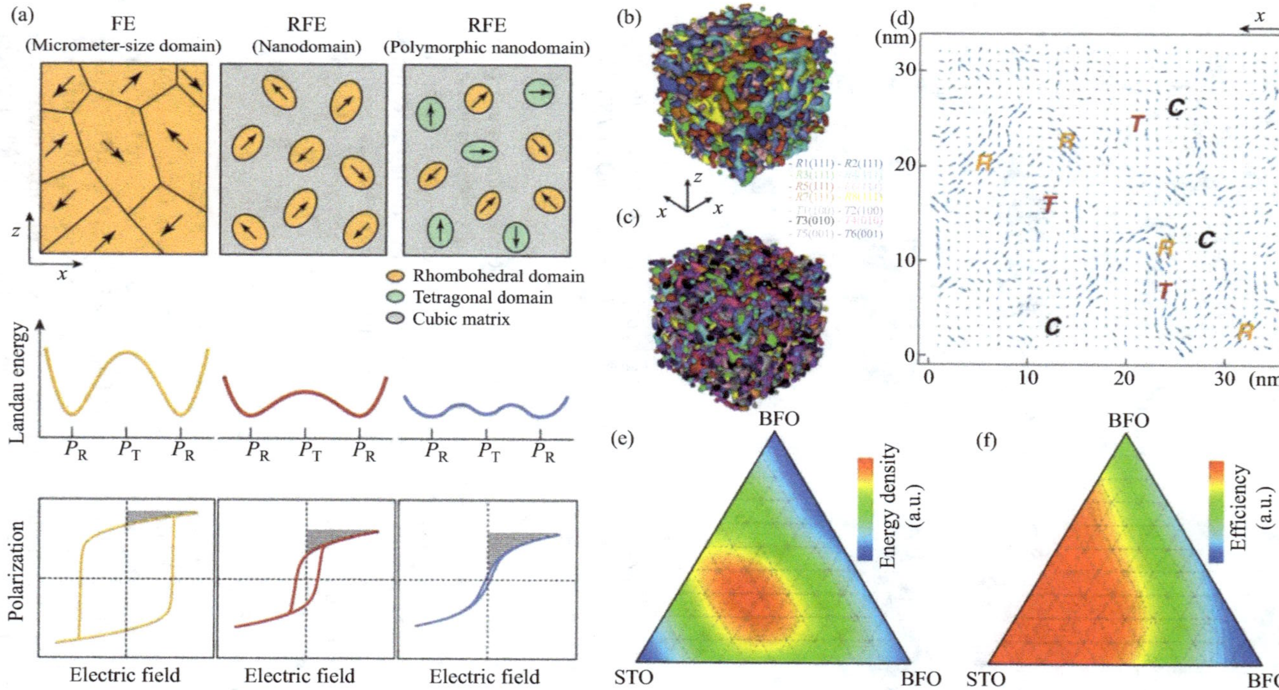

Figure 6.8 *(a) Schematics of domain structures, Landau energy profiles, and P-E loops of a ferroelectric (FE) with micrometer-size domains, a relaxor ferroelectric (RFE) with nanodomains, and an RFE with polymorphic nanodomains. (b) and (c) Simulated three-dimensional domain structures of 0.45BFO-0.55STO with R nanodomains and 0.20BFO-0.25BTO-0.55STO with coexisting R and T nanodomains. (d) Simulated two-dimensional domain structure of 0.20BFO-0.25STO-0.55STO (e) and (f) contour plots of predicted energy density and efficiency of BFO-BTO-STO solid solutions. © 2019 AAAS. Reproduced, with permission, from [32]*

shown in Figure 6.8(a). Also, the phase-field simulations reveal that the BFO-BTO-STO ternary system (0.20BFO-0.25BTO-0.55STO) indeed has a more disordered structure with coexisting R and T nanodomains, as shown in Figure 6.8(c) and (d). Moreover, *P-E* loops of BFO-BTO-STO solid solutions with different compositions have been stimulated and summarized by calculating the energy density and efficiency as concluded in the contour plots in Figure 6.8(e) and (f). One of the most striking features is that the highest U_e lies inside the BFO-BTO-STO triangle, indicating that the BFO-BTO-STO solid solution has great potential of improving the U_e. However, the efficiency of this system is different and more STO content results in higher efficiency due to the reduced hysteresis. Therefore, in order to keep balance of high energy density and high efficiency, a series of compositions (0.55-*x*)BFO-*x*BTO-0.45STO were fabricated. Finally, a high energy density of 112 J/cm^3 and a high-energy efficiency of ~80% are obtained when $x = 0.3$, which is proved to be a successful case of designing a new solid-solution assisted by the phase-field simulations.

Recently, template grain growth has provided a low-cost approach to fabricate textured ferroelectrics with qualities comparable to single crystals. As the template phase usually exhibits properties inferior to the matrix phase, the corresponding dielectric and ferroelectric properties could be largely changed in the polycrystalline-textured structure. Since increasing content of template phase could enhance the quality of matrix phase and meanwhile enlarge the contribution of unwanted template phase, it is crucial to understand the competing effects on the textured microstructure of the grain texture and template volume fraction, so that optimized dielectric and ferroelectric properties could be achieved in this trade-off. Zhou *et al.* used the phase-field method to simulate polycrystalline template-matrix BT-PZT composites with different template phase content, and hence different microstructure and ferroelectric property [33], as shown in Figure 6.9. In their efforts, a hybrid-phase polycrystalline phase-field method is established based on the single-phase method. For the purpose of clear comparison, the hysteresis loop of pure PZT and pure BT with same grain microstructures are also simulated in their work (Figure 6.9(a) and (b)). It is evident that pure BT exhibits smaller residual polarization P_r and coercive field E_c than pure PZT with same grain microstructure, while the template-matrix BT-PZT composite polycrystals show moderate dielectric properties. When small amount of BT is incorporated as template (0.63%), the *P-E* loop profile shows little difference to the profile from pure PZT. Although PZT grains with high quality are induced by increasing BT content (insets in Figure 6.9(b)), the *P-E* loop profile of the composite still deviates away from that of PZT to that of BT, exhibiting lower residual polarization P_r and coercive field E_c as shown in Figure 6.9(b). Therefore, an optimized composition of template-matrix ferroelectric ceramics is suggested in this work, along with the grain microstructure evolution and crystallographic texture development. Moreover, based on the simulation results, the authors have fabricated the representative textured microstructures in experiments to achieve high dielectric properties.

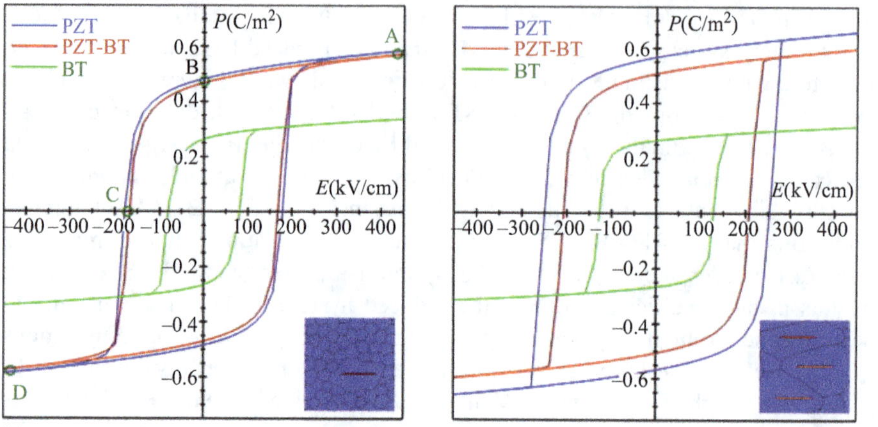

(a) (b)

Figure 6.9 The predicted P-E field hysteresis loop of polycrystals with different template phase volume fractions: (a) 0.63% (b) 2.37%. Three polycrystals are considered with same grain structures and textures in each figure. © 2017 AIP Publishing. Reproduced, with permission, from [33]

6.4.2 Polymer-based nanocomposite

Electrostatic capacitors have been extensively applied in a number of modern electronic and electrical systems due to their specific capability of delivering an ultrahigh power density, low dielectric loss, and high operating voltage. However, the low energy density of the commercially electrostatic capacitor falls short of the increasing demands of reliable, compact, and efficient systems. Therefore, it is imperative to find new solutions to improve the energy density of the electrostatic capacitors.

Upon an applied electric field, the stored energy density U_e could be calculated from electric displacement-electric (*D-E*) field loop as $U_e = \int_{D_{max}}^{0} EdD$. For linear dielectrics, the maximal energy density can be simplified to $U_{max} = 1/2\varepsilon_0\varepsilon_r E_b^2$, where ε_0 is the vacuum permittivity, ε_r is the relative dielectric constant, and E_b is the breakdown strength. Thus, both dielectric constant and breakdown strength are important to high energy density. However, there seems to be an inverse relationship between them in single-phase materials, indicated by the high dielectric constant but low breakdown strength from ceramics and the low dielectric constant, but high breakdown strength from polymers. For the past decade, polymer nanocomposites composed of polymer matrix with high breakdown strength and ceramic nanoparticles with high dielectric constant have been proposed and studied to improve the energy density intensively.

However, due to the large dielectric mismatch between the polymer matrix and ceramic fillers, the local electric field distribution may become inhomogeneous and concentrate around the interfaces, resulting in high electric field at the interface when moderate electric field is applied, and several issues related to this amplified interfacial electric field. The most problematic one is that the interfacial regions with distorted electric field will become the hot spots in the breakdown process, and serve as the starting points to trigger easy propagation of the breakdown paths, decreasing the breakdown strength of the composites. As the maximal energy density is quadratically related to the breakdown strength, the interfacial issues have become the most urgent obstacle to be overcome. However, limited by the experimental characterization techniques, it is hard to characterize the local electric field distribution in this highly non-equilibrium process. As an alternative approach, phase-field method has been extensively applied to simulate the breakdown process, study the underlying breakdown mechanism, and then give beneficial guidance on the experimental design.

Recently, Shen *et al.* developed a continuum phase-field model to study the breakdown process, inspired by the crack propagation process [25]. A high throughput calculation has been performed to study the microstructure effect on dielectric properties, as shown in Figure 6.10. Firstly, a microstructure dataset consisting of 400 polymer nanocomposites with different microstructures was established, as shown in Figure 6.10(d). In this dataset, different microstructures are defined by assigning fillers with different length ratios, and the volume fraction is fixed at 10%. For example, the a_x/a_z and a_y/a_z are assigned to be both 0.067 in the microstructure defined as S1. Therefore, on the basis of this microstructure dataset, corresponding datasets of the breakdown strength, effective dielectric constant, and resultant energy density by a linear approximation of the composites with simulated microstructures could be predicted by high-throughput phase-field calculations, as shown in Figure 6.10(a), (b), and (c), respectively. The most striking features in the predicted datasets is that, with the shape of fillers changing from vertical nanofiber in S1 to parallel nanosheet in S5, the breakdown strength of polymer nanocomposites increases but the effective dielectric constant decreases, consistent with the adversely coupled relationship between the dielectric constant and breakdown strength. Moreover, as illustrated in Figure 6.10(c), the energy density varies monotonically from the vertical nanofiber in S1 to the parallel nanosheet in S5, in accordance with the trend in dataset of the breakdown strength, despite the decreasing effective dielectric constant in these microstructures. Therefore, it is suggested that the polymer nanocomposites filled with parallel nanosheets exhibit much higher energy density than composites with other microstructures, which agrees well with the experimental results.

In addition, based on the understanding of microstructure effects on dielectric properties from those datasets, Shen *et al.* further proposed a design of sandwich-structured nanocomposite filled with parallel nanosheets at the upper and lower layers and vertical nanofibers at the middle layer to improve the energy density. Here, the parallel nanosheets are introduced to enhance the voltage resistance and the vertical nanofibers are responsible for high effective dielectric constant. After optimizing the thickness of each layer, the maximal energy density could reach 2.42 times that of the polymer matrix, which is higher than that of the polymer matrix by 142%, and higher than that of the composites filled with parallel

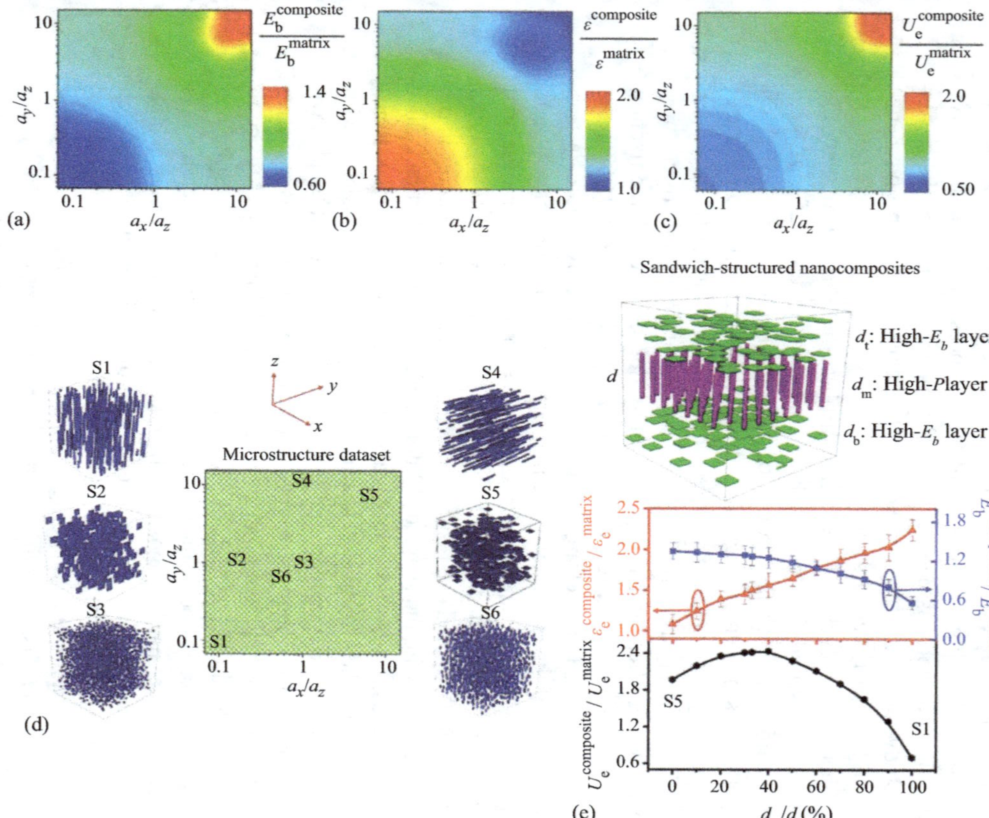

Figure 6.10 Property datasets of (a) breakdown strength, (b) effective dielectric constant, and (c) energy density by high-throughput calculations. (d) Microstructure dataset composed of 400 kinds of microstructures by assigning different length ratios of the nanofillers. (e) A designed sandwich-structured nanocomposite with optimized dielectric properties. © Wiley-VCH. Reproduced, with permission, from [25]

nanosheets by 22.8%. In this work, a phase-field model is built to understand the electrostatic breakdown process in polymer nanocomposites with various microstructures related to aspect ratio of filler. Furthermore, as guided by the calculated data-driven design paradigm, a novel microstructure with concurrently enhanced dielectric constant and breakdown strength is proposed to achieve composite dielectrics for high-energy-density electrostatic capacitors.

Besides the effect of microstructure with different filler shapes, the phase-field method has also been applied to investigate the influence of inhomogeneous distribution of fillers on the breakdown behavior by calculating the evolution of electrical treeing process in nanocomposites with designed out-of-plane gradient and in-plane alignment [3]. A series of 10-layer P(VDF-HFP)-based composites containing BTO nanoparticles and nanofibers with volume fraction fixed at 5% are generated, as shown in Figure 6.11(a) and (b).

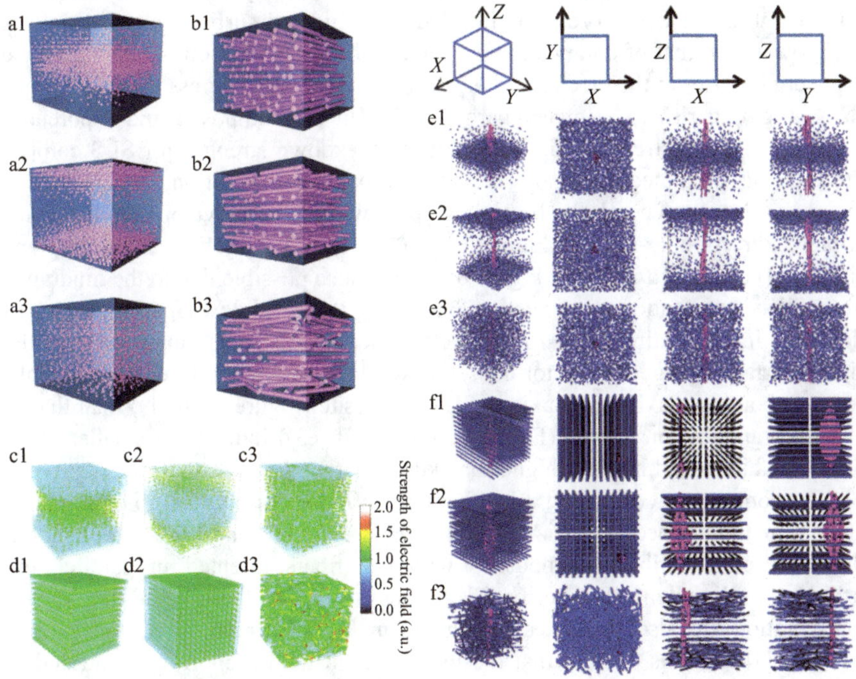

Figure 6.11 Schematic illustration of generated P(VDF-HFP)-based nanocomposites containing (a) nanoparticles and (b) nanofibers with different configurations. Phase-field simulations on electric field distribution in nanocomposites containing (c) nanoparticles and (d) nanofibers with different configurations. The final steady state of the breakdown path evolution in nanocomposites containing (e) nanoparticles and (f) nanofibers with different configurations by phase-field method observed along different projection axis.

For composites containing nanoparticles, the fillers are modulated with composition gradients along the out-of-plane direction. For example, a small-large-small (SLS) composition gradient composite is generated by changing the filler volume fraction from one surface to another in sequence as 0, 2.5, 5, 7.5, 10, 10, 7.5, 5, 2.5, and 0%, as shown in Figure 6.11(a1). Similarly, a large-small-large (LSL) composite can be generated (see Figure 6.11(a2)), along with a composite with random and uniform distribution of BTO nanoparticles (see Figure 6.11(a3)). The local electric field distributions dependent on the generated microstructures are visualized by phase-field method, as shown in Figure 6.11(c). For the composites with gradient composition, the electric field is also generated with gradient features (Figure 6.11(c1) and (c2)), as higher content of BTO nanoparticles will induce more concentrated local electric field. Therefore, in SLS composite, the local electric field in the outer surface layers is highly mitigated (Figure 6.11(c1)), while intensified local electric field is induced in the outer layers of LSL composite (Figure 6.11(c2)). It is worth noticing that the mitigated surficial electric field is even lower than that of composite with random filler distribution, which will raise the energy barrier at the electrode/dielectric interface and suppress the injection of charges, giving rise to a lower charge density in the composite and a potential higher breakdown strength. Enhancement in breakdown strength for SLS composite is further predicted by steady state of the breakdown evolution, as presented in Figure 6.11(e). In general, the breakdown path will mainly develop along the out-of-plane direction (z-axis in Figure 6.11(e)), while propagation along in-plane directions (x-axis and y-axis in Figure 6.11(e)) is also possible due to the hindrance from fillers as mentioned in Section 6.3.2. For the breakdown paths developed along the in-plane directions, more electric energy will be dissipated due to the higher surface energy of the tortuous breakdown paths. Therefore, wider breakdown path along x-axis and y-axis in SLS composite (Figure 6.11(e1)) than that in LSL and random composites (Figure 6.11(e2) and (e3)) indicates a smaller probability of failure and hence a higher breakdown strength.

For composites containing nanofibers, the effect of orientation distribution of filler with high aspect ratio is investigated. Besides a composite with random alignment of nanofibers, composites with nanofibers oriented in parallel and orthotropic (alignment of nanofibers is vertical to that of the neighboring layers) configuration are also generated, as shown in Figure 6.11(b). As visualized in Figure 6.11(d6), it is evident that the local electric field is more concentrated in composite with randomly aligned nanofibers, due to the higher density of connecting points between nanofibers. In contrast, the well-aligned nanofibers help to build more homogeneous local electric field, as revealed by the electric field distribution of nanocomposites filled with aligned nanofibers (Figure 6.11(d4) and (d5)). Moreover, compared to the composite with randomly aligned nanofibers (Figure 6.11(f3)), the tortuosity of breakdown path is dramatically enhanced in fiber-reinforced composites with orthotropic and parallel configurations, indicated by the large widening of the breakdown paths along x-axis or y-axis in Figure 6.11 (f1) and (f2). More notably, the widening of breakdown paths is observed along both x-axis and y-axis in orthotropic composites, while in parallel composite the widening happens mainly along *y*-axis, attributed to the impedance of nanofibers to the propagation of breakdown paths. Therefore, higher breakdown strength in composites filled with orthotropic nanofibers is expected due to their ability to dissipate more electric energy by breakdown strength widening. In addition, composite with optimized configuration, e.g., SLS or orthotropic configurations, could remain the dielectric constant in the same level of the random composite, as the effective contents of the inorganic filler in both composites are the same, which guarantees an enhanced breakdown strength without compromised dielectric constant. Experimental works have also verified the rules from these simulations.

In the above discussions, phenomenological evolutions of breakdown process have been successfully elucidated, while more efforts are devoted to unveil the breakdown mechanisms in dielectrics under realistic situations. Typically, electric breakdown is applied to describe the breakdown process where avalanche multiplication of free charge carriers at high electric field causes irreversible damages to the dielectric. In this case, the insulating dielectrics will be converted into conductors because of the ionization current generated by the collisions between high-energy carriers and the matrix. However, in nanocomposite consisting of polymeric matrix with low thermal conductivity and low Young's modulus, additional breakdown mechanisms, i.e., thermal breakdown and mechanic breakdown, are introduced. The thermal breakdown refers to the thermal instability of the dielectrics under high voltage when Joule heat generated by low-density carriers cannot be effectively dissipated to the surroundings. The mechanical breakdown may take place when shrinkage of dielectrics caused by the electrostriction, piezoelectric effects, and electrostatic attraction between the oppositely charged electrodes overwhelms the dielectrics. An extended phase-field model is built by Shen and co-authors to illustrate the contributions of these breakdown mechanisms [34], as shown in Figure 6.12(a)–(c). Similar intermediate states of pure P(VDF-HFP) with two-needle electrodes at 295, 323, and 363 K is illustrated in Figure 6.12(a1), (b1), and (c1), respectively. At 295 K, the polymer is on the verge of breakdown around 700 kV/mm. When the temperature is elevated to 323 K, the polymer is entirely broken down at 580 kV/mm. If the temperature is further increased to 363 K, the electric field resulting in similar intermediate state will decrease to 415 kV/mm. Therefore, an increase of 68 K in temperature will reduce the breakdown strength from 700 to 415 kV/mm. The distributions of local electric energy density f_{elec}, Joule heat energy density f_{Joule} and strain energy density f_{strain} corresponding to the state in Figure 6.12(a1), (a2), and (a3) are calculated from phase-field simulations (see Figure 6.12(a2)–(a4), (b2)–(b4), (c2)–(c4)). For all kinds of energy densities, the inside and front of the breakdown path exhibits higher value than other regions, leading to the forward growth of the breakdown path. Moreover, the local electric energy density (Figure 6.12(a2), (b2), (c2)) exhibits less dependence on temperature, as the main difference within distributions at different temperature is caused by the different value of applied electric field according to $f_{elec} = 1/2\varepsilon_0\varepsilon_r E^2$. This is consistent with the fact that dielectric constant of P(VDF-HFP) barely changes when temperature is changed from 295 to 363 K. Nevertheless, the thermal energy densities at these three temperatures show large contrast, as shown in Figure 6.12

(a3), (b3), and (c3). The Joule heat energy density is ~2×10^7 J/m^3 at 295 K, whereas the energy density increases to ~10^8 J/m^3 at 363 K, which is ascribed to enhancement in conductivity of P(VDF-HFP) by several orders of magnitude from ~10^{-10} S/m^1 at 295 K to ~10^{-8} S/m^1 at 363 K, revealing the important roles of thermal breakdown under elevated temperature. However, the strain energy density shows little change at different temperature, as the strain energy density increases with the electric field, but decreases with the Young's modulus. Therefore, with the temperature increasing, the breakdown strength is reduced, which leads to a balanced strain energy density. As the decrease in breakdown strength at high temperature is induced by thermal breakdown, it can be suggested that the mechanical breakdown is coupled with thermal breakdown in some extent.

To achieve an analytical expression for breakdown strength as a function of material parameters, Shen *et al.* further performed high-throughput phase-field simulations and machine learning to obtain a predictive model for P(VDF-HFP)-based nanocomposites. Figure 6.12(d) illustrates the strategy proposed to combine machine learning and high-throughput simulations in composites with random nanoparticle distribution and volume fraction fixed at 5%. In the paradigm, the dielectric constant, electrical conductivity, and Young's modulus of fillers are chosen as the variables and fingerprints of the calculated dataset of breakdown strength for the machine learning. Four interactions within these variables and 12 prototypical functions are considered by least-squares regression (LSR) to generate descriptors for regression analysis. To further screen the descriptor, the coefficient of determination (R^2) of the LSR is used as the criterion, and three rounds of LSR and screening are conducted to let a predictive expression of the breakdown strength.

The breakdown strength from the dataset as a function of the dielectric constant, electrical conductivity and Young's modulus is illustrated in Figure 6.12(e). Obviously, during the whole range of all the parameters, the change in breakdown strength versus electrical conductivity ($x_2(\sigma)$) is the most evident. More specifically, when dielectric constant and Young's modulus fixed at 13.5 and 0.982 GPa, the normalized breakdown strength decreases from 1.34 with electrical conductivity of 2.12×10^{-13} S/m^1 to 0.32 with electrical conductivity of 2.12×10^{-17} S/m^1, consistent with the simulations conducted on the pure P(VDF-HFP) in Figure 6.12(a). After three rounds of screening, a simplified practical analytical expression ($R^2 = 0.9093$) with only considering the interaction between electrical conductivity and Young's modulus can be expressed as,

$$\frac{E_b^{\text{composite}}}{E_b^{\text{matrix}}} = 0.9085 - 0.01175 \ln \frac{\varepsilon_{\text{filler}}}{\varepsilon_{\text{matrix}}} - 0.06767 \ln \frac{\sigma_{\text{filler}}}{\sigma_{\text{matrix}}} - 0.01640\left(\frac{Y_{\text{filler}}}{Y_{\text{matrix}}}\right)^{-\frac{1}{2}} + 0.001\left(\frac{Y_{\text{filler}}}{Y_{\text{matrix}}}\right)^{-\frac{1}{2}} \ln \frac{\sigma_{\text{filler}}}{\sigma_{\text{matrix}}} \tag{6.8}$$

Equation (6.8) suggests an enhancement in breakdown strength when fillers with low dielectric constant, low electrical conductivity, and high Young's modulus are incorporated. Moreover, a larger absolute value of the factor before the term $\ln \frac{\sigma_{\text{filler}}}{\sigma_{\text{matrix}}}$ agrees well with the larger change in breakdown strength versus electrical conductivity in Figure 6.12(e).

Additionally, by comparing the predictions with extra phase-field simulations and available experimental measurements, the reliability of the analytical expression is further examined. Along with seven data points simulated but not included in the training dataset (colored solid symbols in Figure 6.12(f), five data points from experimental measurements (colored open symbols in Figure 6.12(f)) are displayed in Figure 6.12(f), and all the test points are dispersed around the solid line representing $E_b^{\text{Simulation/measurement}} = E_b^{\text{Regression}}$, indicating a good agreement between the predictions and the simulated or measured values. However, there are still some factors, including voids and space charges, not considered in this phase-field method leading to the difference between the predicted value from machine learning and the actual value from experiments.

6.4.3 Multilayer dielectric

Although interfaces could be troublesome to handle with, multilayer dielectrics consisting of different compositions in neighboring layers have attracted research interests as the dielectric and chemical mismatch at the mesoscopic interfaces between layers has been mitigated. Therefore, suppressed conduction loss due to the reduced charge migrations by interfaces could contribute to the increase in discharge efficiency and breakdown strength, hence of higher energy density and better reliability of the multilayer capacitors. However, the interface effect in multilayer dielectrics has only been proposed, but hardly verified, due to the experimental difficulties in analyzing this vertically aligned two-dimensional structure. Therefore, phase-field method has been applied to assist the explanation on experimental results obtained in multilayer dielectrics.

To achieve high breakdown strength and high efficiency in polymeric dielectric, Jiang *et al.* has fabricated P(VDF-HFP)/P(VDF-TrFE-CFE) multilayer with 4 layers, 8 layers, and 16 layers, while volume fraction of each polymer is all fixed at 50% [35], as illustrated in Figure 6.13(a). The role of mesoscopic interfaces between the two components is reflected by two experimental phenomena. On the one hand, the multilayer dielectrics exhibit breakdown strength more similar to the value in P(VDF-HFP), instead of getting closer to the average of values in P(VDF-HFP) and P(VDF-TrFE-CFE) (see Figure 6.13(b)). Furthermore, the breakdown strength rises as the layer number increases, and the breakdown strength in composite with 16 layers is even higher than that in P(VDF-HFP). On the other hand, the discharge efficiency in multilayer dielectrics also exhibit deviation from the average of the values in pure components (Figure 6.13(c)). Despite the low efficiency in P(VDF-TrFE-CFE), near all the multilayer composites show higher efficiency than that in P(VDF-HFP), as 4-layered composite quenched at 45 or 60 °C exhibits slightly higher efficiency than P(VDF-HFP); whereas, 8-layered and 16-layered composites exhibit greatly enhanced efficiency, with the highest value of 85.1% obtained in 16-layered composites quenched at 60 °C.

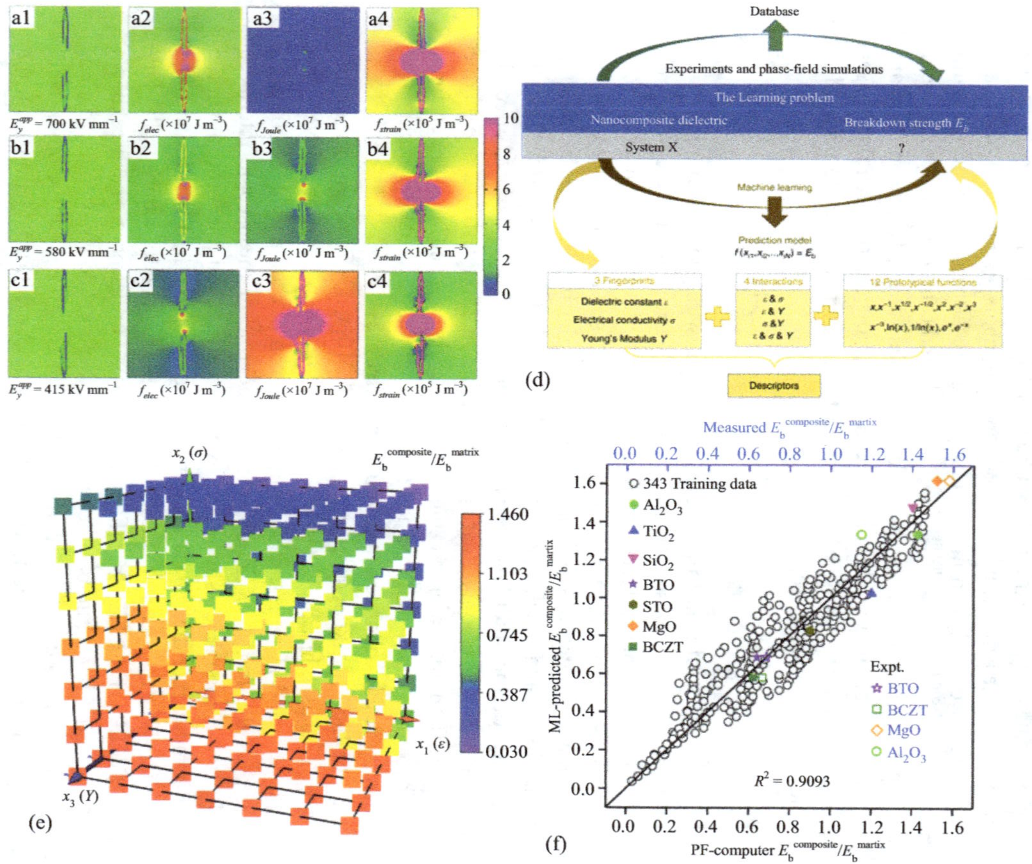

a1
a2
a3
a4
E_y^{app} = 700 kV mm^{-1}
f_{elec} (×10^7 J m^{-3})
f_{Joule} (×10^7 J m^{-3})
f_{strain} (×10^5 J m^{-3})
b1
b2
b3
b4
E_y^{app} = 580 kV mm^{-1}
c1
c2
c3
c4
E_y^{app} = 415 kV mm^{-1}
Database
Experiments and phase-field simulations
The Learning problem
Nanocomposite dielectric
Breakdown strength E_b
System X
?
Machine learning
Prediction model
3 Fingerprints
Dielectric constant ε
Electrical conductivity σ
Young's Modulus Y
4 Interactions
ε & σ
ε & Y
σ & Y
ε & σ & Y
12 Prototypical functions
Descriptors
(d)
x_2 (σ)
$E_b^{composite}/E_b^{matrix}$
1.460
1.103
0.745
0.387
0.030
x_1 (ε)
x_3 (Y)
(e)
Measured $E_b^{composite}/E_b^{martix}$
343 Training data
Al_2O_3
TiO_2
SiO_2
BTO
STO
MgO
BCZT
Expt.
BTO
BCZT
MgO
Al_2O_3
R^2 = 0.9093
ML-predicted $E_b^{composite}/E_b^{matrix}$
PF-computer $E_b^{composite}/E_b^{martix}$
(f)

Figure 6.12 *(a1)–(a4) Breakdown evolution (a1) and distributions of corresponding electrical energy (a2), thermal energy (a3), and strain energy (a4) at 295 K. (b1)–(b4) Breakdown evolution (b1) and distributions of corresponding electrical energy (b2), thermal energy (b3), and strain energy (b4) at 323 K. (c1)–(c4) Breakdown evolution (c1) and distributions of corresponding electrical energy (c2), thermal energy (c3), and strain energy (c4) at 363 K. (d) Machine learning strategy based on high-throughput phase-field simulations. (e) High-throughput simulation results on nanocomposites with changing filler parameters. (f) Comparisons of breakdown strengths between machine learning predictions and phase-field simulations included (black open symbols) or not included (colored solid symbols) in the dataset or experimental measurements (colored open symbols). © 2019 Springer Nature. Reproduced, with permission, from [34]*

Thereafter, phase-field method is applied to explain the nonlinear performance related to dielectric breakdown and discharge efficiency. In this model, the interfaces between P(VDF-HFP) and P(VDF-TrFE-CFE) are considered as a third phase with a dielectric constant of ~500 induced by Maxwell–Wagner–Sillars interfacial polarizations. Figure 6.13(d) depicts the distribution of local electric field along the out-of-plane direction in 8-layered composite. The effective electric fields in P(VDF-HFP) layers is 1.5 times of the applied electric field, due to its lower dielectric constant (~8.6) compared to that of P(VDF-TrFE-CFE) (~35). However, a much lower electric field is simulated at the interfacial regions, with $E_z^{\text{local}}/E_z^{\text{app}}$ ratio of ~0.1, which could serve as trap sites to migrate charged defects, as low local electric field will reduce the energy of passing ions or charges, suppressing the conduction loss. Therefore, more interfaces from more layers in composite could provide more impediment on charge carriers, and thus higher discharge efficiency could be obtained. Moreover, the real-time evolutions of breakdown paths are illustrated in Figure 6.13(e)–(g). The most striking feature is that the breakdown paths become more tortuous and wider in the in-plane direction when there are more layers in the generated composites, which could be attributed to the deceleration of breakdown path at two-dimensional interfaces with low local electric field, consistent with increasing breakdown strength along with increasing layer number in composites. Therefore, the phase-field simulations evidence the blocking effects of mesoscopic interface on breakdown paths and charge carriers in multilayer dielectrics, confirming that multilayer strategy can indeed enhance the efficiency and breakdown strength of the dielectrics.

Following, to further explore nanocomposites with concurrently high energy density and discharge efficiency, Jiang *et al.* proposed an effective route where mesoscopic interfaces in multilayer structure and microscopic interfaces in ceramic/polymer composites are integrated to suppress the charge injection at dielectric/electrode and dielectric/dielectric interfaces [36]. In a P(VDF-HFP)-based nanocomposite with volume fraction of BTO at 10%, the maximal energy density could be enhanced to 30.15 J/cm^3 with an efficiency of ~78%. When conceiving this design idea, the computational simulations give important guidance on how to modulate the structure of nanocomposites. As shown in Figure 6.14, the distributions of local electric field and charge density in nanocomposites with various configurations are simulated by the single electron injection model based on the phase-field theory. Firstly, a nanocomposite filled with high volume fraction of BTO nanoparticles is simulated (Figure 6.14(b) and (c)). When an electric field is applied to the dielectrics, much higher charge density injected from the cathode (see Figure 6.14(b)) is observed in Figure 6.14(c), compared to a moderate charge density in pure P(VDF-HFP) in Figure 6.14(d). This contrast is ascribed to the introduction of randomly distributed BTO nanoparticles, which are less insulating and of high electron mobility, promoting the possibility of charge injection from the electrode and forming faster transport paths for charge carriers. Therefore, modulating the distribution of nanoparticles becomes necessary due to the high leakage current in nanocomposites with uniformly dispersed fillers. Hereby,

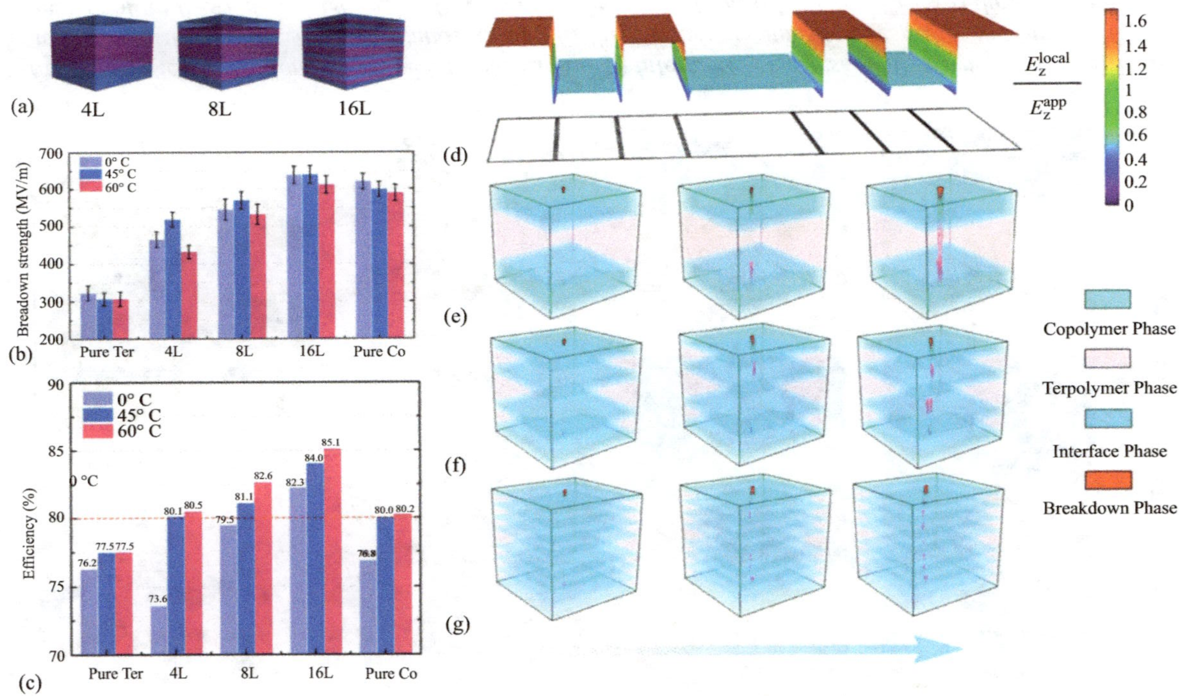

Figure 6.13 *(a) Schematic illustration of multilayer composites with 4, 8, and 16 layers. (b) Breakdown strength of multilayer composites and pure components quenched at 0, 45, and 60 °C. (c) Efficiency of multilayer composites and pure components quenched at 0, 45, and 60 °C. (d) The distribution of local electric field along out-of-plane direction in 8-layered composite upon applying electric field. (e)–(g) Time-dependent three-dimensional views of the breakdown evolutions for 4-layered (e), 8-layered (f), and 16-layered (g) composites. © 2019 Elsevier. Reproduced, with permission, from [35]*

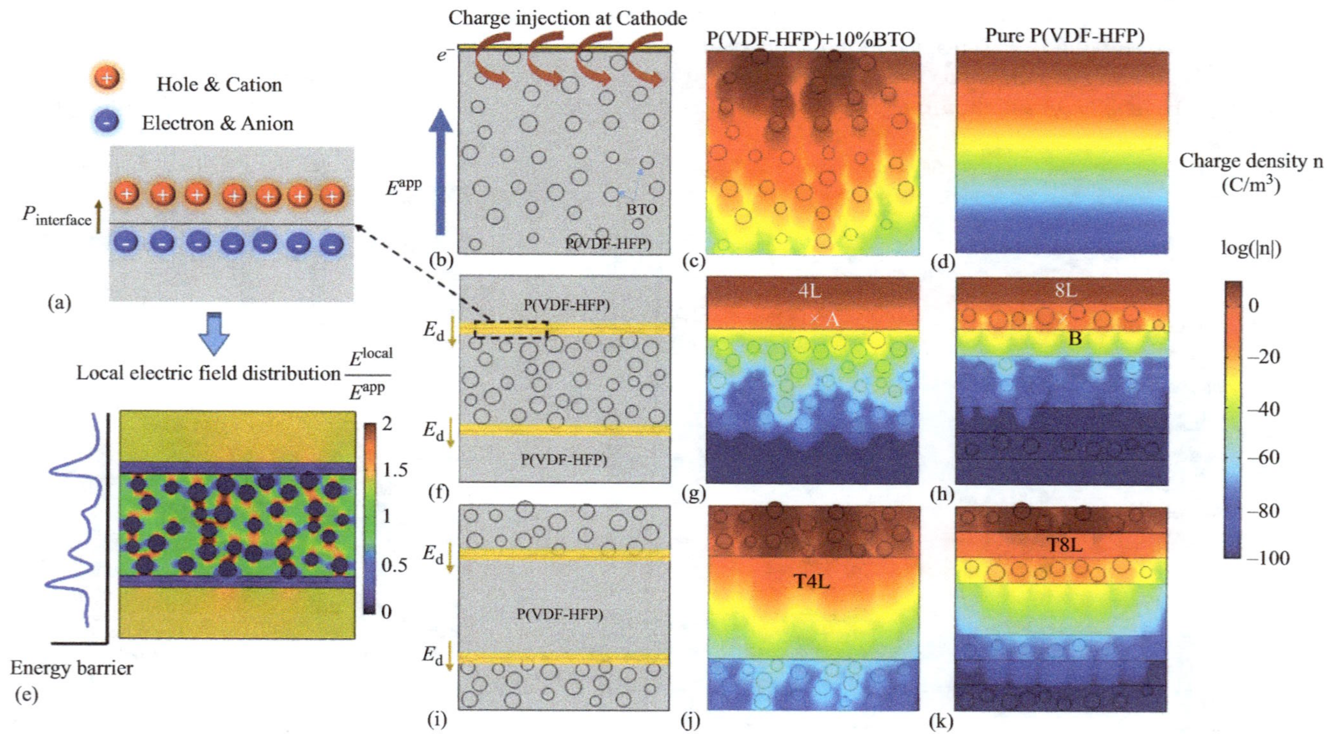

Figure 6.14 *(a) Schematic illustration of interfacial polarization by interfacial charges. (b) 0–3 type P(VDF-HFP)-BTO nanocomposites, (f) the 4L nanocomposites and (i) the T4L nanocomposites, respectively. The distributions of charge density of (c) 0–3 type P(VDF-HFP)-BTO nanocomposites, (d) pure P(VDF-HFP), (g) the 4L nanocomposites, (h) the 8L nanocomposites, (j) the T4L nanocomposites, and (k) T8L nanocomposites. (e) Local electric field distribution and energy barrier of the 4L nanocomposites. © 2019 Elsevier. Reproduced, with permission, from [36]*

nanocomposites denoted as 4L and 8L, with the BTO nanoparticles distributed in the inner layers, and nanocomposites denoted as T4L and T8L, with BTO nanoparticles distributed in outer layers, are generated to study the interface effects on charge behaviors (Figure 6.14(g)–(k)). For all the multilayered nanocomposites, their charge densities are much lower than that of the uniform nanocomposites, comparable with that of the pure P(VDF-HFP), which prove the superiority of the multilayer strategy. Moreover, compared to 4L and 8L composites, higher charge density appears in T4L and T8L composites because of the severe charge injection at the interface between cathode and outer layers filled with BTO. Besides, as the number of stacking layer increases, the charge transport could be suppressed due to the interface effect of polarized ions between different dielectric layers, as shown in Figure 6.14(h) and (k). Strong depolarized field could be induced at the dielectric/dielectric interfaces by the local polarized ions, which may form a higher energy barrier for charge migration as shown in Figure 6.14(e). In consequence, these phase-field simulations predict that the charge behavior and corresponding leakage current and efficiency could be modulated by designing the dielectric/electrode and dielectric/dielectric interfaces, which has also been verified several experiments [37–39].

6.5 Other simulation methods for dielectrics

This chapter mainly focuses on recent applications of phase-field method in dielectrics. However, besides the phase-field method, other computational tools, ranging from density functional theory (DFT) calculation to effective medium theory, have also been successfully applied in dielectrics to address various issues at different length scales. For example, Ramprasad *et al.* used the first-principles computational methodologies to study the impact of surfaces and interfaces on the local properties in nanodieletrics at atomic length [40]. This work is a groundwork due to its thorough analysis of the effects of atomic level defects on electronic and dielectric properties. In addition, based on density functional perturbation theory, Sun *et al.* built an average electron model to calculate the intrinsic breakdown strength of insulating materials [41], which provides a physical insight into the material properties affecting the breakdown process. However, when the simulation scale rises up to micrometers, mesoscopic or macroscopic simulation methods are more suitable to bridge the microstructure and effective properties of a dielectric. For instance, Guseva *et al.* studied the voltage breakdown phenomenon in a polymer nanocomposite with random metal spheres by periodic Monte Carlo method [42]. It has found that in nanocomposites with randomly distributed spheres, the voltage breakdown is strongly dependent on the local microstructure. For breakdown discharge phenomenon, the simplest nontrivial stochastic model developed by Niemeyer *et al.* could naturally leads to fractal structure for the discharge pattern [43]. When calculating the effective properties of a dielectric, effective medium theory may be a good choice, like Li and co-authors did in their work [44], where the effective permittivity, breakdown strength, and electric energy density of nanocomposites are analyzed. Thus, it is important to choose suitable simulation tool for different problems since every method has its own advantages and limitations. Furthermore, we should try to integrate various methods from atomic level to macroscopic scale, to give explanations to more complex phenomena in dielectrics at multiple scales.

In a word, computational simulations have become effective and powerful tools for designing the dielectric materials and related devices. We hope this chapter could give a straight understanding on the role of computational simulation, especially the phase-field method, in the development of capacitor dielectrics, and raise people's attentions on the combination of computational simulations and experiments. We believe that the phase-field method and other methods could play more crucial roles on designing novel dielectrics under the new paradigm of material genome project.

References

[1] J. Jiang, X. Zhang, Z. Dan, *et al.*, 'Tuning phase composition of polymer nanocomposites toward high energy density and high discharge efficiency by nonequilibrium processing'. *ACS Appl Mater Interfaces*. 2017;**9**(35):29717–29731.

[2] Q. Wang and L. Zhu, 'Polymer nanocomposites for electrical energy storage'. *J Poly Sci Part B: Polym Phys*. 2011;**49**(20):1421–1429.

[3] X. Zhang, J. Jiang, Z. Shen, *et al.*, 'Polymer nanocomposites with ultrahigh energy density and high discharge efficiency by modulating their nanostructures in three dimensions'. *Adv Mater*. 2018;**30**(16):e1707269.

[4] L. A. Fredin, Z. Li, M. A. Ratner, M. T. Lanagan, and T. J. Marks, 'Enhanced energy storage and suppressed dielectric loss in oxide core-shell-polyolefin nanocomposites by moderating internal surface area and increasing shell thickness'. *Adv Mater*. 2012;**24**(44):5946–53.

[5] Z. M. Dang, J. K. Yuan, S. H. Yao, and R. J. Liao, 'Flexible nanodielectric materials with high permittivity for power energy storage'. *Adv Mater*. 2013;**25**(44):6334–65.

[6] Prateek, V. K. Thakur, and R. K. Gupta, 'Recent progress on ferroelectric polymer-based nanocomposites for high energy density capacitors: Synthesis, dielectric properties, and future aspects'. *Chem Rev*. 2016;**116**(7):4260–317.

[7] H. Palneedi, M. Peddigari, G.-T. Hwang, D.-Y. Jeong, and J. Ryu, 'High-performance dielectric ceramic films for energy storage capacitors: Progress and outlook'. *Adv Funct Mater*. 2018;**28**(42):1803665.

[8] Z. Pan, L. Yao, J. Zhai, X. Yao, and H. Chen, 'Interfacial coupling effect in organic/inorganic nanocomposites with high energy density'. *Adv Mater*. 2018;**30**(17):e1705662.

[9] X. Zhang, B.-W. Li, L. Dong, *et al.*, 'Superior energy storage performances of polymer nanocomposites via modification of filler/polymer interfaces'. *Adv Mater Interfaces*. 2018;**5**(11):1800096.

[10] S. Peng, X. Yang, Y. Yang, *et al.*, 'Direct detection of local electric polarization in the interfacial region in ferroelectric polymer nanocomposites'. *Adv Mater.* 2019;**31**(21):e1807722.

[11] R. Borgani, L. K. Pallon, M. S. Hedenqvist, U. W. Gedde, and D. B. Haviland, 'Local charge injection and extraction on surface-modified Al_2O_3 nanoparticles in LDPE'. *Nano Lett.* 2016;**16**(9):5934–7.

[12] N. Guo, S. A. DiBenedetto, P. Tewari, M. T. Lanagan, M. A. Ratner, and T. J. Marks, 'Nanoparticle, size, shape, and interfacial effects on leakage current density, permittivity, and breakdown strength of metal oxide−polyolefin nanocomposites: Experiment and theory'. *Chem Mater.* 2010;**22**(4):1567–1578.

[13] Y. Shen, Y. Lin, and Q. M. Zhang, 'Polymer nanocomposites with high energy storage densities'. *MRS Bulletin.* 2015;**40**(9):753–759.

[14] Q. Chen, Y. Shen, S. Zhang, and Q. M. Zhang, 'Polymer-based dielectrics with high energy storage density'. *Ann Rev Mater Res.* 2015;**45**(1):433–458.

[15] J. J. de Pablo, N. E. Jackson, M. A. Webb, L.-Q. Chen, J. E. Moore, D. Morgan. *et al.*, 'New frontiers for the materials genome initiative'. *NPJ Comp Mater.* 2019;**5**(1).

[16] H. Zhao, X. Li, Y. Zhang, L. S. Schadler, W. Chen, and L. C. Brinson, 'Perspective: NanoMine: A material genome approach for polymer nanocomposites analysis and design'. *APL Mater.* 2016;**4**(5):053204.

[17] A. White, 'The materials genome initiative: One year on'. *MRS Bulletin.* 2012;**37**(8):715–716.

[18] R. Ramprasad, R. Batra, G. Pilania, A. Mannodi-Kanakkithodi, and C. Kim, 'Machine learning in materials informatics: Recent applications and prospects'. *NPJ Comp Mater.* 2017;**3**(1).

[19] C. Kim, G. Pilania, and R. Ramprasad, 'From organized high-throughput data to phenomenological theory using machine learning: The example of dielectric breakdown'. *Chem Mater.* 2016;**28**(5):1304–1311.

[20] L.-Q. Chen, 'Phase-field method of phase transitions/domain structures in ferroelectric thin films: A review'. *J Am Cera Soc.* 2008;**91**(6):1835–1844.

[21] L.-Q. Chen, 'Phase-field models for microstructure evolution'. *Ann Rev Mater Res.* 2002;**32**(1):113–140.

[22] I. Steinbach, 'Phase-field models in materials science'. *Model Simul Mater Sci Eng.* 2009;**17**(7):073001.

[23] Y. U. Wang and D. Q. Tan, 'Computational study of filler microstructure and effective property relations in dielectric composites'. *J Appl Phys.* 2011;**109**(10):104102.

[24] Y. U. Wang, D. Q. Tan, and J. Krahn, 'Computational study of dielectric composites with core-shell filler particles'. *J Appl Phys.* 2011;**110**(4):044103.

[25] Z. H. Shen, J. J. Wang, Y. Lin, C. W. Nan, L. Q. Chen, and Y. Shen, 'High-throughput phase-field design of high-energy-density polymer nanocomposites'. *Adv Mater.* 2018;**30**(2).

[26] Z.-H. Shen, J.-J. Wang, J.-Y. Jiang, *et al.*, 'Phase-field model of electrothermal breakdown in flexible high-temperature nanocomposites under extreme conditions'. *Adv Energy Mater.* 2018;**8**(20):1800509.
[27] F. D. Ma and Y. U. Wang, 'Depolarization field effect on dielectric and piezoelectric properties of particulate ferroelectric ceramic-polymer composites'. *J Appl Phys*. 2015;**117**(12):124101.
[28] F. Y. Lin, X. Cheng, L. Q. Chen, and S. B. Sinnott, 'Strain effects on domain structures in ferroelectric thin films from phase-field simulations'. *J Am Cera Soc*. 2018;**101**(10):4783–4790.
[29] Y. Thakur, T. Zhang, C. Iacob, *et al.*, 'Enhancement of the dielectric response in polymer nanocomposites with low dielectric constant fillers'. *Nanoscale.* 2017;**9**(31):10992–10997.
[30] J. Qian, R. Peng, Z. Shen, *et al.*, 'Interfacial coupling boosts giant electrocaloric effects in relaxor polymer nanocomposites: In situ characterization and phase-field simulation'. *Adv Mater.* 2018:1801949.
[31] A. Jain, S. P. Ong, G. Hautier, *et al.*, 'Commentary: The materials project: A materials genome approach to accelerating materials innovation'. *Apl Mater.* 2013;**1**(1):011002.
[32] H. Pan, F. Li, Y. Liu, *et al.*, 'Ultrahigh–energy density lead-free dielectric films via polymorphic nanodomain design'. *Science.* 2019;**365** (6453):578–582.
[33] J. E. Zhou, Y. Yan, S. Priya, and Y. U. Wang, 'Computational study of textured ferroelectric polycrystals: Dielectric and piezoelectric properties of template-matrix composites'. *J Appl Phys*. 2017;**121**(2):024101.
[34] Z.-H. Shen, J.-J. Wang, J.-Y. Jiang, *et al.*, 'Phase-field modeling and machine learning of electric-thermal-mechanical breakdown of polymer-based dielectrics'. *Nat Commun*. 2019;**10**(1):1843.
[35] J. Jiang, Z. Shen, J. Qian, *et al.*, 'Ultrahigh discharge efficiency in multilayered polymer nanocomposites of high energy density'. *Energy Stor Mater.* 2019;**18**:213–221.
[36] J. Jiang, Z. Shen, J. Qian, *et al.*, 'Synergy of micro-/mesoscopic interfaces in multilayered polymer nanocomposites induces ultrahigh energy density for capacitive energy storage'. *Nano Energy.* 2019;**62**:220–229.
[37] Y. Wang, J. Chen, Y. Li, Y. Niu, Q. Wang, and H. Wang, 'Multilayered hierarchical polymer composites for high energy density capacitors'. *J Mater Chem A.* 2019;**7**(7):2965–2980.
[38] J. Chen, Y. Wang, Q. Yuan, *et al.*, 'Multilayered ferroelectric polymer films incorporating low-dielectric-constant components for concurrent enhancement of energy density and charge–discharge efficiency'. *Nano Energy.* 2018;**54**:288–296.
[39] Y. Wang, L. Wang, Q. Yuan, *et al.*, 'Ultrahigh energy density and greatly enhanced discharged efficiency of sandwich-structured polymer nanocomposites with optimized spatial organization'. *Nano Energy.* 2018;**44**:364–370.

[40] N. Shi and R. Ramprasad, 'Local properties at interfaces in nanodielectrics: An ab initio computational study'. *IEEE Trans Dielect Elect In*. 2008;**15** (1):170–177.

[41] Y. Sun, S. Boggs, and R. Ramprasad, 'The intrinsic electrical breakdown strength of insulators from first principles'. *Appl Phys Lett*. 2012;**101** (13):132906.

[42] A. A. Gusev and O. A. Guseva, 'Voltage breakdown in random composites'. *Adv Eng Mater*. 2003;**5**(10):713–715.

[43] L. Niemeyer, L. Pietronero, and H. J. Wiesmann, 'Fractal dimension of dielectric breakdown'. *Phys Rev Lett*. 1984;**52**(12):1033–1036.

[44] J. Li, L. Zhang, and S. Ducharme, 'Electric energy density of dielectric nanocomposites'. *Appl Phys Lett*. 2007;**90**(13):132901.

Chapter 7
Rational design on polymer dielectrics

Yujie Zhu[1]

7.1 Introduction

7.1.1 Introduction on material design

The great inventor Thomas Alva Edison once said, 'None of my inventions came by accident. I see a worthwhile need to be met and I make trial after trial until it comes. What it boils down to is 1% inspiration and 99% perspiration' [1]. It not only shows that he has made great efforts for his invention. But from another perspective, a large number of attempts are indispensable before the final invention is obtained. When Edison experimented with the filament material, after trying it hundreds of times, he finally got an ideal material that met the requirements. This is a traditional experimental methodology derived from 'trial and error'.

The 'trial and error' methodology can be traced back even longer. In ancient China, most materials, such as bronzes and steel [2], were invented under the 'trial and error' shadow. Ancients learned from their experience in failures and attempts, and eventually obtained reasonable manufacturing techniques and applied materials, which became the cornerstones of our modern civilized society. But the progress takes too much time. If finding right materials can only rely on blind 'trial and error', the development of new materials will be deadlocked.

Since the twentieth century, this naive methodology has also been partly inherited. Under the guidance of former experience, some researchers intended to improve the process and material selection continuously, and finally synthesize the required materials. Compared with blind attempts, purposeful synthesis is undoubtedly more efficient. However, many of these details still needed to be tested through trial and error. This would undoubtedly waste some time and resources. At this time, an idea was proposed. If the blueprint of material structure can be drawn in advance by computer, the experiments performed in laboratory will take shortcuts.

With the development of supercomputers [3,4] and the research of quantum mechanics [5], this idea gradually became practical. Moore's Law [6] has remained

[1]State Key Lab of Power Systems, Department of Electrical Engineering, Tsinghua University, Beijing, China

in effect after many years of development. There are tens of billions of transistors on chips [6] today, which provides lots of computing power. In 2019, Google successfully developed a quantum computer [7], which takes the computing power of computers to a higher level. The research of quantum mechanics provides a fundamental look at the structure and properties of materials in the micro perspective. Many quantum mechanical methods, such as discrete Fourier transform (DFT) [8], solve many materials science problems and have been widely applied to study different properties of metals [9], polymers [10] and other materials.

7.1.2 Introduction on polymer dielectric

In this chapter, the polymer dielectric for the energy storage polymer capacitor applications is mainly discussed.

The structure of a polymer capacitor is a stack of polarizable dielectric materials between two conductive metal plates [11]. It can store energy and has the advantage of fast response, compared with lithium batteries, which can quickly release stored energy [11]. On the basis of electrical properties, the dielectric between metal interlayers must have mechanical properties such as flexibility [11]. To ensure its stability during processing and usage, polymer dielectrics with more points replace inorganic dielectrics with high dielectric constants and become the first choice for capacitor dielectrics.

The permittivity and breakdown strengths are two important parameters in deciding the energy density of the polymer dielectric. The maximum energy density of polymer dielectrics is as follows:

$$\text{Maximum energy density} = \frac{1}{2}\varepsilon_0\varepsilon_r E_b^2$$

where ε_0 and ε_r are the dielectric permittivity of the vacuum (8.85×10^{-12} F/m) and the relative permittivity of polymer dielectrics. E_b is the breakdown strength of the medium, which is the maximum electric field that can be applied to a dielectric material without making it conducted. As energy density has a square relationship with the breakdown strength and a linear relationship with the relative permittivity of polymer dielectrics, higher breakdown strength and relative permittivity are required to make the energy density of polymer capacitor higher.

Currently, the polymers used as dielectrics in energy storage film capacitors are at most biaxially oriented polypropylene (BOPP) [12] and polyvinylidene fluoride (PVDF) [13]. BOPP's breakdown field strength exceeds 700 MV/m, low dielectric loss, low permittivity (~2.2) and low operating temperature (85 °C). In contrast, PVDF has a higher permittivity (~10) [12]. Because of the linear relationship between energy density and relative dielectric constant, the energy density of PVDF is much higher than BOPP. However, due to the ferroelectric performance of PVDF, the input and output efficiency of PVDF will be relatively low, and the energy loss will be serious [13].

At present, many studies have tried to find alternatives in the derivatives of BOPP and PVDF [14,15]. At the same time, many studies based on BOPP and

PVDF, such as nano-particle doping [16–18], surface modification [19], etc., are also trying to strengthen the advantages of polymer dielectrics and alleviate their disadvantages. These efforts have achieved great success. But if we can gradually design and select new polymers with the required properties and synthesize them, we will increase the basic level of overall performance to achieve a leap in overall performance. And this is exactly what this chapter wants to introduce. These works are mainly derived from the work within Prof. Rampi Ramprasad's group. Prof. Ramprasad's group is committed to develop and utilize computational and data-driven tools to aid materials design, and has made a lot of research on rational designs of many material properties [20–24] in the field of materials informatics.

7.2 High-throughput calculation on a polymer chemical space

This section mainly introduces how to predict and design material properties through high-throughput calculations [25–28]. After the selected polymer chemical subspace is selected, elements are selected from the subspace to construct a possible chemical structure. Computational prediction of all possible chemical structures properties is done through quantum mechanical calculations. It requires a lot of calculations to support it, and it's the large amount calculations that makes it 'high-throughput' (Figure 7.1).

Before determining the polymer chemistry subspace, a suitable quantum mechanical calculation method needs to be selected to calculate the relevant material properties. The dielectric constant and breakdown field strength of a polymer dielectric material determine the maximum energy density of a polymer capacitor. The breakdown field strength is not easily calculated. There is a

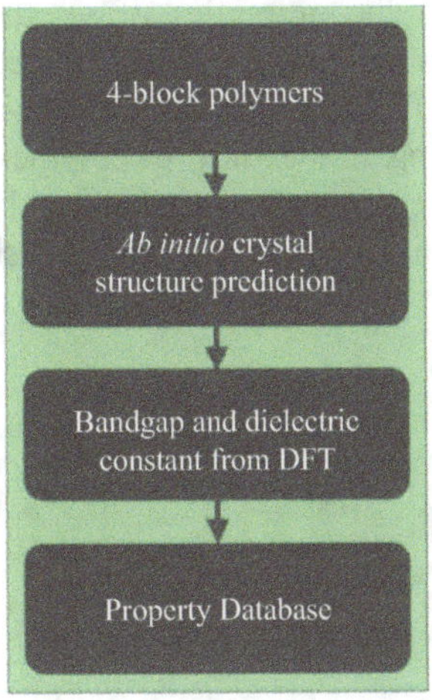

Figure 7.1. Flow chart of high-throughput DFT calculation

correlation between the band gap and the breakdown field strength of a polymer [18]. The dielectric constant and energy band gap of polymer materials are the properties that quantum mechanical calculations need to predict [25].

Therefore, density functional perturbation theory (DFPT) [29] was selected to predict the dielectric constant and band gap of polymer materials. The DFPT method calculates the dielectric constant of a material by studying the response of the system to an electric field [29]. The energy band gap was calculated using the DFT quantum mechanical calculation method of Heyd–Scuseria–Ernzerhof HSE06 [30] mixed electron exchange correlation function. This exchange potential corrects the band gap associated with standard DFT [30].

At the same time, it should be stated that due to the limitation of the calculation cost of the DFT method, although the structure of most polymers is not crystalline, the polymer is considered as a close-packed crystal model during the calculation process [25]. This may cause a certain deviation between the final calculation result and the experimental measurement result, but the calculated results are comparable.

Next, the elements needed for the chemical subspace need to be determined. Among them, most polymers are composed of molecular chains composed of periodic monomers, so some common chemical units can be selected as elements in the chemical subspace. By arranging the chemical units in the combination as the monomers of the molecular chain, the required polymer molecules are constructed.

Elements in the chemical subspace need to satisfy universality and can easily bond with other units. The seven units CH2, NH, CO, C6H4, C4H2S, CS and O are selected as elements of the polymer chemical subspace [25]. As shown in Figure 7.2, the monomers in the n-block organic polymer segment are all selected and linearly linked from the seven elements of the chemical subspace. At the same time, for the convenience of calculation, the number of blocks is limited to four. Excluding chemically infeasible pairings, 284 members remain.

After obtaining a linear polymer arrangement through chemical space, it is then necessary to calculate the properties of all the polymers formed. First, the ground state of the polymer needs to be determined, that is, the state with the minimum energy. Minima hopping method [31] was applied to determine the lowest energy relative packing arrangement of polymer chains in a unit cell. After considering this state as the ground state of the polymer crystal structure, the material properties calculations are implemented using the Vienna Ab initio

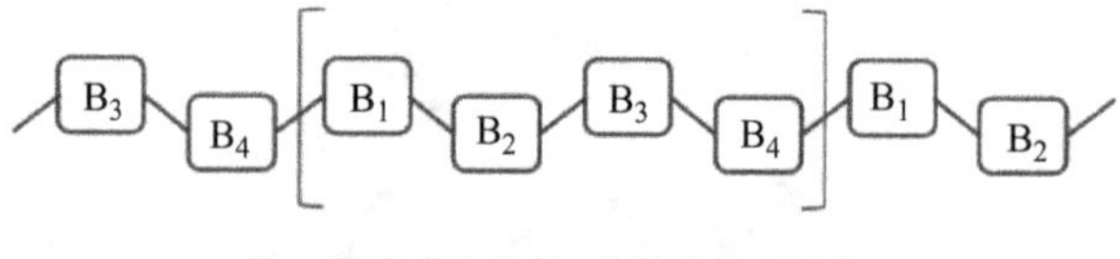

$B_i \in \{NH, CO, C_6H_4, C_4H_2S, O, CH_2\}$

Figure 7.2. Schematic diagram of four-block organic polymer chains and the elements in the chemical subspace

Software Package (VASP) software [32], including dielectric constant and band gap.

The DFPT method divides the calculation of the dielectric constant into two components: the dielectric constant of the electronic part and the dielectric constant of the ionic part [19]. The dielectric constant of the electronic part depends on the polarizability of the atom, while the dielectric constant of the ionic part comes from the infrared (IR) active vibration mode present in the system [31]. The overall dielectric constant is the sum of the two. Figure 7.3 is a plotting of the calculation results of 284 elements. The abscissa is the band gap, and the ordinate is the dielectric constant of each part. It can be clearly seen that there is an inverse relationship between the overall dielectric constant and the band gap. This is due to the inverse correlation between the dielectric constant of the electronic part, which accounts for the main contribution in the overall dielectric constants and band gaps. As the band gap becomes larger, the dielectric constant becomes smaller. Therefore, if you want to get the right energy density, a balance between the two parameters needs to be found.

Considering that there is a positive correlation between energy density, breakdown field strength and dielectric constant, it is considered that polymer materials that participate in the synthesis should be synthesized using polymer materials that have both a high band gap and a dielectric constant. The significance of the diagram is that the material properties can be estimated before laboratory synthesis has been performed. When the two materials have similar band gaps, polymer materials with high dielectric constants tend to have higher energy density. This paved the way for later laboratory synthesis, removing many inappropriate polymer chemical structural components.

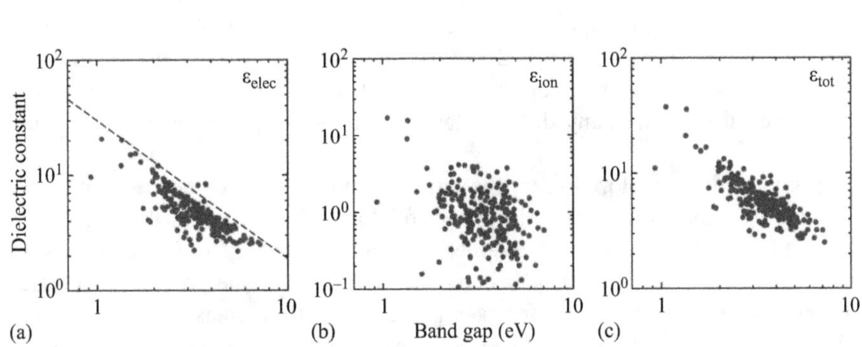

Figure 7.3. *The calculation results of 267 groups of four-block polymers obtained by high-throughput DFT calculations are shown. The abscissa of the three figures is the band gap, and the ordinate is (a) the dielectric constant of the electron, (b) the dielectric constant of the ion and (c) the overall dielectric constant. The dotted line in (a) is an inverse proportional function curve*

7.3 Rational design via machine learning

From Section 7.2, we can find that as the selected polymer chemical subspace elements increase, and the value of n in n-block increases, the computational cost of dielectric polymer DFT will increase explosively. In particular, considering that the calculation cost of using DFT to calculate a single polymer is not low, when calculating the performance of thousands or even tens of thousands of dielectric polymer materials, the explosive cost of high-throughput calculation is unacceptable. Is there any way to solve the problem in rational design?

The field of materials science that uses machine learning (ML) to accelerate materials design is often called materials informatics. Informatics methods have been used in fields such as classification of predicted crystal structure types. A large number of high-throughput calculated databases from Section 7.2 came in handy at this time. The chemical and physical properties of polymers can be related to the properties of polymers through data obtained from existing materials and the ability of computers to learn from laws. The use of ML models to replace DFT calculations which require a large amount of calculation costs [8,9] greatly accelerates the prediction of material properties.

7.3.1 ML model

In layman's terms, ML is to let a computer automatically learn from a limited amount of data to obtain some knowledge (or regularity) and apply these rules to the observed sample. The way to find rules from observed data (samples) and how to use the learned rules (models) to predict unknown or unobservable data is what ML models need to achieve.

In the past decades, machine learning has led to many success in many main domains. Machine learning is the study of computer algorithms that can be automatically improved through experience. Started with overwhelming success of AlexNet [33] on ImageNet [34], deep learning, whose large weight of multiple processing layers structure is optimized through backpropagation [35], becomes most powerful tools in many difficult tasks, such as computer vision and natural language.

Taking a deep exploration on the unreasonable success of deep learning, we can find some explanation why the powerful 'black-box' regains the popularity. First, with the development of Internet and the availability of various sensors, unprecedented amounts of data become the primary engine for the supervised learning. Computer vision area, for example, expediently increased labeled images make it possible to train large neural network. More and more open-sourced benchmark is the best evidence. Second, the advancement of computer hardware, especially Nvidia graphics processing units (GPU), is another reason that enables renewed interest in deep learning. The parallelization [36] can easily speed up network training process more than 100 times. In particular, compared with central processing units (CPU), the parallelization is more suitable for the heavy matrix and heavy computations.

And the improvement of hardware in turn accelerates the research in the deep learning, more capable and novel network architectures are proposed. And the advanced algorithm and novel architecture is the third reason. Residual neural network can help train very deep network, to some degree, easing the problems brought by gradient exploration and gradient vanishment (cite some paper).

However, lack of explanation of behaviour [37] and fragile training [38] make deep learning is not the first choice in many domains, particularly when the data are insufficient for neural networks. Unfortunately, the lack of diverse data makes it unrealistic to train deep neural network in material science. However, it does not mean it is hopeless, some progresses achieved by simple machine learning model encourage more researchers to work. And material science, compared with human cognitive system, can be described with simpler models, which makes the efficient data-driven methods with less data are more welcomed. Kernel Ridge Regression [39], as an efficient non-parametric method, is employed by in many low dimension pattern recognition, either classification and regression. However, it requires empirical feature engineering. On the one hand, feature selection provides a choice to embedding the domain knowledge. The conflict in capacity requirements and difficulties encountered when expanding demand also limit its uses.

Traditional machine learning usually requires the data to be represented as a set of feature vectors. These features are then input to a prediction model and the prediction results are output. An important feature of shallow learning is that it does not involve feature learning, and its features are mainly extracted through artificial experience or feature conversion methods. In practice, the performance of different prediction models is similar, and feature processing has a critical impact on the accuracy of the final system [33]. Because feature processing usually requires human intervention, manual experience is needed to select good features (fingerprint) and ultimately improve the performance of the machine learning system. Therefore, the most critical factor in polymer dielectric calculation is the printing of polymer dielectrics.

7.3.2 Polymer fingerprinting

As can be seen from Section 7.3.1, the key to the rational design of dielectric materials through machine learning is the fingerprint of the dielectric polymer (feature extraction). Fingerprints can then be mapped to target attributes through machine learning methods. Dielectric polymer fingerprints require fingerprints to be intuitive, easy to calculate and easily generalizable to all situations in the same chemical subspace [28]. Considering that the n-block polymer monomers are generated from the seven elements of the chemical subspace, the finite composition of the element units and the periodicity of the structure reminds us that we can digitally represent our polymer to establish associations between fingerprints and attributes through structural features.

Because there are only seven elements in the chemical subspace, the resulting block polymer is composed of seven elements in the chemical subspace anyway, so it is easy to think of the subspace units in the monomer of the polymer element.

The number of occurrences is a feature. M_I can then be represented as a seven-dimensional vector [25,40]. Each of these components represents the number of each block unit repeated in the monomer. But this method has hardly considered the influence of its order of arrangement. It is difficult for M_I feature to correspond one-to-one with the element that obtained the chemical structure. M_I is difficult to consider as the only feature. Therefore, M_{II} and M_{III} were proposed [29]. MII describes the number of adjacent element pairs in a repeating unit. M_{III} describes the number of three-element pairs in the monomer, so M_{II} is a 7×7 matrix formally, and M_{III} is a $7 \times 7 \times 7$ matrix formally. By analogy, more features can be promoted, such as M_{IV}. However, considering that n of the n-block polymer in the training set is equal to 4, the fingerprint feature information of the quadruple unit and above will be less and less, which does not meet the requirements for fingerprint feature extraction. It is a reasonable choice to set the fingerprint characteristics as M_I, M_{II} and M_{III}. It should be noted that the polymer fingerprint does not take into account the spatial crystallinity or other structural factors, so it will not distinguish between two polymers with the same repeating unit but different crystal structures.

After determining the characteristics of the selected structure as its fingerprint, we can conduct a preliminary analysis of the proposed structure [40–44]. Here, Pearson correlation analysis is used to study the correlation between the characteristics M_I and M_{II} and the related properties $\varepsilon_{electron}$, ε_{ion}, ε_{sum} and E_{gap}. Pearson correlation analysis will give a value between -1 and 1, indicating the degree of correlation between the fingerprint element and the block, block pair. As shown in Figure 7.4, the contribution of each element of the block to the band gap and the dielectric constant is almost in the opposite state. For the electronic permittivity and ionic permittivity, each segment unit also has different contributions. In order to improve the band gap, CH_2 units and O units should be introduced. In order to increase the dielectric constant, polar CS and C_4H_2S cells should be introduced. Regarding the effect of block pairs on related properties, it can be found that CH_2-CH_2, CH_2-O, CH_2-NH and CO-O have a strong promotion effect on the band gap performance. And CH_4H_2S-CH_4H_2S and CS-CH_4H_2S have strong positive correlation ability to the overall dielectric constant performance. This also means that from the preliminary analysis of fingerprints, we have summarized artificial experience to guide the subsequent synthesis of high-performance dielectric polymers. We can select more units that have a positive correlation with related parameters and the unit pairs have parameters that affect the target polymer. But most of these experiences stay on the surface. We also need to build a bridge between these fingerprints and target parameters to complete the goal of material performance prediction.

7.3.3 Rational design frameworks

After the fingerprinting of the designed linear block polymer was determined, a Kernel Ridge Regression (KRR) model [39] in machine learning was used to establish a functional relationship between the polymer fingerprint and related

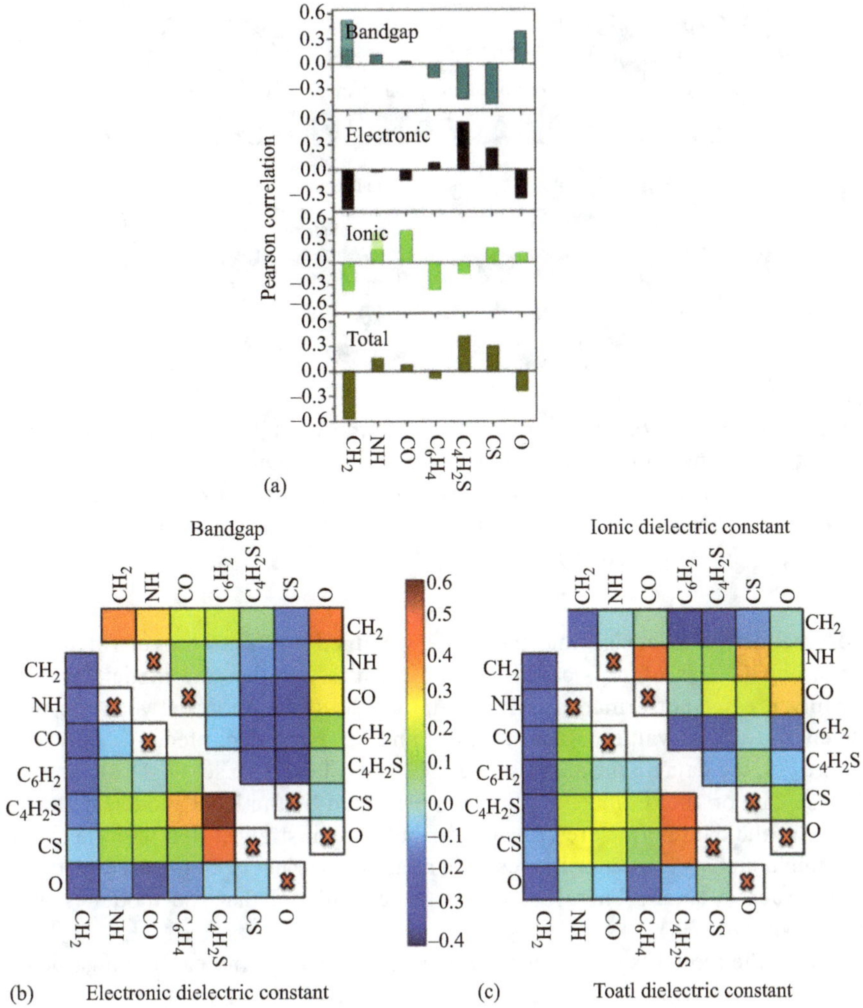

Figure 7.4. *(c) Pearson correlation coefficients between fingerprint M_I and E_{gap}, $\varepsilon_{electron}$, ε_{ion} and ε_{sum}. (d) Pearson correlation coefficients between fingerprint M_{II} and E_{gap}, $\varepsilon_{electron}$, ε_{ion} and ε_{sum} shown in the form of heat maps*

attributes (i.e. E_{gap}, $\varepsilon_{\text{elec}}$ and $\varepsilon_{\text{ionic}}$). By training the four-block polymer data obtained before the machine learning method, a black box of nonlinear function mapping between fingerprint and parameter values is established. After that, this currently trained mapping black box can be used to predict relevant properties of n-block polymers that have not been calculated by DFT (Figure 7.5).

During the training of the model, all known data were split into two parts. Ninety per cent of them were used as training sets. The KRR model is trained by

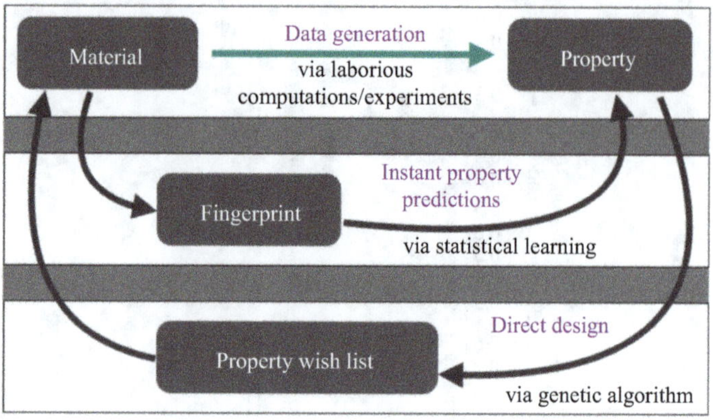

Figure 7.5. Schematic diagram of rational design via ML: (1) the relationship between materials and properties can be obtained through experiments or calculations; (2) fast material performance prediction through statistical learning and (3) direct design of materials with genetic algorithm

cross-validation to ensure that the model obtained will not be over-fitted. The remaining 10% is used as a test set to determine whether the model can reasonably predict performance. After training, the correlation between the predicted and actual values of the training and test sets is plotted. As shown in Figure 7.3, the data on the abscissa are the result of the data set DFT calculation, and the data on the ordinate are the data predicted by the KRR model. It can be seen that the data points are almost distributed on a straight line with the same horizontal and vertical coordinates. The relative error between the predicted and data values is small. Therefore, it can be considered that the model is well trained (Figure 7.6).

After the model is well trained, any n-block polymer molecules constructed in the subspace can be extracted from the fingerprint and then input into the KRR model to obtain possible performance predictions. And this time will be far less than the time required for DFT calculation [25,42–46]. So, a lot of time can be saved. The dielectric constant and band gap properties of a large number of dielectric polymers with different chemical compositions have been predicted. Possible six-block polymers (about 6,000) and eight-block polymers (about 150,000) are extracted in the figure, and their respective fingerprints are determined. Using the ML model of KRR that has been learned before, their properties are estimated. The graph shows the relationship between the predicted total ε_{total} and the predicted E_{gap} value for all six-block polymers and eight-block polymers. Then, eight-block polymers with different predicted values were selected for testing. As shown in Figure 7.7, the dielectric constant and laboratory-measured values calculated by the DFT method are compared with the results predicted by the ML model. It can be seen that the two

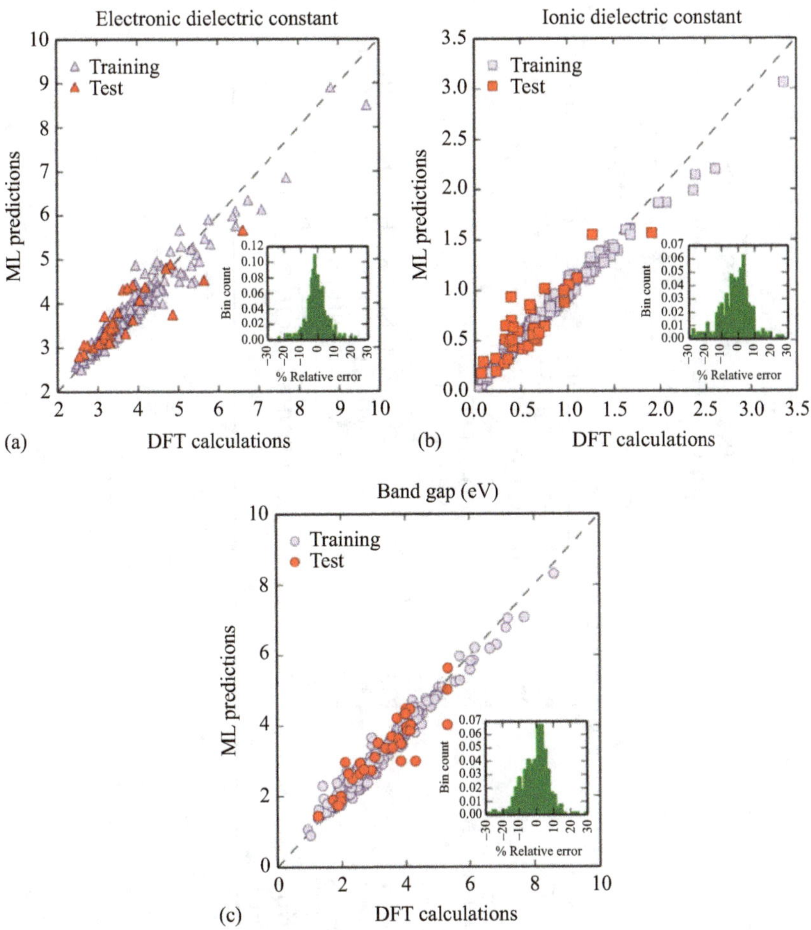

Figure 7.6. Comparison of the prediction results of the material properties by KRR and the calculation results of DFT. The red dots represent the test set, and the grey dots represent the training set. The small picture represents the relative error

parameters have a deep consistency and the method can be applied to a large number of dielectric polymer designs (Table 7.1).

Figure 7.7. (a) Steps of rational design via ML model; (b) dielectric constants and band gaps of four-block polymers, six-block polymers and eight-block polymers; (c) ML of 28 groups of dielectric block polymers compared with the predicted values of the model and DFT direct calculation performance and (d) comparison of the predicted performance values of the nine groups of dielectric block polymer ML models and the directly calculated performance values of DFT and the measured performance in the laboratory.

Table 7.1 Polymer repeat unit labels

Label	Polymer Repeat Unit	Label	Polymer Repeat Unit
P1	CH_2-O-CH_2-O-CH_2-CH_2-CH_2-CH_2	P20	O-C_6H_4-CO-C_4H_2S-CO-NH-O-CO
P2	CH_2-O-CH_2-O-CH_2-CH_2-CH_2-O	P21	CH_2-CH_2-O-CS-NH-CS-C_6H_4-NH
P3	CH_2-NH-CH_2-CH_2-CH_2-O-CH_2-O	P22	C_6H_4-C_6H_4-CH_2-CS-C_4-H_2S-CS-CH_2-O
P4	CH_2-CH_2-O-CO-O-CH_2-CH_2-O	P23	C_6H_4-NH-C_6-H_4-CS-NH-C_4H_2S-CO-NH
P5	CO-O-CH_2-CH_2-CH_2-CH_2-CH_2-O	P24	CO-C_4H_2S-NH-CS-O-C_4H_2S-NH-C_4H_2S
P6	CH_2-CH_2-O-CO-NH-CH_2-CH_2-O	P25	CS-CO-CH_2-CH_2-NH-C_6H_4CS-C_6H_4
P7	CH_2-NH-CO-NH-CH_2-O-CH_2-O	P26	C_6H_4-NH-C4H_2S-C_4H_2S-CS-C_4-H_2S-C_4H_2S-NH
P8	CH_2-CH_2-CH_2-CH_2-NH-CO-CH_2-CH_2	P27	C_4H_2S-C_4H_2S-C_4H_2S-CS-C_4H_2S-NH-CS-NH
P9	CO-NH-O-CH_2-CH_2-CH_2-CH_2-O	P28	C_4H_2S-CS-C_4H_2S-CS-CO-NH-C_6H_4-C_4H_2S
P10	CH_2-O-CO-NH-CH_2-CH_2-NH-CH_2	P29	NH-CO-NH-C_6H_4
P11	CH_2-NH-CH_2-NH-CO-NH-CO-NH	P30	CO-NH-CO-C_6H_4
P12	CH_2-NH-CO-O-NH-CO-NH-O	P31	NH-CS-NH-C_6H_4
P13	CO-NH-CO-O-CO-NH-CH_2-NH	P32	CH_2-CH_2-CH_2-CH_2
P14	CO-NH-CO-NH-CH_2-CH_2-CH_2-NH	P33	NH-CS-NH-C_6H_4-NH-CS-NH-C_6H_4-O-C_6H_4
P15	CO-NH-CO-CH_2-NH-CO-O-NH	P34	NH-CS-NH-C_6H_4-NH-CS-NH-C_6H_4-CH_2-C_6-H_4
P16	C_6H_4-O-CO-CH_2-CO-CH_2-CH_2-O	P35	NH-CS-NH-C_6H_4-NH-CS-NH-C_6H_4
P17	CH_2-CH_2-CO-O-CO-CH_2-C_6H_4-C_6H_4	P36	NH-CS-NH-C_6H_4-NH-CS-NH-$[CH_2]_6$
P18	CO-NH-O-NH-CO-NH-C_4H_2S-NH	P37	NH-CS-NH-C_6H_4-CH_2-C_6H_4
P19	CO-NH-CO-NH-CO-NH-C_4H_2S-NH		

7.4 Conclusion

In this chapter, a rational design of dielectric polymers is introduced. When the number of elements in the chemical subspace is small and there are few candidates for the dielectric polymer to be predicted, we can use high-throughput DFT to calculate the relevant parameters of the dielectric polymer to predict the material properties. When the number of chemical subspace elements increases and the number of candidate dielectric polymers increases, the ML method is used to learn a small amount of material data, replacing the DFT method to accelerate the calculation of material properties. Due to the limitation of the size of the data set, the feature extraction of material attributes (Fingerprinting) has become a very important part of it. After extracting the characteristics of the material structure and learning the material in combination with the NRR ML method, the prediction of the properties of related materials and the rational design of the dielectric polymer can be accelerated.

Of course, the ultimate goal of reasonable material design is to achieve the synthesis and application of high-performance dielectrics, which need to be tested through final experiments. At the same time, it is not necessary to obtain a data set of materials from DFT calculations. A large amount of test data can also be used as a deep learning dataset to predict the performance of materials. Experimental data can also be used to predict characteristics such as electrical conductivity and breakdown field strength of polymer materials that are difficult to calculate by the DFT method.

Acknowledgements

The author is now the visiting student of Prof. Ramprasad's group. The author would like to thank Prof. Jinliang He, Prof. Qi Li, Prof. Rampi Ramprasad, Dr. Huan Doan Tran and Dr. Lihua Chen for their help.

References

[1] Edison T. https://en.wikiquote.org/wiki/Thomas_Edison

[2] Hall M.E. 'Towards an absolute chronology for the Iron Age of Inner Asia', *Antiquity*. 1997;71(274) : 863–874.

[3] Hill, M.D., Jouppi, N.P., Sohi, G. *Readings in Computer Architecture* (The Morgan Kaufmann Series in Computer Architecture and Design). 2000. San Francisco, CA, USA: Morgan Kaufmann Publishers Inc. pp. 40–49.

[4] Eric G.S., David L.F. *Computers: The Life Story of a Technology*. Greenwood Publishing Group 2005. p. 56.

[5] Greiner, W., Müller, B. *Quantum Mechanics Symmetries*, 2nd edn. Springer Science & Business Media: 1994. Chapter 1, p. 52.

[6] Keyes, R.W. The impact of Moore's law. *Solid State Circuits Newsletter*. September 2006 [28 November 2008].

[7] Arute F., Arya K, Babbush R., *et al.* Quantum supremacy using a programmable superconducting processor. Nature. 2019;574(7779): 505–510.

[8] Becke, A.D. 'Perspective: fifty years of density-functional theory in chemical physics'. *The Journal of Chemical Physics*. 2014;140(18): A301.

[9] Assadi, M.H.N., Hanaor, A.H.H. 'Theoretical study on copper's energetics and magnetism in TiO2 polymorphs'. *Journal of Applied Physics*. 2013;113 (23): 233913-1–233913-5.

[10] Zimmerli, U., Parrinello, M., Koumoutsakos, P. 'Dispersion corrections to density functionals for water aromatic interactions'. *Journal of Chemical Physics*. 2004;120 (6): 2693–2699.

[11] Bluhm H. *Pulsed Power Systems: Principles and Applications*. Berlin, Springer: 2006.

[12] Hardy, C.G., Islam, M.S., Gonzalez-Delozier, D., *et al.* Converting an electrical insulator into a dielectric capacitor: end-capping poly- styrene with oligoaniline. *Chemistry of Materials*. 2013;25;799–807.

[13] Song, Y., Shen, Y., Liu, H., Lin, Y., Li, M., Nan, C.-W. 'Improving the dielectric constants and breakdown strength of polymer composites: effects of the shape of the batio$_3$ nanoinclusions, surface modification and polymer matrix'. *Journal of Materials Chemistry*. 2012;22:16491–16498.

[14] Rytoluoto, I., Lahti, K., Karttunen, M., *et al.* 'Large-area dielectric breakdown performance of polymer films-part i: measurement method evaluation and statistical considerations on area-dependence'. *IEEE Transactions on Dielectrics and Electrical Insulation*. 2015; 22(2): 689–700.

[15] Liu B, Chen C, Li T, *et al.* High performance ultrafiltration membrane composed of PVDF blended with its derivative copolymer PVDF-g-PEGMA. *Journal of Membrane Science*. 2013;445:66–75.

[16] Thakur V.K., Gupta R K. Recent progress on ferroelectric polymer-based nanocomposites for high energy density capacitors: synthesis, dielectric properties, and future aspects. *Chemical Reviews*. 2016;116(7):4260–4317.

[17] Li Q, Chen L, Gadinski M R, *et al.* Flexible high-temperature dielectric materials from polymer nanocomposites. Nature. 2015;523(7562):576.

[18] Hu P., Shen Y., Guan Y., Zhang X., Lin Y., Zhang Q., Nan C. Topological-structure modulated polymer nanocomposites exhibiting highly enhanced dielectric strength and energy density. *Advanced Functional Materials*. 2014;24(21): 3172–3178.

[19] Zhou Y., Li Q., Dang B., *et al.* 'A scalable, high-throughput and environmentally benign approach to polymer dielectrics exhibiting significantly improved capacitive performance at high temperatures'. Advanced Materials. 2018;30: 1805672.

[20] Treich G.M., Tefferi M., Nasreen S., *et al.* 'A rational co-design approach to the creation of new dielectric polymers with high energy density'. *IEEE Transactions on Dielectrics and Electrical Insulation*. 2017;24(2):732–743.

[21] Kim C., Chandrasekaran A., Huan T.D., Das D., Ramprasad R. 'Polymer genome: a data-powered polymer informatics platform for property predictions'. *Journal of Physical Chemistry C*. 2018;122:17575–17585.

[22] Chen L., Tran H., Batra R., Kim C., Ramprasad R. 'Machine learning models for the lattice thermal conductivity prediction of inorganic materials'. Computational Materials Science. 2019;170:109155.

[23] Wu C., Li Z., Treich G.M., *et al.* 'Dipole-relaxation dynamics in a modified polythiourea with high dielectric constant for energy storage applications'. Applied Physics Letters. 2019;115:163901.

[24] Patra A., Batra R., Chandrasekaran A., Kim C., Huan T.D., Ramprasad R. 'A multi-fidelity information-fusion approach to machine learn and predict polymer bandgap'. Computational Materials Science. 2020;172(1):109286.

[25] Huan T.D., Mannodi-Kanakkithodi A., Kim C., Sharma V., Pilania G., Ramprasad R. 'A polymer dataset for accelerated property prediction and design'. Scientific Data. 2016;3:16012.

[26] Lorenzini R.G., Kline W.M., Wang C.C., Ramprasad R., Sotzing G.A. 'The rational design of polyurea and polyurethane dielectric materials'. Polymer. 2013;54(14):3529–3533.

[27] Ramprasad R., Batra R., Pilania G., Mannodi-Kanakkithodi A., Kim C. 'Machine learning in materials informatics: recent applications and prospects'. Computational Materials Science. 2017;3:54.

[28] Huan T.D., Mannodi-Kanakkithodi A., Ramprasad R. 'Accelerated materials property predictions and design using motif-based fingerprints'. Physical Review B. 2015;92:014106.

[29] Lin T., Liu X.Y., He C. 'Calculation of infrared/raman spectra and dielectric properties of various crystalline poly (lactic acid) s by density functional perturbation theory (DFPT) method'. *Journal of Physical Chemistry B*. 2012;116(5): 1524–1535.

[30] Heyd J, Scuseria G E. 'Efficient hybrid density functional calculations in solids: assessment of the Heyd–Scuseria–Ernzerhof screened Coulomb hybrid functional'. *Journal of Cmical Physics*. 2004;121(3):1187–1192.

[31] Goedecker S. 'Minima hopping: an efficient search method for the global minimum of the potential energy surface of complex molecular systems'. *Journal of Chemical Physics*. 2004;120(21): 9911–9917.

[32] Kresse G, Hafner J. 'Ab initio molecular dynamics for liquid metals'. Physical Review B. 1993;47(1): 558.

[33] Krizhevsky A, Sutskever I, Hinton G E. 'Imagenet classification with deep convolutional neural networks'. *Advances in Neural Information Processing Systems*. 2012:1097–1105.

[34] Deng J., Dong W., Socher R. *et al.* 'Imagenet: a large-scale hierarchical image database'. 2009 IEEE Conference on Computer Vision and Pattern Recognition. IEEE: 2009. pp. 248–255.

[35] Rumelhart D.E, Hinton G.E., Williams R.J. 'Learning internal representations by error propagation'. California Univ San Diego La Jolla Inst for Cognitive Science, 1985.

[36] Han S, Dally B. 'Efficient methods and hardware for deep learning'. University Lecture, 2017.

[37] Ras G, van Gerven M, Haselager P. 'Explanation methods in deep learning: users, values, concerns and challenges'. *Explainable and Interpretable Models in Computer Vision and Machine Learning*. Cham: Springer; 2018. pp. 19–36.
[38] Akhtar N, Mian A. Threat of adversarial attacks on deep learning in computer vision: a survey. IEEE Access. 2018;6:14410–14430.
[39] An S, Liu W, Venkatesh S. Face recognition using kernel ridge regression. *2007 IEEE Conference on Computer Vision and Pattern Recognition*. IEEE: 2007. pp. 1–7.
[40] Batra R., Tran H.D., Kim C., *et al.* 'General atomic neighborhood fingerprint for machine learning- based methods'. Journal of Physical Chemistry C. 2019;123(25):15859–15866.
[41] Wang C.C., Pilania G., Boggs S.A., Kumar S., Breneman C., Ramprasad R. 'Computational strategies for polymer dielectrics design'. *Polymer*. 2014;55 (4):979–988.
[42] Sharma V., Wang C.C., Lorenzini R.G., *et al.* 'Rational design of all organic polymer dielectrics'. Nature Communications. 2014;5:4845.
[43] Ma R., Baldwin A.F., Wang C.C., *et al.* 'Rationally designed polyimides for high-energy density capacitor applications'. Applied Materials Interfaces. 2014;6(13): 10445–10451.
[44] Mannodi-Kanakkithodi A., Wang C.C., Ramprasad R., 'Compounds based on Group 14 elements: building blocks for advanced insulator dielectrics design'. Journal of Materials Science. 2015;50:801–807.
[45] Mannodi-Kanakkithodi A., Pilania G., Huan T.D., Lookman T., Ramprasad R. 'Machine learning strategy for accelerated design of polymer dielectrics'. Scientific Reports. 2016; 6:20952.
[46] Ma R., Sharma V., Baldwin A.F., *et al.* 'Rational design and synthesis of polythioureas as capacitor dielectrics'. Journal of Materials Chemistry A. 2015;3: 14845.

Chapter 8

Inorganic dielectrics for high-energy-density capacitors

Guangzu Zhang[1,2], Shenglin Jiang[1,2], Muni Yu[1] and Haibo Zhang[2,3]

8.1 Introduction

Since the industrial revolution happened in the eighteenth century, the social productivity has been greatly improved. The human society gets many benefits from the industrial revolution, but simultaneously we have to cope with the energy and environmental crises for the sustainable development. Harvesting powers from renewable and clean energy sources such as wind, solar, tidal, and geothermal energies, and then converting them directly into electricity is promising of alleviating and even completely resolving the energy and environmental crises, the implementation of which rely heavily on the rapid development of energy storage technology, as most of the renewable and clean energy is intermittent in the nature. Nowadays fuel cells, batteries, electrochemical capacitors and electrostatic (dielectric) capacitors are typical devices and technologies for electrical energy storage, distributing in different regions of the energy density–power density coordinate system. As shown in Figure 8.1, unfortunately, up to now no technology is provided with both high energy density and power density [1]. Different from the fuel cells, batteries and electrochemical capacitors whose charging and discharging processes involve complex and time-consuming chemical reactions, capacitors store and release electric energy physically (holding and releasing opposite charges on and from the electrodes separated by dielectrics) with a very high speed, permitting them to have a ultra-high power density in comparison to the other three technologies. The capability of fast charging/discharging makes the capacitors efficient in capturing energy from intermittent sources and providing electricity for

[1]School of Optical and Electronic Information, and Wuhan National Laboratory for Optoelectronics, Huazhong University of Science and Technology, Wuhan, Hubei, China
[2]Engineering Research Center for Functional Ceramics, Ministry of Education, Huazhong University of Science and Technology, Wuhan, Hubei, China
[3]School of Materials Science and Engineering, Huazhong University of Science and Technology, Wuhan, Hubei, China

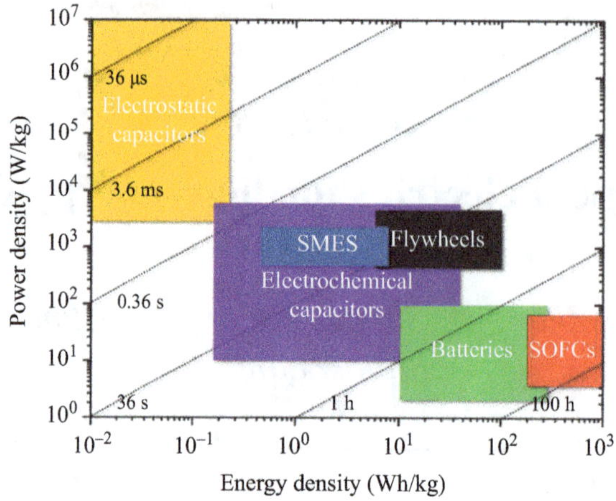

Figure 8.1 The trend of energy density versus power density of fuel cells, batteries, electrochemical capacitors and dielectric capacitors. © 2019 Elsevier. Reprinted, with permission, from [1]

high-power/pulse-power systems. The dielectric materials are the heart of the energy storage capacitors, playing a determining role in the performance of the dielectric energy storage devices. The dielectrics with high energy density and efficiency are essential for the high-power energy system, as high energy density would contribute largely to the reduction of volume, weight and cost of the capacitors and systems, and the high efficiency can significantly depress the heat generated by the electric loss and prevent the capacitors from thermal breakdown. Furthermore, the recent development in the fields, for example, automotive and avionic industries, and deep underground oil and gas explorations, to name a few, require the capacitors which can be operated in elevated-temperature conditions (e.g., >200 °C). Thus, high-temperature dielectric materials with high-temperature stability are highly appreciated as well.

The dielectrics for energy storage can be divided into two categories: inorganic dielectric materials (including single crystals, ceramics, and glasses, and their composites) and organic dielectrics (i.e., polymers). Compared to inorganic dielectric materials, polymer dielectrics are superior in flexibility, facile fabrication process and lightweight. Inorganic dielectric materials have their own advantages as well, for example, they can withstand much higher temperature than most polymers, and there are abundant material systems available with big room for improvement (doping, composition, phase content and microstructure designs) for inorganic dielectrics.

Figure 8.2 schematically illustrates the energy density (U) and efficiency (η) extracted from the polarization-electric field (P–E) loops, and it is intuitive that high polarization and electric breakdown strength (E_b), as well as slim loop with low remanent polarization (P_r) are favorable for the high energy density and

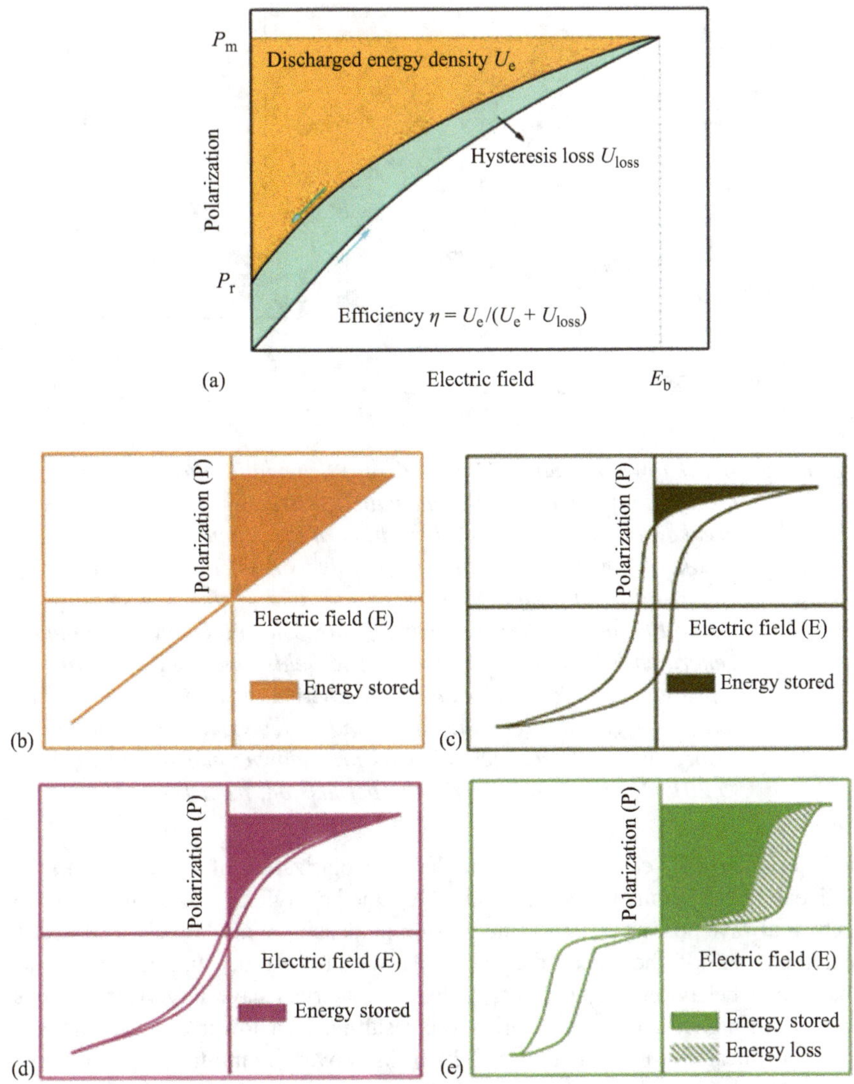

Figure 8.2 (a) Schematic of the energy density (U) and efficiency (η) of a dielectric material extracted from its polarization-electric field loop. (b–e) Typical P–E loops of linear dielectrics (b) and nonlinear dielectric materials in terms of normal ferroelectrics, relaxors and antiferroelectrics (c–e)

efficiency [2,3]. According to different polarization mechanisms, there are two types of inorganic materials for dielectric energy storage, that is linear and nonlinear dielectrics. For nonlinear dielectric materials, for example, ferroelectrics, the spontaneous polarizability provides the materials a high electric polarization that

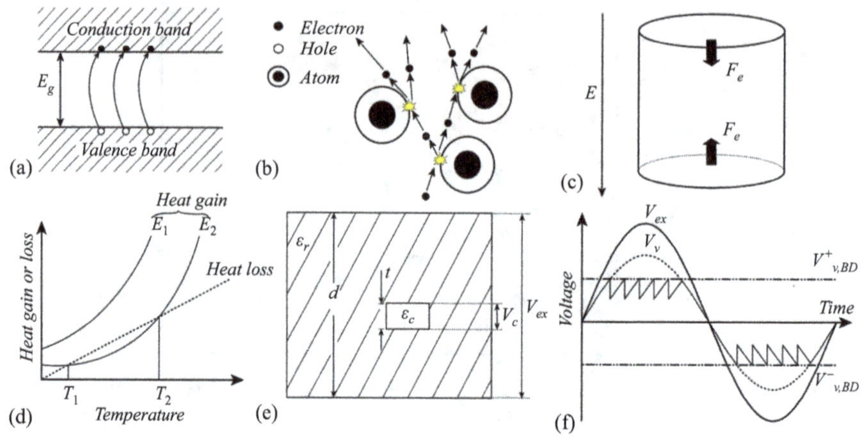

Figure 8.3 Breakdown mechanisms in solids. (a) Schematic illustration of electronic breakdown; (b) schematic illustration of avalanche breakdown; (c) schematic illustration of electromechanical breakdown, where Fe is the compressive stress caused by the electrostatic attraction of the electrodes and (d) thermal stability or instability under different electric fields. The solid curves are the generated heat under different electric fields, and the dashed line represents heat loss; (e) model of a cavity in a solid dielectric. The thickness/permittivity of the cavity and dielectric are t/ε_c and d/ε_r, respectively; (f) sequence of cavity breakdown under AC voltage. © 2019 Elsevier. Reprinted, with permission, from [1]

can be orientated by electric field. Note that the polarization of the ferroelectrics is nonlinear to the applied electric fields. Due to the high polarization, the ferroelectrics always possess a higher energy storage density at a relatively low electric field than that of the linear dielectrics. However, just because of the unique polarization behavior, the nonlinear dielectric materials have larger electric hysteresis with higher remanent polarization, resulting in a lower energy efficiency than their linear counterpart (Figure 8.2(b)). Moreover, owing to the phase transitions (ferroelectric↔ferroelectric, ferroelectric↔antiferroelectric, and ferroelectric↔paraelectric phase transitions) of the ferroelectrics in the operation temperature windows, the temperature stability of the nonlinear dielectrics is inferior to that of nonlinear ones. The linear dielectrics generally have higher breakdown strength than that of nonlinear materials. Although the breakdown process is still unclear, several theoretical models have been proposed and accepted, including intrinsic breakdown, thermal breakdown, electromechanical breakdown, and partial discharge breakdown (Figure 8.3) [1]. Due to the spontaneous polarization effect again, it is believed that the ferroelectrics are more susceptible to the intrinsic and the loss-induced thermal breakdowns than the linear dielectric materials.

The aim of this chapter is to sketch a general picture of the energy storage performance of inorganic dielectrics by reviewing the dielectric characteristics and polarization behavior in different material systems. In addition, the regulation mechanisms for optimizing the comprehensive energy storage performance of the materials are emphasized. At the end of this chapter, we discussed the challenges and future perspectives of the inorganic dielectrics for capacitors and energy storage technology.

8.2 Linear dielectrics

Theoretically, the electric polarization of linear dielectrics increases linearly with the electric fields (Figure 8.2(b)), and their dielectric constant is independent of the electric fields. Therefore, the energy storage density can be simply estimated in the following equation:

$$U_e = \frac{1}{2}\varepsilon\varepsilon_0\varepsilon_r E^2 \tag{8.1}$$

where ε_0 is the dielectric permittivity in vacuum ($\sim$8.85 $\times$ 10^{-12} F/m), ε_r is the relative dielectric constant and E is the electric field applied on the dielectric materials. Compared to nonlinear dielectric materials, the linear dielectrics have higher breakdown strength. Without the spontaneous polarization effect, the electric displacement of the linear dielectrics is much lower than the nonlinear materials with the same electric field. However, also because they have no spontaneous polarization, there is no electric hysteresis in ideal linear dielectric materials in ideal condition, endowing them with a low dielectric loss and high energy efficiency (the stored energy can be fully released).

8.2.1 Glasses

Due to the amorphous nature associated with the elimination of grain boundaries, high-quality glasses have very few defects, contributing largely to their extremely high breakdown strength that is comparable to or even surpass that of dielectric polymers. Attractively, the glasses have the self-healing potential during the breakdown, as the energy generated by the dielectric breakdown is sufficient to burn out the electrode surrounding the local failure spot, preventing an electrical short across the positive and negative electrodes.

Aluminosilicate, aluminoborate, and aluminophosphate with alkali elements (e.g., Na, K)-alkaline earth (e.g., Mg, Ca, Sr, and Ba) network modifiers are typical dielectric glasses for energy storage. With the highly polarizable alkaline earth oxides, the glasses possess a permittivity of about 7 at room temperature. Alkali elements are needed for providing the glass-forming ability; however, the high ion mobility of alkali elements deteriorate the inherent high resistivity and breakdown strength of the glasses. Furthermore, the ion mobility of the alkali elements depends heavily on the type and concentration of the alkaline earth in the glass composition. For example, in the Na-alkaline earth (Mg and Ba)-aluminosilicate system, the divalent alkaline earth can bridge two AlO_4 charge compensation centers which have to be deformed for the sodium ion transport (Figure 8.4) [4]. The alkaline-earth ion size and field strength on the local structure would significantly affect the activation energy for the ionic conduction and determine the resistivity and breakdown strength of the glasses. With an optimized composition, this glass system has a breakdown strength around 4,200 kV cm^{-1}, and an energy storage density of 5 J cm^{-3}.

In order to further enhance the breakdown strength, alkali-free glasses have been developed and now many alkali-free glass compositions are commercially available [5]. The general compositions are SiO_2 50%–60%, B_2O_3 10%–14%, Al_2O_3 11%–15%, BaO 2%–24%, CaO 5% and SrO 5% (by % weight). Without the alkali elements, the breakdown strength of the glasses reaches up to an extremely high level of 12,000 kV cm^{-1} with a mild decrease in dielectric constant (~6 for the alkali-free glasses) compared to the glasses with alkali elements. As a result, the energy storage density dramatically increases to ~35 J cm^{-3}, seven times that of the glasses containing alkali elements. In addition, because of the elimination of the alkali elements, the resistivity of the glasses enhances greatly, yielding a low dielectric loss and a high energy efficiency (<95%). It is noted that the high-temperature performance of the alkali-free glasses is excellent [6]. The glass-transition temperatures of the alkali-free glasses are higher than 500 °C, well above the operation temperature range for energy storage and conversion applications, and because of their amorphous characteristics (without phase transition during the whole operation temperature window), the alkali-free glasses exhibit a very stable energy storage properties ranging from 20 °C to 180 °C.

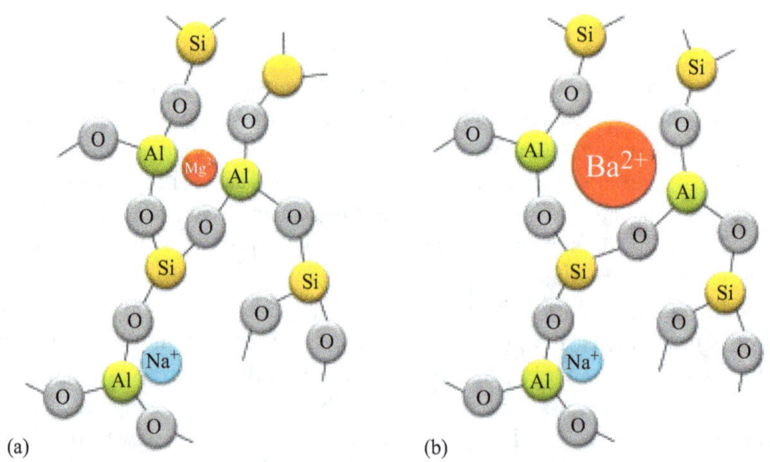

Figure 8.4 Schematic illustration in two dimensions of possible atomic structure in glass containing (a) magnesium and (b) barium. © 2016 John Wiley and Sons. Reprinted, with permission, from [4]

8.2.2 *Ceramics*

In stark contrast to the glasses that are amorphous, the ceramics are polycrystals that possess higher dielectric constant than that of glasses. Alkaline earth zirconates and hafnates are representative materials for linear ceramics for dielectric energy storage [7]. In the $Ca(Zr_{1-x}Ti_x)O_3$ (CZT), for example, $CaTiO_3$ is regarded as an incipient ferroelectric material with high dielectric constant and $CaZrO_3$ is a linear dielectric material with high resistivity (large band gap > 3.8 eV) and breakdown strength. By balancing the permittivity and dielectric breakdown strength with various Zr:Ti ratios, the $Ca(Zr_{1-x}Ti_x)O_3$ solid-solution ceramics shows an energy storage density of ~4 J cm^{-3}. Even with an incipient ferroelectric content, the $Ca(Zr_{1-x}Ti_x)O_3$ solid-solution ceramics show a linear polarization behavior. Similar polarization behavior has also been observed in the $Ca(Ti_xHf_{1-x})O_3$ (CTH) system. The energy storage properties of the ceramics can be further improved by elements doping, a straightforward and effective method that has been extensively used for electric property regulation in functional materials. Mn doping can effectively improve the breakdown strength of CZT and CTH, as Mn can change from Mn^{2+} to Mn^{3+} and Mn^{4+} with various oxygen activities to compensate the oxygen vacancies, which minimize the ionic conductivity of the ceramics to a very low level. Moreover, co-doped with Mg and Dy, the CTH shows an optimized energy density of ~10 J cm^{-3}. The CZT and CTH ceramics can store energy at high temperature exceeding 200 °C. For instance, at high termperature, the breakdown strength of $Ca(Zr_{0.8}Ti_{0.2})O_3$ can maintain at a high level beyond 1,500 kV cm^{-1} with a polarization of 6.5 μC cm^{-2} and a slim *P–E* loop, showing an energy storage density of 4 J cm^{-3}. Although with good energy performance at high temperature, few efforts have been devoted to the linear dielectric ceramics, as their polarization is much lower than that of ferroelectric ceramics and breakdown strength is inferior to that of glasses.

In summary, linear inorganic dielectrics, including the glasses and ceramics, are promising candidates for high-field and high-temperature applications. At the end of this section, it is emphasized that the energy storage density calculated by (8.1) with low-field dielectric constant may be overestimated as the polarization may saturate under high electric field, and more seriously, even linear dielectric materials may have open hysteresis loops with high remanent polarization at high field, which are attributive to the high ionic/electronic conductions and/or space charges.

8.3 Nonlinear dielectrics

As shown in Figure 8.2(c)–(e), the electric polarization of ferroelectrics (including antiferroelectric materials) is nonlinear to the electric field, which are typical and well-known nonlinear dielectric materials for energy storage. However, not all ferroelectrics are suited for dielectric energy storage. Even with a high polarization, the normal ferroelectric materials working below their Curie temperatures show a squire ferroelectric hysteresis with high remanent polarization, the recoverable energy density and efficiency of which are poor (Figure 8.2(c)). Differently, the relaxors, antiferroelectrics and normal ferroelectrics working with paraelectric phase have relatively small remanent polarization, whose energy efficiency is more favorable than that of normal ferroelectrics, having been extensively investigated for power energy storage (Figure 8.2 (d)–(e)). Moreover, ceramics and glass-ceramics are popular, while the ferroelectric single crystals, at present, have limited potential for practical energy storage applications, due to the high cost and complicated fabrication processes [8]. Composition design (including elements doping) and microstructure modification are effective approaches for energy storage performance improvement, and diverse optimization mechanisms have been established.

8.3.1 Ferroelectrics working at paraelectric state

The paraelectric materials usually have a Curie temperature below the operating temperature for energy storage. In the paraelectric phase, although the polarization of the materials is lower than they with a ferroelectric phase, which, nevertheless, is much higher than that of the linear dielectrics. In addition, without electric dipoles, the paraelectric dielectrics show a slim electric hysteresis loop with a low remanent polarization, yielding a relatively high energy efficiency.

$SrTiO_3$ is a typical ferroelectric material with a low Curie temperature around -271 °C, which has been widely studied for energy storage [9,10]. At room temperature, $SrTiO_3$ ceramic possesses a polarization of ~12 μC cm^{-2} and a Weibull-statistic breakdown strength of ~300 kV cm^{-1}. As working with a paraelectric state, $SrTiO_3$ also has a very low remanent polarization below 1 μC cm^{-2}. The pristine $SrTiO_3$ ceramics consequently possess an energy density of 1.5 J cm^{-3} and an efficiency about 90%.

The polarization behavior can be modified by changing the composition. The Curie temperature of $SrTiO_3$ increases linearly with the increasing content of Ba in the $(Sr_xBa_{1-x})TiO_3$ solid solution, leading to a higher polarization at room temperature relative to that of the pure $SrTiO_3$ [1,11]. $(Sr_{0.6}Ba_{0.4})TiO_3$ is a typical composition for the solid solution whose Curie point is located at ~-60 °C with a polarization of 15 μC cm^{-2} @ 200 kV cm^{-1}, and a high energy density of 1.2 J cm^{-3} is achieved with the relatively low electric field. However, the $(Sr_{0.6}Ba_{0.4})TiO_3$ has a larger hysteresis than that of the $SrTiO_3$, as they are working closer to the ferroelectric state, resulting in their inferior energy efficiency. Besides, it is reported that substituting Ca for Sr decreases the cell volume and rise the binding energy between the ions, leading to the enhancement in the breakdown strength. Doped with 2 mol % Ca, the breakdown strength of $SrTiO_3$ increases 30% without compromising polarization, and therefore the energy density increases to ~2 J cm^{-3}. However, the enhanced hysteresis caused by Ca doping results in the decreased energy efficiency ($\eta = 75\%$). Partially replacing Sr with Mn, Bi, Dy, Nd, etc., and tuning the Sr: Ti ratio improves the energy performance of $SrTiO_3$ as well. A significant improvement has been demonstrated in a complex composition $(1-x)$ $SrTiO_3-x(Bi_{0.48}La_{0.02}Na_{0.48}Li_{0.02}Ti_{0.98}Zr_{0.02}O_3)$, which is ascribable to the cell volume shrink and chemical stability enhancement caused by the A and B sites co-substitution. With $x = 0.1$, the modified $SrTiO_3$ exhibits a high recoverable energy storage density of 2.59 J/cm^3 and a high efficiency 85% simultaneously.

Microstructure is another important factor that determines the energy storage characteristics. By using the spark plasma sintering (SPS) method, the pure and doped $SrTiO_3$ and $Ba_{0.4}Sr_{0.6}TiO_3$ with grain sizes ranging from 300 nm to 5 μm have been successfully prepared [12]. With the decrease of grain size, the polarization increases as the content of grain boundary increases, leading to an enhancement of interfacial polarization, which, however, impairs the breakdown strength simultaneously. Moreover, due to the SPS, the air pores could be effectively eliminated from the samples, which is conducive to the high breakdown strength and energy efficiency.

8.3.2 Relaxors

As the same crystallographic sites of the cells occupied by different cations with various valences, the polar nanoregions are formed in the relaxors instead of macroscopic domains in the normal ferroelectrics, leading to very different polarization behaviors (Figure 8.2(d)). For example, the diffused and smeared phase transition peak appears in the relaxors, and the dielectric constant decreases and the temperature corresponding to the maximum permittivity shifts toward higher temperature as the frequency increases. Most importantly, the dipoles of relaxors can be aligned to follow the electric fields with a negligible remanent polarization, exhibiting slim hysteresis loops that contribute to the low energy loss and high energy efficiency. In recent years, massive efforts in the scientific and industrial communities have been made to optimize the energy storage performance in relaxors and many reviews surveyed the progress in this field [13,14], so here we just highlight representative compositions and strategies for high-performance energy storage in relaxors.

$Pb(Mg_{1/3}Nb_{2/3})O_3$–$PbTiO_3$ (PMN–PT), $Pb(Zn_{1/3}Nb_{2/3})O_3$ (PZN), (Pb,La)(Zr, Ti)O_3 (PLZT), $BaTiO_3$-based solid solutions [including BTO–$BiScO_3$ (BSO), BTO–$SrTiO_3$ (STO), $BaTiO_3$–$BiFeO_3$ (BFO)–$La(Mg_{1/2}Ti_{1/2})O_3$ (LMT), BTO–BFO–$Ba(Mg_{1/3}Nb_{2/3})O_3$ (BaMN) and BTO–$Bi(Zn_{2/3}Nb_{1/3})O_3$ (BZN)], $(K_{0.5}Na_{0.5})NbO_3$ (KNN), $(Na_{0.5}Bi_{0.5})TiO_3$ (NBT), etc. are relaxors that have been widely investigated for energy storage [1]. By tuning the compositions, the energy storage performance of the relaxors can be significantly improved. For example, in the BTO-based material system, composited with 15 mol% BZN, the BTO–BZN shows an energy storage density of 0.8 J/cm^3 with an extremely high energy efficiency of 93.5%, both of which are two times that of BTO with only 5 mol% BZN. The improved energy storage features are attributive to the elevated activation energy of the relaxation process [15]. Besides, the energy storage density increases from 0.4 J/cm^{-3} in BTO–BFO to 1.6 J cm^{-3} in 6 mol% LMT-doped samples, with a significant enhancement in energy efficiency (form 30% to 80%) as well. Similar result was obtained in the BTO–BFO doped with 6 mol% BMN [16].

KNN and NBT-based solid solution relaxors are also studied for their high energy storage property [18,19]. Doped with adequate contents of $Sr(Sc_{0.5}Nb_{0.5})O_3$(SSN) and STO, high energy density of 2.02 J cm^{-3} and 3.67 J cm^{-3} were reported in KNN with an energy efficiency beyond 70%. More interestingly, doped with 10 mol% $Bi(Mg_{2/3}Nb_{1/3})O_3$ (BiMN), the energy storage density can be enhanced to 4.08 J cm^{-3}, which is attributed to the high polarization orientated from the strong hybridization between the Bi 6*s* and O 2*p* orbitals. However, the drawback of the composition is the relatively low energy efficiency (~62%). NBT exhibits normal ferroelectric feature (square *P–E* loop) at room temperature which, intriguingly, can be tuned to show relaxor and paraelectric polarization behaviors by composition design. For example, with the increased content of $(Na0_{.73}Bi_{0.09})NbO_3$ (NBN), the 0.92NBT–0.08BTO changes from an ergodic relaxor state to a paraelectric material with a reduced remanent polarization. The sample with 20% NBN possesses an energy storage density of 1.36 J cm^{-3} and a preferred energy efficiency. The energy storage density can be further upgraded in the BNT–BNB by coping with Li, La, Zr and Nb. With the precisely designed composition, the BNT-based ceramics demonstrate a high energy storage density of 2.04 J cm^{-3} @ ~180 kV cm^{-1}. It is noted that due to the smeared phase transition that occurs at high temperature, the KNN- and NBT-based dielectrics generally exhibit a favorable temperature stability of energy storage property, which are important materials for capacitors being operated under extreme temperature environment.

In addition to composition design, microstructure plays another important role in the energy storage property. A reprehensive work is the ceramic consists of core-satellite BTO@STO grains [17]. In stark contrast to the homogeneously doped BTO, the STO wraps the BTO cores as a layer of shells, which makes a remarkable suppression of grain size and results in the significant enhancement in breakdown strength (Figure 8.5). Furthermore, the local compositionally graded structure arising from the controlled sintering and modulated diffusion of Sr enables the ceramic composite to exhibit a relaxor behavior with a relatively high polarization and low energy loss simultaneously. This study proposed a new strategy for designing relaxor dielectrics for energy storage application.

Above we discussed the energy storage properties of relaxor in bulk ceramics. Compared to bulks, it is accepted that preferred energy storage performance can be acquired in films, as with the same composition, the films have the similar polarization but much higher breakdown strength than that of the bulks. Avalanche breakdown is one type of intrinsic breakdown in dielectrics, which can be triggered by a chain reaction that the electrons gain energy exceeding the ionization potential by colliding with lattice atoms repeatedly under electric fields (Seitz F.: On the theory of electron multiplication in crystals) [20]. The avalanche breakdown would occur when the avalanche exceeds a critical size with ~10^{12} electrons, corresponding to 40 times of ionizing collisions by an initiating electron. The critical distance in a dielectric material for

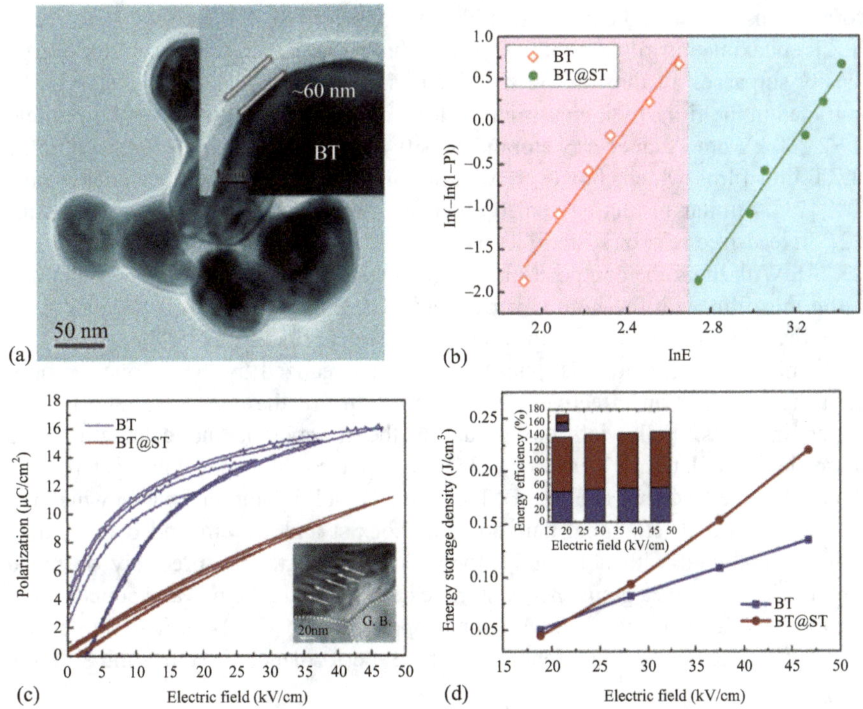

Figure 8.5 *(a) TEM images of ceramic with core-satellite BTO@STO grains. (b) Weibull distribution for the BT and BT@ST ceramics. The ceramics consists of core-satellite BTO@STO grains have higher breakdown strength (218 kV cm^{-1}) than that of the BT ceramics (113 kV cm^{-1}). Polarization-electric field hysteresis loops (c) and energy storage density (d) for the BT and BT@ST ceramics. The inset in (c) shows the domain patterns for the BT@ST ceramic, indicated by arrows, where GB is grain boundary, and the inset in (d) shows the energy efficiency for the BT and BT@ST ceramics. © 2015 Royal Society of Chemistry. Reprinted, with permission, from [17]*

the ionizing collision chain reaction to trigger the avalanche breakdown can be estimated by

$$E_{\text{ava}} = \frac{k}{\ln\left(\frac{\alpha_0 d}{40}\right)} \tag{8.2}$$

where k and α_0 are constants determined by the dielectric materials. Therefore, it is understandable that thin films can withstand higher electric fields (>1,000 kV cm^{-1}) than the bulks (typically <100 kV cm^{-1}) and possess more favorable energy storage features. With a high electric field of 1000 kV cm^{-1}, a recoverable energy storage density of 13.7 J cm^{-3} with a high efficiency of 88.2% has been achieved in PLZT epitaxial thin films with 300 nm in thickness. As known to us, the energy density surpasses all those in bulk dielectrics. The energy storage density can be upgraded in the films by increasing the electric field further. With a very high field of ~2,200 kV cm^{-1}, an energy storage density of ~30 J cm^{-3} can be obtained in the PLZT thin films which, however, with a compromised energy efficiency below 80% [21]. Similar results have also been reported in PMN–PT relaxor thin films [22]. In lead-free relaxor system, let us taking BTO–$Bi(Mg_{1/2}Ti_{1/2})O_3$, for example [23,24]. With the same composition, a high energy density of 37 J cm^{-3} is attained in the thin film with the high electric field of 1900 kV cm^{-1}, which is about 20 times that obtained in bulks with the limited electric field of 224 kV cm^{-1}.

Similar to bulk materials, phase engineering realized by the rational composition designing is an effective approach to optimize the energy storage performance in films. Following the guide of the thermodynamic calculation and phase-field simulation (Figure 8.6) [25], high polarization with low hysteresis is obtained in the lead-free $BiFeO_3$–$BaTiO_3$–$SrTiO_3$ solid-solution films in which the rhombohedral and tetragonal nanodomains coexist and are embedded in a cubic matrix. Combined with a favorable breakdown strength, the precisely designed composition exhibits a giant energy storage density of 112 J cm^{-3} and concurrently a high energy efficiency of ~80%. More importantly, this work provides a general strategy for achieving high energy storage performance in the solid-solution inorganics.

Notably, much more flexible approaches can be utilized to regulate the polarization behavior and energy storage performance in thin films than in bulks. For instance, a desirable energy storage density exceeding 50 J cm^{-3} is gained in the $Ba_{0.7}Ca_{0.3}TiO_3$–$BaZr_{0.2}Ti_{0.8}O_3$ (BCT–BZT) thin films with a multilayer-structured design that contribute largely to the high breakdown strength, which has been rationalized by electric tree growth-obstruction model. In the multilayered structure, the electric trees are blocked by the BCT–BZT interfaces which construct obstructers for preventing the breakdown. As a result, the BCT–BZT consists of 8 layers thin films can survive at a high electric field of 4,500 kV cm^{-1}, being 1.5 times that of the film with only two layers, shows the high energy storage density of 52.4 J cm^{-3} that is about two times that of the two-layer films. Besides, improvement in polarization and breakdown strength can be realized in thin film with texturing [26]. With a (111) texture orientation, the $Pb_{0.8}Ba_{0.2}ZrO_3$ (PBZ) thin film shows a recoverable energy storage density of 40 J cm^{-3} along with an efficiency of 64%. Another representative work is reported in the epitaxial BZT thin films fabricated by means of RF-magnetron sputtering. By creating an adaptive nanodomain morphology via phase engineering capitalizing on the misfit strain between the substrates and the films (Figure 8.7), an unprecedented energy storage density of 166 J cm^{-3} and efficiency of 96% are obtained in the thin film [27]. Additionally, the phase-engineered films exhibit an excellent frequency and temperature stability in broad ranges.

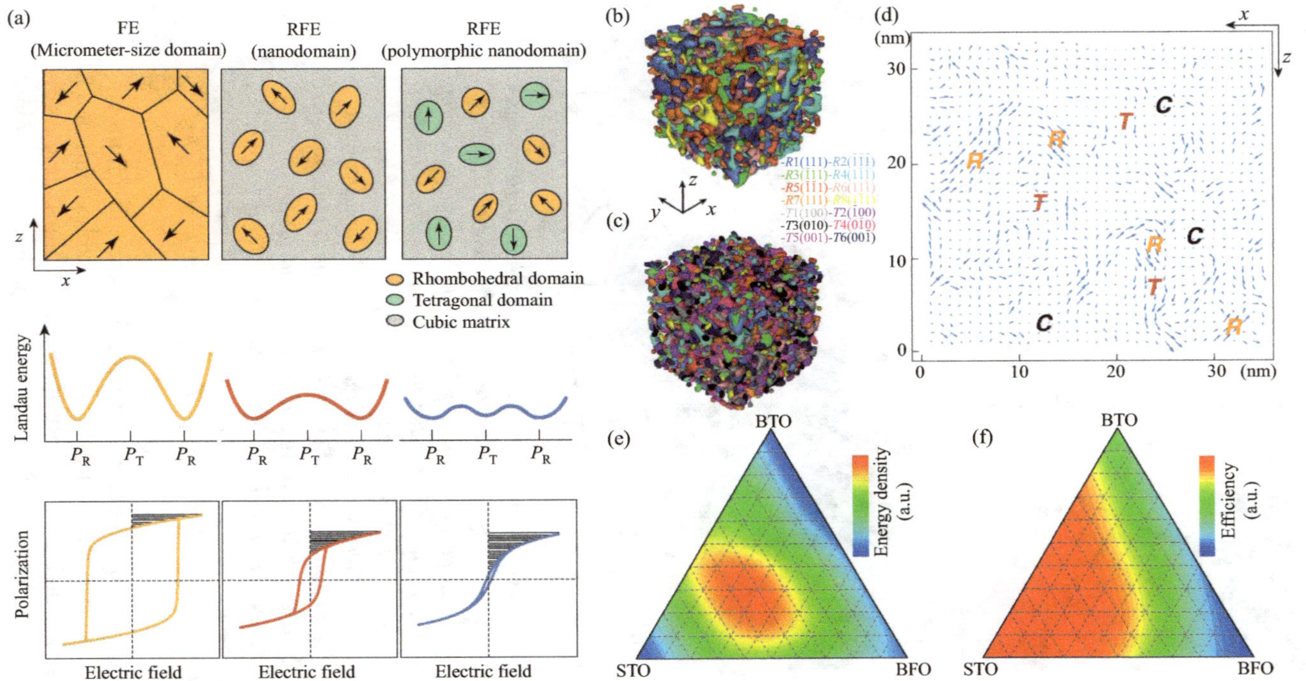

Figure 8.6 *Design of new RFEs with enhanced energy performance via polymorphic nanodomain design. (a) Comparative display of Landau energy profiles and P–E loops of an FE with micrometer-size domains, an RFE with nanodomains, and an RFE with polymorphic nanodomains. The P_R schematics represent the polarization states along the rhombohedral (R) [111] and ½111directions, and P_T along the intermediate tetragonal (T) ½001direction. The shadowed area in the P–E loops represents the energy density. (b and c) Phase-field-simulated three-dimensional domain structures of (b) 0.45BFO–0.55STO with R nanodomains and (c) 0.20BFO–0.25BTO–0.55STO with coexisting R and T nanodomains. The size for simulation is 32 nm × 32 nm × 32 nm. The different colors represent different polarization directions of the nanodomains, as shown by the inset notations. For clarity, the cubic (c) matrix is not depicted. (d) Simulation of the two-dimensional domain structure of 0.20BFO–0.25BTO–0.55STO showing the R and T nanodomains in the C matrix. (e–f) Contour plots of the simulated (e) energy density and (f) efficiency of BFO–BTO–STO solid solutions. Reprinted, with permission, from [25]*

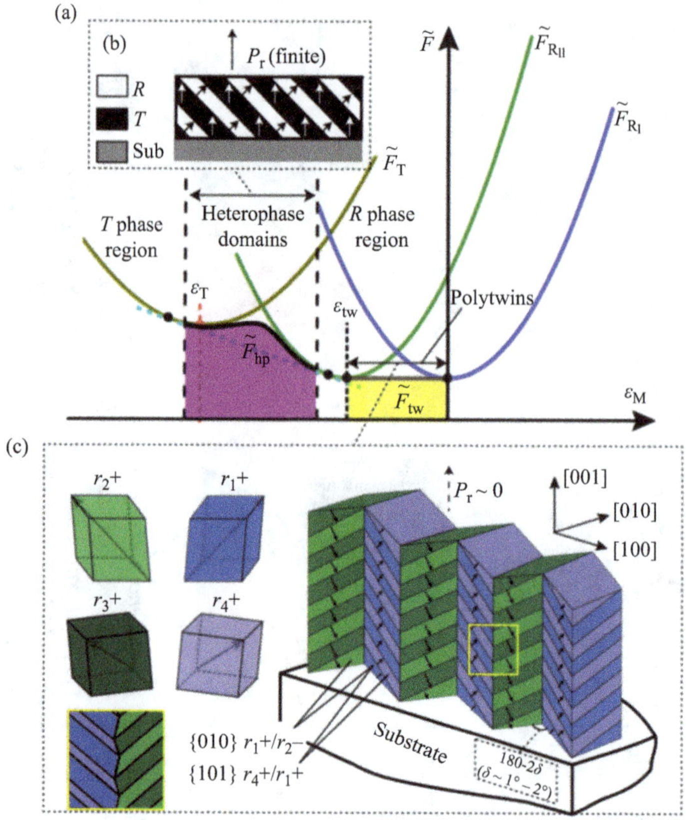

Figure 8.7 *(a) Free energies versus misfit strain for a film with a single-domain rhombohedral phase, a single-domain tetragonal phase, T/R heterophase polydomains and rhombohedral polytwins. Stable regions of the polydomain structures are determined by the common tangent method applied to the free energy curves of single-domain films. ε_T and ε_{tw} are the spontaneous strains of the T and RII domains with respect to the RI domain. Schematics of the polydomain structures are shown in (b) for a T/R heterophase polydomain, and (c) for hierarchical rhombohedral polytwins formed in a (001)-oriented epitaxial ferroelectric film with a bulk rhombohedral phase.* *© 2017 Springer Nature. Reprinted, with permission, from [27]*

Although thin films exhibit superb energy storage performance, thin films are still far away for practical application, for their low volume, complex fabrication process, difficulty in scalability and high cost. Relatively, thick films are a compromise between bulks and thin films. With the same composition, thick films possess the similar polarization but higher breakdown strength to the bulks, even though the breakdown strength is inferior to that of thin films. More importantly, thick films can be manufactured with mature and low-cost tape-casting process. With a multilayer structure, the multiplayer ceramic capacitors (MLCC) with high energy density dielectrics exhibit a great potential for power electronics applications. At around 800 kV cm^{-1}, the PLZT thick films shows an energy storage density of 12.4 J cm^{-3} and an efficiency of 75% [28]. With a core-shell grain structure, a sizable energy storage density of 10.5 J cm^{-3} along with an efficiency of 87% are reported in the MLCC composed of BFO–BTO relaxor thick films with interdigital electrodes. Similar results are reported in the $0.87BaTiO_3$–$0.13Bi(Zn_{2/3}(Nb_{0.85}Ta_{0.15})_{1/3})O_3$ (BTO–BZNT) MLCC [29]. With an electric field, the BTO–BZNT shows an energy storage density of 10.12 J cm^{-3} along with a high efficiency of 89.4%, which simultaneously exhibits preferred temperature stability, that is the energy storage density varies $<\pm 5\%$ as the temperature ranges from −75 °C to 150 °C at 40 MV m^{-1}.

8.3.3 Antiferroelectrics

As the adjacent dipoles orientate antiparallelly at zero electric field which can be aligned to the same direction with a high field (transit from the antiferroelectric state to the ferroelectric state, that is AFE–FE transition), antiferroelectrics (AFEs) exhibit unique double hysteresis *P–E* loops (the threshold electric fields correspond to the AFE–FE transition are E_F and E_A, respectively). Thus, at zero electric field, the antiferroelectric materials have no macroscopic polarization. The double hysteresis *P–E* loops with low remanent polarization are highly beneficial to deliver a high recoverable energy storage density (Figure 8.2(e)). In early stage, antiferroelectric effect is mostly studied in $PbZrO_3$ (PZO)-based system. Actually, it is practically impossible to observe the double hysteresis loop in pristine PZO at room temperature as its E_F exceeds the breakdown strength. However, under the guide of Goldschmidt tolerance theory that predicts the phase stability in the perovskites as a function of the ion radii, the polarization behavior can be modified by composition design [30]. Energy storage properties are widely reported in the (Pb, La)(Zr, Sn, Ti)O_3 (PLZST) system, which is very sensitive to the composition. For example, in the $(Pb_{0.87}Ba_{0.1}La_{0.02})(Zr_xSn_{0.95-x}Ti_{0.05})O_3$ (PBLZST) ceramics, the ferroelectric state with x = 0.6 can be transformed to antiferroelectric state with x = 0.625, and the polarization is maximized at x = 0.675, due to the larger cell volume for the relative ion displacement between the B site cations and O^{2-} with the increase content of Zr4$^+$ whose ion radius is larger than that of Ti^{4+}. By tuning the Zr:Sn ratio, an energy storage density of 2.05 J cm^{-3} and an efficiency of ~70% is achieved in the bulk antiferroelectric ceramics. Moderate Y doping can further improve the energy storage property of the PBLZST, as the substitution of Y^{3+} for Pb^{2+} could depress the formation of oxygen vacancies, leading to an enhanced polarization. With 0.75 mol% Y doping, the energy storage density of PBLZST can be enhanced to 2.05 J cm^{-3} (Figure 8.8) [31]. The temperature stability of the PBLZST can also be improved by composition design [31,32]. Cooperated with 55 wt% $Pb_{0.97}La_{0.02}(Zr_{0.93}Sn_{0.05}Ti_{0.02})O_3$ (PLZST), a orthorhombic phase whose polarization is

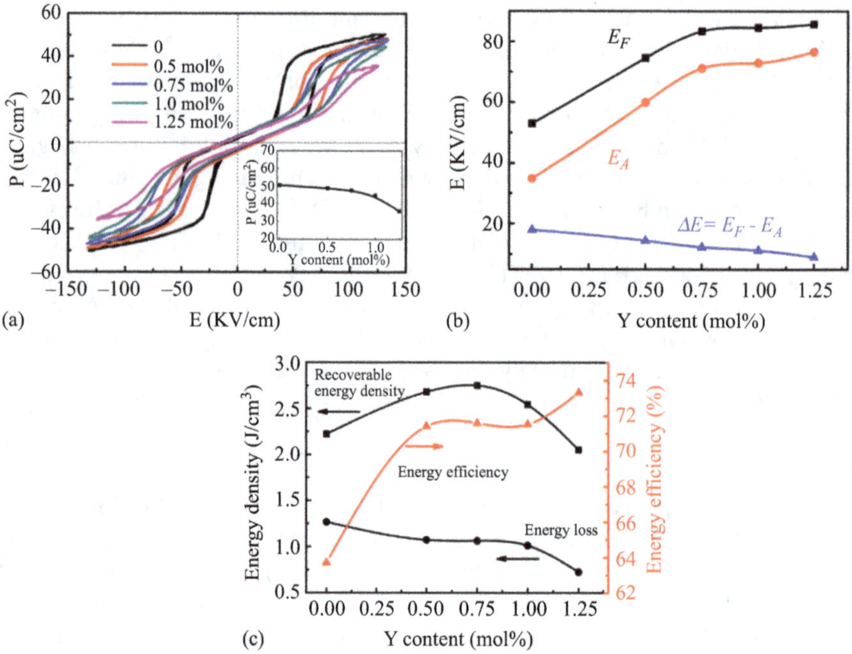

Figure 8.8 (a) The hysteresis loops and (b) the saturation polarization and (c) electric fields for the antiferroelectric–ferroelectric and ferroelectric–antiferroelectric phase transitions (E_F and E_A, respectively) and ΔE of the PBLZST-x ceramics. The recoverable energy density, energy loss and energy efficiency of the samples with different contents of Y. © 2014 Elsevier. Reprinted, with permission, from [31]

independent of temperature from room temperature to around 150 °C, the PBLZST–PBZST composite shows an energy storage density of 3.2 J cm^{-3} and efficiency of 75% which varies within 15% during the temperature range from 20 °C to 150 °C [33].

Recently, lead-free antiferroelectric materials for energy storage have attracted much attention for addressing the environmental issue of lead-containing materials [34]. The most popular lead-free antiferroelectric material recently is $AgNbO_3$ (ANO). At room temperature, the pristine ANO shows the ferroelectric phase which transfers to antiferroelectric phase at 67 °C (Figure 8.9). The energy storage performance of ANO also relies heavily on the composition modifications, for example A-site Bi^{3+} and B-site Ta^{5+} doping can improve the energy storage property significantly. The doped Bi ions interact with the Ag vacancies and stabilize the AFE in the ANO, resulting in a high energy storage density of 2.6 J cm^{-3} @ ~210 kV cm^{-1} with an energy efficiency of 86% [35,36]. The AFE of ANO can also be stabilized with the doped Ta ions with reduced polarizability and electronegativity. In addition, the introduction of Ta leads to an increased E_A, which enlarges the area that represents the recoverable energy storage density as schematically illustrated in Figure 8.2(e). Consequently, the optimized energy storage density of 4.2 J cm^{-3} with a high-temperature stability is achieved in the doped ANO.

As we discussed above, the FE phase can be induced with electric field that exceeds the E_F, and in many cases, the induced FE state is very stable, leading to a low E_A and wide square double hysteresis loops that is detrimental to the energy efficiency. So the concept of relaxor-antiferroelectric materials has been proposed to anticipate the antiferroelectrics to exhibit a slim double hysteresis loop with a reduced E_F–E_A. In $(Na_{0.43}Bi_{0.43}Al_{0.06})TiO_3$ (BNAT), the introduction of Ba^{2+} can induce a relaxor-antiferroelectric behavior like double hysteresis loop in the material [37], which provides a practicable strategy for designing the chemical composition for high energy storage antiferroelectrics.

Besides composition design, the energy storage performance is closely associated with the microstructure. For example, higher breakdown strength can be achieved in ceramics with smaller grains (higher grain boundary density) and lower porosity (higher density). To this end, hot-press sintering (HPS) and spark plasma sintering (SPS) have been used to fabricate antiferroelectric ceramics during the sintering processes of which the high mechanical stress is directly applied on the samples to eliminate the pores effectively (Figure 8.10) [38,39]. Furthermore, the high-speed heating process of HPS and SPS can depress the anomalous grain growth during sintering. High energy storage density of 6.4 J cm^{-3}, one of the highest energy densities in bulk ceramics, with an efficiency of 60% is achieved in the $(Pb_{0.858}Ba_{0.1}La_{0.02}Y_{0.008})(Zr_{0.65}Sn_{0.3}Ti_{0.05})O_3$-$(Pb_{0.97}La_{0.02})(Zr_{0.9}Sn_{0.05}Ti_{0.05})O_3$ composition fabricated by SPS, with an acceptable energy efficiency of 60.6%. Similarly, antiferroelectric films show superior energy storage performance to that of the bulks. For example, the favorable energy storage density of 9.5 J cm^{-3}, together with a desirable energy efficiency of 92% is demonstrated in $(Na_{0.5}Bi_{0.5})TiO_3$–$0.45(Sr_{0.7}Bi_{0.2})TiO_3$ (NBT–0.45SBT) MLCC, which is a promising candidate for practical high-power energy storage applications [40].

8.4 Glass ceramics

Glass ceramics are produced to combine the features of high polarization of crystals and high breakdown strength of amorphous phase. To this end, two types of glass ceramics are investigated: one is introducing a small content of glasses into ceramics to fill the grain boundaries (glass-ceramic composites), for the purpose of eliminating pores and enhance the breakdown strength in the materials with a minimal compromise in polarization. In the other one, the high-polarization crystal phases are crystalized during the annealing process, which is wrapped by the amorphous glasses.

For glass-ceramic composites, paraelectric BST with a Ba:Sr ration of 4:6 is usually used as the host material doped with glasses. Cooperated with a glass with the composition of BaO–20.3% SiO_2–26.7% B_2O_3–0.7% Al_2O_3–1.95% ZrO_2–

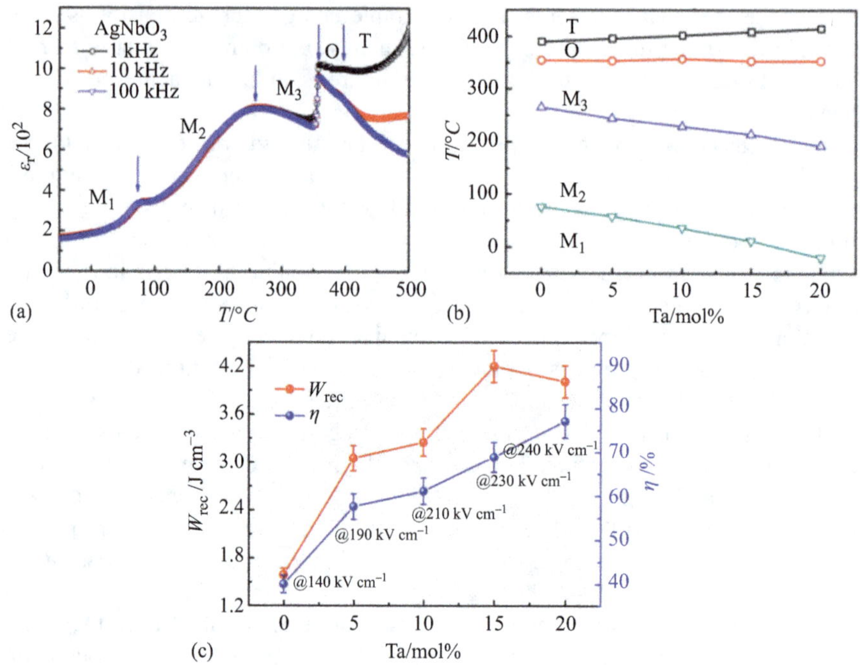

Figure 8.9 (a) Temperature and frequency dependence of dielectric constant of $AgNbO_3$ ceramic; (b) phase diagram of $Ag(Nb_{1-x}Ta_x)O_3$ determined from dielectric constant and (c) energy storage performance of the Ag $(Nb_{1-x}Ta_x)O_3$ ceramics prior to their breakdown. © 2017 John Wiley and Sons. Reprinted, with permission, from [34]

1.94% SrO (mol%), the breakdown strength of the BST can be enhanced from 120 kV cm^{-1} to 240 kV cm^{-1} as the glass content increases from 0 wt% to 20 wt%, which contributes to the increased energy storage density (0.9 J cm^{-3} with 5 wt% glass @ 200 kV cm^{-1}) [41]. Similar results have been reported in BST with BaO–B_2O_3–SiO_2–Na_2CO_3–K_2CO_3 glass and BTO with BaO–SrO–TiO_2–Al_2O_3–SiO_2–BaF_2 glass (Figure 8.11) [42]. However, as the polarization decreases with the added glass in this type of glass ceramics, the energy storage density is still relatively low.

Comparatively, most attentions are paid to synthesize the glass ceramics consist of glass matrixes embedded with crystal phases. BTO and BST are still popular lead-free crystals in this type of glass ceramics. Annealed at 850 °C–1200 °C, BST can be crystalized in the glass ceramic in the Al_2O_3, SiO_2, B_2O_3, LiF, BaF_2, and P_2O_5 glass system [43]. Even though with a relatively low polarization because of the limited content of crystal, the sample possesses much higher breakdown than that of the glass-ceramic composites due to their low defect density owing to the different crystal-amorphous ratios and crystallization processes. Applied with a high field of 800 kV cm^{-1}, the specimen shows an energy storage density of

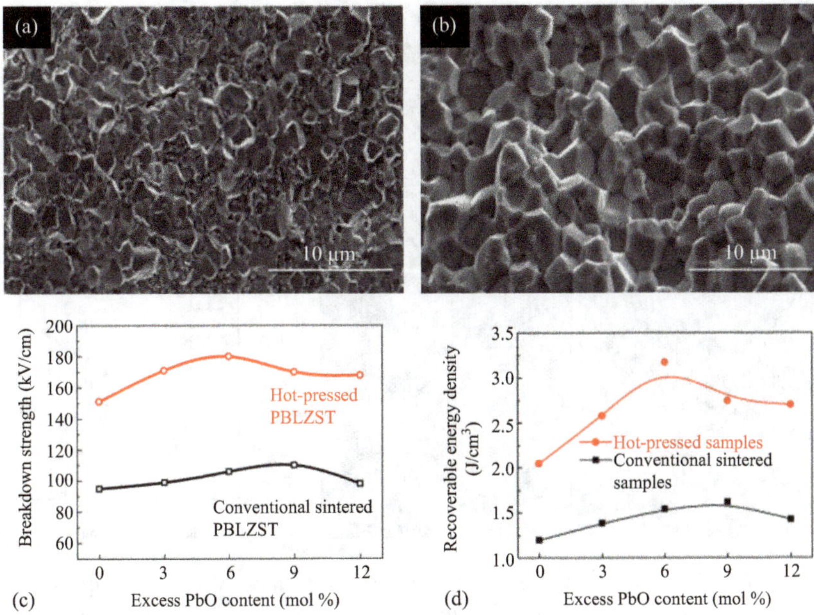

Figure 8.10 (a-b) Cross-sectional SEM images of PBLZST AFE ceramics are sintered by the conventional and hot-press methods, respectively. It is obvious that the hot-pressed samples are denser than that of the conventionally sintered one, whose grain size is more homogeneously as well. (c) The breakdown strength of the PBLZST ceramics is sintered by the conventional and hot-press sintering methods with different contents of excess PbO. The hot-pressed samples have higher breakdown strength due to their optimized microstructures. (d) Comparison of the recoverable energy density of the PBLZST ceramics is sintered by the conventional and hot-pressing methods. © 2015 John Wiley and Sons. Reprinted, with permission, from [38]

0.8 J cm^{-3}. Much higher energy density has been evidenced in the BaO–K_2O–Nb_2O_5–SiO_2 (BKNS) glass ceramic system that shows a quasi-linear polarization behavior [44]. Annealed to 750 °C–1000 °C, the $BaNb_2O_6$ and $Ba_2KNb_5O_{15}$ with a tungsten bronze structure and the K_3NbO_4 with perovskite structure are obtained in the amorphous glass matrix. With a very high breakdown strength exceeding 2,400 kV cm^{-1}, the samples exhibit a considerable energy storage density of 15 J cm^{-3}, which surpass all those of bulk relaxors and antiferroelectrics. Even higher energy storage density can be obtained in the lead-containing glass ceramic system. For example, an energy storage density of 17 J cm^{-3} has been achieved in the 15.4Na_2O–15.4PbO–23.1Nb_2O_5–46.2SiO_2 glass ceramics embed with $NaNb_3$ and $Pb_2Nb_2O_7$ crystals [45].

Figure 8.11 (a) Polarization-electric field (P–E) hysteresis loops for the BKNS glass ceramic at different temperatures. (b-c) Discharge energy density and efficiency versus the applied electric field of the BKNS. © 2012 Elsevier. Reprinted, with permission, from [42]

8.5 Conclusions and outlooks

In this chapter, the energy storage properties in inorganic dielectrics are surveyed. Overall, relaxors and antiferroelectrics have high energy density at relatively low electric fields due to their high polarization, while linear dielectrics and ceramic glasses, especially which are crystalized by the annealing process, can be operated with a much higher electric field to achieve sizable energy storage density. In addition, linear dielectrics exhibit lower losses, higher energy efficiency and more favorable temperature stability compared to that of nonlinear dielectrics. Compared to bulk materials, films with the same composition have much higher energy storage density due to the similar polarization but significantly larger breakdown strength.

Until now, most attentions are focused on how to further upgrade the energy storage density in the materials. However, it is suggested that more efforts should be made to push the material closer to practical application. For example, only high energy storage density is in vain as high energy loss and low efficiency are unacceptable as large amount of heat will be generated during the operation process and burn the dielectrics and capacitors. Moreover, dielectrics which can withstand elevated temperature with high-temperature stability are highly desired for the applications in extreme environments. Besides the properties of the materials themselves, optimization in capacitor structure design is important as

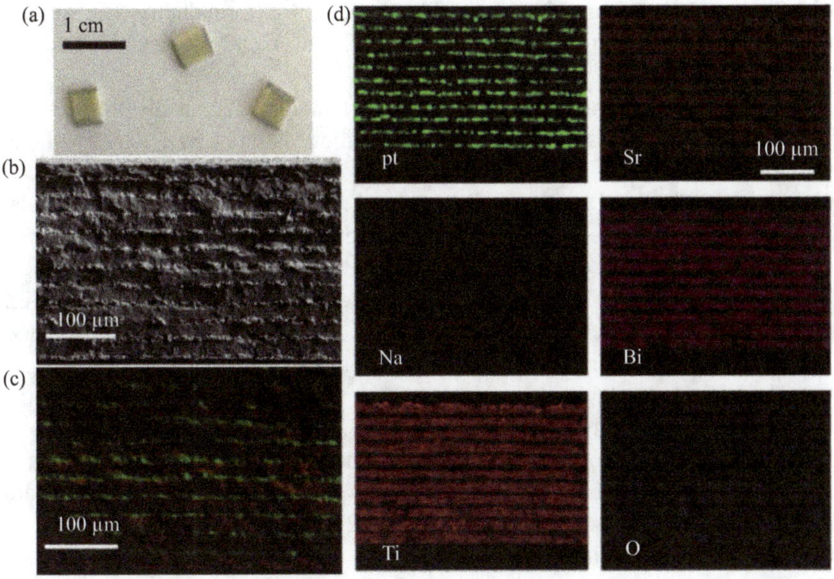

Figure 8.12 Picture and microstructures of multilayer ceramic. (a) A picture of the prepared NBT MLCC. (b) An SEM image of the MLCC, in which the ceramics and electrodes can be observed. (c) An SEM/energy-dispersive X-ray spectroscopy (EDX) image with all elements (Pt, Sr, Na, Bi, Ti and O) overlapping. (d) SEM/EDX images of the distributions of specific elements. The thickness of a single ceramic layer is measured according to the image of Ti. © 2018 John Wiley and Sons. Reprinted, with permission, from [40]

well. For example, good structure design of interdigital electrode can accelerate the heat transfer in the capacitor, preventing the heat breakdown of the dielectrics. Additionally, the development in fabrication technology, such as MLCC manufacturing for high-quality energy storage capacitors is anticipated (Figure 8.12) [40].

References

[1] Yang, L., Kong, X., Li, F., *et al.* 'Perovskite lead-free dielectrics for energy storage applications'. Prog. Mater. Sci. 2019, vol. 102, pp. 72–108

[2] Pan, H., Ma, J., Ma J., *et al.* 'Giant energy density and high efficiency achieved in bismuth ferrite-based film capacitors via domain engineering'. Nat. Commun. 2019, vol. 9, pp. 1813

[3] Satyanarayan, P., Aditya, C., Rahul V. 'Enhanced electrical energy storage density in mechanical confined antiferroelectric ceramic'. Ferroelectrics. 2015, vol. 486, pp. 114–125

[4] Amma, S., Lanagan, M., Kim, S., Pantano, C. 'Ionic conductivity in sodium–alkaline earth–aluminosilicate glasses'. J. Am. Ceram. Soc. 2016, vol. 99, pp. 1239–1247
[5] Smith, N., Rangarajan, B., Lanagan, M., Pantano, C. 'Alkali-free glass as a high energy density dielectric material'. Mater. Lett. 2009, vol. 63, pp. 1245–1248
[6] Gao, J., Kwon, D., Perini, S., Long, J., Zhang, S., Lanagan, M. 'Glass dielectrics in extreme high-temperature environment'. J. Am. Ceram. Soc. 2016, vol. 99, PP. 4045–4049
[7] Lee, H., Kim, J., Lanagan, M., Trolier-McKinstry, S., Randall, C. 'High-energy density dielectrics and capacitors for elevated temperatures: Ca(Zr, Ti)O_3'. J. Am. Ceram. Soc. 2013, vol. 96, pp. 1209–1213
[8] Hao, X. H. 'A review on the dielectric materials for high energy-storage application'. J. Adv. Dielect. 2013, vol. 3, pp. 1330001
[9] Zhang, G., Liu, H., Yao, Z. Cao, M. H., Hao, H. 'Effects of Ca doping on the energy storage properties of (Sr, Ca)TiO_3 paraelectric ceramics'. J. Mater. Sci.: Mater. Electron. 2015, vol. 26, pp. 2726–2732
[10] Rajesh, V. S., Daniel, S. K., George, A. R., Stephen, J. L. 'Strontium zirconate and strontium titanate ceramics for high-voltage applications: Synthesis, processing, and dielectric properties'. J. Am. Ceram. Soc. 2001, vol. 84, pp. 1648–50
[11] Song, Z., Liu, H., Zhang, S., *et al.* 'Effect of grain size on the energy storage properties of $(Ba_{0.4}Sr_{0.6})TiO_3$ paraelectric ceramics'. J. Eur. Ceram. Soc. 2014, vol. 34, pp. 1209–1217
[12] Huang, Y., Wu, Y., Qiu, W., Li, J., Chen, X. 'Enhanced energy storage density of $Ba_{0.4}Sr_{0.6}TiO_3$–MgO composite prepared by spark plasma sintering'. J. Eur. Ceram. Soc. 2015, vol. 35, pp. 1469–1476
[13] Li, F., Zhai, J., Shen, B., Zeng, H. 'Recent progress of ecofriendly perovskite-type dielectric ceramics for energy storage applications'. J. Adv. Dielect. 2018, vol. 8, pp. 1830005
[14] Zou, K., Dan, Y., Xu, H., *et al.* 'Recent advances in lead-free dielectric materials for energy storage'. Mater. Res. Bull. 2019, vol. 113, pp. 190–201
[15] Wu, L., Wang, X., Li, L. 'Lead-free $BaTiO_3$–$Bi(Zn_{2/3}Nb_{1/3})O_3$ weakly coupled relaxor ferroelectric materials for energy storage'. RSC Adv. 2016, vol. 6, pp. 14273–14282
[16] Zheng, D., Zuo, R. 'Enhanced energy storage properties in La $(Mg_{1/2}Ti_{1/2})O_3$-modified $BiFeO_3$–$BaTiO_3$ lead-free relaxor ferroelectric ceramics within a wide temperature range'. J. Eur. Ceram. Soc. 2017, vol. 37, pp. 413–418
[17] Wu, L., Wang, X., Gong, H., Hao, Y., Shen, Z., Li, L. 'Core-satellite $BaTiO_3$@$SrTiO_3$ assemblies for a local compositionally graded relaxor ferroelectric capacitor with enhanced energy storage density and high energy efficiency'. J. Mater. Chem. C, 2015, vol. 3, pp. 750–758
[18] Qu, B., Du, H., Yang, Z. 'Lead-free relaxor ferroelectric ceramics with high optical transparency and energy storage ability'. J. Mater. Chem. C, 2016, vol. 4, pp. 1795–1803

[19] Yang, Z., Du, H., Qu, S., *et al.* 'Significantly enhanced recoverable energy storage density in potassium–sodium niobate-based lead free ceramics'. J. Mater. Chem. A, 2016, vol. 4, pp. 13778–13785

[20] Seitz, F. 'On the theory of electron multiplication in crystals'. Phys. Rev. 1949, vol. 76, pp. 1376–1393

[21] Hu, Z., Ma, B., Liu, S., Narayanan, M., Balachandran, U. 'Relaxor behavior and energy storage performance of ferroelectric PLZT thin films with different Zr/Ti ratios'. *Ceram. Int.* 2014, vol. 40, pp. 557–562.

[22] Wang, X., Zhang, L., Hao, X., An, S. 'High energy-storage performance of 0.9 Pb $(Mg_{1/3}Nb_{2/3})O_3$–0.1 $PbTiO_3$ relaxor ferroelectric thin films prepared by RF magnetron sputtering'. Mater. Res. Bull. 2015, vol. 65, pp. 73–79.

[23] Kwon, D.-K., Lee, M. H. 'Temperature-stable high-energy-density capacitors using complex perovskite thin films'. IEEE Trans. Ultrason. Ferroelectr. Freq. Control, 2012, vol. 59, pp. 1894–1899

[24] Hu, Q., Jin, L., Wang, T., Li, C., Xing, Z., Wei, X. 'Dielectric and temperature stable energy storage properties of 0.88 $BaTiO_3$–0.12 $Bi(Mg_{1/2}Ti_{1/2})O_3$ bulk ceramics'. J. Alloy. Compd. 2015, vol. 640, pp. 416–420

[25] Pan, H., Li, F., Liu, Y., *et al.* 'Ultrahigh–energy density lead-free dielectric films via polymorphic nanodomain design'. Science, 2019, vol. 365, pp. 378–582

[26] Peng, B., Zhang, Q., Li, X., *et al.* 'Large energy storage density and high thermal stability in a highly textured (111)-oriented $Pb_{0.8}Ba_{0.2}ZrO_3$ relaxor thin film with the coexistence of antiferroelectric and ferroelectric phases'. ACS Appl. Mater. Interfaces, 2015, vol. 7, pp. 13512–13517

[27] Cheng, H., Ouyang, J., Zhang, Y.-X., *et al.* 'Demonstration of ultra-high recyclable energy densities in domain-engineered ferroelectric films'. Nat. Commun. 2017, vol. 8, pp. 1999.

[28] Zhang, L., Hao, X., Yang, J., An, S., Song, B. 'Large enhancement of energy-storage properties of compositional graded $(Pb_{1-x}La_x)(Zr_{0.65}Ti_{0.35})O_3$ relaxor ferroelectric thick films'. Appl. Phys. Lett. 2013, vol. 103, pp. 113902

[29] Zhao, P., Wang, H., Wu, L., *et al.* 'High-performance relaxor ferroelectric materials for energy storage applications'. Adv. Energy Mater. 2019, vol. 9, pp. 1803048

[30] Steven C., T. 'Mapping comparison of Goldschmidt's tolerance factor with perovskite structural conditions'. Ferroelectrics, 2014, vol. 470, pp. 13–27

[31] Zhang, L., Jiang, S., Zeng, Y., *et al.* 'Y doping and grain size co-effects on the electrical energy storage performance of $(Pb_{0.87}Ba_{0.1}La_{0.02})(Zr_{0.65}Sn_{0.3}Ti_{0.05})O_3$ anti-ferroelectric ceramics'. Ceram. Int. 2014, vol. 40, pp. 5455–5460

[32] Jiang, S., Zhang, L., Zhang, G., *et al.* 'Effect of Zr: Sn ratio in the lead lanthanum zirconate stannate titanate anti-ferroelectric ceramics on energy storage properties. '. Ceram. Int. 2013, vol. 39, pp. 5571–5575

[33] Liu, P., Li, M. Y., Zhang, Q., *et al.* 'High thermal stability in PLZST anti-ferroelectric energy storage ceramics with the coexistence of tetragonal and orthorhombic phase'. J. Eur. Ceram. Soc. 2018, vol. 38, pp. 5396–5401

[34] Zhao, L., Liu, Q., Gao, J., Zhang, S., Li, J.-F. 'Lead-free antiferroelectric silver niobate tantalate with high energy storage performance'. Adv. Mater. 2017, vol. 29, pp. 1701824

[35] Tian, Y., Jin, L., Zhang, H., *et al.* 'Phase transitions in bismuth-modified silver niobate ceramics for high power energy storage'. J. Mater. Chem. A, 2017, vol. 5, pp. 17525–17531

[36] Han, K., Luo, N., Mao, S., Zhuo, F., Liu, L., Peng, B., Chen, X., Hu, C., Zhou H., Wei Y. 'Ultrahigh energy-storage density in A-/B-site codoped AgNbO3 lead-free antiferroelectric ceramics: Insight into the origin of antiferroelectricity'. J. Mater. Chem. A. 2019, vol. 7, p. 26293

[37] Borkar, H., Singh, V. N., Singh, B. P., Tomar, M., Gupta, V., Kumar, A. 'Room temperature lead-free relaxor–antiferroelectric electroceramics for energy storage applications'. RSC Adv. 2014, vol. 4, pp. 22840–22847

[38] Zhang, G., Zhu, D., Zhang, X., *et al.* 'High-energy storage performance of $(Pb_{0.87}Ba_{0.1}La_{0.02})(Zr_{0.68}Sn_{0.24}Ti_{0.08})O_3$ antiferroelectric ceramics fabricated by the hot-press sintering method'. J. Am. Ceram. Soc. 2015, vol. 98, pp. 1175–1181

[39] Zhang, L., Jiang, S., Fan, B., & Zhang, G. 'Enhanced energy storage performance in $(Pb_{0.858}Ba_{0.1}La_{0.02}Y_{0.008})(Zr_{0.65}Sn_{0.3}Ti_{0.05})$ O_3–$(Pb_{0.97}La_{0.02})(Zr_{0.9}Sn_{0\ 05}Ti_{0.05})O_3$ anti-ferroelectric composite ceramics by Spark Plasma Sintering'. J. Alloy. Compd. 2015, vol. 622, pp. 162–165

[40] Li, J., Li, F., Xu, Z., Zhang, S. 'Multilayer lead-free ceramic capacitors with ultrahigh energy density and efficiency'. Adv. Mater. 2018, vol. 30, pp. 1802155

[41] Zhang, Q., Wang, L., Luo, J., Tang, Q., Du, J. 'Improved energy storage density in barium strontium titanate by addition of $BaO–SiO_2–B_2O_3$ glass'. J. Am. Ceram. Soc. 2009, vol. 92, pp. 1871–1873

[42] Wang, X., Zhang, Y., Song, X., *et al.* 'Glass additive in barium titanate ceramics and its influence on electrical breakdown strength in relation with energy storage properties'. J. Eur. Ceram. Soc. 2012, vol. 32, pp. 559–567

[43] Gorzkowski, E. P., Pan, M. J., Bender, B., Wu, C. C. M. 'Glass-ceramics of barium strontium titanate for high energy density capacitors'. J. Electroceram. 2007, vol. 18, pp. 269–276

[44] Wang, H., Liu, J., Zhai, J., Shen, B. 'Ultra high energy-storage density in the barium potassium niobate-based glass-ceramics for energy-storage applications'. J. Am. Ceram. Soc. 2016, vol. 99, pp. 2909–2912

[45] Luo, J., Du, J., Tang, Q., Mao, C. 'Lead sodium niobate glass-ceramic dielectrics and internal electrode structure for high energy storage density capacitors'. IEEE Trans. Dielectr. Electr. Dev. 2008, vol. 55, pp. 3549–3554

Chapter 9
Ceramic dielectrics for microwave communication

Zhifu Liu[1], *Jing Guo*[2] *and Hong Wang*[3]

9.1 Introduction

People have never been relying on wireless communication facilities as much as they do nowadays. Mobile phone, wireless local area network (LAN), intelligent transport systems, Internet of things (IoT) systems, etc. have been "implanted" into our life deeply. As the fundamental building blocks of the facilities for electromagnetic wave receiving, transmitting, and transferring, dielectric materials played an important role as milestones in the evaluation of wireless communication technology. The first mobile phone could only be made in 1973 by Martin Cooper, a Motorola researcher, after the breakthrough of the first temperature-stable low-loss barium tetra titanate ($BaTi_4O_9$) ceramic in 1971 [1] and its application in band-pass filters in 1972 [2]. The mass production of temperature-stable dielectric resonators for base station in early 1980s made commercialization of mobile phone possible in 1983. Furthermore, the invention of a series of high-performance dielectric ceramics and the ceramic components/modules production technology pushed the miniaturization and multifunctionalization of mobile phone.

On the other hand, the evolution of wireless communication technology also stimulated the innovation of microwave dielectric materials and device technology. The current 4G technology came into play around 2011 with the networks working frequency in the 700 MHz–2.7 GHz range. The 5G network is already on the stage of commercialization to fulfill the faster data transferring rates needed for ultra HD video streaming, ultra-reliable low latency in modern wireless communication including the Internet of things (IoT) [3]. Sub-6 GHz bands (3 GHz-6 GHz) have been assigned for the first used 5G cellular bands and massive multiple-input and

[1]CAS Key Laboratory of Inorganic Functional Materials and Devices, Shanghai Institute of Ceramics, Chinese Academy of Sciences, Shanghai, China
[2]State Key Laboratory for Mechanical Behavior of Materials and School of Materials Science and Engineering, Xi'an Jiaotong University, Xi'an, China
[3]Department of Materials Science and Engineering and Shenzhen Engineering Research Center for Novel Electronic Information Materials and Devices, Southern University of Science and Technology, Shenzhen, China

multiple-output (MIMO) technology has been applied to improve the throughput, latency, and cell efficiency. Since the limit band source in sub-6 GHz spectrum, high-frequency bands such as 24.25–27.5 GHz (n258), 26.5–29.5 GHz (n257), and 37–40 GHz (n260) have been assigned for 5G wireless communication bands [4]. To cover all the communication bands, the base station, the mobile phone and other terminals need to be designed to access all the working spectra. Dozens of narrow-band channels need to be integrated with complicated circuits without interference. These give big challenges not only to the communication systems design but also the dielectric materials.

From the point view of dielectric materials, ceramic dielectrics for 5G dielectric resonators should have low dielectric constant to increase the signal speed, ultra-low dielectric loss to improve the selectivity of devices, low-temperature variation of relative dielectric constant for the reliable operation, high thermal conductivity for fast heat dissipation, and low coefficient of thermal expansion to prevent failure during the harsh conditions [5,6]. The development of high-performance dielectric ceramics with dielectric constant lower than 20 is one of the main streams of the dielectric material research. A series of low dielectric constant materials such as Li–Zn–Ti–O, aluminates, and sillicates-based ceramics have been developed [7]. Furthermore, exploring low temperature cofired ceramic (LTCC)/ultra-low temperature cofired ceramic (ULTCC) materials for millimeter-wave applications never stopped. New process for fabrication low loss LTCC materials have been put forward [8]. A series of ULTCC materials with low loss, for example, molybdates, tungstates have been developed [9]. These materials will be strong support of millimeter-wave facilities of the upcoming 5G era. This chapter will briefly summarize the ceramic microwave dielectric materials developed in the past decades.

9.2 Brief history of the development of microwave ceramic dielectrics

In 1970s, the rapid growth in semiconductor technology has provided enormous miniaturization for active microwave components. But the miniaturization of passive microwave components such as resonators and filters has lagged behind. Although initially miniaturization has been realized by the development of microstrip resonators on alumina or plastic substrates, the quality factor (Q) of these devices is too low to replace the conventional bulky and expensive cavity resonators. To reduce the size of passive microwave components like dielectric resonators, filters for base station was the main task at the early stage of wireless communication development. This reduction depends on the phenomenon that the wavelength of an electromagnetic wave is shortened by a factor of $1/\varepsilon^{1/2}$ (ε is the dielectric constant) in dielectrics compared with a wavelength in free space. The dimensions of a dielectric resonator are on the order of one wavelength in the dielectric material. So, engineers have searched high dielectric constant microwave ceramics for long time.

Titania (TiO_2) has a high dielectric constant of 100 and high Q value up to 17,000 (@3 GHz). In 1968, Cohn and Harrison reported the ceramic dielectric resonator filter based on TiO_2 ceramics, respectively [10]. But the high-temperature coefficient of the resonant frequency (τ_f) of about 450 ppm/K prevented its practical application in the base station. The Ba–Ti–O material system was intensively investigated in early 1970s. A real breakthrough in dielectric resonator ceramic technology occurred in 1971 when the first temperature-stable low-loss $BaTi_4O_9$ dielectric ceramic was developed by researchers from Raytheon Company. The $BaTi_4O_9$ dielectric ceramics showed a dielectric constant of about 38, τ_f value ~−50 ppm/K, and Q value of about 2,500 at 11 GHz [1]. In 1972, the same group of people reported the temperature-stable bandpass filters using dielectric resonators made from $BaTi_4O_9$ ceramics and the temperature coefficient measured approaches that of an Invar cavity after compensation [2]. Later, a new barium titanate, $Ba_2Ti_9O_{20}$, ceramics with the improved performance was reported by Bell Laboratories. The single-phase $Ba_2Ti_9O_{20}$ ceramics have a dielectric constant of 39.8, Q value of 8,000 at 4 GHz, and τ_f value of -24 ± 2 ppm/K [11]. The Q value could be improved to 9,300 after heat-treated in oxygen-rich atmosphere [12]. The τ_f value could be tuned to near zero without bad effect on dielectric properties by substituting Ti with a small amount of Sn [13]. Another breakthrough in high performance dielectric ceramic material was made by researchers from Murata Manufacturing Company Japan. They developed the Zr–Sn–Ti–O based dielectric ceramics. The temperature coefficients could be tuned between +10 and −10 ppm/K by adjusting the ratio of Zr/Sn. The optimized Zr–Sn–Ti–O can have a dielectric constant of 38, Q value of 7,000, and τ_f value of 0 ppm/K [14].

Perovskites are another group of materials have been extensively investigated because of their high dielectric constants, high-quality factors and low-temperature coefficients. Many perovskite-type microwave dielectric ceramics, such as (Mg, Ca)TiO_3, Ca(Zr,Ti)O_3, and (Ba,Sr)(Zr,Ti)O_3 have been reported. The complex perovskites, $Ba(M^{2+}{}_{1/3}M^{5+}{}_{2/3})O_3$($M^{2+}$ = Mg, Zn, Co, Ni; M^{5+} = Nb, Ta), ceramics are the most attractive materials [15]. $Ba(Zn_{1/3}Ta_{2/3})O_3$ (BZT), $Ba(Mg_{1/3}Ta_{2/3})O_3$ (BMT), and $Ba(Zn_{1/3}Nb_{2/3})O_3$(BZN)-based ceramics with excellent microwave dielectric performance have been obtained and widely used in the electronics industry. Kawashima *et al.* first reported the microwave dielectric properties of the $Ba(Zn_{1/3}Ta_{2/3})O_3$ (BZT) ceramic, which can have a dielectric constant of about 30, Q value of 14,000 at 12 GHz, and a τ_f value of 0 ± 0.5 ppm/K [16]. BMT ceramic as a temperature stable microwave dielectric material was first reported by Nomura *et al.* The material showed a dielectric constant of 25, Q value of 16,800 at 10 GHz, and τ_f value of 2.7 ppm/K doped with 1 mol% Mn as sintering aid [17]. The Q value of BMT ceramic can be improved to about 40,000 (at 10 GHz) by adding 0.5% $Ba(Mg_{1/2}W_{1/2})O_3$ (BMW) [18]. Single-phase BZN ceramics have a dielectric constant of 41, Q value of 5,400 at 10 GHz, and TCF of about 30 ppm/K, while, the $Ba(Co_{0.7}Zn_{0.3})_{1/3}(Nb_{2/3})O_3$ (BCZN) ceramic shows a dielectric constant of 33.5, an improved Q value of 11,000 at 6.5 GHz, and a τ_f value of 0 ppm/K [19]. Considering the lower price of Nb than Ta, the excellent properties of BCZN makes it a cheaper replacement of BZT and BMT ceramics for dielectric resonator applications.

At the beginning of 1990s, miniaturization of microwave passive components for mobile phone become a main task for electrical engineers. Since the large effect of metallization on the Q value of components, the Q value of microwave dielectric ceramics is no longer the bottle neck of their application. Materials researchers focused on developing high dielectric constant microwave dielectric ceramics to solve the components miniaturization problem. One kind of classical high dielectric constant microwave dielectric materials are the tungsten bronze-type perovskite structure $Ba_{6-3x}Re_{8+2x}Ti_{18}O_{54}$ (Re = Nd, Sm, Pr, La) system. The microwave dielectric properties of the material can be tuned by the x value and Re element. High dielectric constant of 78–80 can be obtained with $Q{\times}f$ value of 8,800–9,720 GHz and can be obtained from the $Ba_{6-3x}Sm_{8+2x}Ti_{18}O_{54}$ ceramics when x is in the range of 0.5–0.7 [20]. M-phase Li–Nb–Ti–O ceramics are another high dielectric constant microwave dielectric material system. The dielectric properties of the $Li_{1-x-y}Nb_{1-x-3y}Ti_{x-4y}O_3$ ceramics with composition located in the M-phase field found the close relation of the dielectric properties with composition. The observed ε_r values were quite high and ranged from 78 ($x = 0.15$, $y = 0$) to 55 ($x = 0.1$, $y = 0.175$), and generally decreases with increasing x and y. The quality factor increases with the titanium content, reaching a $Q{\times}f$ value of 9,000 (at 6 GHz) for the sample with $x = 0.1$, $y = 0.175$. Furthermore, the τ_f value also exhibits systematic changes with composition and can be tuned to zero in the middle of the M-phase field [21].

These classical dielectric ceramic materials largely promoted the commercialization of miniaturized ceramic dielectric resonators, the base station for wireless communication, and the wide application of mobile phones. On the other hand, investigation on microwave dielectric materials year by year also benefits the understanding of the effect of chemical composition, crystal phase, microstructure, defects on dielectric constant, Q value, temperature coefficient, and other physical parameters. The established materials preparation, dielectric performance characterization, microwave components design and fabrication methods are important engines accelerating the evolution of the microwave dielectric ceramics.

9.3 Ceramic dielectrics for millimeter-wave wireless communication

To fulfill the demands of the fast huge data wireless transformation, the wireless communication frequency is expanding to millimeter-wave region. Except that the millimeter-wave bands 24–40 GHz have been assigned for 5G wireless communication, a standard of the millimeter-wave communication by the 60 GHz zone converges to "IEE802.11ad." The 77 GHz band was assigned for industrial radar sensors. Presently, it is uncertain exactly what type of filters will be used for handsets or for base stations in the millimeter-wave region. Waveguide filters or any kinds of architectures may be used for 5G millimeter-wave filters. It is known that for a given frequency, the size of a dielectric filter decreases in proportion to the square root of the dielectric constant. In the millimeter-wave region, components with a high dielectric constant material would be extremely small so that machining these ceramics to target dimensions would be extremely challenging. So, except for the low dielectric loss and good temperature stability, ceramic dielectrics with low dielectric constant are expected for millimeter-wave applications. Materials intended for millimeter-wave application were summarized as below.

9.3.1 Silicates

Silicates are one of the most abundant compounds on earth. Because of their low dielectric constant and low dielectric loss due to their basic structure units of silicon tetrahedron [SiO_4] containing strong covalent bond with 50% covalence, silicates are suitable for millimeter-wave dielectrics. A series of silicates, such as cordierite ($Mg_2Al_4Si_5O_{18}$), forsterite (Mg_2SiO_4), willemite (Zn_2SiO_4), and Anorthite ($CaAl_2Si_2O_8$)-based ceramic materials have been fabricated and reported in recent years. Many of them showed excellent microwave dielectric properties.

9.3.1.1 Cordierite based ceramics

Cordierite ($Mg_2Al_4Si_5O_{18}$) is regarded as a good candidate as microwave and millimeter-wave dielectric substrates. According to the arranging order of [SiO_4] and [AlO_4] tetrahedrons in [$(Si_4Al_2)O_{18}$] hexagonal rings, cordierite has two stable polymorphs, that is, orthorhombic cordierite phase with [$(Si,Al)O_4$] tetrahedrons arranging orderly and hexagonal indialite phase with [$(Si,Al)O_4$] tetrahedrons arranging disorderly. In orthorhombic cordierite structures, a series of six-member inner [$(Si_4Al_2)O_{18}$] and external [Mg_6O_{36}] rings are parallel to *a–b* planes and stack along the *c*-axis. Alternate layers of [$(Si_4Al_2)O_{18}$] hexagonal rings are connected by [MgO_6] octahedrons, one [SiO_4] and two [AlO_4] tetrahedrons [22]. Pure cordierite phase ceramic prepared by solid-state reaction has a dielectric constant of 6.2, $Q \times f$ value of 40,000 GHz, and τ_f value of –24 ppm/K. Comparatively, the cordierite glass ceramics exhibit a much higher $Q \times f$ value of 200,000 GHz. This may be caused by the worse density of the cordierite ceramic than glass-ceramic [23], while the octahedrons and tetrahedrons in the cordierite structure provide large space for microwave dielectric properties modification.

Partially substitution of Mg by Ni can improve the $Q \times f$ values of cordierite ceramics. The $Q \times f$ value of $Mg_{1.8}Ni_{0.2}Al_4Si_5O_{18}$ ceramics can be improved from 39,900 to 99,110 GHz. Ohsato *et al.* indicated that the origin of improvement of $Q \times f$ value comes from changing to the high symmetry, which changes from a distorted ring to an equilateral hexagonal ring according to the substitution Ni for Mg as shown in Figure 9.1 [24]. The replacement of Mg by rare earth elements can also affect the symmetry of cordierite crystal and therefore, the microwave dielectric properties. [$(Mg_{1-x}Ln_x)_6O_{36}$] external hexagonal rings in cordierite ceramics tend to be distorted as Ln^{3+} (Ln = La, Sm) ions substitutes Mg^{2+} ions, and the shapes of [$(Si_4Al_2)O_{18}$] inner hexagonal rings are alternatively tuned from

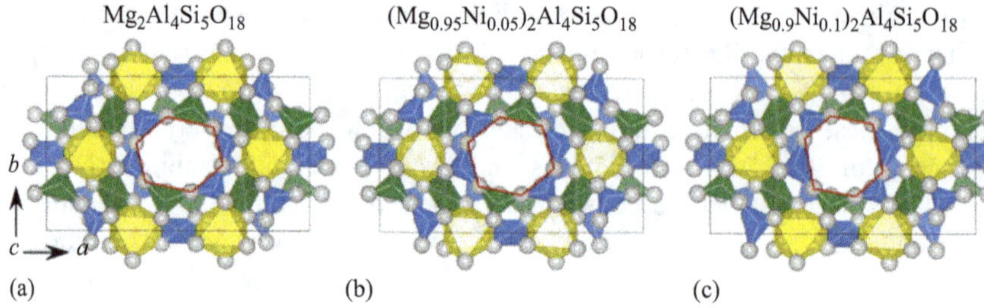

Figure 9.1 Crystal structure on Ni-substituted cordierite $(Mg_{1-x}Ni_x)_2Al_4Si_5O_{18}$ ($x = 0$, 0.05, 0.1, and 0.15). Reproduced from [24]

elliptical to circular. Compared with pure $Mg_2Al_4Si_5O_{18}$ cordierite ceramics, the improvement of $[(Si_4Al_2)O_{18}]$ ring centro-symmetry and the uniform distribution of grain sizes obviously enhance the $Q\times f$ value of $(Mg_{1-x}Ln_x)_2Al_4Si_5O_{18+x}$ cordierite type ceramics. The optimum microwave dielectric properties with a dielectric constant of 6.7, a $Q\times f$ value of 78,500 (@14.3 GHz), and τ_f value of −22 ppm/K were obtained by $(Mg_{0.95}La_{0.05})_2Al_4Si_5O_{18.05}$ ceramics [25]. To obtain near-zero temperature coefficient, Wu *et al.* prepared the $(1-x)Mg_2Al_4Si_5O_{18}$–$xTiO_2$ $[(1-x)$MAS–xT] ($0 \leq x \leq 0.35$) composite ceramics. A dielectric constant of 6.8, $Q\times f$ value of 37,800 GHz, and τ_f value of −0.2 ppm/K were obtained from the 0.75MAS-0.25T ceramic [22].

9.3.1.2 Forsterite based ceramics

Forsterite (Mg_2SiO_4) belongs to the Pbnm space group and consists of isolated SiO_4 tetrahedra surrounded by MgO_6 octahedra occupying sites of two different symmetries ($Mg(1)O_6$ and $Mg(2)O_6$). The $Mg(1)O_6$ site is slightly smaller and flattened along its threefold axis. The Mg(2) O_6 site is slightly larger and considerably more distorted. Tsunooka *et al.* reported the low dielectric constant and ultrahigh $Q\times f$ value ~270,000 GHz of the Mg_2SiO_4 ceramics [26]. The low dielectric constant of Mg_2SiO_4 ceramic is formed by SiO_4 tetrahedral framework with 45% ionic bond and 55% covalent bond. However, the Mg_2SiO_4 ceramics possess negative temperature coefficients of resonance frequency of −67 ppm/K. The sintering temperature of the Mg_2SiO_4 ceramics is also very high, which prohibits their practical application.

Works have been carried out to modify the dielectric performance and fabrication process to fulfill the requirement of the applications. The covalent bond reduced the dielectric constant because of the rattling effect reduced. Ando *et al.* investigated the effect of raw material on the dielectric properties of the forsterite ceramics. The Mg_2SiO_4 ceramics made from amorphous silica with a fine grain size of 0.25 mm brought higher $Q\times f$ value and a wider process window for sintering temperature than that of ceramics made from an irregular crystalline form with a coarse grain size of 0.82 mm [27]. To improve the microwave dielectric properties of Mg_2SiO_4 ceramics, various ionic substitution, such as divalent cations Cu^{2+}, Mn^{2+}, Ca^{2+} and Zn^{2+} ions and tetravalent Ti^{4+} ions were substituted for

Mg^{2+} and Si^{4+} ions, respectively. The Cu^{2+} ion presented a solid solution with the Mg_2SiO_4 phase and preferentially occupy Mg(1) site in the $Mg_{2-x}Cu_xSiO_4$ ceramics. The distortion of MgO_6 octahedron was modified by Cu^{2+} ions, resulting in a positive change in the temperature coefficient values. Excellent microwave dielectric properties with dielectric constant of 6.35, high $Q\times f$ value of ~188,500 GHz and near-zero τ_f value of –2.0 ppm/K were achieved at $x = 0.08$ [28]. TiO_2 was also applied to tune the τ_f value of the forsterite ceramics. Small amount of TiO_2 addition did not significantly affect the dielectric properties but decreased the sintering temperature [29]. Improved forsterite ceramics with τ_f value of 0 ppm/K, dielectric constant of 11 and $Q\times f$ value of 82,000 GHz could be obtained when they contain 24 wt% of TiO_2 and sintered at 1,200 °C for 2 h [30].

9.3.1.3 Feldspar based ceramics

The aluminosilicate feldspars are the most abundant minerals in the earth's crust. The feldspar crystal structure is composed of a 3D framework of corner-shared tetrahedra that are centered by Si^{4+} and Al^{3+}. Large alkali (Na^+, K^+) or alkaline-earth ions (Ca^{2+}, Sr^{2+} and Ba^{2+}) in the interstices balance the charge caused by the different charges of the tetrahedral Al^{3+} and Si^{4+} ions. The symmetry of the lattice depends on the effective size of the large cations (Na^+, Ca^{2+} and Ba^{2+}) and the distribution of the tetrahedral ions (Al^{3+} and Si^{4+}). The feldspars with smaller cations (Na^+ and Ca^{2+}) are triclinic, whereas the larger cations (Sr^{2+} and Ba^{2+}) tend to support the monoclinic symmetry [31]. Krzmanc *et al.* found that the $Q\times f$ values of the $MeAl_2Si_2O_8$ (Me = Ca, Sr and Ba) materials decrease in the following order: $BaAl_2Si_2O_8 > SrAl_2Si_2O_8 > CaAl_2Si_2O_8$. Monoclinic $BaAl_2Si_2O_8$ exhibits an order-higher $Q\times f$ value (40,000–90,000 GHz) compared to the triclinic $CaAl_2Si_2O_8$ (6,000–11,000 GHz) [32]. Krzmanc *et al.* also reported the microwave dielectric properties of the $Sr_xBa_{1-x}Al_2Si_2O_8$ ($0 \leq x \leq 1$) and $Ca_yBa_{1-y}Al_2Si_2O_8$ ($0 \leq y \leq 0.05$) ceramics. The substitution of Ba with Ca or Sr promotes the transformation of hexacelsian to monocelsian. The $Q\times f$ value of the $Sr_{0.05}Ba_{0.95}Al_2Si_2O_8$ ($x = 0.05$) was only 5% lower than the Q·f value of the $BaAl_2Si_2O_8$, which exhibited $Q\times f$ values of 70,600 and 77,700 GHz when sintered at 1,500 °C for 12 and 40 h, respectively. The improvement in the $Q\times f$ value with the prolongation of the heat-treatment time can be attributed to an increase in the Al:Si ordering [33].

The high sintering temperature up to 1,500 °C is the drawback of the $BaAl_2Si_2O_8$ based ceramics. The sintering temperature can be reduced by substituting Ba or Al with other elements. Krzmanc *et al.* reported the microwave dielectric properties of the $Na_xCa_{1-x}Al_{2-x}Si_{2+x}O_8$ plagioclase solid solutions ($0 \leq x \leq 1$). They found that slow-cooled ceramics exhibited higher $Q\times f$ values than fast-cooled ones. The highest $Q\times f$ value of 17,600 GHz was obtained for the slow-cooled $Na_{0.8}Ca_{0.2}Al_{1.2}Si_{2.8}O_8$ ceramic sintered at 1,100 °C, which has a dielectric constant of 5.8 and τ_f value of –7 ppm/K [34]. Song *et al.* presented a new type of plagioclase $BaZnSi_3O_8$ ceramic, which can be sintered at 1,100 °C and possesses good microwave dielectric properties with a dielectric constant of 6.60, a $Q\times f$ value of 52,401 GHz (at 15.4 GHz) and τ_f value of –24.5 ppm/K [35].

9.3.1.4 Other silicates based ceramics

Except for the above materials, there are many other silicates showed excellent microwave dielectric performances. The wollastonite possesses two polymorphs: (1) a low-temperature form composed of a chain structure of SiO_4 tetrahedra and (2) a high-temperature form (pseudo-wollastonite: α-$CaSiO_3$) with three-membered ternary rings of SiO_4 tetrahedra. High-temperature form α-wollastonite $CaSiO_3$ has excellent microwave dielectric properties with a dielectric constant of 6.82, a $Q\times f$ value of 42,200 GHz, and relative low τ_f value of -18.9 ppm/K [36]. The $Q\times f$ value can be improved by coupling with $SrSiO_3$ to form solid solutions. The lowest dielectric constant of 6.62 and highest $Q\times f$ value of 66,700 GHz with τ_f value of -40 ppm/K were obtained from the $Ca_{1-x}Sr_xSiO_3$ solid solutions when $x = 0.8$ [37]. Guo *et al.* reported the excellent millimeter-wave dielectric properties of the willemite (Zn_2SiO_4) ceramics, which has a dielectric constant of 6.6, a $Q\times f$ value of 219,000 GHz and a τ_f value of -61 ppm/K. With 11 wt% of TiO_2 addition, a dielectric constant of 9.3, a $Q\times f$ value of 113,000 GHz, and a τ_f value of 1.0 ppm/K were obtained [38]. Diopside ($CaMgSi_2O_6$) is one of the silicates with a chain SiO_4 structure and a low dielectric constant. The best ceramic of diopside with dielectric constant 7.6, $Q\times f$ value of 121,381 GHz and τ_f value of -66 ppm/K was obtained by Ohsato *et al.* [39]. Kan *et al.* investigated the microwave dielectric property of the $Ca_3(Zr_{1-x}Sn_x)Si_2O_9$(CZSS) solid solutions with cuspidine structure. The dielectric constant of CZSS solid solutions slightly decreased from 10.6 to 8.4, whereas the $Q\times f$ value increased from 54,791 to 93,296 GHz with the increase of x. The highest $Q\times f$ value of 93,296 GHz was obtained at $x = 1$, where the dielectric constant and temperature coefficient of resonant frequency were 8.4 and -64.7 ppm/K, respectively [40]. Table 9.1 lists some typical silicate ceramics with excellent microwave dielectric properties.

9.3.2 Rock-salt structure ceramics

The NaCl or rock salt structure is one of the most common and well known structure types. The regular octahedrons coordinated for both cations and anions are edge shared through three-dimensional network. Superstructure phases which display ordering among cation sites might form when two or three types of cation occupied the sodium sites in a nonrandom manner. It is well known that the Q values of the complex perovskites are proportional to the cations ordering degree. In addition, the temperature coefficients of resonant frequency of perovskites are usually tuned by changing the tilting angle of oxygen octahedrons since the oxygen octahedrons are corner shared in perovskites. However, the oxygen octahedrons should not tilt like in perovskites since oxygen octahedrons are edge shared in rock salt structure. Therefore compounds with rock salt structure may exhibit a different tuning mechanism. Furthermore, low cost could be expected due to the cheaper raw materials and relatively lower sintering temperature resulted from low melting lithium constituent. So, lithium contained oxides with rock salt structure are of great interest for the generation of new high Q microwave dielectrics. A series of lithium containing rock salt structure ceramic materials such as Li_2TiO_3,

Table 9.1 Silicate ceramics and their dielectric properties

Material	Dielectric constant	$Q\times f$ (GHz)	τ_f (ppm/K)	Reference
$Na_{0.8}Ca_{0.2}Al_{1.2}Si_{2.8}O_8$	5.8	17,600	−7	36
$Mg_2Al_4Si_5O_{18}$	6.2	40,000	−24	24
$Mg_{1.92}Cu_{0.08}SiO_4$	6.35	188,500	−2.0	30
$BaAl_2Si_2O_8$	6.36	44,800	−46.9	34
$BaZnSi_3O_8$	6.6	52,401	−24.5	37
Zn_2SiO_4	6.6	219,000	−61	40
$Ca_{0.2}Sr_{0.8}SiO_3$	6.62	66,700	−40	39
$(Mg_{0.95}La_{0.05})_2Al_4Si_5O_{18.05}$	6.7	78,500	−22	27
Mg_2SiO_4	6.79	30,000	−65	28
$0.75Mg_2Al_4Si_5O_{18}$–$0.25TiO_2$	6.8	37,800	−0.2	24
α-$CaSiO_3$	6.82	42,200	−18.9	38
$Sr_{0.05}Ba_{0.95}Al_2Si_2O_8$	7	73,500	−24	34
$CaAl_2Si_2O_8$	7.4	11,000	−130	34
$BaGa_2Si_2O_8$	7.6	62,300	−30	34
$CaMgSi_2O_6$	7.6	121,381	−66	41
$Ca_3SnSi_2O_9$	8.4	93,296	−64.7	42

Li_2MgTiO_4, $Li_2Mg_3BO_6$ (B = Ti, Zr, Sn), and $Li_3Mg_2NbO_6$ have been reported showing excellent microwave dielectric properties.

Rock salt structure Li_2TiO_3 crystals contains the above mentioned supercells, which make them show excellent performance. Yuan *et al.* reported the excellent microwave dielectric properties of Li_2TiO_3 ceramics, which had the dielectric constant of 22, $Q\times f$ value of 63,500 GHz, and τ_f value of 20.3 ppm/K [41]. They improved the density of the Li_2TiO_3 ceramics by introducing excessive amount of lithium. Optimized microwave dielectric properties were obtained from the $Li_{2.08}TiO_3$ ceramics, which have dielectric constant of 24.6, $Q\times f$ value of 66,000 GHz, and τ_f value of 22.1 ppm/K. The $Q\times f$ value of Li_2TiO_3 ceramics can be significantly improved to 106,226 GHz (at 8.8 GHz) by a small amount of MgO addition, although the ordering degree decreased with the addition of MgO. The τ_f value could be adjusted to 3.56 ppm/K simultaneously [42]. Bian *et al.* found that small amount of substitution for Ti^{4+} by $(Mg_{1/3}Nb_{2/3})^{4+}$ slightly decreased the dielectric constant, while considerably improved the $Q\times f$ value. The temperature coefficient of resonant frequency changed from positive into negative value. Excellent combined microwave dielectric properties with a dielectric constant of ~21.0, $Q\times f$ value of ~200,000 GHz and τ_f value of −1 ppm/K could be obtained from the $Li_2Ti_{0.76}(Mg_{1/3}Nb_{2/3})_{0.24}O_3$ ceramic after optimizing the calcining temperature [43]. Zhang *et al.* investigated the Al^{3+} and Nb^{5+} codoped Li_2TiO_3 ceramics [$Li_2Ti_{1-x}(Al_{0.5}Nb_{0.5})_xO_3$]. The complex substitution of Al^{3+} and Nb^{5+} produced a significant effect on the microstructural morphology, which resulted in the obviously enhanced $Q\times f$ values. Excellent microwave dielectric properties of dielectric constant ~21.2, $Q\times f$ value ~181,800 GHz and τ_f value ~12.8 ppm/K were achieved in the $Li_2Ti_{0.85}(Al_{0.5}Nb_{0.5})_{0.15}O_3$ ceramic [44].

The dielectric properties of the Li_2MgTiO_4 ceramics were first investigated by Kishore *et al.* [45]. Tseng *et al.* reported the excellent microwave dielectric performance of Li_2MgTiO_4 ceramic sintered at 1,360 °C, with a dielectric constant of ~17.25, $Q\times f$ value of ~97,300 GHz (at 9.86 GHz) and τ_f value of ~−27.2 ppm/K [46]. The effects of isovalent substitution $(Mg_{1/3}Ta_{2/3})^{4+}$ at the Ti-site on the sintering behaviors, microstructures, and microwave dielectric properties of Li_2MgTiO_4 ceramics have been investigated. With the $(Mg_{1/3}Ta_{2/3})^{4+}$ content increasing, the quality factor $Q\times f$ first increased and decreased thereafter, while the dielectric constant almost linearly decreased and the τ_f values shifted to positive value. The best composition appeared to be $Li_2MgTi_{0.6}(Mg_{1/3}Ta_{2/3})_{0.4}O_4$, which showed excellent microwave dielectric properties with a dielectric constant of ~15.73, $Q\times f$ value of ~184,000 GHz, and τ_f value of ~−12.54 ppm/K [47]. Substitution of Ti^{4+} with $(Mg_{1/3}Nb_{2/3})^{4+}$ has the similar effect as that of $(Mg_{1/3}Ta_{2/3})^{4+}$ substitution. But $(Mg_{1/3}Nb_{2/3})^{4+}$ addition leads to a more significant shift of τ_f value towards zero. The $Li_2MgTi_{0.7}(Mg_{1/3}Nb_{2/3})_{0.3}O_4$ ceramics sintered at 1,400 °C could achieve an excellent combination of microwave dielectric properties with a dielectric constant of ~16.19, $Q\times f$ value of ~160,000 GHz, and τ_f value of ~−3.14 ppm/K [48].

Fu *et al.* reported the microwave dielectric properties of the $Li_2Mg_3BO_6$ (B = Ti, Zr, Sn) materials [49]. Single-phase rock-salt structure $Li_2Mg_3TiO_6$, $Li_2Mg_3SnO_6$ and $Li_2Mg_3ZrO_6$ ceramics were obtained. The atmosphere-protective sintering method could effectively overcome the lithium volatilization. As a result, there was excellent microwave dielectric performance, including dielectric constant values of 10.12–14.47, $Q\times f$ values of 114,000–153,000 GHz and τ_f value of 11.13 to 17.47 ppm/K. Ultrahigh-Q $Li_{2(1+x)}Mg_3ZrO_6$ microwave dielectric ceramics were successfully prepared by means of atmosphere-controlled sintering through simultaneously adopting double crucibles and sacrificial powder. Desirable microwave dielectric properties with dielectric constant of 12.8, $Q\times f$ value of 307,319 GHz (@0.88 GHz), and τ_f value of −35 ppm/K were achieved from the $Li_{2.12}Mg_3ZrO_6$ ceramic [50]. To tune the τ_f value to zero, partially substitution for Mg^{2+} in the $Li_2Mg_3ZrO_6$ structure with Ca^{2+}, Ni^{2+}, Zn^{2+}, and Mn^{2+} was investigated. Although second phases appeared in the ceramics, the $Li_2(Mg_{0.95}Zn_{0.05})_3TiO_6$ ceramics exhibited excellent microwave dielectric properties of dielectric constant ~14.6, $Q\times f$ value ~158,000 GHz (at 9.11 GHz), and a near-zero τ_f value of 3.2 ppm/K [51].

Yuan *et al.* first reported that the $Li_3Mg_2NbO_6$ ceramics exhibited microwave dielectric properties with a dielectric constant of 16.8, $Q\times f$ value of 79,643 GHz and τ_f value of 27.2 ppm/K [41]. Similar with other lithium-containing rock-salt structure ceramics, small amount excessive lithium can improve the quality factor of the $Li_3Mg_2NbO_6$ ceramic. The $Li_{3.04}Mg_2NbO_6$ ceramic possessed extraordinarily excellent microwave dielectric properties: dielectric constant ~15.8, $Q\times f$ value ~150,000 GHz (at 9.85 GHz), and τ_f value ~−29 ppm/K [52]. The dielectric properties can be modified by partial substitution Mg^{2+} with Ca^{2+}, Ni^{2+}, Zn^{2+}, or

Co^{2+}. Zuo *et al.* reported that the $Li_3(Mg_{0.92}Zn_{0.08})_2NbO_6$ ceramics possessed a higher quality factor of 142,311 GHz, while the τ_f value showed a downward trend with the increasing content of Zn^{2+} [53]. The $Li_3(Mg_{0.95}Ca_{0.05})_2NbO_6$ ceramic exhibited appropriate microwave dielectric properties with a dielectric constant ~15.62, a $Q\times f$ value ~96,160 GHz, and τ_f value ~−18.49 ppm/K [54]. The $Li_3(Mg_{0.98}Co_{0.02})_2NbO_6$ ceramic sintered at 1,300 °C exhibited excellent microwave dielectric properties of dielectric constant ~15.22, $Q\times f$ value ~127,600 GHz and τ_f value ~−3.64 ppm/K, which made the ceramics as promising candidates for millimeter-wave applications [55].

9.3.3 Spinel structure ceramics

Compounds with the general formula AB_2O_4[A = Mg, Zn, Co, Ni, Cu; B = Al, Ga, Fe] are referred as spinels, which characterized by two symmetrically distinct polyhedra: an octahedron and a tetrahedron, occupied by divalent and trivalent cations. In the normal spinel structure, tetrahedral sites are occupied by divalent cations whereas octahedral sites are occupied by trivalent cations, whereas "inverse" spinels have formula $B(AB)O_4$ with A and B randomly distributed among the octahedral sites. Intermediate cation distributions are very common and a general formula can be written as $(A_{1-x}B_x)(A_xB_{2-x})O_4$, where x is the inversion parameter and is usually taken as the fraction of B cation on the tetrahedral site. Spinel-structured ceramics are expected to be suitable candidates for millimeter wave communication applications because of their microwave dielectric properties correspond to the intermediate spinel structures [56]. Some reported spinel structure microwave dielectric ceramics and their performance were summarized in Table 9.2.

$MgAl_2O_4$ is the most common spinel material. The microwave dielectric properties of a spinel-structured $MgAl_2O_4$ ceramic have been reported by Zheng *et al.*, who found that the dielectric constant and $Q\times f$ value of the ceramic were 7.9 and 68,900 GHz, respectively [57]. $MgAl_2O_4$ ceramics obtained by microwave sintering of high-purity ultrafine powders have sufficiently high transparency in the millimeter-wave range with $Q\times f$ value of 320,000 GHz at very high frequency of 296 GHz [58]. In general, $MgAl_2O_4$ is considered to have a normal spinel structure with the general formula AB_2O_4, in which the A^{2+} cations occupy the tetrahedral sites and the B^{3+} cations occupy the octahedral sites. However, ^{27}Al nuclear magnetic resonance (NMR) and Raman spectroscopy analyses of $MgAl_2O_4$ have revealed that the Al^{3+} cations can occupy both tetrahedral and octahedral sites, and the Al^{3+} cations position strongly depends on the fabrication process [59]. It has been suggested that the cation distribution in the $MgAl_2O_4$ lattice affects the covalency of cation-oxygen bonds and the microwave dielectric properties of the ceramics. The introduction of cation vacancies in $MgAl_2O_4$ could facilitate the preferential occupation of the tetrahedral sites by Al^{3+} cations [60]. Takahashi *et al.* reported the dielectric properties enhancement of the $Mg_{0.7}Al_{2.2}O_4$ and $Mg_{0.4}Al_{2.4}O_4$ ceramics with cation vacancies synthesized using the molten salt method. The $Mg_{0.7}Al_{2.2}O_4$ ceramics showed a dielectric constant of 7.7 and a $Q\times f$

Table 9.2 Microwave dielectric ceramics with rock-salt and spinel structures

Material	Dielectric constant	$Q\times f$ (GHz)	τ_f (ppm/K)	Reference
$Mg_{0.4}Al_{2.4}O_4$	7.5	232,301	−60	63
$Mg_{0.7}Al_{2.2}O_4$	7.7	201,111	−60	63
$MgAl_2O_4$	7.9	68,900	−73	59
$Zn_{0.4}Al_{2.4}O_4$	8.2	202,468	–	66
$ZnAl_2O_4$	8.5	56,300	−79	65
$Li_2Mg_3SnO_6$	8.8	123,000	−32	51
$ZnAl_{1.9}(Zn_{0.5}Ti_{0.5})_{0.1}O_4$	9.1	115,800	−78	67
$Li_2Mg_3ZrO_6$	12.6	86,000	−36	51
$0.83ZnAl_2O_4$–$0.17TiO_2$	12.67	99,500	–	65
$Li_{2.12}Mg_3ZrO_6$	12.8	307,319	−35	52
Mg_2TiO_4	14.51	161,570	−49.3	69
$Li_2(Mg_{0.95}Zn_{0.05})_3TiO_6$	14.6	158,000	3.2	53
$0.93Mg_2TiO_4$–$0.07CaTiO_3$	15	35,000	−2	75
$Li_2Mg_3TiO_6$	15.2	152,000	−39	51
$Li_3(Mg_{0.98}Co_{0.02})_2NbO_6$	15.22	127,600	−3.64	57
$(Mg_{0.95}Zn_{0.05})_2TiO_4$	15.48	275,300	−34	70
$Li_3(Mg_{0.95}Ca_{0.05})_2NbO_6$	15.62	96,160	−18.49	56
$(Mg_{0.95}Co_{0.05})_2TiO_4$	15.7	286,000	−52.5	71
$Li_2MgTi_{0.6}(Mg_{1/3}Ta_{2/3})_{0.4}O_4$	15.73	184,000	−12.54	49
$Li_{3.04}Mg_2NbO_6$	15.8	150,000	−29	54
$Li_2MgTi_{0.7}(Mg_{1/3}Nb_{2/3})_{0.3}O_4$	16.19	160,000	−3.14	50
$Li_3Mg_2NbO_6$	16.8	79,643	27.2	43
$Li_3(Mg_{0.92}Zn_{0.08})_2NbO_6$	17.2	142,311	−23.2	55
Li_2MgTiO_4	17.25	97,300	−27.2	48
0.76 Li_2TiO_3–$0.24MgO$	19.2	106,226	3.56	44
$Li_2Ti_{0.76}(Mg_{1/3}Nb_{2/3})_{0.24}O_3$	21	200,000	−1	45
$Li_2Ti_{0.85}(Al_{0.5}Nb_{0.5})_{0.15}O_3$	21.2	181,800	12.8	46
Li_2TiO_3	22	63,500	20.3	43

value of 201,111 GHz. The $Mg_{0.4}Al_{2.4}O_4$ ceramics showed a dielectric constant of 7.5 and a $Q\times f$ value of 232,301 GHz. The preferential occupation of tetrahedral site by Al^{3+} cation contributes to the enhancement of $Q\times f$ value [61]. Zn substitution for Mg can also affect the distribution of Al^{3+} in $MgAl_2O_4$ ceramics. The effects of Zn substitution for Mg on the microwave dielectric properties and cation distribution of the $Mg_{1-x}Zn_xAl_2O_4$ ceramics have been investigated. ^{27}Al NMR measurement revealed that the distribution of Al^{3+} cations in the octahedral site changed with the Zn content. The highest $Q\times f$ value of 222,600 GHz was obtained for the $Mg_{0.25}Zn_{0.75}Al_2O_4$ ceramics [57].

$ZnAl_2O_4$ is also a member of spinel family and naturally occurring as mineral gahnite. A fully dense $ZnAl_2O_4$ ceramic showed a Young's modulus of 242 GPa and a thermal conductivity of about 20–25 W/m·K [62]. Surendran *et al.* reported that $ZnAl_2O_4$ ceramic has a low dielectric constant of 8.5, high $Q\times f$ value of

56,300 GHz, and a τ_f value of −79 ppm/K. The high negative τ_f value could be tailored to zero by forming a composite with TiO_2. The composite $0.83ZnAl_2O_4$–$0.17TiO_2$ ceramics showed nearly zero τ_f value with excellent microwave dielectric properties (dielectric constant ~12.67, $Q\times f$ value ~99,500 GHz) [63]. Similar to the $MgAl_2O_4$ ceramics, excessive Al^{3+} in the $ZnAl_2O_4$ ceramic can lead to the preferential distribution of Al^{3+} in the spinel lattice and enhance the microwave dielectric properties. The dielectric constant of the spinel- structured $Zn_{1-3x}Al_{2+2x}O_4$ ($x = 0$–0.2) ceramics slightly decreased from 8.5 to 8.2 with increasing x, while the $Q\times f$ value increased from 127,532 to 202,468 GHz due to the preferential distribution of Al^{3+} and the introduction of cation vacancies [64]. Lan *et al.* reported that partial substitution of $[Zn_{0.5}Ti_{0.5}]^{3+}$ for Al^{3+} effectively lowered the sintering temperature of the $ZnAl_2O_4$ ceramics and remarkably increased the quality factor values. An enhanced $Q\times f$ value of 115,800 GHz with dielectric constant of 9.1 were obtained in the $ZnAl_{1.9}(Zn_{0.5}Ti_{0.5})_{0.1}O_4$ ceramics sintered at 1,400 °C in oxygen atmosphere [65].

Mg_2TiO_4 crystallizes in a special spinel structure. Generally, nearly half of Mg^{2+} ions occupy the octahedral site. Mg_2TiO_4 is composed of $[Mg_{(1)}/Ti_{(1)}O_4]$ tetrahedron and $[Mg_{(2)}/Ti_{(2)}O_6]$ octahedron. Octahedrons are connected with each other by sharing one of the edges while tetrahedrons are connected with octahedrons by sharing one of the vertexes [66]. Mg_2TiO_4 spinel ceramic acts as a promising material due to its typical microwave dielectric properties with dielectric constant of ~14.51, $Q\times f$ value of ~161,570 GHz and τ_f value of about −49.3 ppm/K [67]. Like the above spinel ceramics, partial substitution of Mg or Ti by Zn, Ni, Sn, Mn, Co, or even $(Mg_{1/3}Sb_{2/3})$ complex ions can largely improve the $Q\times f$ values [68–73]. For example, the $(Mg_{0.95}Zn_{0.05})_2TiO_4$ and $(Mg_{0.95}Co_{0.05})_2TiO_4$ ceramics showed $Q\times f$ value of 275,300 GHz (with dielectric constant of 15.48 and τ_f value of −34 ppm/K) and 286,000 GHz (with dielectric constant of 15.7 and τ_f value of −52.5 ppm/K), respectively [68,69]. But, these substitutions have little effects on the τ_f value. Near-zero τ_f value could be obtained from the $0.93Mg_2TiO_4$–$0.07CaTiO_3$ ceramics which has a dielectric constant of ~15, a $Q\times f$ value of 350,00 GHz, and τ_f value of ~−2 ppm/K [73] (Table 9.2).

9.3.4 Other dielectric ceramics for millimeter-wave applications

$A^{2+}B^{6+}O_4$ (A^{2+}: Ca, Pb, Ba; B^{6+}: Mo, W) ceramics with Scheelite structure are another kind of materials suitable for millimeter wave applications. $A^{2+}B^{6+}O_4$ (A^{2+}: Ca, Pb, Ba; B^{6+}: Mo, W) scheelite compound crystallize in tetragonal symmetry with four molecules per unit cell (space group: I41/a). As shown in Figure 9.2, A-site and B-site ions are positioned at the corner and face-center of the primitive unit cell, respectively, while the oxygen ion is located at the corner of a tetrahedron that also includes two A-site ions and one B-site ion. Microwave dielectric properties of $A^{2+}B^{6+}O_4$ (A^{2+}: Ca, Pb, Ba; B^{6+}: Mo, W) ceramics were investigated. The dielectric constant of $A^{2+}B^{6+}O_4$ scheelite compounds was not only dependent on the ionic polarizability but also the packing fraction resulting from the effective ionic size and unit cell volume. With the increase of packing

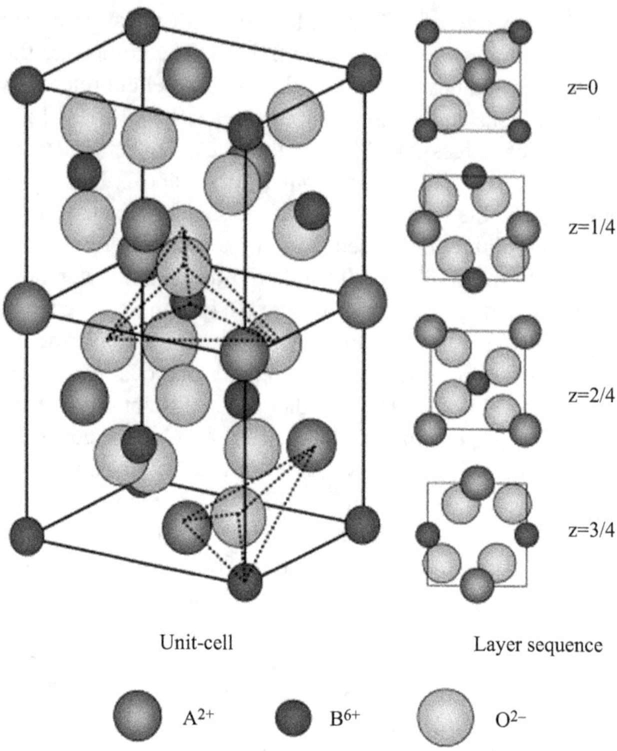

Figure 9.2 Crystal structure of $A^{2+}B^{6+}O_4$ scheelite compound. Reproduced from [74]

fraction from 54% ($BaBO_4$) to 62% ($CaBO_4$), the $Q\times f$ value increased from 21,620 GHz of $BaWO_4$ to 76,990 GHz of $CaWO_4$ [74]. The τ_f value of the specimens decreased with increasing of oxygen bond valences.

The dielectric ceramics $A_4M_2O_9$ (where A = Mg^{2+}, Co^{2+}, Zn^{2+} and Ni^{2+}; M = Ta^{5+} and Nb^{5+}) with corundum type structure also attracted much interest because of theirs high $Q\times f$ value and low dielectric loss. Among these materials, the $Mg_4Nb_2O_9$ ceramics were mostly investigated because the dielectric constant and $Q\times f$ values of $Mg_4Nb_2O_9$ ceramics are comparable to those of sintered Al_2O_3 ceramics [75]. With M (M = Zn, Mn and Co) substitution for Mg, the $Q\times f$ value values of $(Mg_{4-y}M_y)Nb_2O_9$ solid solutions decreased with increasing y, whereas those of $Mg_4(Nb_{2-x}Ta_x)O_9$ solid solutions increased from 210,000 to 347,000 GHz [76]. The $Mg_{0.7}Co_{0.3}Ta_2O_9$ ceramic shows a near-zero τ_f value with a dielectric constant of 13.79 and a $Q\times f$ value of 56,876 GHz [77]. A small amount of V substitution for Nb can effectively reduce the sintering temperature. The $Mg_4(Nb_{2-x}V_x)O_9$ ceramic at $x = 0.0625$ can be sintered at 1,025 °C and shows good dielectric properties with dielectric constants and the $Q\times f$ values of 11.6 and 160,256 GHz, respectively [78]. The dependences of the microwave dielectric properties on the isovalent substitution of $(Li^{1+}{}_{1/5}W^{6+}{}_{4/5})^{5+}$, $(Mg^{2+}{}_{1/4}W^{6+}{}_{3/4})^{5+}$, $(Al^{3+}{}_{1/3}W^{6+}{}_{2/3})^{5+}$ and $(Ti^{4+}{}_{1/2}W^{6+}{}_{1/2})^{5+}$ at the Nb^{5+}-sites of $Mg_4Nb_2O_9$ ceramics have also been investigated. With increasing isovalent substitution at the Nb^{5+}-sites, the dielectric constant decreased due to a higher dielectric polarizability of Nb^{5+} compared with those of the substitution ions. The τ_f values were also affected to below −60 ppm/K [79,80].

As shown above, some rock-salt and spinel structure titanates, niobates and tungstates have dielectric constant of 10–20 with high $Q\times f$ values. Silicates and aluminates always show dielectric constant lower than 10. Many of them can be good candidates for millimeter-wave applications. However, most of them have negative intrinsic temperature coefficient. Much work still needs to be carried out to tune their temperature coefficient to zero or near zero to fulfill the needs of millimeter-wave applications. The low price, high performance of many silicates and aluminates give much driving force to push their application in millimeter-wave devices in near future.

9.4 Low/ultralow-temperature cofired ceramic dielectrics

Low-temperature cofired ceramic (LTCC) technology is a kind of the multilayered technologies, which are generally applied in the industry for most functional ceramic circuits, passive and active components, actuators and modules productions. The history of multilayered technology can be traced back to 1940s when people tried to reduce the size of ceramic capacitors [81]. In 1970s, high-temperature co-fried ceramic (HTCC) technology was developed to fabricate ceramic substrates or board for semiconductor chip packaging or integration. But the metal electrode materials W, Mo and their alloys have poor electrical performance and cannot be used for high-frequency circuits and components. So, low-temperature cofirable ceramic materials with sintering temperature of 850–1,000 °C were developed so that these materials can be cofired with excellent conductors Ag, Au, or Cu in 1980s. LTCC technology provides a pathway to integrate conductive, dielectric and/or resistive components and enables rapid signal transmission between different electronic components. These advantages make LTCC technology more and more attractive for researchers in scientific research institutes and industries. In recent years, the application of mobile facilities is very common in daily life. There are emerging request on developing cost-effective, ecological process for electrical components production. Ultra-low temperature cofirable ceramics (ULTCC) can be sintered at temperature lower than 600 °C so that the cheap aluminum metal electrode could be used in components or circuits developments. The sintering temperature of ULTCC materials can even be reduced to less than 400 °C [9]. This reveals the possibility to allow the ceramic process compatible with silicon-based semiconductor chip fabrication process and to integrate the ceramic components with IC chips directly. More recently, the new developed cold sintering process (CSP) can sinter a series functional ceramics at

Table 9.3 Cofiring technologies, corresponding cofiring temperatures, and typical metal electrode materials. ST: sintering temperature; CSP: cold sintering process; RT: room temperature process

Multilayered technologies		**ST (°C)**	**Metal electrodes**	
			Material	**Melting point (°C)**
HTCC		1,600–1,800	W	3,410
			Mo	2,610
			Mo–Mn	1,246–1,500
LTCC		850–1,000	Cu	1,083
			Au	1,063
			Ag	960
ULTCC	ULTCC-II	400–700	Al, Ag, Au, or Cu	660
	ULTCC-I	<400	Almost all metal electrodes and some polymers	
	CSP and RT	≤ 300		

temperatures lower than 300 °C [82]. The cold sintering process opens a new window for fabricating ceramic components on flexible polymer substrates for wearable electronics applications. Table 9.3 shows different cofiring technologies, corresponding sintering temperatures and applicable electrode materials. The following context will review the development of related ceramic materials of these cofiring technologies.

9.4.1 LTCC dielectric materials

LTCC dielectrics should have a low sintering temperature (typically $\leq$950 °C) to cofire with metal electrodes. The performance of LTCC dielectrics were always restricted by two factors: (1) essential properties: low dielectric loss, low thermal expansion, good thermal resistance, high mechanical strength, etc. and (2) processing demand: sintering at low temperature. However, for most cases, these two requirements are contradictory, and the optimum is at a critical point to balance property condition and processing demand [83]. So, to reduce the sintering temperature and at the same time, keep good dielectric performance is the main theme of LTCC dielectric materials development.

The traditional LTCC materials usually employ glass additions to meet the requirement of low sintering temperature. The typical glasses include ZnO–B_2O_3, ZnO–B_2O_3–BaO, ZnO–B_2O_3–La_2O_3, ZnO–B_2O_3–SiO_2, MgO–B_2O_3–SiO_2, CaO–B_2O_3–SiO_2, SrO–B_2O_3–SiO_2, BaO–B_2O_3–SiO_2, Al_2O_3–B_2O_3–SiO_2, PbO–B_2O_3–SiO_2, MgO–Al_2O_3–SiO_2, La_2O_3–B_2O_3–TiO_2, Li_2O–B_2O_3, Li_2O–B_2O_3–SiO_2, Li_2O–B_2O_3–CaO–Al_2O_3, Li_2O–B_2O_3–SiO_2–CaO–Al_2O_3, etc. [84]. Generally, the glasses have low relative dielectric constants in the range of 4–25 and low dielectric losses. The dielectric properties of some typical glasses are shown in Table 9.4. The LTCC materials with glasses are categorized into two groups. One is the glass-ceramic, where the stating materials are pure glass powders. Another one is the glass plus ceramic composite, where the stating materials are composed of glass

Table 9.4 Dielectric properties of some typical glasses. Reproduced from [84]

Glass systems	Dielectric constant	Dielectric loss (1 MHz)
$ZnO–B_2O_3$	4.2	0.003
$ZnO–B_2O_3–BaO$	6.9	0.009
$Li_2O–B_2O_3–SiO_2$	6–9	0.004–0.6
$Li_2O–B_2O_3–CaO–Al_2O_3$	8.3	0.0027
$Li_2O–B_2O_3–SiO_2–CaO–Al_2O_3$	8–9	0.002–0.004
$BaO–B_2O_3–SiO_2$	7–10	0.00025–0.01
$ZnO–B_2O_3–SiO_2$	7–8	0.009–0.011
$PbO–B_2O_3–SiO_2$	9–20	0.009–0.02
$ZnO–B_2O_3–SiO_2–Bi_2O_3$	21	0.01

powders and crystalline ceramic powders [83]. In the first category, the glasses are first densified and then crystallized to ceramics during the sintering process. The $CaO–B_2O_3–SiO_2$ system-based Ferro A6 series materials are the typical examples of the glass-ceramic LTCC dielectrics. The dielectric properties of the final glass-ceramic products are dominated by the degree of crystallization. So the glass-ceramic type LTCC materials could have low dielectric loss. But sintering process may have large effect on the performance of this kind of material because the crystallization of glass-ceramics relies on the thermal history.

In the case of the second category, glasses are added into ceramics, and the glass plus ceramic composites are densified by liquid phase sintering. During the sintering process, the ceramic particles are wetted by glasses and the mass transportation is enhanced by the liquid phase. Therefore, the sintering temperatures are lowered by the addition of proper amount of glasses and the dielectric properties are controlled by both glasses and ceramics. Dupont 951 material is one typical glass plus ceramic composite system. But, the glass plus ceramic composite LTCC systems always show relative high dielectric loss because of the random glass network, especially at high frequency. So, sometimes, crystallizable glasses were also used as the sintering aids to reduce the amount of glass phase in the final state to achieve low loss LTCC materials. Dupont 9k7 material is a typical example of crystallizable glass plus ceramic material system. The sintered body of Dupont 9k7 material contains multiple crystalline phases with very small amount of residual glass [85]. But, similar to the glass-ceramic type LTCC material, the performance of the material containing crystallizable glass is very sensitive to sintering process.

To overcome the drawback of LTCC material with crystallizable glass, Liu *et al.* proposed a solid-state synthesis process to prepare LTCC materials by calcining the raw material of crystallizable glass at high temperature [8]. The resulting mixture contains multiple crystallized phases with small amount glass, and can be sintered at temperature lower than 950 °C. The performance could be tuned by adjusting the phase composition through chemical composition. The sintered ceramics show very low dielectric loss and are suitable for millimeter-wave applications. Nakajima *et al.* developed cofirable LTCC materials with different dielectric constants using the solid-state synthesis method by calcining the mixture of oxides and carbonates [86]. They did not use B_2O_3 but the mixture of Bi_2O_3, Na_2CO_3, K_2CO_3, CuO, and Mn_3O_4 to reduce the sintering temperature. The LTCC material with the main phase of $SrAl_2Si_2O_8$ prepared by this method can have a dielectric constant of 5.7, a high $Q\times f$ value of 100,000 GHz, and a near-zero τ_f value of −4.8 ppm/°C.

Researchers are trying to develop glass-free LTCC materials to avoid the effect of glass phase. Low melting point oxides or fluorides have been used as sintering aids to reduce the sintering temperature of ceramics. $BaCuB_2O_5$ has a low intrinsic sintering temperature of about 810 °C and been used as sintering aids for different material systems such as TiO_2 [87], $BaTi_4O_9$ [88], $Ba(Zn_{1/3}Nb_{2/3})O_3$ [89], etc. Since the $BaCuB_2O_5$ itself has high $Q\times f$ value up to 50,000 GHz, the LTCC materials obtained by using $BaCuB_2O_5$ as sintering aids normally show good performance. Recently, Shigeno *et al.* reduced the sintering temperature of Al_2O_3 to 925 °C by introducing small amount of $4CuO–TiO_2–4Nb_2O_3$(CTN) [90]. The Al_2O_3+CTN ceramic could have a dielectric constant of ~10, an excellent $Q\times f$ value of 34,000 GHz (@12 GHz), and a τ_f value of −51 ppm/K. Although some oxide sintering aids may introduce liquid phase during the sintering process, the final state of these additives are crystalline phase. This kind of glass-free state is very helpful for obtaining high dielectric performance. Surely, there are also many materials have low intrinsic sintering temperature and can be cofired with Ag or other low firing electrode material. Many ultra-low temperature cofired ceramics (ULTCCs) are glass-free intrinsic low temperature sintering materials.

9.4.2 ULTCC dielectric materials

Ultra-low temperature cofired ceramics (ULTCCs) are ceramics that can be sintered at temperatures lower than 700 °C. Generally, the ULTCCs can be categorized as ULTCC-II for sintering temperature in the range of 700–400 °C and ULTCC-I for sintering temperature less than 400 °C. The low sintering temperature makes it possible to cofire ceramics with metal electrodes with low melting points, high conductivities, and low prices. Especially, ULTCC-I materials may be cofired with almost all the normal metal electrodes. The much lower sintering temperature enables the cofiring of ceramics and organic materials, making it promising for the applications of flexible electronics. Besides, it could also reduce the sintering time and save the energy of the sintering process, leading to green manufacturing. Therefore, ULTCC materials have attracted more and more attention, and the published papers are increasing dramatic. Most of ULTCC materials are composed of low-melting-point oxides, such as MoO_3 (795 °C), TeO_2 (733 °C), V_2O_5 (690 °C), Bi_2O_3 (817 °C), PbO (886 °C), B_2O_3/H_3BO_3 (450/171 °C), P_2O_5 (340 °C), Li_2CO_3 (723 °C), etc. As typical examples, some of the molybdates, tungstates, tellurates, vanadates, and glass-ceramics are discussed below.

9.4.2.1 Molybdates

Molybdates are compounds with low melting points, and a large number of them show low sintering temperatures that are suitable for ULTCC applications.

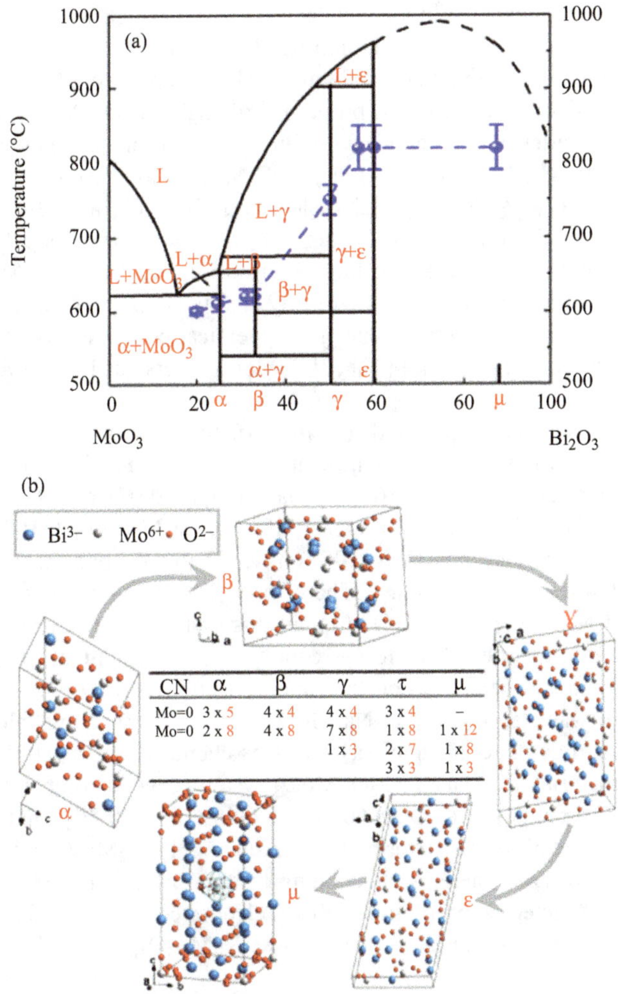

Figure 9.3 The phase diagram and sintering temperatures of Bi_2O_3–MoO_3 system. Reproduced from [91]

Zhou *et al.* reported the microwave dielectric properties of Bi_2O_3–MoO_3 system, including Bi_2O_3–$4MoO_3$, $Bi_2Mo_3O_{12}$, Bi_2O_3–$2.2MoO_3$, $Bi_2Mo_2O_9$, Bi_2MoO_6, $1.3Bi_2O_3$–MoO_3, $3Bi_2O_3$–$2MoO_3$, and $7Bi_2O_3$–MoO_3 [91]. Figure 9.3 illustrates the phase diagram and densification temperatures for Bi_2O_3–MoO_3 system. Many low melting-point compounds can be formed in this system. Some compounds have ultra-low sintering temperatures below 700 °C and are suitable for ULTCC applications, such as Bi_2O_3–$4MoO_3$ (600 °C), $Bi_2Mo_3O_{12}$ (610 °C), Bi_2O_3–$2.2MoO_3$ (620 °C) and $Bi_2Mo_2O_9$ (620 °C). Among them, $Bi_2Mo_3O_{12}$ exhibits the highest $Q{\times}f$ value of 21,800 GHz with a dielectric constant of 19 and a τ_f value of −215 ppm/K. $Bi_2Mo_2O_9$ shows the highest dielectric constant of 38 with a $Q \times f$ value of 12,500 GHz and a τ_f value of +31 ppm/K. Both $Bi_2Mo_3O_{12}$ and $Bi_2Mo_2O_9$ are chemically compatible with Al electrode materials [91]. Zhou *et al.* also studied the structures and microwave dielectric properties of Li_2O–Bi_2O_3–MoO_3 ternary system and found a series of compounds with ultra-low sintering temperatures, such as $Li_2Mo_4O_{13}$ (520 °C), Li_2MoO_4 (540 °C), $LiBiMoO_5$ (660 °C), $(Li_{0.5}Bi_{0.5})MoO_4$ (560 °C), $Li_3BiMo_3O_{12}$ (540 °C), $Li_8Bi_2Mo_7O_{28}$ (540 °C), and Bi_2O_3 (680 °C) [92]. Among them, Li_2MoO_4 has the highest $Q \times f$ value of 46,000 GHz with a dielectric constant of 5.5 and a τ_f value of −160 ppm/K. $(Li_{0.5}Bi_{0.5})MoO_4$ shows the highest dielectric constant of 44.4 with a $Q \times f$ value of 3,200 GHz and a τ_f value of +245 ppm/K. Li_2MoO_4 ceramic has good chemical compatibility with both Ag and Al electrode materials, while $(Li_{0.5}Bi_{0.5})MoO_4$ reacts easily with Al electrode material.

Recently, Zhang *et al.* reported that most of the compounds in Na_2O–MoO_3 system show ultra-low sintering temperatures and good microwave dielectric properties, such as $2Na_2O$–MoO_3 (565 °C), Na_2MoO_4 (660 °C), $Na_2Mo_2O_7$ (575 °C), Na_2O–$3MoO_3$ (510 °C), $3Na_2O$–$10MoO_3$ (510 °C), $Na_6Mo_{11}O_{36}$ (510 °C), Na_2O–$4MoO_3$ (505 °C), and Na_2O–$5MoO_3$ (505 °C) [93]. Table 9.5 presents the dielectric properties of these compositions. There are three single phases including Na_2MoO_4, $Na_2Mo_2O_7$ and $Na_6MoO_{11}O_{36}$, in this system, exhibit excellent microwave dielectric properties. $Na_2Mo_2O_7$ ceramic has the optimal $Q \times f$ value of 62,400 GHz, and its dielectric constant and τ_f value are 12.9 and −72 ppm/K, respectively. It was also found that Na_2MoO_4 is chemically compatible with both Ag and Al electrode materials and $Na_2Mo_2O_7$ is chemically compatible with only Al electrode material. Similar to Na_2O–MoO_3 system, several compounds in K_2O–MoO_3 system, such as $K_2Mo_2O_7$ (460 °C), $K_2Mo_3O_{10}$ (520 °C), and $K_2Mo_4O_{13}$ (540 °C), also have ultra-low sintering temperatures [94]. $K_2Mo_2O_7$ with a triclinic crystal structure can be sintered well at a temperature lower than 500 °C. It exhibits good microwave dielectric properties with a dielectric constant of 7.5, a $Q \times f$ value of 21,995 GHz, and a τ_f value of −63 ppm/K. $K_2Mo_3O_{10}$ sintered at 520 shows

Table 9.5 Microwave dielectric properties of ULTCC materials in Na_2O–MoO_3 system. Reproduced from [93]

Composition	ST(°C)	Phase	*f*(GHz)	ε_r	$Q \times f$(GHz)	τ_f(ppm/K)
$2Na_2O$–MoO_3	565	α+Na_2O	12.531	5.8 ± 0.1	19,900 ± 400	−115 ± 2
Na_2MoO_4	660	α	13.394	4.1 ± 0.2	35,000 ± 1,000	−76 ± 3
$Na_2Mo_2O_7$	575	β	8.415	12.9 ± 0.1	62,400 ± 1,200	−72 ± 1
Na_2O–$3MoO_3$	510	β+γ	9.180	10.3 ± 0.1	41,000 ± 2,000	−74
$3Na_2O$–$10MoO_3$	510	β+γ	9.456	10.2 ± 0.1	57,400 ± 2,000	−68
$Na_6Mo_{11}O_{36}$	510	γ	9.764	9.9 ± 0.1	57,000 ± 2,500	−61 ± 2
Na_2O-$4MoO_3$	505	γ+MoO_3	9.752	9.5 ± 0.2	33,800 ± 1,700	−57 ± 2
Na_2O-$5MoO_3$	505	γ+MoO_3	9.919	9.4 ± 0.1	22,100 ± 400	−59 ± 2

ST, sintering temperature; ε_r, relative dielectric constant.

slightly higher $Q\times f$ value of 35,830 GHz with a dielectric constant of 5.6, and a τ_f value of −92 ppm/K. Both $K_2Mo_2O_7$ and $K_2Mo_3O_{10}$ have good chemical compatibility with Ag and Al electrode materials.

Wang and Zhou groups studied the microwave dielectric properties of Ag_2O–MoO_3 system materials, including $Ag_2Mo_4O_{13}$, $Ag_2Mo_2O_7$, and Ag_2MoO_4 [95,96]. Figure 9.4 presents the phase diagram of Ag_2O–MoO_3 system and crystal structures of typical Ag–Mo based compounds. It is seen that the melting points of $Ag_2Mo_4O_{13}$(536 °C), $Ag_2Mo_2O_7$(516 °C), and Ag_2MoO_4(576 °C) are lower than 600 °C. Ag_2MoO_4 sintered at 450 °C shows good microwave dielectric properties with a dielectric constant, $Q\times f$ value, and τ_f value of 8.08, 17,000 GHz and −133 ppm/K, respectively [95]. It is chemically compatible with Ag electrode material

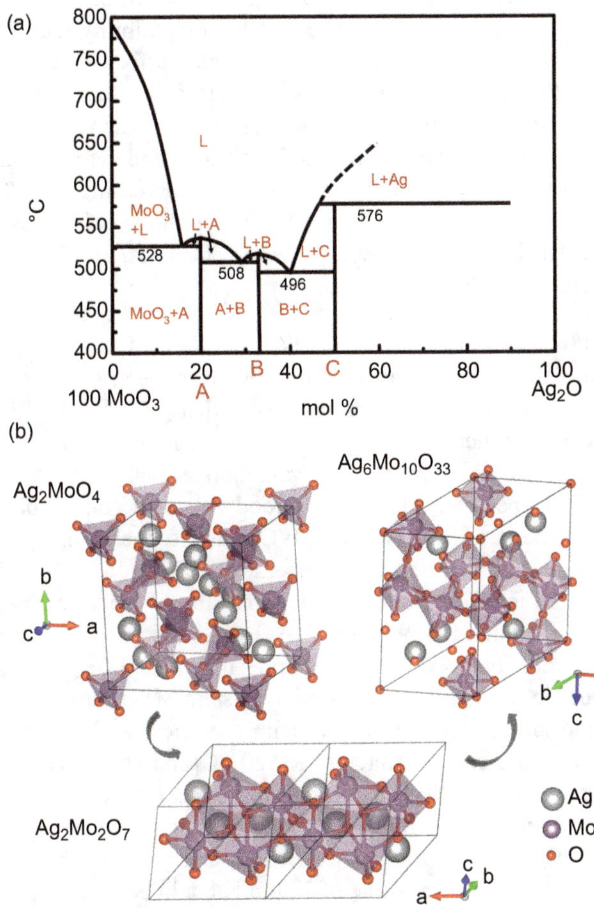

Figure 9.4 (a) The phase diagram of Ag_2O–MoO_3 system (A: $Ag_2Mo_4O_{13}$; B: $Ag_2Mo_2O_7$; Ag_2MoO_4) and (b) crystal structures of Ag–Mo based compounds. Reproduced from [96]

but reacts seriously with Al electrode material. $Ag_2Mo_2O_7$ sintered at 460 °C for 4 h has the highest $Q\times f$ value of 25,300 GHz with a dielectric constant of 13.3 and a τ_f value of −142 ppm/K [96]. $Ag_6Mo_{10}O_{33}$ sintered at 500 °C for 4 h exhibits a dielectric constant, $Q\times f$ value, and τ_f value of 14.00, 8,500 GHz and −50 ppm/K, respectively [96]. Xue *et al.* reported that the negative τ_f value of Ag_2MoO_4 can be tailored by the addition of $Ag_{0.5}Bi_{0.5}MoO_4$ [97]. All the $(1-x)Ag_2MoO_4$–$xAg_{0.5}Bi_{0.5}MoO_4$ $(0 \leq x \leq 1)$ composites show high densities at an ultra-low sintering temperature of less than 540 °C. The optimal microwave dielectric properties are obtained for $0.35Ag_2MoO_4$–$0.65Ag_{0.5}Bi_{0.5}MoO_4$ ceramic sintered at 520 °C with a dielectric constant of 23.9, a $Q\times f$ value of 16,200 GHz and a near-zero τ_f value of −2.4 ppm/K. $NaAgMoO_4$ with a spinel-like structure can be sintered at an ultra-low temperature of 400 °C [98], and shows the optimal microwave dielectric properties with a dielectric constant of 7.9, a $Q\times f$ value of 33,000 GHz, and a τ_f value of −120 ppm/K. $NaAgMoO_4$ is chemically compatibility with both Ag and Al electrode materials, which makes it potential material for ULTCC applications. $(Na_xAg_{2-x})MoO_4$ $(0 \leq x \leq 2)$ ceramics are solid solutions with spinel structure. All the compounds can be densified at a temperature lower than 420 °C [99]. With x increasing, the dielectric constant decreases from 8 to 5, and the τ_f value shifts from −140 to −60 ppm/K. The $Q\times f$ value increases with x increasing from 0 to 1.2, and then decreases with x for x larger than 1.2. The optimal microwave dielectric properties of dielectric constant = 8.1, $Q\times f$ =44,800 GHz, and $\tau_f = -82$ ppm/K were obtained with $x = 1.2$.

9.4.2.2 Tungstates

A lot of tungstates were reported as good microwave dielectric materials, but very few of them show sintering temperatures lower than 600 °C. Li_2WO_4 ceramic (rhombohedral crystal structure) sintered at 650 °C has good microwave dielectric properties with a dielectric constant, $Q\times f$ value, and τ_f value of 5.5, 62,000 GHz and −146 ppm/K, respectively. It shows a good chemical compatibility with Ag and Al electrode materials [100]. $(AgBi)_{0.5}WO_4$ has a scheelite monoclinic crystal structure and can be sintered well at a temperature of 580 °C [101]. The sintered $(AgBi)_{0.5}WO_4$ ceramic shows a dielectric constant of 35.9, a $Q\times f$ value of 1,300 GHz, and a τ_f value of −69 ppm/K. $(AgBi)_{0.5}WO_4$ reacts with Ag but has a good chemical compatibility with Al electrode materials. Xie *et al.* studied the microwave dielectric properties of PbO–WO_3 system. There are two compounds with low sintering temperatures in this system, specifically $PbWO_4$ and Pb_2WO_5 [102]. Pb_2WO_5 ceramic can be sintered at 520 °C and shows good microwave dielectric properties with a dielectric constant of 16.4, a $Q\times f$ value of 14,800 GHz, and a τ_f value of 95 ppm/K. As for $PbWO_4$, the sintering temperature is 620 °C, a little higher than that of Pb_2WO_5. A dielectric constant of 21.4, a higher $Q\times f$ value of 43,000 GHz and a near-zero τ_f value of 7 ppm/K could be obtained from $PbWO_4$ ceramic sintered at optimal condition. Both $PbWO_4$ and Pb_2WO_5 are chemically compatible with Ag and Al electrode materials.

Wang's group reported the microwave dielectric properties of NaO–WO_3 system [103]. There are three stable compounds in this system, including Na_2WO_4,

$Na_2W_2O_7$ and $Na_2W_4O_{13}$. Na_2WO_4 ceramic has the lowest sintering temperature of 565 °C and the best microwave dielectric properties with a dielectric constant, $Q{\times}f$ value, and τ_f value of 5.6, 124,200 GHz and −63 ppm/K, respectively. It shows good chemical compatibility with Ag and Al electrode materials. $Na_2W_2O_7$ ceramic can be sintered well at 650 °C. It has a dielectric constant of 8.1, a $Q{\times}f$ value of 22,600 GHz, and a τ_f value of −78 ppm/K. The sintering temperature of $Na_2W_4O_{13}$ ceramic is the highest (740 °C) in the NaO–WO_3 system. It shows a dielectric constant, $Q{\times}f$ value, and τ_f value of 11.2, 11,000 GHz and −65 ppm/K, respectively. With Ag substitution, the sintering temperature of Na_2WO_4 ceramic can be reduced from 565 °C to 510–520 °C [104]. The $(Na_{0.9}Ag_{0.1})_2WO_4$ ceramic sintered at 520 °C shows good microwave dielectric properties with a dielectric constant, $Q{\times}f$ value, and τ_f value of 5.8, 97,600 GHz and −70 ppm/K, respectively.

9.4.2.3 Tellurates

Maeda *et al.* first reported the sinterability and microwave dielectric properties of tellurate based compounds, including $TiTe_3O_8$, $SnTe_3O_8$, $ZrTe_3O_8$, and $(Ti_{0.8}Sn_{0.2})Te_3O_8$ [105]. All these materials have a trirutile-type formula of AB_2O_6, where A and B ions occupy the locations of Ti ions in the rutile structure in order. The porosities of $TiTe_3O_8$, $SnTe_3O_8$, $ZrTe_3O_8$, and $(Ti_{0.8}Sn_{0.2})Te_3O_8$ ceramics sintered at 700 °C are 12%, 12%, 4% and 10%, respectively, indicating that they can be sintered well at a low temperature of 700 °C. $TiTe_3O_8$ ceramic shows a dielectric constant of 36, a Q value of 3,400 (@4 GHz), and a τ_ε value of −220 ppm/K. $SnTe_3O_8$ ceramic has similar microwave dielectric properties with a dielectric constant, Q value (@4 GHz), and τ_ε value of 32, 3,300 and −380 ppm/K, respectively. $ZrTe_3O_8$ shows the highest dielectric constant of 68, the lowest Q value of 240 (@4 GHz), and the largest negative τ_ε value (−570 ppm/K). The $(Ti_{0.8}Sn_{0.2})Te_3O_8$ ceramic has the highest Q value of 5,400 (@4 GHz) together with a dielectric constant of 41 and τ_ε value of −310 ppm/K.

Later on, several groups developed more tellurate based microwave dielectric materials with low sintering temperatures, such as TeO_2, $Bi_2Te_2O_8$, $BaTe_4O_9$, $BaTe_2O_6$, $Zn_2Te_3O_8$, etc. [106–111]. Udovic *et al.* reported the microwave dielectric properties of TiO_2–TeO_2 system [106]. They claimed that $TiTe_3O_8$ is the only binary compound stable in air in the TiO_2–TeO_2 system. $TiTe_3O_8$ sintered at 720 °C in air shows a density higher than 95% and good microwave dielectric properties with a dielectric constant of 50, a $Q{\times}f$ value of 30,600 GHz, and a positive τ_f value of 133 ppm/K at 5 GHz. TeO_2 ceramic sintered at 640 °C for 15 h has a porosity of 20%, but exhibits not bad microwave dielectric properties with a dielectric constant, $Q{\times}f$ value and τ_f value of 19.3, 30,000 GHz, and −119 ppm/K, respectively. The negative τ_f value of TeO_2 can be tailored by the addition of $TiTe_3O_8$. The $0.15TiTe_3O_8$–$0.85TeO_2$ ceramic shows a near-zero τ_f value and good $Q{\times}f$ value of 22,000 GHz. The negative τ_f value of TeO_2 can be also tailored by the addition of $CaTiO_3$ and $SrTiO_3$ [112]. TeO_2–10 wt% $CaTiO_3$ can be sintered well in the temperature range of 630–670 °C with a dielectric constant of 28.6, a $Q{\times}f$ value of 15,600 GHz, and a near-zero τ_f value of −3.24 ppm/K. In the case of $SrTiO_3$ addition, the sintering temperature is slightly lower (610 °C) and the microwave dielectric properties are similar: dielectric constant = 28.7, $Q \times f$ =12,400 GHz, and $\tau_f = -3.81$ ppm/K. It is noted that there are impure phases ($TiTe_3O_8$ and $CaTe_2O_5/SrTe_2O_5$) formed in both TeO_2–$CaTiO_3$ and TeO_2–$SrTiO_3$ systems, which may deteriorate the microstructures and physical properties.

Kwon *et al.* reported the microwave dielectric properties of BaO–TeO_2 system [108–110]. The compounds rich in Te have lower sintering temperature. Among all the compounds in BaO–TeO_2 system ($BaTe_4O_9$, $BaTe_2O_6$, $BaTeO_3$, Ba_2TeO_5), $BaTe_4O_9$ shows the lowest sintering temperature of 550 °C [111]. The monoclinic $BaTe_4O_9$ ceramic has good microwave dielectric properties with a dielectric constant of 17.5, a $Q \times f$ value of 54,700 GHz, and a τ_f value of −90 ppm/K. The chemical compatibility test reveals that $BaTe_4O_9$ is easy to react with Ag but compatible with Al electrode material, making it promising for ULTCC applications. The negative τ_f value of $BaTe_4O_9$ ceramic can be tailored by $TiTe_3O_8$ ceramic [113]. The 60$BaTe_4O_9$–40$TiTe_3O_8$ (wt%) sintered at 575 °C shows a near-zero τ_f value of 3 ppm/K, as well as a dielectric constant of 25 and a $Q \times f$ value of 19,300 GHz.

Sebastian and co-workers reported the microwave dielectric properties of ATe_3O_8 [A = Sn, Zr, Zn] ceramics [114,115]. $Zn_2Te_3O_8$ ceramic with a monoclinic crystal structure can be sintered well at 585 °C. It has good microwave dielectric properties of dielectric constant = 16.2, $Q \times f$ = 66,000 GHz, and $\tau_f = -60$ ppm/K and chemical compatible with both Ag and Al electrode materials [114]. With the addition of 4 wt% TiO_2, the τ_f value is improved to near zero (−8.7 ppm/K) but the $Q \times f$ value drops to 27,000 GHz. With 30 wt% $TiTe_3O_8$ addition, the τ_f value can be tailored to −8.7 ppm/K with a dielectric constant of 19.8 and high $Q \times f$ value of 50,000 GHz [116]. In summary, some of the tellurates have low sintering temperatures, good microwave dielectric properties, and chemical compatibly with metal electrode materials, making it feasible for the ULTCC technology. However, the toxicity of Te may limit their commercial applications.

9.4.2.4 Vanadates

Valant and Suvorov first reported the sinterability and microwave dielectric properties of $BiVO_4$ ceramic, which has a dielectric constant of 68, a $Q \times f$ value of 6,500 GHz, and a τ_f value of −260 ppm/K [117]. However, $BiVO_4$ reacts with Ag seriously at the sintering temperature of 750 °C, which limits its application as an LTCC material. Later on, a number of researchers developed more vanadate-based microwave dielectric materials, such as $Mg_3(VO_4)_2$, $R_2V_2O_7$ (R = Ca, Sr, Ba), $NaCa_2Mg_2V_3O_{12}$, $Ca_5A_4(VO_4)_6$ (A = Zn, Mg), etc. [118–121]. Most of them have sintering temperatures lower than 950 °C but higher than 700 °C, which are not suitable for ULTCC applications. Recently, a few vanadates with low firing temperature ($\leq$ 600 °C) were reported, including BaV_2O_6, $Ba_3V_4O_{13}$, etc. [122,123]. The BaV_2O_6 ceramic can be densified at a low temperature of 550 °C for 1 h [122]. It shows good microwave dielectric properties with a dielectric constant, $Q \times f$ value, and τ_f value of 11.2, 42,800 GHz and 28 ppm/K, respectively. BaV_2O_6 is chemically compatible with Al electrode material. CaV_2O_6 has a monoclinic structure with the space group of C2/m, and its sintering temperature is a little higher than that of BaV_2O_6. CaV_2O_6 sintered at 675 °C shows a relative density of 95.5%, and excellent microwave dielectric properties with a dielectric constant of 10.2, a $Q\times f$ value of 123,000 GHz, and a τ_f value of 60 ppm/K [124]. However, the chemical compatibility of CaV_2O_6 with metal electrode materials was not reported. Pure $Ba_3V_4O_{13}$ phase with a monoclinic crystal structure can be obtained at a temperature $\geq$ 580 °C [123]. When the sintering temperature is lower than $\leq$ 560 °C, two other phases including orthorhombic BaV_2O_6 and triclinic $Ba_2V_2O_7$ are observed. The $Ba_3V_4O_{13}$ ceramic sintered at 600 °C for 1 h shows good microwave dielectric properties with a dielectric constant of 9.6, a $Q\times f$ value of 56,100 GHz, and a τ_f value of 42 ppm/K and has good chemical compatibility with Al electrode material [123]. It is noted that the compounds that are rich in V or Nb are easily to react with Ag [117]. The reactivity of vanadates towards Ag limits the application in LTCC technology. However, Al is a possible candidate electrode to replace Ag for vanadates with ultra-low sintering temperatures.

9.4.2.5 Glass-ceramics/glass+ceramic

There are some glass-ceramic or glass+ceramic composites with low sintering temperatures suitable for ULTCC applications. Chen *et al.* reported that the sintering temperature of Al_2O_3 and $BaTiO_3$ ceramics can be reduced to 450 °C with the addition of $27B_2O_3$–$35Bi_2O_3$–$6SiO_2$–32ZnO (mol%, BBSZ for short) glass [125,126]. After sintering at 450 °C, a second phase of $Bi_{24}Si_2O_{40}$ is formed in both BBSZ–Al_2O_3 and BBSZ–$BaTiO_3$ composites, and thus physical properties are dominated by the amount of BBSZ, Al_2O_3/$BaTiO_3$, and $Bi_{24}Si_2O_{40}$. The $50BaTiO_3$–50BBSZ (wt%) composite sintered at 450 °C exhibits a dielectric constant of 136 and a low dielectric loss of 0.02 at 100 MHz. With higher amount of BBSZ addition up to 90 wt%, the dielectric loss decreases to 0.011 and the dielectric constant drops to 63. The $30Al_2O_3$–70BBSZ (vol%) composite sintered at 450 °C has a lower dielectric constant of 20.8 and dielectric loss of 0.006.

9.4.3 Cold sintering and RT sintering of ULTCC materials

Cold sintering process (CSP) is a low temperature sintering technique for ceramic materials developed by Randall and co-workers [82]. It was inspired by the natural phenomena of geological and biological mineralization, such as the formation of sedimentary rock and the agglomeration of table salt, where an aqueous solution dominates the densification and grain growth [127,128]. During cold sintering process, a small amount of aqueous solution, such as pure water, weak acid, or weak base, is added to the ceramic powders, serving as the transient liquid phase to increase the driving force for the densification. The general process of cold sintering is shown in Figure 9.5. At the beginning, the ceramic powders are homogenously wetted by the transient liquid phase, and then the wetted powders are placed in a steal die with certain temperature and pressure. The temperature and pressure are usually in the range of RT-300 °C and 0–500 MPa, respectively. With the assistance of the liquid phase and external pressure, the particles undergo dissolution-rearrangement process, followed by the

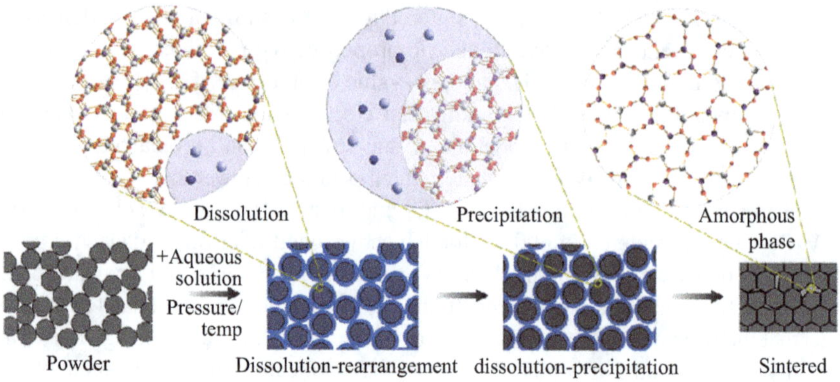

Figure 9.5 Schematic showing the general process of cold sintering. Reproduced from [82]

dissolution-precipitation process. The contact areas of particles show higher chemical potential, and thus the dissolved matters diffuse through the liquid and precipitate at the locations away from the contact areas during the dissolution-precipitation process [82]. This process drives the mass transportation and fills in the pores leading to densification of ceramics.

Since dissolution and precipitation play a critical role in cold sintering process, the pathways under cold sintering process can be categorized based on the dissolution characteristic of ceramic powders, as shown in Figure 9.6. In the case of the congruent dissolution, which is the simplest condition, the particles can be easily cold sintered to dense ceramics. The dissolution process is dominated by the composition of ceramics, the particle size and distribution, the type of solvents, temperature, and pressure. The transient solvent evaporates gradually during cold sintering process, leading to the precipitation of ionic species and/or atomic clusters from the supersaturated solution. If this process is too fast, there may be glassy or amorphous phases formed at the grain boundaries. For example, $Na_2Mo_2O_7$ microwave dielectric ceramic, prepared by cold sintering process at 120 °C for 15 min with pure water as the liquid phase, shows a thick amorphous phase at some grain boundaries (less than 10%) and clear grain–grain interfaces at most of the grain boundaries, as shown in Figure 9.7. However, with or without amorphous phases formed, the ceramic particles with congruent dissolution can be sintered to dense, robust bodies under the cold sintering condition.

The incongruent dissolution is more complex than the congruent dissolution, where the powders may be not easily densified using cold sintering process. Due to the incongruent dissolution, a passive surface may appear at the interface of ceramic particles and liquid phases, preventing the precipitation of ionic species and/or atomic clusters from the liquid. For example, $BaTiO_3$ is difficult to be densified using the simple cold sintering condition with pure water. This issue can be solved by adding solutions that are rich of solutes with dominant dissolution rates. The

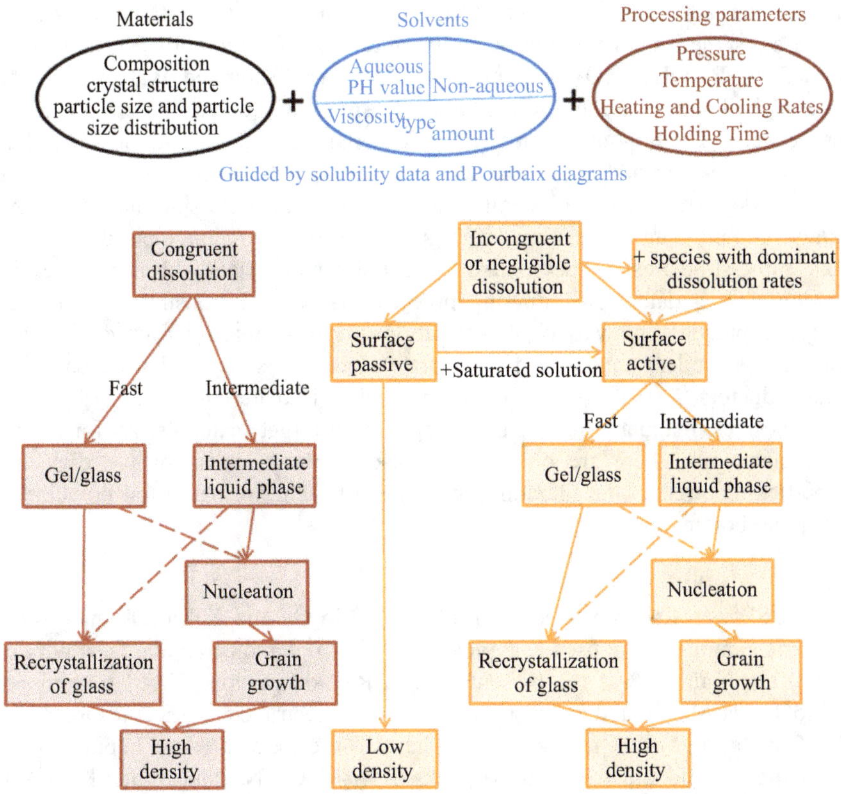

Figure 9.6 Flowchart illustrating the critical parameters for cold sintering process and the pathways under cold sintering. Reproduced from [128]

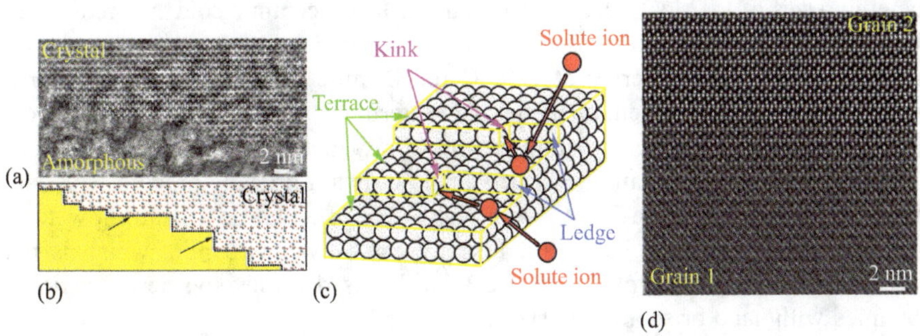

Figure 9.7 Interfaces of $Na_2Mo_2O_7$ cold-sintered at 120 °C and the Terrace–Ledge–Kink (TLK) model for the crystal growth. (a) HAADF-STEM micrograph for the interface of crystal and amorphous phases. (b) Schematic showing the terraces and ledges at the interface. (c) Schematic of the TLK model. (d) HAADF-STEM micrograph of the clear grain boundary. Reproduced from [82]

added solutes act as active layers that limit the incongruent dissolution process, and further assist the densification through dissolution-precipitation process. In the case of $BaTiO_3$, $Ba(OH)_2/TiO_2$ suspension together with saturated $BaTiO_3$ solution serve as the liquid phase and initiate the cold sintering process. If there are active surfaces at the solid–liquid interfaces under incongruent dissolution condition, there is no need to add additional solutions to limit the incongruent dissolution process. As for materials with negligible dissolution, the cold sintering conditions are also complex, and the pathways are similar to that of incongruent dissolution. Thus, same strategies can be applied to cold sinter materials with negligible dissolution. Up to date, cold sintering process has been demonstrated to densify inorganic materials in a variety of applications [129], such as microwave dielectric materials, cathodes, anodes and solid state electrolytes in solid state batteries, semiconductors, ferroelectric materials, piezoelectric materials, refractories, ceramic adhesives, magnetic materials, and sputtering target materials. As for microwave dielectrics, there are only a few cases with good microwave dielectric properties reported in the literature, and most of them are based on molybdates, NaCl, and borates.

9.4.3.1 Molybdates

When CSP was first proposed, Li_2MoO_4, $Na_2Mo_2O_7$ and $K_2Mo_2O_7$ microwave dielectric ceramics were used as typical examples to demonstrate the feasibility of CSP [82]. Figure 9.8(a)–(c) presents the X-Ray diffraction (XRD) patterns of Li_2MoO_4, $Na_2Mo_2O_7$ and $K_2Mo_2O_7$ microwave dielectric ceramics cold sintered at 120 °C with water as the mediate liquid phase. There are no obvious impure phases detected after cold sintering. All the peaks of Li_2MoO_4, $Na_2Mo_2O_7$ and $K_2Mo_2O_7$ can be indexed to rhombohedral (R-3), triclinic (P-1), and orthorhombic (Cmca) structures, respectively. As seen in Figure 9.8(d)–(f), dense microstructures can be observed for all the cold sintered samples, demonstrating that Li_2MoO_4, $Na_2Mo_2O_7$ and $K_2Mo_2O_7$ can be sintered well under cold sintering condition. It is noted that the grain sizes of Li_2MoO_4, $Na_2Mo_2O_7$ and $K_2Mo_2O_7$ ceramic cold sintered at 120 °C for 15 min are similar to those of raw powders, indicating that the grain growth is limited using cold sintering at 120 °C for 15 min. Generally, the mass transportation during the conventional thermal sintering at high temperatures triggers the densification as well as the grain growth of raw materials. The grain growth of ceramics under CSP typically described as the Ostwald ripening process. The cold sintering of Li_2MoO_4, $Na_2Mo_2O_7$ and $K_2Mo_2O_7$ ceramics is an ultra-low temperature sintering at 120 °C with a fast process only for 15 min, which is not sufficient for the grain growth. So, CSP can be used to develop new functional ceramics with nano or submicron sizes.

Li_2MoO_4 ceramic cold sintered at 120 °C under a pressure of 350 MPa for 15 min shows a high relative density of 95.7%, and good microwave dielectric properties with a dielectric constant, $Q\times f$ value, and τ_f value of 5.61, 30,500 GHz and −174 ppm/K, respectively [82,130]. $Na_2Mo_2O_7$ ceramic can be cold sintered well at 120 °C under a pressure of 350 MPa for 15 min with a relative density of 93.7%. The cold sintered $Na_2Mo_2O_7$ ceramic has a dielectric constant of 13.4, a

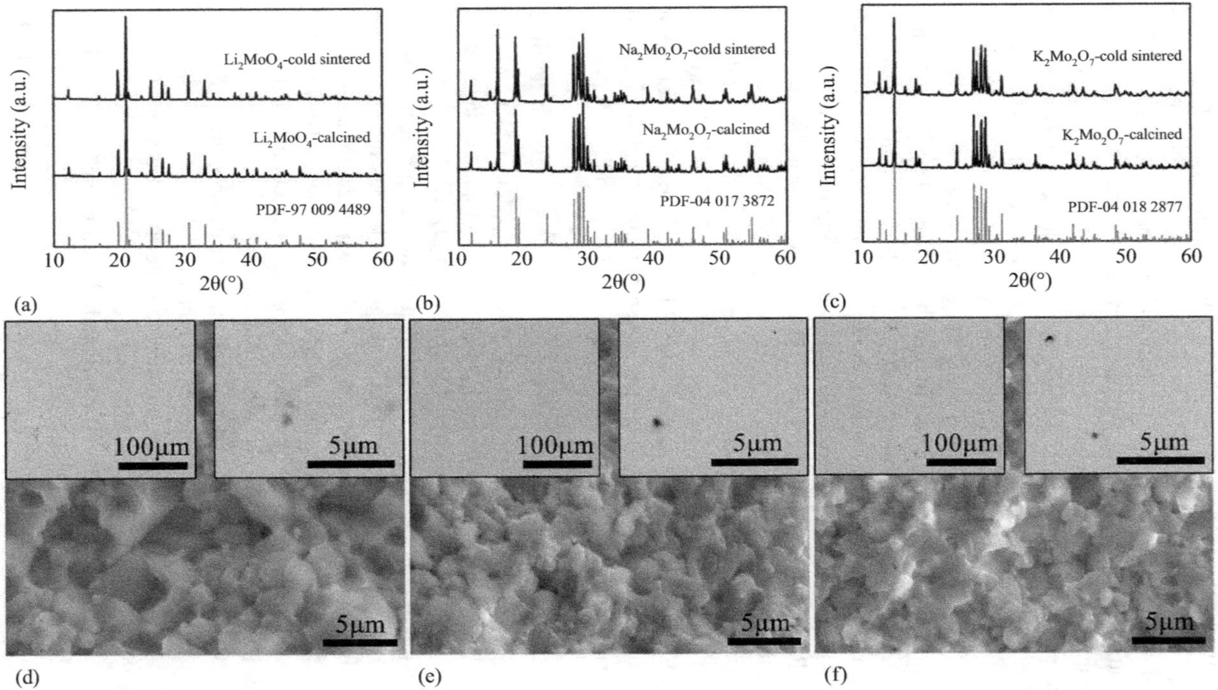

Figure 9.8 *XRD patterns (a–c) and SEM micrographs (d–f) of Li_2MoO_4 (a, d), $Na_2Mo_2O_7$ (b, e), and $K_2Mo_2O_7$ (c, f) microwave dielectric ceramics cold sintered at 120 °C for 15 min. Microstructures of polished surfaces are shown in the insets of SEM micrographs. Reproduced from [82,129]*

$Q\times f$ value of 14,900 GHz, and a τ_f value of −90 ppm/K [82,130]. The $K_2Mo_2O_7$ ceramic cold sintered at 120 °C with 350 MPa for 15 min shows a relative density, dielectric constant, $Q\times f$ value, and τ_f value of 94.1%, 9.8, 16,000 GHz, and −66 ppm/K, respectively [82,130]. Overall, the Li_2MoO_4, $Na_2Mo_2O_7$, $K_2Mo_2O_7$, and $(LiBi)_{0.5}MoO_4$ ceramics cold sintered at 120 °C show comparable microwave dielectric properties to those prepared by conventional thermal sintering at higher temperatures, indicating that cold sintering opens a new era for the development of microwave dielectric materials and devices [130].

It is also reported by Guo *et al.* that $(LiBi)_{0.5}MoO_4$ ceramic can be cold sintered to 87.8%–88.5% of theoretical density at 120 °C with water as the mediate liquid phase [130]. Considering the heating process, the experimental methods of cold sintering of $(LiBi)_{0.5}MoO_4$ are divided into two groups. One approach is that the wetted $(LiBi)_{0.5}MoO_4$ powders were put in a preheated die (120 °C) and then pressed at the target sintering temperature of 120 °C (CSP-120, for abbreviation). The other approach is that the wetted powders were placed in a room temperature die and then pressed at room temperature, followed by a heating process with a rate of 5–6 °C/min (CSP-RT-120, for abbreviation). The target sintering temperature and holding time of these two methods are the same, which are 120 °C and 15 min, respectively. Using the approach of CSP-120, $(LiBi)_{0.5}MoO_4$ shows a relative density of 88.5%, a dielectric constant of 33.7, a $Q\times f$ value of 2,300 GHz, and a τ_f value of +184 ppm/K. With the approach of CSP-RT-120, $(LiBi)_{0.5}MoO_4$ has a similar density of 88.5%. However, the dielectric constant is improved to 37.1, and the $Q\times f$ value and τ_f value are decreased a little to 1,700 GHz, and +179 ppm/K, respectively.

Due to the low sintering temperature, cold sintering process makes it feasible to cofire ceramics with other materials, such as metal materials and polymers [82,131]. Guo *et al.* successfully fabricated $Na_2Mo_2O_7$–PEI composites using cold sintering at 120 °C [132]. The cold sintered $(1-x)Na_2Mo_2O_7$–xPEI ($0 \leq x \leq 50$ vol%) composites show relative densities higher than 90%, and good microwave dielectric properties with dielectric constant in the range of 7.1–13.4 and $Q\times f$ value of 10,000–15,000 GHz. In addition, the dielectric breakdown strength is improved dramatically from 55.1 MV/m to 107.5 MV/m with the addition of 10–20 vol% PEI. Furthermore, a device prototype of integrated multilayer circuits with ring resonators based on $Na_2Mo_2O_7$–PEI–Ag was prepared using cold sintering cofired technology at 120 °C. The details of process and microstructures are shown in Figure 9.9. No obvious delamination or warping can be observed in the multilayer ring resonators. All these results demonstrate that cold sintering process provides a pathway for the development of microwave devices.

Reaney and co-workers studied several molybdate based composites, such as $Na_{0.5}Bi_{0.5}MoO_4$–Li_2MoO_4, Li_2MoO_4–$BaFe_{12}O_{19}$, and $(Bi_{0.95}Li_{0.05})(V_{0.9}Mo_{0.1})O_4$–$Na_2Mo_2O_7$ [133–136] using cold sintering process. $Na_{0.5}Bi_{0.5}MoO_4$–Li_2MoO_4 composite can be cold sintered to 92.6%–96.4% of theoretical density at 150 °C under a pressure of 200 MPa for 30 min with water as the mediate liquid phase [133]. With the amount of Li_2MoO_4 increasing, the dielectric constant and τ_f value decrease gradually while the $Q\times f$ value increases dramatically. The optimal

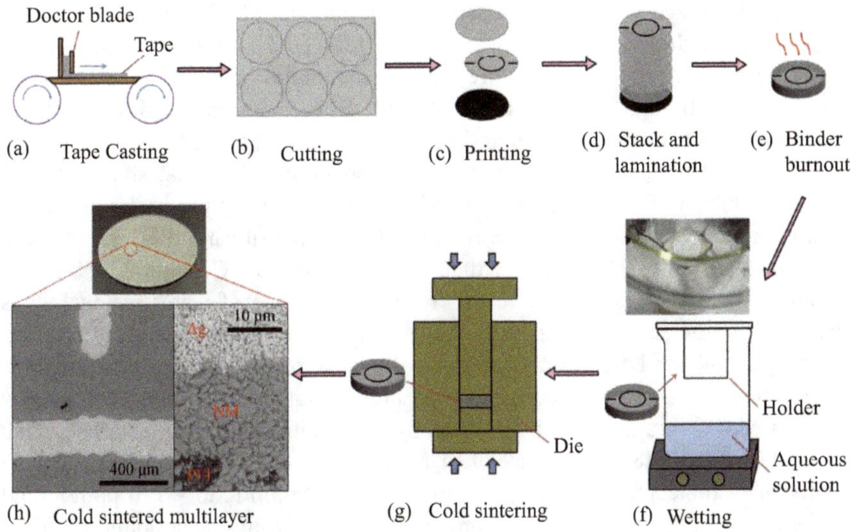

Figure 9.9 Schematic illustrating the CSP for the fabrication of $Na_2Mo_2O_7$–PEI–Ag multilayered ring resonator (a–g), and the photograph as well as SEM micrographs of surfaces of $Na_2Mo_2O_7$–PEI–Ag multilayered ring resonator sintered at 120 °C (h). Reproduced from [131]

microwave dielectric properties are obtained from ceramic with 20 wt% Li_2MoO_4: dielectric constant = 17.4, $Q{\times}f$ = 7,470 GHz, and $\tau_f = -4.7$ ppm/K. Based on the $Na_{0.5}Bi_{0.5}MoO_4$–Li_2MoO_4 composite, a dielectric graded radial index lens was designed and successfully fabricated using cold sintering process. The simulated results reveal that the directivity can be improved by 6–8 dB in the frequency range of 26–40 GHz using the designed lens, and has approximately 78% aperture efficiency at 34 GHz. All these results demonstrate that cold sintering is a promising technology for RF applications. $(Bi_{0.95}Li_{0.05})(V_{0.9}Mo_{0.1})O_4$–20 wt%$Na_2Mo_2O_7$ composites are densified well using cold sintering process at 150 °C under 200 MPa for 30 min, with the optimal microwave dielectric properties of dielectric constant = 40, $Q{\times}f$ = 4,000 GHz, and $\tau_f = -4$ ppm/K [135]. To tailor the permeability of Li_2MoO_4 ceramic, small amount of $BaFe_{12}O_{19}$ was added to form Li_2MoO_4–$BaFe_{12}O_{19}$ composites. The composites cold sintered at 120 °C shows high relative densities of 94%–97% [136]. With the amount of $BaFe_{12}O_{19}$ increasing from 5% to 15%, the permeability tested in the X band (8 GHz) increases slightly from 0.94 to 1.02. The dielectric constant and $Q{\times}f$ values are in the range of 5.6–5.8 and 16,000–22,000 GHz, respectively. The cold sintered Li_2MoO_4–$BaFe_{12}O_{19}$ composites with the combination of good dielectric constant and permeability are suitable for the application of dual mode or enhanced bandwidth microstrip path antennas [136].

Kahari *et al.* reported a room temperature method for the densification of Li_2MoO_4 ceramic [137]. The Li_2MoO_4 powders were first wetted by a small amount of deionized water and pressed under a pressure of 130 MPa. Afterwards, the pellets were dried at room temperature or 120 °C. The XRD analysis reveals that all the peaks of room temperature fabricated samples belong to pure Li_2MoO_4 phase and no obvious hydrates are observed. Both the samples dried at room temperature and 120 °C show similar densities in the range of 87%–93%, which are comparable to that sintered at 540 °C using conventional thermal sintering. The dielectric constant of Li_2MoO_4 samples dried at room temperature is also similar to that dried at 120 °C (4.6–5.2). However, the $Q\times f$ value of samples dried at room temperature (10,200 GHz) is lower than that dried at 120 °C (14,000 GHz). The deterioration of $Q\times f$ value may result from the presence of water residual. It is noted that the dielectric properties of Li_2MoO_4 ceramics prepared by room temperature method can be optimized by controlling the particle size of raw powders, applied pressures and post-processing time [138]. Moreover, the physical properties of Li_2MoO_4 ceramics can be tailored by the addition of other materials, such TiO_2 and $BaTiO_3$ [138]. In the case of Li_2MoO_4–TiO_2 composites, the temperature coefficient of dielectric constant (τ_ε) shifts from 320 ppm/K to −170 ppm/K and the dielectric constant increases from 5.1 to 10.1, with the amount of TiO_2 increasing from 0 to 30 wt% [139]. The Li_2MoO_4–TiO_2 composites prepared by room temperature method did not react with Ag electrode material. The chemical compatibility of Li_2MoO_4–TiO_2 with Ag electrode material as well as the low processing temperature makes it feasible to utilize nano-Ag inks in the embedded electrodes.

9.4.3.2 NaCl based composites

Sebastian and co-workers studied the microwave dielectric properties of NaCl, NaCl–Al_2SiO_5 and NaCl–Al_2O_3 prepared by cold sintering process using water as the mediate liquid phase [140,141]. The NaCl ceramic cold sintered at 120 °C under a pressure of 200 MPa shows a density, dielectric constant, $Q\times f$ value, and τ_f value of 2.1 g/cm^3, 5.22, 12,000 GHz and −36 ppm/K, respectively. Al_2SiO_5 (sillinanite) is easily decomposed to $Al_6Si_2O_{13}$ (mullite) and SiO_2 in the form of glass or cristobalite at a sintering temperature above 1,525 °C, and thus its sintering temperature is limited to 1,525 °C using conventional thermal sintering. The Al_2SiO_5 ceramic sintered at 1,525 °C has a density of 2 g/cm^3, a dielectric constant of 4.43, a $Q\times f$ value of 41,800 GHz, and a τ_f value of −17 ppm/K. With NaCl addition, the sintering temperature of Al_2SiO_5 ceramic is reduced to 120 °C using cold sintering process. The cold sintered NaCl–Al_2SiO_5 composites show a high density of 2.39 g/cm^3 and good microwave dielectric properties with a dielectric constant, $Q\times f$ value, and τ_f value of 4.52, 22,350 GHz and −24 ppm/K, respectively. 50NaCl-50Al_2O_3 (wt%) composites can be cold sintered well at 120 °C with a high relative density of 96%. There are no visible impurity phases in the cold sintered composites and no obvious weight loss up to 790 °C, indicating that there is no moisture residual in the samples [141]. The cold sintered 50NaCl-50Al_2O_3 composites show good microwave dielectric properties with a dielectric constant of 6.53 and tanδ of 0.007 at 1 MHz, and a dielectric constant of 6.04, dielectric loss of 0.002, and τ_f value of +32 ppm/K, at 5 GHz [141].

Hong *et al.* studied the cold sintering mechanism of NaCl ceramics, and found that dissolution-precipitation process together with pressure-induced plastic deformation dominate the densification of NaCl during cold sintering process [142]. As the mediate liquid phase, water is critical in the cold sintering process of NaCl ceramics, which could improve the microstructural homogeneity and further enhance the Qf values. Dense (94.3%–94.6%) NaCl ceramics with optimal microwave dielectric properties (dielectric constant = 5.55; $Q \times f = 49{,}600$ GHz; $\tau_f = -173$ ppm/K) are obtained using cold sintering process at room temperature under a pressure of 300 MPa with 4 wt% water as the liquid phase. Later on, Li *et al.* reported the densities, microstructures, and microwave dielectric properties of NaCl ceramics cold sintered with different amounts of water additions at room temperature and 120 °C [143]. In the case of cold sintering at room temperature, the relative density of NaCl decreases from 99.3% to 93.4% with increasing the amount of water from 0 to 10 wt%. The evaporation of residual water results in the mass loss and volume increase, leading to the decrease of densities. The dielectric constant shows a decrease trend, which is similar to the density data. However, the $Q \times f$ values increases from 27,600 GHz to 496,00 GHz with water from 0 to 4 wt% and fluctuates with water from 4 to 10 wt%. The increase of $Q \times f$ values results from the improvement of microstructural homogeneity during the dissolution-precipitation process activated by liquid phase. Therefore, the amount of water should be optimized to obtain ceramics with both high densities and good microwave dielectric properties. Furthermore, using cold sintering process at 120 °C with water amount in the range of 0–10 wt%, the densities and $Q \times f$ values can be improved to 98.8%–99.6% and 48,100–56,000 GHz, respectively. The improvement of densities and $Q \times f$ values is attributed to the decrease of the amount of water residual and the promotion of mass transportation at 120 °C.

9.4.3.3 Borates

Recently, the microstructures and microwave dielectric properties of HBO_2 sintered at a temperature below 200 °C were reported [144]. It was found that a large amount of pores, residual H_3BO_3 and amorphous B_2O_3 exist in the sintered HBO_2 samples, and the relative density of HBO_2 is only 70.3%. However, the HBO_2 ceramic shows good microwave dielectric properties with a dielectric constant of 2.12, a $Q \times f$ value of 32,700 GHz, and a τ_f value of −43 ppm/K. In 2019, Hong *et al.* reported that HBO_2-II ceramic can be sintered to high relative densities of 94.7% using cold sintering process at 120 °C with dehydrated ethanol as the mediate liquid phase [145]. A 10 wt% dehydrated ethanol serves as the liquid phase to promote the dissolution and precipitation process, and further drives the densification of HBO_2. In the case of cold sintering at 30 °C, the density of HBO_2-II can be improved from 77.5% to 84.5%, with the applied pressure increasing from 200 to 500 MPa. If the temperature for cold sintering is increased to 120 °C, the relative density can be increased to 94.7% with the pressure of 500 MPa. HBO_2-II ceramic prepared by cold sintering process at 120 °C under a pressure of 500 MPa for 10 min shows the best microwave dielectric properties with a dielectric constant of 4.21, a $Q \times f$ value of 47,500 GHz, and a τ_f value of −70 ppm/K.

9.5 Summary and outlook

The above review on ceramic microwave dielectrics shows that large progress has been made in the past more than 50 years. A series of ceramic dielectric materials have been developed and commercialized. Some new concepts like ULTCC material and cold sintering process have been developed. These new materials and new process make it possible to promote the application of ceramic dielectrics in more emerging fields. While, although large amount of materials with different element compositions, phase compositions, forms and microstructures have been explored, to develop dielectric materials with low dielectric loss and near-zero temperature coefficient for higher frequency applications is still the main theme of ceramic microwave dielectric study. It is still difficult to establish the composition–structure–property relationship to guide new material design due to the complicity of the impact factors of dielectric performance. But, the huge experimental and theoretical data on microwave dielectric ceramics make it possible for researchers to use the emerging technology like machine learning to explore high-performance new microwave dielectric materials.

The continuous evolution of wireless communication and information technology are the main driving force of the development of ceramic microwave dielectrics. The commercialization of 5G and the initiation of 6G investigation will bring big market for ceramic millimeter-wave dielectric materials. Millimeter-wave applications need high-precision shape machining and metallization. The ceramic dielectrics should have excellent machinability and compatibility with metal electrodes. Thin film techniques, photolithography, direct writing, and 3D printing may be used to achieve fine and high precision metallization patterns. Thin film and thick film hybrid integration technology would be used for developing millimeter-wave LTCC components and modules. On the other hand, the ULTCC plus CSP or RT sintering technology allows the integration of ceramic dielectrics-based passive components using the mild semiconductor and organic materials compatible process, which will open the way to develop ceramic-silicon hybrid modules, flexible electronics using ceramic microwave dielectrics.

References

[1] Masse D.J., Pucel R.A., Readey D.W., Maguire E.A., Hartwig C.P. "A new low-loss high-*k* temperature compensated dielectric for microwave applications." *Proceedings of the IEEE*. Nov. 1971, pp.1628–1629.

[2] Masse D.J., Pucel R.A. "A temperature-stable bandpass filter using dielectric resonators." *Proceedings of the IEEE*. Nov. 1972, pp.730–731.

[3] https://www.qualcomm.com/media/documents/files/whitepaper-making-5g-nr-a-reality.pdf

[4] https://www.3gpp.org/specifications/releases

[5] Hill M.D., Cruickshank D.B. "Ceramic materials for 5G wireless communication systems." *Am. Ceram. Soc. Bull.* 2019, vol. 98(6), pp. 20–25.

[6] Reaney I.M., Iddles D. "Microwave dielectric ceramics for resonators and filters in mobile phone networks." *J. Am. Ceram. Soc.* 2006, vol.89 (7), pp. 2063–72.
[7] Ohsato H. "Microwave materials with high Q and low dielectric constant for wireless communications." *Mater. Res. Soc. Symp. Proc.* 2005, vol. 833, pp. 55–62.
[8] Liu Z.F., Li Y.X. Low temperature co-fired ceramic material and preparation method therefor. US Patent 9718696, 2017.
[9] Sebastian M.T., Wang H., Jantunen H. "Low temperature co-fired ceramics with ultra-low sintering temperature: A review." *Curr. Opin. Solid State Mater. Sci.* 2016, vol. 20(3), pp. 151–170.
[10] Cohn S.B. "Microwave band-pass filters containing high-Q dielectric resonators." *IEEE Trans. Microw. Theory Tech.* 1968, vol. 16(4), pp.218–227.
[11] O"Bryan H.M., Thomson J., Plourde J.K. "A new BaO–TiO_2 compound with temperature-stable high dielectric constant and low microwave loss." *J. Am. Ceram. Soc.* 1974, vol. 57(10), pp.450–453.
[12] O"Bryan H.M., Thomson J. *Apparatus for processing microwave electrical energy*. US Patent 4337446, 1982.
[13] O"Bryan H.M., Plourde J.K., Thomson J. *Dielectric for microwave applications*. US Patent 4563661, 1986.
[14] Katsube M., Ishikawa Y., Tamura H., Tomono K. *Dielectric ceramic composition for high frequencies*. US Patent 4102696, 1978.
[15] Wersing W. "Microwave ceramics for resonators and filters." *Curr. Opin. Solid State Mater. Sci.* 1996, vol. 1(5), pp. 715–731.
[16] Kawashima S., Nishida M., Ueda I., Ouchi H. "$Ba(Zn_{1/3}Ta_{2/3})O_3$ ceramics with low dielectric loss at microwave frequencies." *J. Am. Ceram. Soc.* 1983, vol. 66(6), pp. 421–423.
[17] Nomura S., Toyama K., Kaneta K. "$Ba(Mg_{1/3}Ta_{2/3})O_3$ ceramics with temperature-stable high dielectric constant and low microwave loss." *Jpn. J. Appl. Phys.* 1982, vol. 21(10A), pp. 624–626.
[18] Nomura S., Konishi Y. *Low-loss microwave dielectric materials*. US patent 4487842.
[19] Endo K., Fujimoto K., Murakawa K. "Dielectric properties of ceramics in Ba $(Co_{1/3}Nb_{2/3})O_3$ - $Ba(Zn_{1/3}Nb_{2/3})O_3$ solid solution." *J. Am. Ceram. Soc.* 1987, vol. 70 (9), pp. 215–218.
[20] Ohsato H. "Microwave dielectrics with perovskite-type structure" in Pan L. K.(Ed.). Perovskite materials: synthesis, characterisation, properties, and applications. IntechOpen; 2016. pp. 281–330.
[21] Liu Z.F, Wang Y.L., Wu W.J., Li Y.X. "Li–Nb–Ti–O microwave dielectric ceramics." *J. Asian Ceram. Soc.* 2013, vol. 1(1), pp. 2–8.
[22] Wu S., Song K., Liu P., *et al.* "Effect of TiO_2 doping on the structure and microwave dielectric properties of cordierite ceramics." *J. Am. Ceram. Soc.* 2015, vol. 98 (6), pp. 1842–1847.
[23] Ohsato H., Kim J.S., Kim A.Y., Cheon C.I., Chae K.W. "Millimeter-wave dielectric properties of cordierite/indialite glass ceramics." *Jpn. J. Appl. Phys.* 2011, vol. 50 (9S2), p. 09NF01.

[24] Ohsato H., Kagomiya I., Terada M., Kakimoto K. "Origin of improvement of Q based on high symmetry accompanying Si–Al disordering in cordierite millimeter-wave ceramics." *J. Eur. Ceram. Soc.* 2010, vol. 30(2), pp. 315–318.

[25] Song K., Liu P., Lin H., *et al.* "Symmetry of hexagonal ring and microwave dielectric properties of $(Mg_{1-x}Ln_x)_2Al_4Si_5O_{18+x}$ (Ln=La, Sm) cordierite-type ceramics." *J. Eur. Ceram. Soc.* 2016, vol. 36 (5), pp. 1167–1175.

[26] Ohsato H., Tsunooka T., Sugiyama T., Kakimoto K., Ogawa H. "Forsterite ceramics for millimeterwave dielectrics." *J. Electroceram.* 2006, vol. 17 (2–4), pp. 445–450.

[27] Ando M., Himura K., Tsunooka T., Kagomiya I., Ohsato H. "Synthesis of high-quality forsterite." *Jpn. J. Appl. Phys.* 2007, vol. 46(10S), pp. 7112–7116.

[28] Lai Y., Tang X., Huang X., *et al.* "Phase composition, crystal structure and microwave dielectric properties of $Mg_{2-x}Cu_xSiO_4$ ceramics." *J. Eur. Ceram. Soc.* 2018, vol. 38(4), pp. 1508–1516.

[29] Tsunooka T., Androu M., Higashida Y., Sugiura H., Ohsato H. "Effects of TiO_2 on sinterability and dielectric properties of high-Q forsterite ceramics." *J. Eur. Ceram. Soc.* 2003, vol. 23(14), pp. 2573–2578.

[30] Tsunooka T., Ando M., Suzuki S., Yasufuku Y., Ohsato H. "Research & developments for millimeter-wave dielectric forsterite with low dielectric constant, high Q, and zero temperature coefficient of resonant frequency." *Jpn. J. Appl. Phys.* 2013, vol. 52(9S1), p. 09KH02.

[31] Mccauley R.A. "Polymorphism and dielectric electric properties of Ba- and Sr-containing feldspars." *J. Mater. Sci.* 2000, vol. 35(15), p. 3939–3942.

[32] Krzmanc M.M., Jančar B., Suvorov D. "The influence of tetrahedral ordering on the microwave dielectric properties of $Sr_{0.05}Ba_{0.95}Al_2Si_2O_8$ and $BaM_2M'_2O_8$ (M= Al, Ga, M′= Si, Ge) ceramics." *J. Eur. Ceram. Soc.* 2008, vol. 28(16), pp. 3141–3148.

[33] Krzmanc M.M., Valant M., Suvorov D. "The synthesis and microwave dielectric properties of $Sr_xBa_{1-x}Al_2Si_2O_8$ and $Ca_yBa_{1-y}Al_2Si_2O_8$ ceramics." *J. Eur. Ceram. Soc.* 2007, vol. 27(2–3), pp. 1181–1185.

[34] Krzmanc M.M., Valant M., Jancar B., Suvorov D. "Sub-solidus synthesis and microwave dielectric characterization of plagioclase feldspars." *J. Am. Ceram. Soc.* 2005, vol. 88(9), pp. 2472–2479.

[35] Song X.Q., Lu W.Z., Wang X.C., *et al.* "Sintering behaviour and microwave dielectric properties of $BaAl_{2-2x}(ZnSi)_xSi_2O_8$ ceramics." *J. Eur. Ceram. Soc.* 2018, vol. 38(4), pp. 1529–1534.

[36] Kagomiya I., Suzuki I., Ohsato H. "Microwave Dielectric Properties of $(Ca_{1-x}Sr_x)SiO_3$ Ring Silicate Solid Solutions." *Jpn. J. Appl. Phys.* 2009, vol. 48(9S1), p. 09KE02.

[37] Ohsato H., Suzuki I., Kagomiya I.. "Crystal structure and microwave dielectric properties of α-$(Ca_{1-x}Sr_x)SiO_3$ (x=1 and 0.8) ring silicates for millimeter-wave applications." *Mater. Res. Bull.* 2017;96:115–120.

[38] Guo Y., Ohsato H., Kakimoto K.I. "Characterization and dielectric behavior of willemite and TiO_2-doped willemite ceramics at millimeter-wave frequency." *J. Eur. Ceram. Soc.* 2006, vol. 26(10–11), pp. 1827–1830.

[39] Ohsato H., Terada M., Kawamura K. "Fabrication conditions of diopside for millimeterwave dielectrics." *Jpn. J. Appl. Phys.* 2012;51(9S1):09LF02.

[40] Kan A., Ogawa H., Ohsato H. "Synthesis and crystal structure-microwave dielectric property relations in Sn-substituted $Ca_3(Zr_{1-x}Sn_x)Si_2O_9$ solid solutions with cuspidine structure." *Jpn J. Appl. Phys.* 2007, vol. 46(10S), pp. 7108–7111.

[41] Yuan L.L., Bian J.J. "Microwave dielectric properties of the lithium containing compounds with rock salt structure." *Ferroelectrics.* 2009, vol. 387 (1), pp. 123–129.

[42] Bian J.J, Dong Y.F. "Sintering behavior, microstructure and microwave dielectric properties of $Li_{2+x}TiO_3$ ($0 \leq x \leq 0.2$)." *Mater. Sci. Eng. B.* 2011, vol. 176(2), pp. 147–151.

[43] Bian J.J., Zhang X.H. "Structural evolution, grain growth kinetics and microwave dielectric properties of $Li_2Ti_{1-x}(Mg_{1/3}Nb_{2/3})_xO_3$." *J. Eur. Ceram. Soc.* 2018, vol. 38(2), pp. 599–604.

[44] Zhang T., Zuo R., Zhang J. "Structure, microwave dielectric properties, and low-temperature sintering of acceptor/donor codoped $Li_2Ti_{1-x}(Al_{0.5}Nb_{0.5})_xO_3$ ceramics." *J. Am. Ceram. Soc.* 2016, vol. 99(3), pp. 825–832.

[45] Kishore M.S., Marinel S., Pralong V., Caignaert V., D'Astorg S., Raveau B.. "The rock salt oxide Li_2MgTiO_4: Type I dielectric and ionic conductor." *Mater. Res. Bull.* 2006, vol. 41(7), pp. 1378–1384.

[46] Tseng Y.W., Chen J.Y., Kuo Y.C., Huang C.L. "Low-loss microwave dielectrics using rock salt oxide Li_2MgTiO_4." *J. Alloy. Compd.* 2011;509 (33):L308–L310.

[47] Yang C.H., Liu Q.Q., Bi J.X., Tao W.H., Wu H.T. "Effects of $(Mg_{1/3}Ta_{2/3})$-substitution for Ti-site on the microwave dielectric properties of Li_2MgTiO_4 ceramics." *Ceram. Int.* 2018, vol. 44(6), pp. 5982–5987.

[48] Yang C.H., Wu H.T. "Phase composition, Raman spectra, infrared spectra and dielectric properties of $Li_2MgTi_{1-x}(Mg_{1/3}Nb_{2/3})_xO_4$ ceramics at microwave frequency." *Ceram. Int.* 2018, vol. 44 (8), pp. 9255–9262.

[49] Fu Z., Liu P., Ma J., Chen X., Zhang H. "New high Q low-fired $Li_2Mg_3TiO_6$ microwave dielectric ceramics with rock salt structure." *Mater. Lett.* 2016, vol. 164, pp. 436–439.

[50] Song J., Zhang J., Zuo R. "Ultrahigh Q values and atmosphere-controlled sintering of $Li_{2(1+x)}Mg_3ZrO_6$ microwave dielectric ceramics." *Ceram. Int.* 2017, vol. 43(2), pp. 2246–2251.

[51] Zhang P., Xie H., Zhao Y., Xiao M. "Microwave dielectric properties of low loss $Li_2(Mg_{0.95}A_{0.05})_3TiO_6$ ($A{=}Ca^{2+}$, Ni^{2+}, Zn^{2+}, Mn^{2+}) ceramics system." *J. Alloy Compd.* 2016, vol. 689, pp. 246–249.

[52] Wang G., Zhang H., Xu F., *et al.* "Investigation and characterization on crystal structure and enhanced microwave dielectric properties of

non-stoichiometric $Li_{3+x}Mg_2NbO_6$ ceramics." *Ceram. Int.* 2018, vol. 44(16), pp. 20539–20544.

[53] Zhang T., Zuo R. "Effect of Li_2O–V_2O_5 addition on the sintering behavior and microwave dielectric properties of $Li_3(Mg_{1-x}Zn_x)_2NbO_6$ ceramics." *Ceram. Int.* 2014, vol. 40(10), pp. 15677–15684.

[54] Zhao Y., Zhang P. "Microstructure and microwave dielectric properties of low loss materials $Li_3(Mg_{0.95}A_{0.05})_2NbO_6$ (A=Ca^{2+}, Ni^{2+}, Zn^{2+}, Mn^{2+}) with rock-salt structure." *J. Alloy Compd.* 2016, vol. 658, pp. 744–748.

[55] Xing C.F., Bi J.X., Wu H.T. "Effect of Co-substitution on microwave dielectric properties of $Li_3(Mg_{1-x}Co_x)_2NbO_6$ ($0.00 \leq x \leq 0.10$) ceramics." *J. Alloy Compd.* 2017, vol. 719, pp. 58–62.

[56] Ma Y., Liu X. "Kinetics and thermodynamics of Mg–Al disorder in $MgAl_2O_4$-spinel: a review." *Molecules*. 2019, vol. 24(9), p. 1704.

[57] Zheng C.W., Wu S.Y., Chen X.M., Song K.X. "Modification of $MgAl_2O_4$ microwave dielectric ceramics by Zn substitution." *J. Am. Ceram. Soc.* 2007, vol. 90(5), pp. 1483–1486.

[58] Egorov S.V., Sorokin A.A., Ilyakov I.E., *et al.* "Terahertz dielectric properties of polycrystalline $MgAl_2O_4$ spinel obtained by microwave sintering and hot pressing." *J. Infrared Millimet. Terahertz Waves*. 2019, vol. 40, pp. 447–455.

[59] Takahashi S., Kan A., Ogawa H. "Microwave dielectric properties and crystal structures of spinel-structured $MgAl_2O_4$ ceramics synthesized by a molten-salt method." *J. Eur. Ceram. Soc.* 2017, vol. 37(3), pp. 1001–1006.

[60] Takahashi S., Ogawa H., Kan A. "Electronic states and cation distributions of $MgAl_2O_4$ and $Mg_{0.4}Al_{2.4}O_4$ microwave dielectric ceramics." *J. Eur. Ceram. Soc.* 2018, vol. 38(2), pp. 593–598.

[61] Takahashi S., Kan A., Ogawa H. "Microwave dielectric properties and crystal structures of $Mg_{0.7}Al_{2.2}O_4$ and $Mg_{0.4}Al_{2.4}O_4$ ceramics with defect structures." *J. Am. Ceram. Soc.* 2017, vol. 100, pp. 3497–3504.

[62] Laaga N.J., Snela M.D., Magusinb P.C., With G. "Structural, elastic, thermophysical and dielectric properties of zinc aluminate ($ZnAl_2O_4$)." *J. Eur. Ceram. Soc.* 2004, vol. 24 (8), pp. 2417–2424.

[63] Surendran K.P., Santha N., Mohanan P., Sebastian M.T. "Temperature stable low loss ceramic dielectrics in $(1-x)ZnAl_2O_4$–$x TiO_2$ system for microwave substrate applications." *Eur. Phys. J. B.* 2004, vol. 41(3), pp. 301–306.

[64] Takahashi S., Kan A., Ogawa H. "Microwave dielectric properties and cation distributions of $Zn_{1-3x}Al_{2+2x}O_4$ ceramics with defect structures." *J. Eur. Ceram. Soc.* 2017, vol. 37(9), pp. 3059–3064.

[65] Lan X.K., Li J., Zou Z.Y., *et al.* "Improved sinterability and microwave dielectric properties of $[Zn_{0.5}Ti_{0.5}]^{3+}$-doped $ZnAl_2O_4$ spinel solid solution." *J. Am. Ceram. Soc.* 2019, vol. 102(10), pp. 5952–5957.

[66] Li H., Xiang R., Chen X., *et al.* "Intrinsic dielectric behavior of Mg_2TiO_4 spinel ceramic." *Ceram Int.* https://doi.org/10.1016/j.ceramint.2019.10.143.

[67] Li H., Zhang P., Yu S., Yang H., Tang B., Li F., Zhang S. "Structural dependence of microwave dielectric properties of spinel structured Mg_2

$(Ti_{1-x}Sn_x)O_4$ solid solutions: Crystal structure refinement, Raman spectra study and complex chemical bond theory." *Ceram. Int.* 2019, vol. 45(9), pp. 11639–11647.

[68] Huang C.L., Chen J.Y. "Low-loss microwave dielectric ceramics using $(Mg_{1-x}Mn_x)_2TiO_4$ (x=0.02–0.1) Solid Solution." *J. Am. Ceram. Soc.* 2009, vol. 92(3), pp. 675–678.

[69] Huang C.L., Liu S.S. "Low-loss microwave dielectrics in the $(Mg_{1-x}Zn_x)_2TiO_4$ Ceramics." *J. Am. Ceram. Soc.* 2008;91(10):3428–3430.

[70] Huang C.L., Chen J.Y. "High-Q microwave dielectrics in the $(Mg_{1-x}Co_x)_2TiO_4$ ceramics." *J. Am. Ceram. Soc.* 2009, vol. 92(2), pp. 379–383.

[71] Kim S.H., Kim E.S. "Intrinsic factors affecting the microwave dielectric properties of $Mg_2Ti_{1-x}(Mg_{1/3}Sb_{2/3})_xO_4$ ceramics." *Ceram. Int.* 2016;42 (13):15035–15040.

[72] Kim S.H., Kim E.S. "Effect of substituting Ti sites with $(Mg_xNb_{0.8-0.4x})^{4+}$ on the microwave dielectric properties of Mg_2TiO_4 ceramics." *Ceram. Int.* 2017, vol. 43, pp. S321–S325.

[73] Belous A, Ovchar O, Durilin D, Krzmanc MM, Valant M, Suvorov D. "High-Q microwave dielectric materials based on the spinel Mg_2TiO_4." *J. Am. Ceram. Soc.* 2006, vol. 89 (11), pp. 3441–3445.

[74] Kim E.S., Chun B.S., Freer R., Cernik R.J. "Effects of packing fraction and bond valence on microwave dielectric properties of $A^{2+}B^{6+}O_4$ (A^{2+}: Ca, Pb, Ba; B^{6+}: Mo, W) ceramics." *J. Eur. Ceram. Soc.* 2010, vol. 30(7), pp. 1731–1736.

[75] Kim J.H., Kim E.S. "Effect of isovalent substitution on microwave dielectric properties of $Mg_4Nb_2O_9$ ceramics." *J. Electron. Mater.* 2019, vol. 48(4), pp. 2411–2417.

[76] Kan A., Ogawa H., Yokoi A., Nakamura Y. "Crystal structural refinement of corundum- structured $A_4M_2O_9$ (A=Co and Mg, M=Nb and Ta) microwave dielectric ceramics by high-temperature X-ray powder diffraction." *J. Eur. Ceram. Soc.* 2007, vol. 27(8–9), pp. 2977–2981.

[77] Ogawa H., Kan A., Ishihara S., Higashida Y. "Crystal structure of corundum type $Mg_4(Nb_{2-x}Ta_x)O_9$ microwave dielectric ceramics with low dielectric loss." *J. Eur. Ceram. Soc.* 2003, vol. 23(14), pp. 2485–2488.

[78] Mohanty P., Keshri S., KumarSinha M., Gupta V.R. "Study on microwave dielectric properties of corundum type $(Mg_{1-x}Co_x)_4Ta_2O_9$ (x=0–0.6) ceramics for designing a microwave low pass filter." *Ceram. .Int.* 2016, vol. 42 (5), pp. 5911–5920.

[79] Kan A., Ogawa H., Yokoi A., Ohsato H. "Low-temperature sintering and microstructure of $Mg_4(Nb_{2-x}V_x)O_9$ microwave dielectric ceramic by V Substitution for Nb." *Jpn J. Appl. Phys.* 2003, vol. 42(9S), pp. 6154–6157.

[80] Kim J.H., Kim E.S. "Microwave dielectric properties of $Mg_4Nb_2O_9$-based ceramics with $(B_xW_{1-x})^{5+}$ substitutions at Nb^{5+} sites (B = Li, Mg, Al, Ti)." *Ceram. Int.* 2017, vol. 43, pp. S339–S342.

[81] Imanaka Y.. *Multilayered Low Temperature Cofired Ceramic Technology*. Springer Science+ Business Media, Inc. New York, NY 10013, USA;2005.

[82] Guo J., Guo H.Z., Baker A.L., *et al.* "Cold sintering: a paradigm shift for processing and integration of ceramics." *Angew. Chem.-Int. Edit.*, 2016, vol. 55(38), pp. 11457–11461.

[83] Zhou J. "Towards rational design of low-temperature co-fired ceramic (LTCC) materials." *J. Adv. Ceram.* 2012, vol.1(2), pp. 89–99.

[84] Sebastian M.T., Jantunen H. "Low loss dielectric materials for LTCC applications: a review." *Int. Mater. Rev.* 2008, vol. 53(2), pp. 57–90.

[85] Dai, S.X., Casias, Adrian L. Process evaluation of DuPont(TM) GreenTape (TM) 9K7 LTCC. presentation at the ceramic interconnect and ceramic microsystems technologies (CICMT 2011), April 5–7, 2011, San Diego, USA.

[86] Nakajima H., Sakaguchi K. *Ceramic substrate and method for production thereof*, US Patent 10383220, 2019.

[87] Gu Y.J., Lei L.W., Huang J.L., *et al.* "A novel low-fired, temperature-stable, and low-cost $(1-x)BaCu(B_2O_5)-xTiO_2$ microwave dielectric ceramic." *J. Eur. Ceram. Soc.* 2019, vol. 39, pp. 1137–1141.

[88] Lim J.B., Kim M.H., Kim J.C., Nahm S., Paik J.H., Kim J.H. "Effect of $BaCu(B_2O_5)$ Additive on the sintering temperature and microwave dielectric properties of $BaTi_4O_9$ ceramics." *Jpn J. Appl. Phys.* 2006, vol. 45(8), pp. 242–244.

[89] Kim M.H., Lim J.B., Kim J.C., Nahm S., Paik J.H., Kim J.H. "Synthesis of $BaCu(B_2O_5)$ ceramics and their effect on the sintering temperature and microwave dielectric properties of $Ba(Zn_{1/3}Nb_{2/3})O_3$ Ceramics." *J. Am. Ceram. Soc.* 2006, vol. 89(10), pp. 3124–3128.

[90] Shigeno K., Kojima E., Fujimori H. "Improvement in the low-temperature sintering performance and characteristics of alumina with $CuO–iO_2–Nb_2O_5$ additive by controlling the firing atmosphere." *J. Jpn Soc. Powd. Powd. Metall.* 2016, vol. 63, pp. 701–7015.

[91] Zhou D., Wang H., Pang L.X., Randall C.A., Yao X. "$Bi_2O_3–MoO_3$ binary system: an alternative ultralow sintering temperature microwave dielectric." *J. Am. Ceram. Soc.*, 2009, vol. 92(10), pp. 2242–2246.

[92] Zhou D., Randall C.A., Wang H., Pang L.X., Yao X. "Microwave dielectric ceramics in $Li_2O–Bi_2O_3–MoO_3$ system with ultra-low sintering temperatures." *J. Am. Ceram. Soc.*, 2010, vol. 93(4), pp. 1096–1100.

[93] Zhang G.Q., Wang H., Guo J., He L., Wei D.D., Yuan Q.B. "Ultra-low sintering temperature microwave dielectric ceramics based on $Na_2O–MoO_3$ binary system." *J. Am. Ceram. Soc.*, 2015, vol. 98(2), pp. 528–533.

[94] Zhang G.Q., Guo J., He L., *et al.* "Preparation and microwave dielectric properties of ultra-low temperature sintering ceramics in $K_2O–MoO_3$ binary system." *J. Am. Ceram. Soc*, 2014, vol. 97(1), pp. 241–245.

[95] Zhou D., Li W.B., Pang L.X., *et al.* "Sintering behavior and dielectric properties of ultra-low temperature fired silver molybdate ceramics." *J. Am. Ceram. Soc.* 2014, vol. 97(11), pp. 3597–3601.

[96] Zhang G.Q., Guo J., Wang H., "Ultra-low temperature sintering microwave dielectric ceramics based on $Ag_2O–MoO_3$ binary system." *J. Am. Ceram. Soc.* 2017, vol. 100(6), pp. 2604–2611.

[97] Xue X., Yuan X.F., Ma R., Yuan Q.B., Wang H. "Temperature stable $0.35Ag_2MoO_4$- $0.65Ag_{0.5}Bi_{0.5}MoO_4$ microwave dielectric ceramics with ultra-low sintering temperatures." *J. Eur. Ceram. Soc.* 2019, vol. 39 (13), pp. 3744–3748.

[98] Zhou D., Pang L.X., Qi Z.M., Jin B.B., Xi Y. "Novel ultra-low temperature co-fired microwave dielectric ceramic at 400 degrees and its chemical compatibility with base metal." *Sci. Rep.* 2014, vol. 4, pp. 5980–5985.

[99] Zhou D., Li J., Pang L.X., Wang D.W., Reaney I.M. "Novel water insoluble (Na_xAg_{2-x}) MoO_4 $(0 \leq x \leq 2)$ microwave dielectric ceramics with spinel structure sintered at 410 degrees." *J. Mater. Chem. C*, 2017, vol. 5(24), pp. 6086–6091.

[100] Zhou D., Randall C.A., Pang L.X., *et al.* "Microwave dielectric properties of Li_2WO_4 ceramic with ultra-low sintering temperature." *J. Am. Ceram. Soc.* 2011, vol. 94(2), pp. 348–350.

[101] Zhou D., Pang L.X., Xie H.D., *et al.* "Crystal structure and microwave dielectric properties of an ultralow-temperature-fired (AgBi)(0.5)WO_4 ceramic." *Eur. J. Inorg. Chem.* 2014, vol. 2014(2), pp. 296–301.

[102] Xie H.D., Xi H.H., Chen C., Zhou D. "Microwave dielectric properties of two low temperature sintering ceramics in the PbO–WO_3 binary system." *Ceram. Int.* 2015, vol. 41(8), pp. 10287–292.

[103] Yuan X.F., Xue X., Wang H. "Preparation of ultra-low temperature sintering ceramics with ultralow dielectric loss in Na_2O–WO_3 binary system." *J. Am. Ceram. Soc.* 2019, vol. 102(7), pp. 4014–4020.

[104] Yuan X.F., Xue X., Jin F., Wang H. "High-*Q* $(Na_{1-x}Ag_x)_2WO_4$ (x=0.1, 0.2) ceramics with ultra-low sintering temperature." *J. Eur. Ceram. Soc.* 2019, vol. 39(14), pp. 4156–4159.

[105] Maeda M., Yamamura T., Ikeda T. "Dielectric characteristics of several complex oxide ceramics at microwave-frequencies." *Jpn. J. Appl. Phys. Part 1*.1987, vol. 26, pp. 76–79.

[106] Udovic M., Valant M., Suvorov D. "Dielectric characterization of ceramics from the TiO_2–TeO_2 system." *J. Eur. Ceram. Soc.* 2001, vol. 21(10–11), pp. 1735–1738.

[107] Udovic M, Valant M, Suvorov D. "Phase formation and dielectric characterization of the Bi_2O_3–TeO_2 system prepared in an oxygen atmosphere." *J. Am. Ceram. Soc.* 2004, vol. 87(4), pp. 591–597.

[108] Kwon D.K., Lanagan M.T., Shrout T.R. "Microwave dielectric properties and low-temperature cofiring of $BaTe_4O_9$ with aluminum metal electrode." *J. Am. Ceram. Soc.* 2005, vol. 88(12), pp. 3419–3422.

[109] Kwon D.K., Lanagan M.T., Shrout T.R. "Synthesis of $BaTiTe_3O_9$ ceramics for LTCC application and its dielectric properties." *J. Ceram. Soc. Jpn.* 2005, vol. 113(1315), pp. 216–219.

[110] Kwon D.K., Lanagan M.T., Shrout T.R. "Microwave dielectric properties of BaO–TeO_2 binary compounds." *Mater. Lett.* 2007, vol. 61(8–9), pp. 1827–1831.

[111] Wang S.F., Wang Y.R., Hsu Y.F., Tsai J.S. "Densification, microstructure and microwave dielectric properties of ultra-low fire $BaTe_4O_9$–$TiTe_3O_8$ ceramic composites." *J. Eur. Ceram. Soc.* 2010, vol. 30(7), pp. 1737–1741.

[112] Wang S.F., Huang C.Y., Liu Y.L. "Effects of $CaTiO_3$ and $SrTiO_3$ additions on the microstructure and microwave dielectric properties of ultra-low-fire TeO_2 ceramics." *J Am Ceram Soc*. 2010, vol. 93(10), pp. 3272–3277.

[113] Jiao X.Q., Zhong C.W., Zhang S.R., Liu X.S., Li B. "Microwave dielectric properties of BaO–TiO_2–TeO_2 ternary system." *J. Mater. Sci.* 2010, vol. 45 (12), pp. 3331–3335.

[114] Subodh G., Sebastian M.T. "Glass-free $Zn_2Te_3O_8$ microwave ceramic for LTCC applications." *J. Am. Ceram. Soc.* 2007, vol. 90(7), pp. 2266–2268.

[115] Subodh G., Sebastian M.T. "Microwave dielectric properties of ATe_3O_8 (A = Sn, Zr) ceramics." *Jpn J. Appl. Phys.* 2008, vol. 47(10), pp. 7943–7946.

[116] Wang S.F., Hsu Y.F., Wang Y.R., Sung C.C. "Ultra-low-fire $Zn_2Te_3O_8$–$TiTe_3O_8$ ceramic composites." *J. Am. Ceram. Soc.* 2011, vol. 94(3), pp. 812–816.

[117] Valant M., Suvorov D. "Chemical compatibility between silver electrodes and low-firing binary-oxide compounds: Conceptual study." *J. Am. Ceram. Soc.* 2000, vol. 83(11), pp. 2721–2729.

[118] Umemura R., Ogawa H., Kan A. "Low temperature sintering and microwave dielectric properties of $(Mg_{3-x}Zn_x)(VO_4)_2$ ceramics." *J. Eur. Ceram. Soc.* 2006, vol. 26(10–11), pp. 2063–2068.

[119] Joung M.R., Kim J.S., Song M.E., Nahm S., Paik J.H. "Formation process and microwave dielectric properties of the $R_2V_2O_7$ (R=Ba, Sr, and Ca) ceramics." *J. Am. Ceram. Soc.* 2009, vol. 92(12), pp. 3092–3094.

[120] Fang L., Xiang F., Su C.X., Zhang H. "A novel low firing microwave dielectric ceramic $NaCa_2Mg_2V_3O_{12}$." *Ceram. Int.* 2013, vol. 39(8), pp. 9779–9783.

[121] Yao G.G., Liu P., Zhang H.W. "Novel series of low-firing microwave dielectric ceramics: $Ca_5A_4(VO_4)_6$ (A^{2+}=Mg, Zn)." *J. Am. Ceram. Soc.*2013, vol. 96(6), pp. 1691–1693.

[122] Neelakantan U.A., Kalathil S.E., Ratheesh R. "Structure and microwave dielectric properties of ultralow-temperature cofirable BaV_2O_6 ceramics." *Eur. J. Inorg. Chem.*2015, vol. 2, pp. 305–310.

[123] Kalathil S.E., Neelakantan U.A., Ratheesh R. "Microwave dielectric properties of ultralow-temperature cofirable $Ba_3V_4O_{13}$ ceramics." *J. Am. Ceram. Soc.* 2014, vol. 97(5), pp. 1530–1533.

[124] Yao G.G., Pei C.J., Xu J.G., Liu P., Zhou J.P., Zhang H.W. "Microwave dielectric properties of CaV_2O_6 ceramics with low dielectric loss." *J. Mater. Sci.-Mater. Electron.* 2015, vol. 26(10), pp. 7719–7722.

[125] Chen M.Y., Juuti J., Hsi C.S., Chia C.T., Jantunen H. "Dielectric $BaTiO_3$–BBSZ glass ceramic composition with ultra-low sintering temperature." *J. Eur. Ceram. Soc.* 2015, vol. 35(1), pp. 139–144.

[126] Chen M.Y., Juuti J., Hsi C.S., Chia C.T., Jantunen H. "Dielectric properties of ultra-low sintering temperature Al_2O_3–BBSZ glass composite." *J. Am. Ceram. Soc.* 2015, vol 98(4), pp. 1133–1136.

[127] Guo J., Zhao X.T., Beauvoir T.H., *et al.* "Recent progress in applications of the cold sintering process for ceramic-polymer composites." *Adv. Funct. Mater.* 2018, vol. 28(39), pp. 1801724.

[128] Guo H.Z., Baker A., Guo J., Randall C.A. "Cold sintering process: a novel technique for low-temperature ceramic processing of ferroelectrics." *J. Am. Ceram. Soc.* 2016, vol. 99(11), pp. 3489–3507.

[129] Guo J., Floyd R., Lowum S., *et al.* "Cold sintering: progress, challenges, and future opportunities." *Ann. Rev. Mater. Res.* 2019, vol. 49, pp. 275–295.

[130] Guo J., Baker A.L., Guo H.Z., Lanagan M., Randall C.A. "Cold sintering process: a new era for ceramic packaging and microwave device development." *J. Am. Ceram. Soc.* 2017, vol. 100(2), pp. 669–677.

[131] Guo J., Berbano S.S., Guo H.Z., Baker A.L., Lanagan M.T., Randall C.A. "Cold sintering process of composites: bridging the processing temperature gap of ceramic and polymer materials." *Adv. Funct. Mater.* 2016, vol. 26 (39), pp. 7115–7121.

[132] Guo J., Pfeiffenberger N., Beese A., *et al.* "Cold sintering $Na_2Mo_2O_7$ ceramic with Poly(ether imide) (PEI) polymer to realize high-performance composites and integrated multilayer circuits." *ACS. Appl. Nano Mater.*2018, vol. 1(8), pp. 3837–3844.

[133] Wang D.W., Zhou D., Zhang S.Y., *et al.* "Cold-sintered temperature stable $Na_{0.5}Bi_{0.5}MoO_4$–Li_2MoO_4 microwave composite ceramics." *ACS Sustain. Chem. Eng.*. 2018, vol. 6(2), pp. 2438–2444.

[134] Faouri S.S., Mostaed A., Dean J.S., *et al.* "High quality factor cold sintered Li_2MoO_4–$BaFe_{12}O_{19}$ composites for microwave applications." *Acta Mater.* 2019, vol. 166, pp. 202–207.

[135] Wang D.W., Zhang S.Y., Zhou D., *et al.* "Temperature stable cold sintered $(Bi_{0.95}Li_{0.05})(V_{0.9}Mo_{0.1})O$-$4Na_2Mo_2O_7$ microwave dielectric composites." *Mater.* 2019, vol. 12(9), p. 1370.

[136] Wang D.W., Zhou D., Song K.X., Feteira A., Randall C.A., Reaney I.M. "Cold-sintered C0G multilayer ceramic capacitors." *Adv. Electron .Mater.* 2019, vol. 5(7), p.1900025.

[137] Kahari H., Teirikangas M., Juuti J., Jantunen H. "Dielectric properties of lithium molybdate ceramic fabricated at room temperature." *J. Am. Ceram. Soc.* 2014, vol. 97(11), pp. 3378–3379.

[138] Kahari H., Teirikangas M., Juuti J., Jantunen H. "Improvements and modifications to room-temperature fabrication method for dielectric Li_2MoO_4 Ceramics." *J. Am. Ceram. Soc.* 2015, vol. 98(3), pp. 687–689.

[139] Kahari H., Teirikangas M., Juuti J., Jantunen H. "Room-temperature fabrication of microwave dielectric Li_2MoO_4–TiO_2 composite ceramics." *Ceram. Int.* 2016, vol. 42(9), pp. 11442–11446.

[140] Induja I.J, Sebastian M.T. "Microwave dielectric properties of mineral sillimanite obtained by conventional and cold sintering process." *J. Eur. Ceram. Soc.* 2017, vol. 37(5), pp. 2143–2147.

[141] Induja I.J., Sebastian M.T. "Microwave dielectric properties of cold sintered Al_2O_3–NaCl composite." *Mater. Lett.* 2018, vol. 211, pp. 55–57.
[142] Hong W.B., Li L., Cao M., Chen X.M. "Plastic deformation and effects of water in room-temperature cold sintering of NaCl microwave dielectric ceramics." *J. Am. Ceram. Soc.* 2018, vol. 101(9), pp. 4038–4043.
[143] Li L., Hong W.B., Yang S., Yan H., Chen X.M. "Effects of water content during cold sintering process of NaCl ceramics." *J. Alloy Compd.* 2019, vol. 787, pp. 352–357.
[144] Pang L.X., Zhou D., Li W.B., Yue Z.X. "High quality microwave dielectric ceramic sintered at extreme-low temperature below 200 degrees and co-firing with base metal." *J. Eur. Ceram. Soc.* 2017, vol. 37(9), pp. 3073–3077.
[145] Hong W.B., Li L., Yan H., Chen X.M. "Cold sintering and microwave dielectric properties of dense HBO_2-II ceramics." *J. Am. Ceram. Soc.* 2019, vol. 102(10), pp. 5934–5940.

Chapter 10
Ceramic dielectrics for MLCCs

Ziming Cai[1], *Chaoqiong Zhu*[1], *Limin Guo*[2], *Longtu Li*[1] *and Xiaohui Wang*[1]

10.1 Introduction

Multilayer ceramic capacitors (MLCCs), characterized by their high capacitance and compactness, are in high demand due to the rapid development of modern electronics [1,2]. The typical structure of an MLCC is shown in Figure 10.1(a). Due to the clamor for ever-smaller sizes, improved performance, and lower component costs in the electronic industry, there is a strong drive to develop MLCCs with thinner dielectric layers [3,4]. The layer thickness in an MLCC can be 1 μm or less (see Figure 10.1(b)), with the attendant ceramic grain size (GS) shrinking below 100 nm. The $BaTiO_3$ (BT) ceramic is widely used in MLCCs due to its large dielectric constant (*K*) and low loss. It is known that the dielectric constant of ferroelectrics, including BT, significantly depends on the grain size. This phenomenon is called the size effect, and many researchers have explored the size dependence of the dielectric constant of BT ceramics [5–8]. The drive to decrease the dielectric layer thicknesses of MLCCs is expected to continue for the next decade, which makes it important to understand the mechanism for the size effect of BT ceramics. Consequently, it is of great practical interest to prepare dense nanograin BT ceramics and investigate their dielectric and ferroelectric characteristics. The findings will provide useful guidance when designing next-generation MLCCs.

Section 10.2 of this chapter begins with a discussion of the size effect for grain sizes down to 5 nm [9]. Whether nanograin BT can provide an elevated *K* is obviously an important practical issue that has repeatedly been questioned in the literature. In the coarse-grain regime, *K* has a well-known maximum at GS = 1 μm, below which *K* decreases [10]. In our work from 2006, we found a ferroelectric polydomain state during the microscopic characterization of BT samples with an average GS as small as 8 nm that were obtained by a two-step sintering method

[1]State Key Laboratory of New Ceramics and Fine Processing, School of Materials Science and Engineering, Tsinghua University, Beijing, China
[2]School of Science, Beijing University of Posts and Telecommunications, Beijing, China

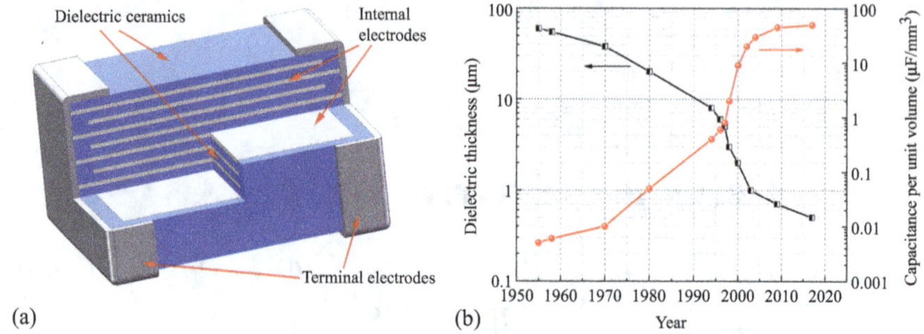

Figure 10.1 Schematic of an MLCC (a) and temporal trends in dielectric layer thickness and capacitance per unit volume (b)

[11]. Robust *K*-temperature (*T*) performance, a high dielectric constant (~1400), low dielectric loss, and multiple dielectric/ferroelectric transitions were clearly observed for this ceramic. This implies that a ceramic capacitor with a layer thickness of 40–50 nm (corresponding to five to seven 8-nm grains) is viable. A phenomenological thermodynamic theory that also considers stress effects caused by surface bond contraction was proposed by Sun *et al.* [12] as a means to predict the critical grain size. This theory predicts the relationships between the changing trends in phase transitions and the critical grain size for the disappearance of the ferroelectric phases, including two opposite relationships between the grain size and two types of stresses. The grain size that leads to the disappearance of ferroelectric phases at room temperature is found to be about 5.2 nm. When the grain size is less than 5.2 nm, ferroelectric phases are completely absent across the entire temperature range. Most recently, to determine whether it is feasible to fabricate sub-5 nm ceramics, we generated ultrafine BT nanocrystalline powder with grain sizes of 3–5 nm using a one-step solvothermal method [13]. A dense BT ceramic with a grain size of 5 nm was then obtained by two-step sintering, which is currently the finest dense bulk nanograin material on record. This is reassuring for the future of the MLCC industry, and augurs well for the conventional method of fabricating MLCCs (tape casting of green layers, screen printing of electrodes, and co-firing of BT/electrode stacks comprising hundreds of layers). With incremental changes and advances, this method should still be useable for the foreseeable future.

Presently in the MLCC industry, however, there is an urgent demand for dielectric ceramics with grain sizes of < 200 nm that can meet the need for dielectric layer thicknesses of < 1 μm, given that at least five grains across each dielectric layer are required to ensure MLCC reliability [14–17]. However, the preparation of dielectric layer ceramics with such small grain sizes is a huge challenge, since the grains easily grow to micrometer sizes at high sintering temperature (~1200 °C), that is, the grain size (< 200 nm) is hard to control. Also, base metal electrodes (BMEs) such as Ni and Cu are preferred for MLCCs in order to lower costs and facilitate mass production [18,19], implying that ceramic dielectrics such as BT need to be nonreducible to ensure that they can be co-sintered with Ni or Cu electrodes under a reducing atmosphere. However, $BaTiO_3$-based ceramics readily become semiconducting when sintered under a reducing atmosphere. Thus, further adjustment of the components of $BaTiO_3$-based ceramics is needed to suppress their reducibility [20–23].

In Section 10.3 of this chapter, the chemical coating method for the preparation of core–shell-structured dielectric ceramics is reviewed. This technique causes the doping elements to distribute uniformly in the shell part of the grain, leading to high dielectric constant and stability and fine grains. The effects of grain size on the dielectric properties are then further summarized. After that, the two-step sintering method – which refines the grains and reduces shell occupation – is introduced, and its mechanism is discussed. The fabrication of a temperature stable BME-MLCC based on a nanograin $BaTiO_3$-based ceramic is then described with excellent performance.

Aside from dielectric capacitor applications, this chapter also explores the use of MLCCs for energy-storage applications. In contrast to batteries and electrochemical capacitors, dielectric ceramics possess ultrahigh power density and can release the stored energy very rapidly (within in a microsecond) to create intense power pulses. This fast charge–discharge capability is crucial in pulsed power systems such as femtosecond lasers, medical defibrillators, hybrid electric vehicles, high-voltage pulse-power circuits, and kinetic energy weapons [24–28]. However, the energy density of a dielectric capacitor is much lower than that of an electrochemical capacitor or battery. Theoretically, the energy storage density (J) of a dielectric capacitor can be calculated via

$$J = \int_0^{D_{\max}} E\mathrm{d}D \approx \int_0^{P_{\max}} E\mathrm{d}P \tag{10.1}$$

where E, D, and $D_{\max}$ are the electric field, electric displacement, and maximum electric displacement at the breakdown strength, respectively. P and $P_{\max}$ are the electric polarization and maximum polarization at the breakdown strength. In general, there are four kinds of dielectric materials [26]: linear dielectrics, ferroelectrics, relaxor ferroelectrics, and antiferroelectrics, for which the P–E relations and the corresponding discharge energy density (W_{dis}) and energy loss (W_{loss}) are illustrated in Figure 10.2. It is clear that the energy density is dominated by the electric polarization and the breakdown strength. Ferroelectric polymer-based and ceramic-based dielectric materials are widely used in dielectric capacitors due to their large electric polarizations. Polymer-based dielectrics possess high breakdown strength, resulting in a high discharge energy density (~20–30 J cm^{-3}) [29–33]. However, their low polarization induces low dielectric constant (ε) and low energy efficiency, limiting their applications. They are also problematic to use at high temperature because they show reduced breakdown strength and greatly increased energy loss as the temperature ramps up. In spite of their low energy-storage density (~2–3 J cm^{-3}), ferroelectric ceramic-based dielectrics (such as

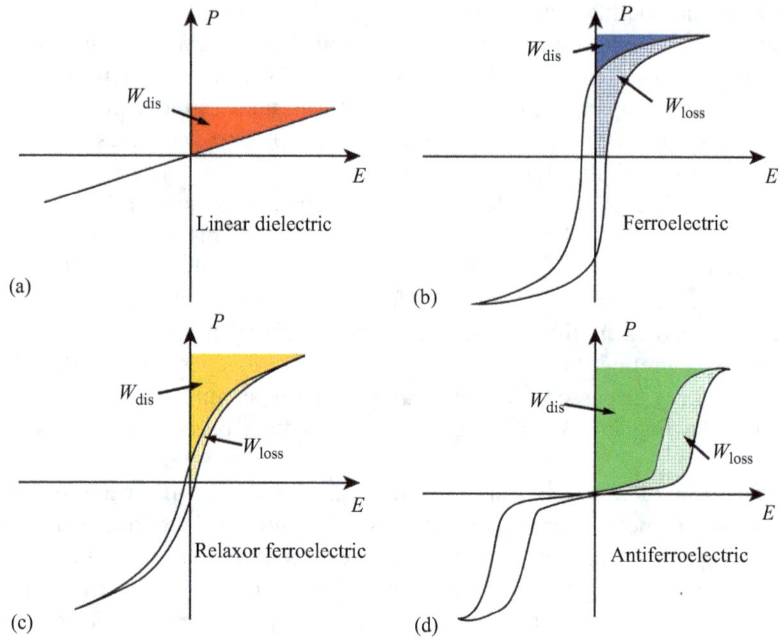

Figure 10.2 Diagrams showing the P–E relation and the corresponding discharge energy density (W_{dis}) and energy loss (W_{loss}) for (a) linear dielectrics, (b) ferroelectrics, (c) relaxor ferroelectrics, and (d) antiferroelectrics. Reproduced from [26]

$BaTiO_3$) exhibit high electric polarization (i.e., high dielectric constant) and good temperature stability, so they find considerable use, especially in high-temperature applications [34–38]. However, their low dielectric strength and large hysteresis lead to low energy density and low energy efficiency, which limits the energy-storage applications of ferroelectric ceramics. In recent decades, a great deal of effort has been directed into improving the energy-storage performance of ferroelectric ceramics, for example by modifying their chemical or electrical components through doping [39–41], generating core–shell structure via chemical coating [42,43], refining the grain size through two-step sintering [4,44–47], and so on. Also, compared to bulk ceramics, the dielectric breakdown strength of MLCCs is dramatically enhanced due to their significantly decreased dielectric layer thickness [48–51]. Therefore, MLCCs exhibit unusually high energy-storage density, making them particularly promising candidates for high-power energy-storage systems.

In Section 10.4 of this chapter, the most recently reported MLCCs with greatly improved energy-storage densities (> 10 J cm^{-3}, comparable to or even larger than those of thin-film capacitors) are reviewed. Then, the process of optimizing the sintering method to achieve high-temperature energy storage MLCCs is discussed, along with a special high-temperature X9R-type energy-storage MLCC. After that, an advanced design strategy for energy-storage MLCCs based on phase-field simulations is summarized. Finally, we draw some conclusions and consider the prospects for MLCCs.

10.2 Effect of grain size on the dielectric properties of barium titanate ceramics

10.2.1 Preparation of nanograin $BaTiO_3$ ceramics by two-step sintering

The development of new methods of synthesizing monodisperse BT nanocrystals has prompted investigations of the size effect. The hydrothermal synthesis of ceramic powders with strictly controlled parameters (concentration, pH, and temperature) has been widely reported, especially for the preparation of products with strong single-phase character at low temperature [52,53]. This method has been used to prepare highly disperse BT powders with particles that are smaller than several tens of nanometers. However, aqueous hydrothermal systems mostly produce cubic-phase BT powders. To obtain the tetragonal phase capable of ferroelectricity, high-temperature annealing, crystallite growth, and powder aggregation are inevitable. Thus, nonaqueous reaction approaches are considered to be advantageous for controlling crystallization and particle growth. We have synthesized monodisperse BT nanocrystals with a tetragonal structure using a one-step solvothermal method [13]. These nanocrystals have an average size of about 3–5 nm, with a narrow size distribution. The high-resolution (HR) transmission electron microscopy (TEM) image presented in Figure 10.3(a) demonstrates the high crystallinity of the nanoparticles. A tetragonality (c/a ratio $\approx$ 1.0069) can also be discerned from the HRTEM micrograph in Figure 10.3(a) based on the lattice fringe. The nanoparticle size distribution (average size: 5 nm) is shown in Figure 10.3(b).

Two-step sintering (TSS) was invented by Wang and Chen [44] to allow the fabrication of dense nanograin ceramics, including both oxides and nonoxides

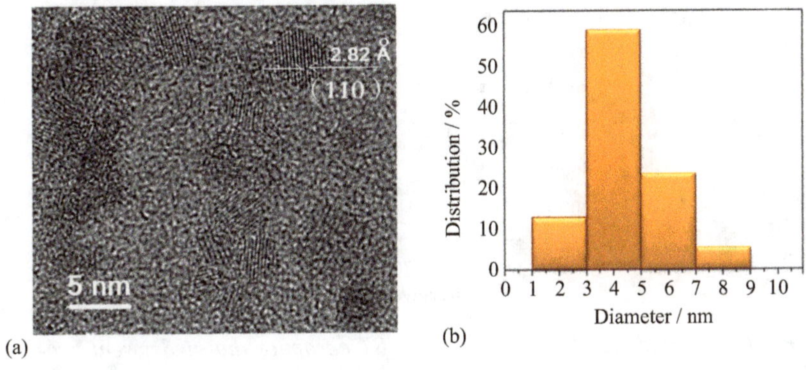

Figure 10.3 HRTEM image (a) and size distribution (b) of 5-nm BT nanoparticles. Reproduced from [9]

[54–58]. This technique relies on suppressing grain-boundary migration while maintaining grain-boundary diffusion. In this method, the porous material is first sintered at a high temperature (T_1) to eliminate the pores, which are thermodynamically unstable, leading to a relative density of 70%–80%. The partially sintered material is then cooled and held at a lower temperature (T_2) for a long time to inhibit grain growth but not densification. There is also a "kinetic window" to achieve densification. The remarkable feature of this procedure is that there is no apparent grain growth during the second densification step because the four-grain junctions are "frozen," resisting grain boundary migration, even though diffusion is still proceeding [47]. This is rather unique and completely unlike the case in normal sintering in which final stage densification is always accompanied by rapid grain growth. The dramatic contrast between the development of microstructure in BT ceramics during normal sintering (NS) and that during TSS can be seen in Figure 10.4, where several grain size–density trajectories are depicted. It is clear that grain growth during the second sintering step ($T_2 < T_1$) was completely suppressed.

Recently, using the 3- to 5-nm nanopowders of BT (obtained by solvothermal synthesis) mentioned above, and aided by cold isotropic pressing (CIP) at an ultrahigh pressure (3~6 GPa), we realized a dense BT nanoceramic with a grain size of 5~8 nm, again via TSS (see Figure 10.5(a)) [59]. Employing the two-step sintering strategy without applying a high pressure during CIP, a series of high-density bulk BT ceramics with grain sizes ranging from micrometers to 30 nm were also prepared [59–61]. Table 10.1 lists the samples ($BaTiO_3$-1 to $BaTiO_3$-18) obtained from some of the successful two-step sintering experiments that used 10-nm, 30-nm, and 100-nm BT powders. These experiments all achieved a high density ($\geq$ 96%) without grain growth in the second step. Samples with larger grains were also obtained by increasing T_2 to allow some grain growth ($BaTiO_3$-19: GS = 4.32 μm and $BaTiO_3$-20: GS = 8.61 μm in Table 10.1).

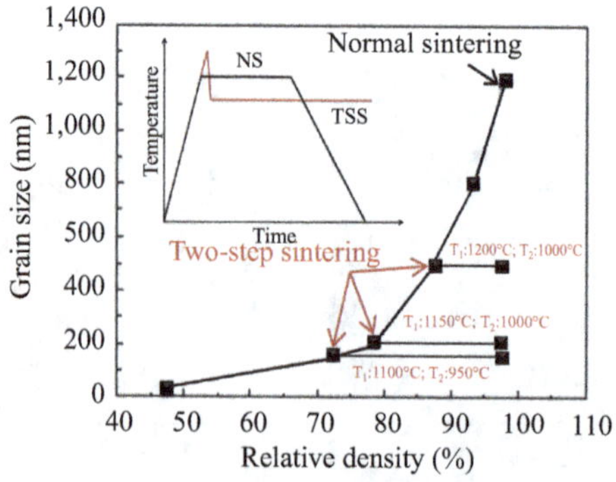

Figure 10.4 Grain size versus density for BT ceramics that underwent two-step sintering and normal sintering. Reproduced from [9]

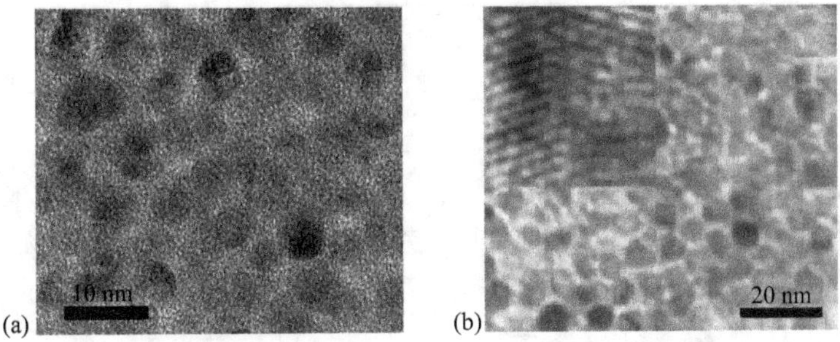

Figure 10.5 *(a,b) TEM and HRTEM (inset in (b)) images of BT ceramics with grain sizes of 5 nm and 8 nm. Reproduced from [9]*

Table 10.1 *Two-step sintering results for BT ceramics (initial powder sizes of 10 nm (*), 30 nm (**), or 100 nm (***))*

Sample	ρ_0 (%)	**After first sintering step**				**After second sintering step**			
		T_1 (°C)	t_1 (h)	ρ_1 (%)	GS_1 (nm)	T_2 (°C)	t_2 (h)	ρ_2 (%)	GS_2 (nm)
$BaTiO_3$-1*	61	950	0	86	33	900	2	98.0	35
$BaTiO_3$-2*	46	980	0	78	68	900	4	97.0	70
$BaTiO_3$-3**	46	1,100	0	73	148	900	20	96.2	150
$BaTiO_3$-4**	46	1,100	0	73	148	950	20	97.1	150
$BaTiO_3$-5**	46	1,150	0	78	200	900	20	96.3	200
$BaTiO_3$-6**	46	1,150	0	78	200	1,000	20	97.2	200
$BaTiO_3$-7**	46	1,180	0	83	296	950	20	97.0	300
$BaTiO_3$-8**	46	1,180	0	83	296	1,000	20	97.2	300
$BaTiO_3$-9**	46	1,200	0	87	495	850	20	96.0	500
$BaTiO_3$-10**	46	1,200	0	87	595	900	20	96.3	500
$BaTiO_3$-11**	46	1,200	0	87	495	1,000	10	97.2	500
$BaTiO_3$-12**	46	1,200	0	87	495	1,100	10	97.0	500
$BaTiO_3$-13**	46	1,230	0	90	795	850	20	97.5	800
$BaTiO_3$-14**	46	1,230	0	90	795	1,000	20	98.0	800
$BaTiO_3$-15***	48	1,310	0	88	990	800	24	96.4	990
$BaTiO_3$-16***	48	1,310	0	88	990	850	24	96.4	1,020
$BaTiO_3$-17***	48	1,340	0	90	1,950	900	24	98.0	1,970
$BaTiO_3$-18***	48	1,340	0	90	1,950	950	24	97.4	2,210
$BaTiO_3$-19***	48	1,340	0	90	1,950	1,050	24	96.8	4,320
$BaTiO_3$-20***	48	1,340	0	90	1,950	1,100	24	96.5	8,610

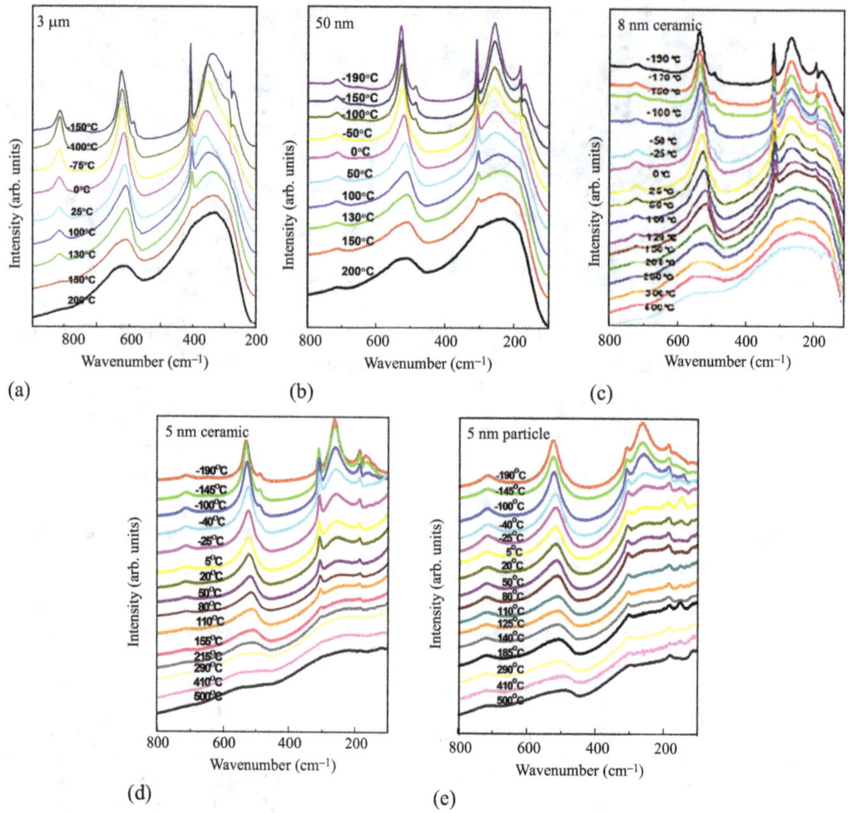

Figure 10.6 Raman spectra of BT as a function of temperature from −190 °C to 500 °C: (a) 3-µm ceramic, used as a reference; (b) 50-nm ceramic; (c) 8-nm ceramic; (d) 5-nm ceramic; and (f) 5-nm powder. Reproduced from [9]

10.2.2 Microstructure and dielectric properties of nanograin $BaTiO_3$ ceramics

10.2.2.1 Raman spectra

Raman spectroscopy can provide very sensitive measurements of the local crystal symmetry in powders, polycrystals, and single crystals of materials such as $PbTiO_3$ and $BaTiO_3$ [62–64]. In the following, we describe our observations of Raman spectroscopy of BT powders and nanoceramics obtained using a confocal microscopic Raman spectrometer (RM2000, Renishaw, UK). Raman spectra of a sample with GS = 3 µm were collected first and used as a reference. As shown in Figure 10.6(a), the spectra demonstrate distinct C/T/O/R dielectric transitions in the temperature range −190 to 500 °C, such spectra feature distinct C/T/O/R dielectric transitions. Concerning the above spectra, while symmetry-characteristic spectra (e.g., 200 °C for the C phase, 100 °C for the T phase, 0 °C for the O phase, and

-150 °C for the R phase) might be designated, the sensitivity of BT Raman spectra to both optical and microstructural details (e.g., polarization, mode–mode interference, orientation, single-domain vs. multidomain vs. polycrystal) demands a more cautious interpretation. It is apparent that the sharp peaks at 310 cm^{-1} and 715 cm^{-1} are forbidden for C symmetry, the intensity of the broad band at 240–270 cm^{-1} drops discontinuously during the T/O transition, and the sharp multiple peaks at 170–190 cm^{-1} are a characteristic feature of the R phase, although they are still present as a weaker, more diffuse feature for both O and T symmetries. The above information allowed us to identify the structural transition in the BT ceramics with GS = 50 nm (Figures 6(b)). The presence of R symmetry at -190 to -150 °C (a double peak at 170–190 cm^{-1}) and the T/O transition at 50–100 °C (a discontinuous shift of 240–270 cm^{-1}) are sufficient to establish the existence of all three (T/O/R) symmetries. Please note that the nominally "C" spectrum at 200 °C still has weak features at 310 cm^{-1} and 715 cm^{-1}, indicating a T symmetry remnant (Figure 10.6(b)). Therefore, the C/T transition is rather diffuse in the nanocrystal. Figure 6(c) and (d) shows the Raman spectra of another two BT ceramics with GS = 8 and 5 nm, respectively. All of the peak evolutions mentioned above are seen in the spectra of coarse-grain ceramics. Furthermore, the nominally "C" spectra at 310 cm^{-1} and 715 cm^{-1} still have weak features at 290 °C, indicating that the T remnant can survive above the Curie temperature. This suggests that all polar phases observed in coarse-grain ceramics are also found in nanoceramics at least those with grains as fine as 5 nm. They should all undergo R/O/T/C phase transitions as the temperature rises [65,66]. In particular, the Raman spectroscopic behavior of a 5-nm BT powder (Figure 10.6(e)) is similar to that of a dense 5-nm ceramic, as is the behavior of the 8-nm BT ceramic reported in our previous study [11].

10.2.2.2 High-resolution synchrotron X-ray diffraction

Quantitative phase analysis of nanograin BT ceramics is hampered by the broadening of diffraction peaks when using conventional diffraction techniques. Therefore, high-resolution synchrotron X-ray diffraction was performed to study phase evolution and coexistence in these particles (GS: 5–100 nm) in the range 150–450 K [67]. Our results indicate that, as the temperature rises, nanograin BT continues to present the same phase sequence but the phase boundaries become more diffuse and more than one phase may coexist between 200 K and 450 K. The diffraction patterns were collected at APS 11-BMB (Argonne National Laboratory) with a resolution of $< 2 \times 10^4$ $\Delta Q/Q$. The temperature was controlled to within 1 K. All diffraction data were analyzed through Rietveld refinement using the General Structure Analysis System (GSAS) software package [68], including the lattice parameters, atomic positions, thermal parameters, and the fractions of the R, O, T, and C phases. Table 10.2 gives the phase fractions of BT nanopowders with particle sizes of 5–100 nm at various temperatures. Except for the R phase at 150 K, all other diffraction patterns can be fitted with two or three phases: R+O, R+O+T, O+T, and O+T+C, in order of increasing temperature. Compared to coarse-grain ceramics and coarse powders, the R, O, T, and C phases extend and coexist over a

Table 10.2 Phase fractions (in %) of BT nanopowders as functions of the particle size and temperature

Particle size (nm)	Temperature (K)						
	150	**200**	**250**	**300**	**350**	**400**	**450**
100	R (100)	R+O (64+36)	R+O+T (25+25+50)	O+T (25+75)	O+T (20+80)	O+T+C (32+54+14)	O+T+C (28+24+48)
50	R (100)	R+O (60+40)	R+O+T (33+17+50)	O+T (38+62)	O+T (33+67)	O+T+C (33+44+23)	O+T+C (17+33+50)
40	R (100)	R+O (58+42)	R+O+T (22+13+65)	O+T (47+53)	O+T (38+62)	O+T+C (21+44+35)	O+T+C (15+33+52)
20	R (100)	R+O (56+44)	R+O+T (21+8+71)	O+T (50+50)	O+T (41+59)	O+T+C (20+40+40)	O+T+C (6+31+63)
10	R (100)	R+O (63+37)	R+O+T (45+19+36)	O+T (38+62)	O+T (32+68)	O+T+C (30+44+26)	O+T+C (18+33+49)
5	R (100)	R+O (80+20)	R+O+T (50+25+25)	O+T (33+67)	O+T (23+77)	O+T+C (32+48+20)	O+T+C (11+42+47)

wide temperature range. This trend is consistent with observations of nanograin ceramics. Also, these nanopowders remain ferroelectric across a wide temperature range, making them potentially suitable for practical ferroelectricity-related applications (Figure 10.7).

10.2.2.3 Dielectric properties

In the coarse grain regime, the dielectric constant K is well known to have a maximum at GS $\approx$ 1 μm [10]. Whether nanograin BT can provide an elevated K has been repeatedly questioned in the literature due to a lack of reliable data for a large range of grain sizes. Therefore, using high-purity two-step-sintered ceramic samples with grain sizes ranging from nanometers to micrometers, we investigated this issue. To remove the influence of moisture, which is especially strong at small grain sizes, the dielectric measurements reported below were all performed in a vacuum environment after the samples had first been baked in situ.

As shown in Figure 10.8, all the temperature spectra of the dielectric constant show a distinct C/T phase transition. In comparison, the low-temperature transitions are much weaker and broader. For the smaller grain sizes ($\leq$ 100 nm), the diffuse humps superimposed on the sloping background at low temperature may be partly due to the tail of the C/T transition. Accurate values for the transition temperatures can be obtained at the minimal of d^2K/dT^2 (this procedure largely removes the distortion caused by the sloping background). As shown in Figure 10.9, over the GS range 50–800 nm, the C/T transition temperature steadily decreases as the GS decreases (by about 12 °C for the smallest GS). In contrast, the temperatures of the T/O transition and the O/R transition steadily increase with decreasing GS (by about 20 °C for the smallest GS). The C/T transition temperature (115 °C) for the sample with GS = 50 nm agrees well with that observed for SPS

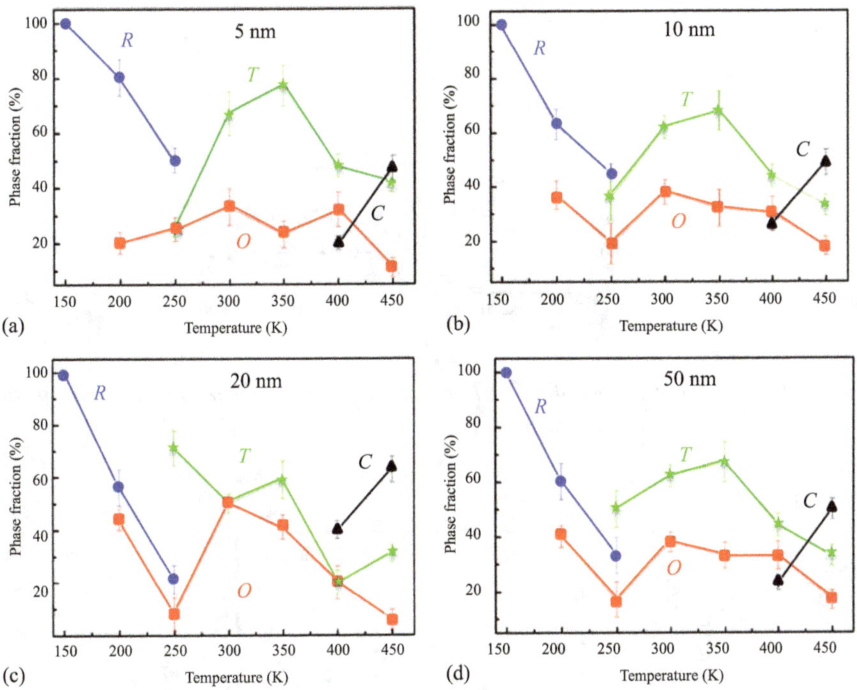

Figure 10.7 Plots showing how the phase fractions of various BT nanopowders evolve as a function of temperature. Particle size: (a) 5 nm, (b) 10 nm, (c) 20 nm, and (d) 50 nm. Reproduced from [9]

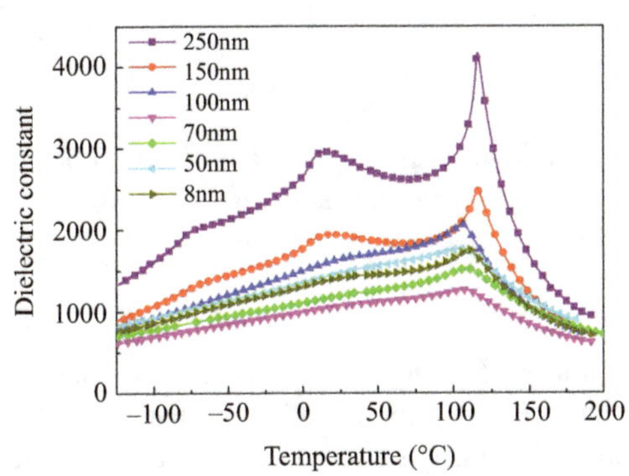

Figure 10.8 Temperature spectra of the dielectric constant for various BT ceramics during cooling. Reproduced from [9]

BT ceramics [69]. Although the T/O/R transitions have not been reported previously for GS < 300 nm, the observed increase in transition temperature with decreasing grain size is similar to that seen for coarse-grain BT (1.1–53 μm) [7]. Moreover, identical temperature spectra for the relative dielectric constant with

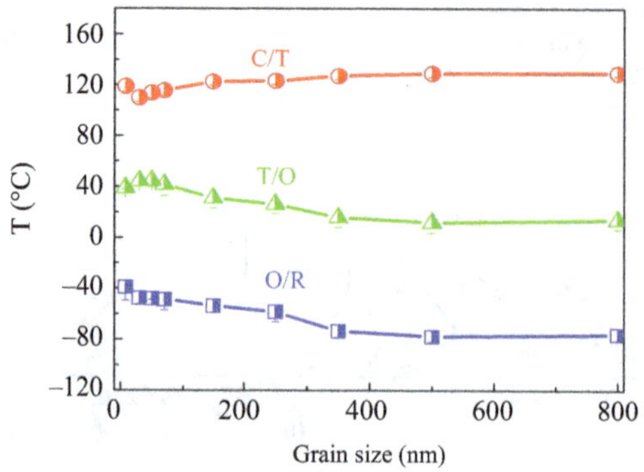

Figure 10.9 Temperatures of the C/T, T/O, and O/R phase transitions versus grain size. Reproduced from [9]

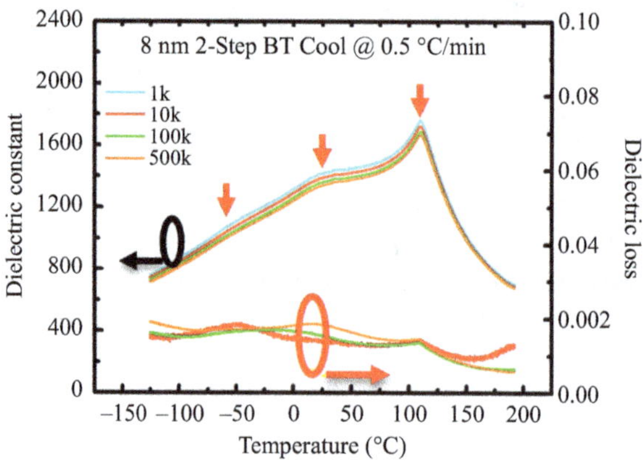

Figure 10.10 Temperature dependence of the dielectric constant of 8-nm BT ceramics at different frequencies (T_c = 118 °C). Reproduced from [9]

C/T/O/R transitions during cooling and heating processes were observed (not shown), indicating that the behavior is rather robust. Therefore, it can be summarized that nanograin BT ceramics show similar trends in structural distortions to those of coarse-grain ceramics, even though all transitions are very broad when the GS is less than 50 nm.

In our study, the 8-nm BT ceramic showed excellent dielectric properties, with three discernible peaks (indicated by the arrows in Figure 10.10) corresponding to the three phase transitions. These peaks are clearly more diffuse than those of the sample with GS = 150 nm (Figure 10.8). More interestingly, the frequency dependence of these peaks is very weak below T_C and nonexistent above T_C. The diffuse nature of these peaks cannot therefore be attributed to a broad distribution of polarization relaxation times, that is, it is not a relaxor. The small dielectric loss (typically <2%) is consistent with the high resistivity (7.8×10^{11} Ω cm), high density microstructure revealed by the TEM micrograph in Figure 10.5(b). On the other hand, despite these robust dielectric properties, the switchable polarization measured by integrating the depolarization current after poling (under 0.1 MV/m during cooling from 500 K to 77 K) is very small in the nanoceramics (Figure 10.11), indicating that the ferroelectricity is clamped, making temperature-dependent spontaneous depolarization difficult.

The technological driving forces for the development of nanograin MLCC ceramics are component miniaturization and device performance. The figure of merit of a capacitor can be defined as the capacitance versus the unit volume, scaled by K/t^2, where t is the layer thickness of the capacitor. In capacitors with ultrathin dielectric layers, dielectric breakdown is a concern, so the charge stored per unit volume at the breakdown electrical field E_b, that is, KE_b/t, may be used as an alternative figure of merit. Note that both of these measures improve with decreasing thickness of the dielectric layer. The feasibility of thinner layers strongly depends on the presence of very small ceramic grains (typically, t/d = 5–10, where d = GS) to ensure reliable properties and smooth layer interfaces. Therefore, the figure of merit scales with K/d, since E_b and t/d are almost constant. As described above, the BT ceramic with GS = 8 nm displays robust K-temperature performance, with very low dielectric loss and clear evidence of multiple dielectric/ferroelectric transitions. This nanograin ceramic exhibits a higher figure of merit at 25 °C than state-of-the-art MLCC ceramics with GS = 100 nm (Figure 10.12). Such results are greatly encouraging from the perspective of the continuing development of the MLCC industry toward the goal of nanograin ceramics.

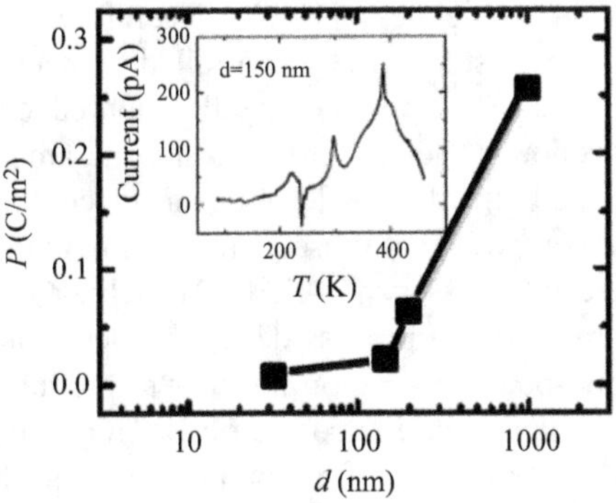

Figure 10.11 Effect of the grain size on the switchable polarization of BT ceramics. Inset: depolarization current with peaks at transition temperatures. Reproduced from [9]

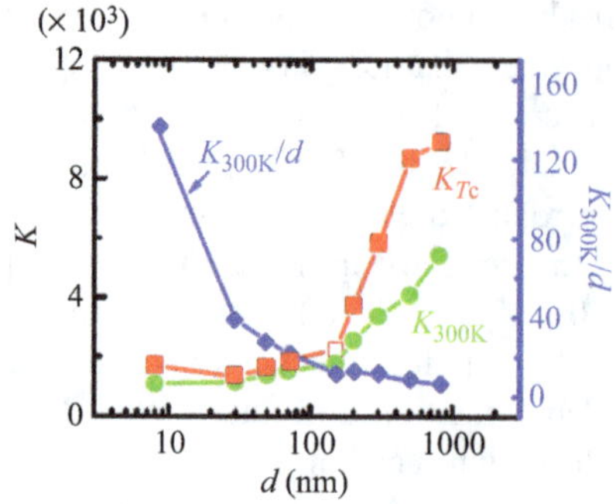

Figure 10.12 Relative dielectric constant (K) and figure of merit (K/d) of BT ceramics with different grain sizes (d = GS). Reproduced from [9]

10.3 Nonreducible $BaTiO_3$-based ceramics and ultrathin dielectric layer MLCCs

10.3.1 High-performance $BaTiO_3$-based ceramics prepared via the chemical coating method

Driven by the demands of the electronics industry for miniaturization, improved performance, and lower component costs, the development of multilayer ceramic capacitors (MLCCs) with base-metal internal electrodes (BME) and thinner dielectric layers is underway [2,70]. Barium titanate ($BaTiO_3$) is a ceramic material that is widely used in MLCCs due to its high dielectric constant and low loss. For large-capacity BME-MLCCs to fulfill the requirements of the X7R /X5R specification (i.e., the change in capacitance should be less than ±15% across the temperature range from –55 to 125 °C / 85 °C), the thickness of the dielectric layer should be < 1 μm and the grain size needs to be < 200 nm, and preferably much smaller, considering the reliability of electronic devices that at least 5 grains are needed across a dielectric thickness [15,71]. Therefore, the uniformity of the powder and the grain size must be strictly controlled to meet the demands of next-generation BME-MLCCs. Also, to achieve the nonreducible sintering of the dielectric layer, various donor and acceptor dopants and amphoteric trivalent rare-earth dopants (Y, Ho, and Dy) [21,72,73] can be introduced to ameliorate the effect of reducing atmosphere (as well as a subsequent step involving the application of a reoxidizing atmosphere if necessary [74–76]). To deliver enhanced temperature-stable dielectric performance, the process design further calls for these dopants to be localized at the grain-boundary region in order to produce a core–shell microstructure where the core is the pure ferroelectric $BaTiO_3$ and the shell a nonferroelectric solid solution. With such fine particles, it is difficult to obtain a homogeneous dopant distribution around the particles using the solid-state method and mechanical ball-milling, especially when the $BaTiO_3$ particle size decreases to the nanometer scale. Furthermore, the ultrafine nanoparticles and submicron particles easily form agglomerates, which causes abnormal grain growth during sintering, producing a nonoptimal dopant distribution in the ceramic grains and poor dielectric properties [77].

Chemical coating is an alternative to the conventional solid-state mixing method that can produce a highly homogeneous dopant shell containing all of the additive elements [78–82]. Agglomeration can be avoided and a uniform distribution of additives can be achieved irrespective of the $BaTiO_3$ particle size, leading to perfect core–shell microstructure control and dielectric performance.

10.3.1.1 Advantages of the chemical coating method

Based on the principle of chemical coprecipitation, the chemical coating method involves adjusting the pH such that each doping element is encouraged to precipitate from the solution to form a continuous coating layer on the surface of the $BaTiO_3$, which improves the uniformity of the dopant dispersion. The microstructures and properties of nanosized $BaTiO_3$ based ceramics prepared by chemical coating are significantly better than those using the conventional solid-state mixing method [80,81,83,84].

Figure 10.13 presents the surface morphology of a chemical coated $BaTiO_3$ particle as characterized by TEM. A uniform, continuous coating layer with a thickness of 2~3 nm is shown. When compared to powders prepared by the solid-state method, the chemical coated powders exhibit increased sintering activity due to the uniform distribution of the dopants. Samples prepared by chemical coating also show enhanced bulk density. Figure 10.14 shows the microstructural development of samples prepared by the chemical coating method (CC) and the solid-state method (SS) at a series of sintering temperatures. With increasing sintering temperature, the density of the ceramic samples increases without significant grain growth, except at the maximum sintering temperature, 1285 °C. The grains are a little finer and the grain distribution is narrower and more uniform for the CC samples (CC-1230: 125 nm, SS-1230:138 nm).

The dielectric properties and insulation resistivity at room temperature of samples prepared using the two methods are presented in Figure 10.15. At a particular sintering temperature, both the dielectric constant and the resistivity of the

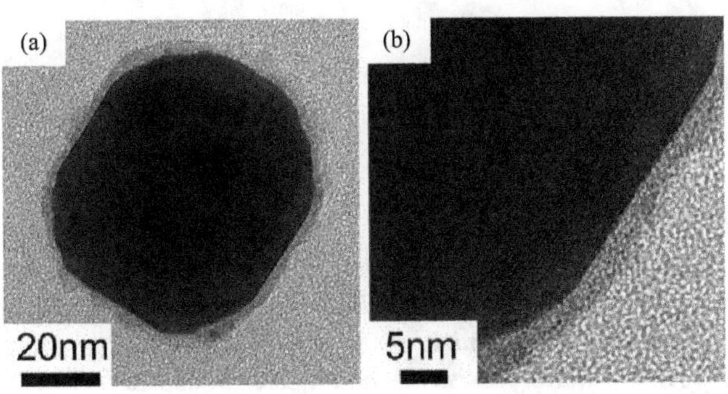

Figure 10.13 Surface morphology of additives uniform coated $BaTiO_3$ powder

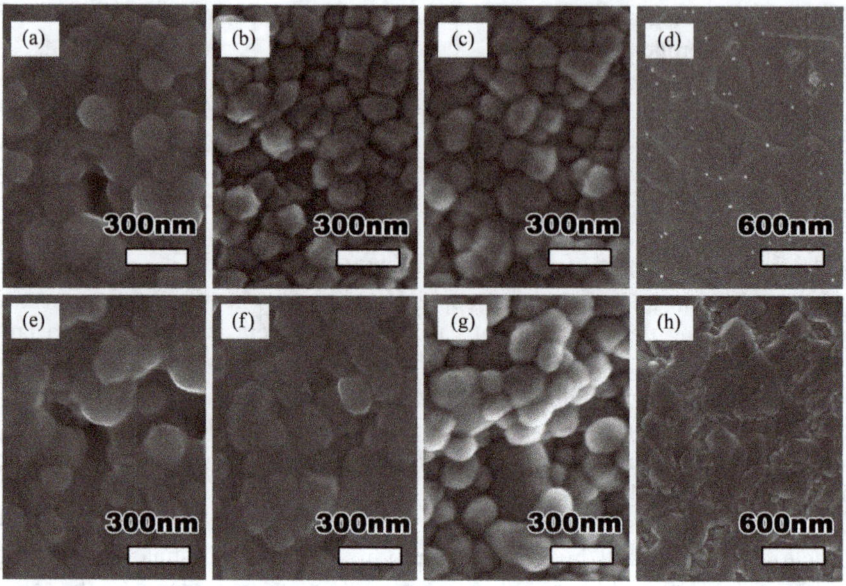

Figure 10.14 Morphologies of ceramic samples prepared by either the chemical coating method (CC samples) or the solid-state method (SS samples) and sintered at different temperatures [83]. Sintering temperatures of the CC samples: (a) 1210 °C, (b) 1230 °C, (c) 1260 °C, and (d) 1285 °C. Sintering temperatures of the CC samples: (e) 1210 °C, (f) 1230 °C, (g) 1260 °C, and (h) 1285 °C

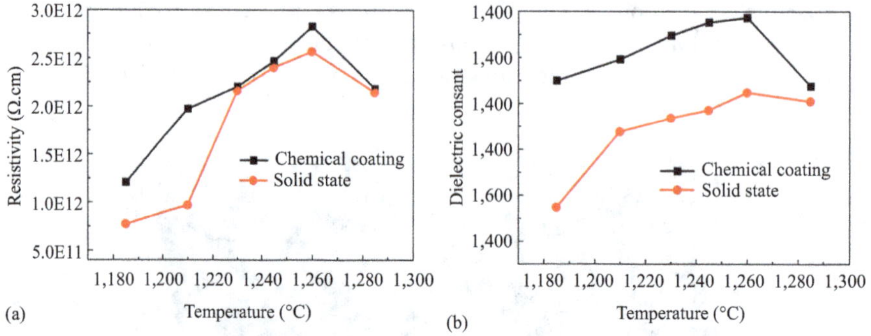

Figure 10.15 Plots of insulation resistivity (a) and dielectric constant (b) versus sintering temperature for $BaTiO_3$-based ceramic samples prepared by either the chemical coating method or the solid-state method. The insulation resistivity and dielectric constant were measured at room temperature. Reproduced from [83]

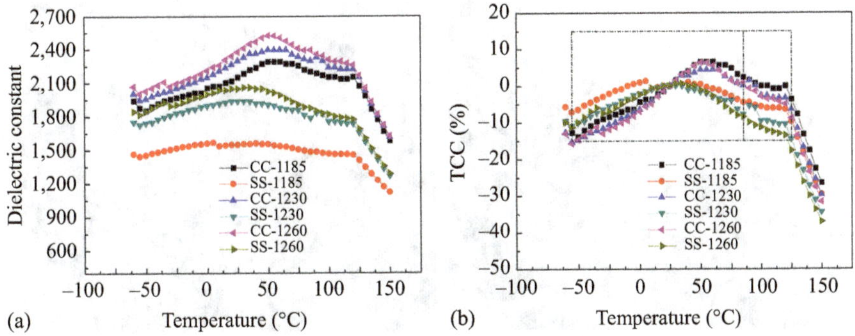

Figure 10.16 Temperature dependence of the dielectric constant (a) and the temperature coefficient of capacitance (TCC) (b) for $BaTiO_3$-based ceramic samples sintered at different temperatures. Reproduced from [83]

CC samples are higher than those for the SS samples. For both the CC and the SS samples, the dielectric constant and insulation resistivity initially gradually increase with sintering temperature due to the increase in density, but they then decrease due to the destruction of the microstructure for the samples sintered at 1285 °C. The temperature dependence of the dielectric constant and the temperature coefficient of capacitance (TCC) for the samples sintered at 1185, 1230, and 1260 °C are shown in Figure 10.16. In this temperature range, the *K–T* curves of the CC samples are located above those for the SS samples. All of these samples meet EIA X7R specification.

The core–shell structure is key to achieving the temperature-stable capacitance and the reliable performance of MLCCs [85–87]. To explore this structure, energy-dispersive X-ray spectroscopy (EDS) analysis of the distributions of the additive elements in grains of the CC sample and the SS sample was performed, as shown in Figure 10.17. A grain of considerable size in each sample was selected for investigation. Figure 17(c) and (d) shows that upon moving from the grain boundary to the grain core, the dopant contents in the grains of the CC and SS samples gradually decrease to a very low level. It can be concluded that the dopants, such as Y, Si, and Mg, are mainly localized in the shell regions of the $BaTiO_3$ grains and form a solid solution of a paraelectric phase, while the core regions are composed of almost pure $BaTiO_3$, which remains a ferroelectric phase. However, the dopants diffuse more deeply (~30 nm) into the grain of the SS sample than into the grain of the CC sample (~19 nm). That is, the core portion of the grain in the CC sample is larger, which explains the observed TCC behavior at high temperature and why the dielectric peak in the TCC curve at approximately 125 °C is higher for the CC samples than for the SS samples. During the formation of the core–shell structure, dopant interactions (such as Mg^{2+} preventing the diffusion of Y^{3+} into the core during sintering [88]) are much stronger in the CC sample (with the dopants coated around the $BaTiO_3$ grains) than in the SS sample. This results in rather different

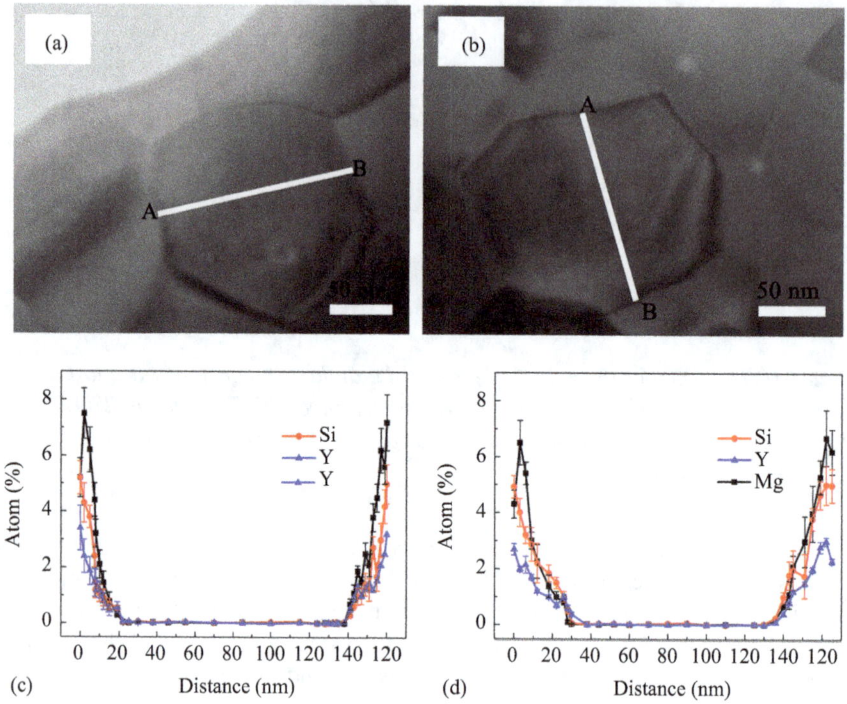

Figure 10.17 TEM images (a,b) and EDS profiles (along the line AB in each TEM image) of the dopant distributions (c,d) for $BaTiO_3$-based ceramics sintered at 1230 °C prepared by the chemical coating method (a,c) and the solid-state method (b,d). Reproduced from [83]

shell thicknesses for the grains in the CC and SS samples, which in turn influences their dielectric behavior.

Furthermore, a large difference is observed in the highly accelerated lifetime test (HALT) results. Figure 10.18 reveals that for the CC samples, the resistivity degradation includes two stages: a rapid decay stage and a stable stage; in other words, the insulation resistivity decreases rapidly at first and then remains almost unchanged, with no obvious resistivity degradation occurring with increasing time. For the SS samples, the resistivity degradation includes three stages: a rapid decay stage, a transition stage, and a slow decay stage. That is, the insulation resistivity first decreases rapidly and then the decay rate decreases with increasing time; the transition is a very short process. The insulation resistivity is much lower and the degradation is more severe for the SS samples than the CC samples. It can therefore be concluded that the electrical properties of the CC samples are much better than those of the SS samples.

To establish a link between microstructure and electrical properties for the samples prepared using different methods, and to obtain reliable values for the resistance of the grains and grain boundaries, a complex impedance analysis was

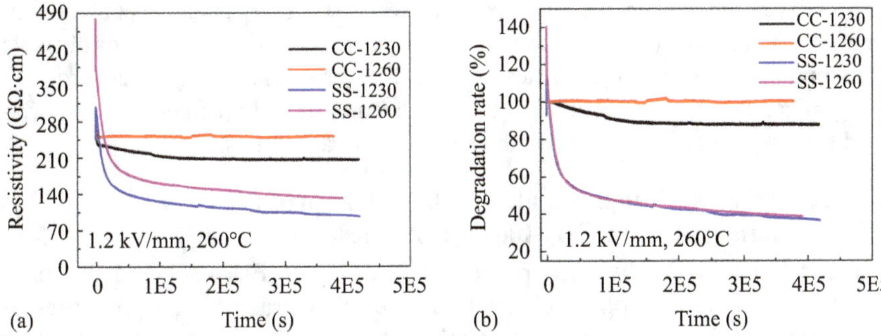

Figure 10.18 Resistivity (a) and degradation (b) versus time for various $BaTiO_3$-based ceramics, as measured at an elevated temperature of 260 °C and a DC bias of 1,200 V/mm. Reproduced from [83]

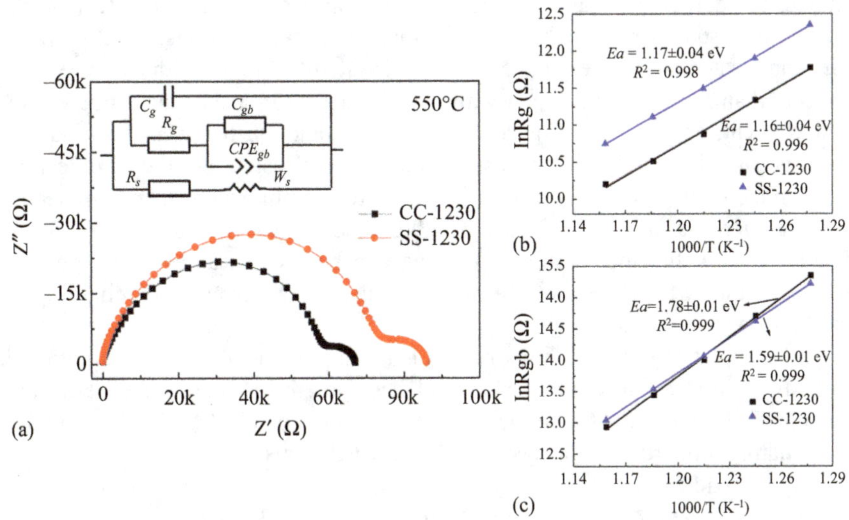

Figure 10.19 (a) Complex impedance plane plots for samples CC-1230 and SS-1230, as measured at 550 °C (inset: the effective electrical equivalent circuit model containing electronic and ionic components). Also shown are the Arrhenius plots and the fitting results for (b) the grain and (c) the grain boundary for samples CC-1230 and SS-1230. Reproduced from [83]

performed (Figure 10.19). As observed in Figure 10.19(b), the activation energy of the grain bulk differs only slightly between the two types of samples. However, the activation energy of the grain boundary (see Figure 10.19(c)) shows a much greater difference between the two types of samples. The higher activation energy for grain boundaries in the CC sample than in the SS sample may be due to more uniform and smaller grain sizes in the CC sample as well as a lower chemical defect concentration at grain boundaries in this sample. Grain boundaries are a major obstacle to the migration of oxygen vacancies [74,89]. These results are in good accord with the significantly improved reliability that has been observed for fine-grain $BaTiO_3$-based ceramic capacitors based on their HALT characteristics.

10.3.1.2 Effects of dopants on the dielectric properties of nanograin $BaTiO_3$-based ceramics

In order to improve the dielectric properties of nanograin $BaTiO_3$-based ceramics, the effects of different doping elements on the microstructure and properties of these ceramics have been systematically studied. The doping modification mechanism is of particular interest.

Due to synergistic effects in the Ho-Dy co-doping system [20], the Ho^{3+} ions and Dy^{3+} ions preferentially occupy different positions in the perovskite structure of $BaTiO_3$, forming the donor doping and acceptor doping, respectively, and then generate the defect association. Defect association has a beneficial effect on the insulation resistivity and reliability of the material system. For instance, the resistivity and the activation energy are better for a system co-doped with two rare-earth elements than for a system doped with a single acceptor species or a single rare-earth element, as shown in Figure 10.20. The interaction between the co-doped elements can inhibit not only grain growth but also diffusion into the core, leading to a thinner shell and finally exhibit a flatter temperature dependence dielectric constant curve as shown in Figure 10.20(a) and longer life time as shown in Figure 10.20(b). In other words, co-doping can be used to adjust the core-shell thickness ratio, which greatly affects the insulation resistance and reliability of modified $BaTiO_3$-based ceramics.

The element Mg plays an important role in suppressing abnormal grain growth [90]. Mg acts as an acceptor dopant and occupies the B site in the perovskite structure of $BaTiO_3$, facilitating the capture of conduction electrons and optimizing the insulating properties of the system. Mn^{2+}, which is also an acceptor dopant, can partially be oxidized to Mn^{3+} during the reoxidation step, which consumes some oxygen vacancies and avoids any loss of resistivity [22]. Thus, the inclusion of Mn as a dopant is the key to achieving nonreducible $BaTiO_3$.

10.3.2 Grain size effect for nonreducible $BaTiO_3$-based ceramics

The electrical performance of a dielectric ceramic is greatly influenced by its grain size, which derives from the induced changes in internal stress, crystal structure, tetragonality, and phase transitions of the ceramic [91]. This phenomenon is called the grain size effect, and it has attracted increasing attention in recent years due to the demand for miniaturized MLCCs with high capacitance. As the dielectric layer thickness in MLCCs decreases below 1 μm, and each layer needs to be at least five grains thick to ensure reliability and good electrical performance, it has become

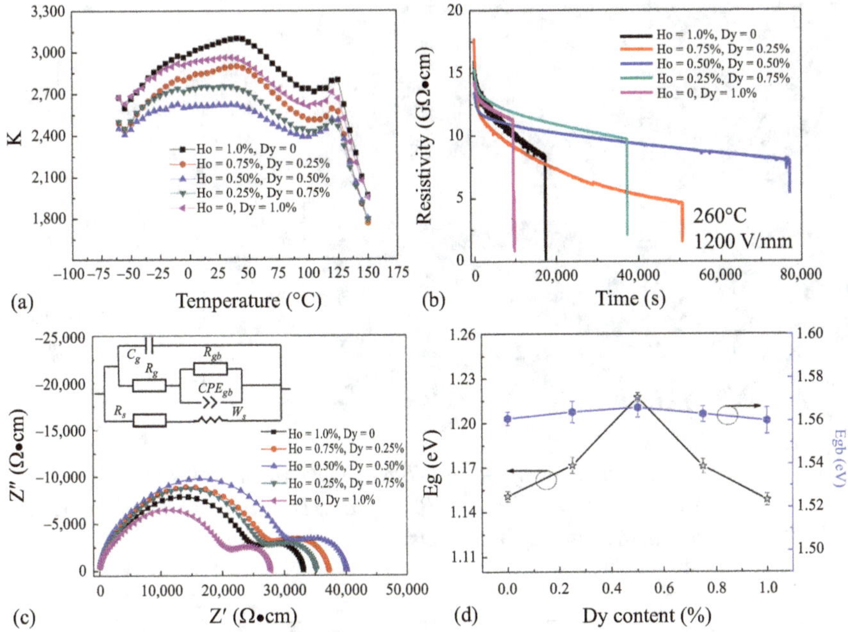

Figure 10.20 Doping effect on the temperature-dependent dielectric constant (a), HALT (b), and impedance (c) of $BaTiO_3$-based ceramics. The activation energy of grain (Eg) and grain boundary (Egb) calculated from (c) are plotted in (d)

increasingly important to study the effect of grain size on the reliability and electrical performance of nanograin $BaTiO_3$-based ceramics.

Our group [92] prepared fine-grain $BaTiO_3$-based ceramics with various grain sizes (around 100 nm to 500 nm) and a core–shell structure via the chemical coating method in a reducing atmosphere ($H_2/N_2/H_2O$, $P_{O2} = 10^{-11}$ to 10^{-13} Pa) in order to systematically study the effect of grain size on the dielectric and the reliability of the ceramic.

Figure 10.21 presents micrographs of polished sections of thermally etched $BaTiO_3$-based ceramic samples with mean grain sizes of 136 ± 50, 238 ± 75, 323 ± 133, and 529 ± 245 nm (denoted BTP1, BTP2, BTP3, and BTP4, respectively). No obvious pores were observed, and the grain-size distributions for the samples were narrow and uniform, demonstrating the superiority of the chemical coating method described above.

10.3.2.1 Size-dependent dielectric properties

The effect of the grain size on the dielectric constant of these $BaTiO_3$-based ceramics is consistent with that observed for other $BaTiO_3$ ceramics [93–95], since it is the grain core (i.e., $BaTiO_3$) that plays the dominant role in this respect. Figure 10.22 shows the temperature dependence of the dielectric constant for the

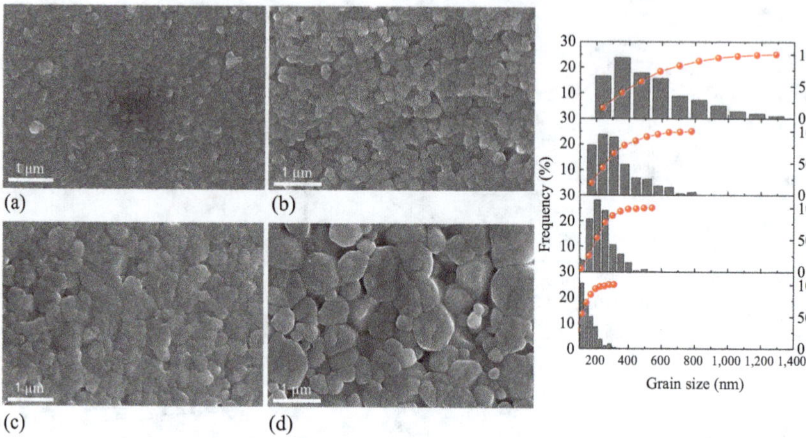

Figure 10.21 Microstructures of $BaTiO_3$-based ceramic samples with various grain sizes: (a) BTP1, (b) BTP2, (c) BTP3, and (d) BTP4. (e) Grain-size distributions for the samples. Reproduced from [92]

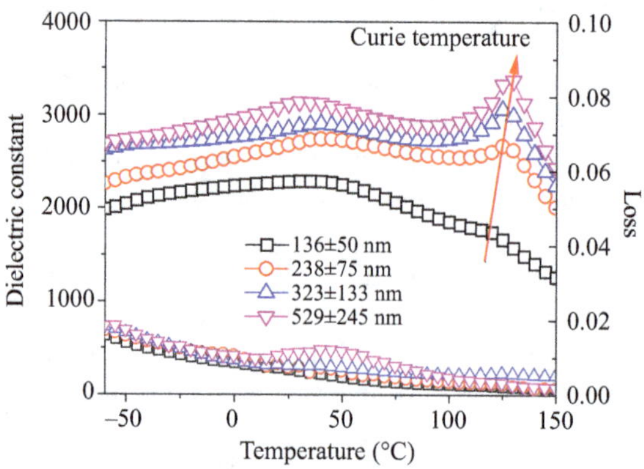

Figure 10.22 Temperature dependence of the dielectric constant and loss for $BaTiO_3$-based ceramics with various grain sizes. Reproduced from [92]

$BaTiO_3$-based ceramics with various grain sizes, as well as the temperature dependence of the loss. All of the samples exhibit typical core–shell behavior, with two evident peaks: one for the shell contribution around the ambient temperature and another for the core contribution near the Curie temperature. As the grain size increases, the dielectric constant increases over the whole test temperature range. In addition, the Curie temperature shifts to a higher temperature, indicating a coarsening of the core.

The capacitance–voltage (*C–V*) characteristics of the ceramic are also highly dependent on the grain size. The *C–V* curves at ambient temperature and the DC

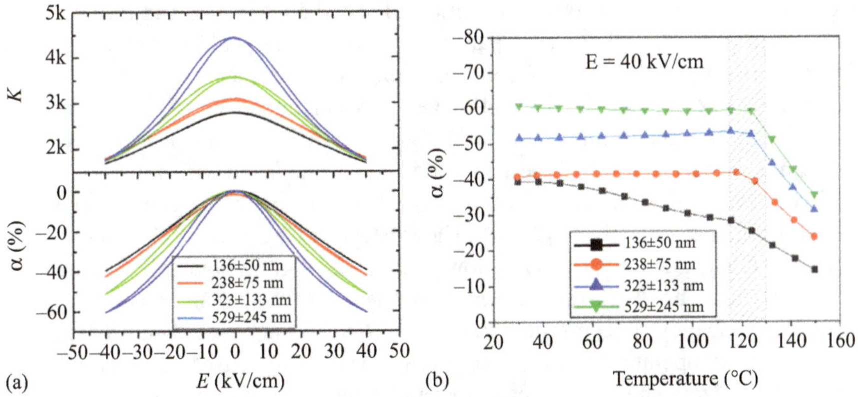

Figure 10.23 Capacitance–voltage (C–V) curves at ambient temperature (a) and the temperature dependence of the DC bias stability under a fixed bias field of 40 kV/cm (b) for the $BaTiO_3$-based ceramics with various grain sizes. Reproduced from [92]

bias stability under a fixed bias field of 40 kV/cm for the $BaTiO_3$-based ceramics are depicted in Figure 10.23(a). Here, the DC bias stability is defined as

$$\alpha = (K_E - K_0)/K_0 \times 100\%, \tag{10.2}$$

where K_0 and K_E are the relative dielectric constant under bias fields of 0 and E, respectively. As the DC bias field increases, the dielectric constant decreases due to the suppression of the anharmonic interactions of Ti^{4+} ions in Ti–O octahedrons [96]. Fine-grain ceramics yield better bias-field stabilities than coarse-grain ceramics; the relative dielectric constant of the coarse-grain ceramics drop rapidly as the DC bias field is increased. This difference in stability derives from differences in the grain boundary ratios, ferroelectric contributions, and oxygen vacancies in the fine-grain and coarse-grain ceramics. The DC bias stability of ceramics is also dependent on the temperature, and this dependence under a fixed bias field of 40 kV/cm for the $BaTiO_3$-based ceramics with various grain sizes is shown in Figure 10.23(b). The DC bias stability curves flatten and the absolute value of the DC bias stability drops as the temperature increases (especially at temperatures higher than the Curie temperature). Additionally, the fine-grain ceramics show smaller absolute values of the DC bias stability at all temperatures than the coarse-grain ceramics, indicating better DC bias-field stability.

10.3.2.2 Size-dependent reliability

Thermally stimulated depolarization current (TSDC) analysis can be employed to characterize the relaxation behavior and the concentrations of defects (dipoles, trapped charge, and mobile ions). Figure 24(a)–(d) shows TSDC curves for the $BaTiO_3$-based ceramics with various grain sizes. The curves exhibit an α peak around ambient temperature and a β peak near the Curie temperature, which originate from the diffuse phase transition of the shell part and the ferroelectric/paraelectric transition of the core part, respectively. The peak γ originating from dipoles can decrease the effective concentration of oxygen vacancies to some extent, since the acceptor dopants in the shell develop associations with oxygen vacancies, especially for the fine-grain ceramics. At higher temperatures, a δ peak and even a ζ peak can be discerned. Both of these become more pronounced with increased T_P and thus were assigned to the relaxation of oxygen vacancies inside grains and across grain boundaries, respectively [97,98]. The existence of a ζ peak for the coarse-grain ceramic BTP4 is due to its relatively low abundance of grain boundaries, which can block oxygen vacancy relaxation. Generally speaking, weak δ and ζ peaks imply a low oxygen vacancy concentration. Figure 10.24(e) shows a plot of current strength (J_{max}) versus temperature (T_m) for the δ peak for the $BaTiO_3$-based ceramics with various grain sizes. This plot indicates that a high temperature is required to generate current from the relaxation of oxygen vacancies inside fine grains. Around the δ peak, the depolarization current can be approximately expressed as

$$\ln[J(T)] \approx C - E_a/kT - (T/T_m)^4 \exp(E_a/kT_m - E_a/kT) \tag{10.3}$$

where C is a geometric constant and E_a is the activation energy associated with the relaxation of oxygen vacancies inside the grains. Figure 10.24(f) shows the activation energy obtained by fitting the δ peak. A higher E_a indicates that more energy is required to activate the relaxation of oxygen vacancies inside the grains.

The reliability of a device is one of the most important performance indices monitored during mass production. Highly accelerated lifetime testing (HALT) was therefore adopted to investigate the effect of the grain size on the reliability of the insulation resistivity. The results shown in Figure 10.25 indicate that the insulation resistance degradation behavior of the ceramic samples under a high electric field and elevated temperature improves dramatically with decreasing grain size. This is because the grain boundaries play the dominant role in the formation of Schottky barriers [15,100].

Furthermore, a complex impedance study was carried out to study the electrical microstructure of the $BaTiO_3$-based ceramics with various grain sizes (118–462 nm), and the results are presented in Figure 10.26. The results show that the fine-grain ceramics with core–shell grain structure have higher grain and grain boundary activation energies, whereas their proportion of ionic conductivity was lower, suggesting greater insulating reliability. This can be explained by noting that the fine-grain ceramics will have a higher proportion of the grain boundaries, and these boundaries block oxygen vacancy migration [101,102]. These results provide the theoretical foundations for the application of nanograin ceramics in the next generation of ultrathin-layer BME-MLCCs.

10.3.3 Nonreducible $BaTiO_3$-based ceramics prepared by two-step sintering method

Reliability, including high electrical and mechanical strength and a narrow strength distribution, is commonly associated with fine-grain ceramics with a narrow

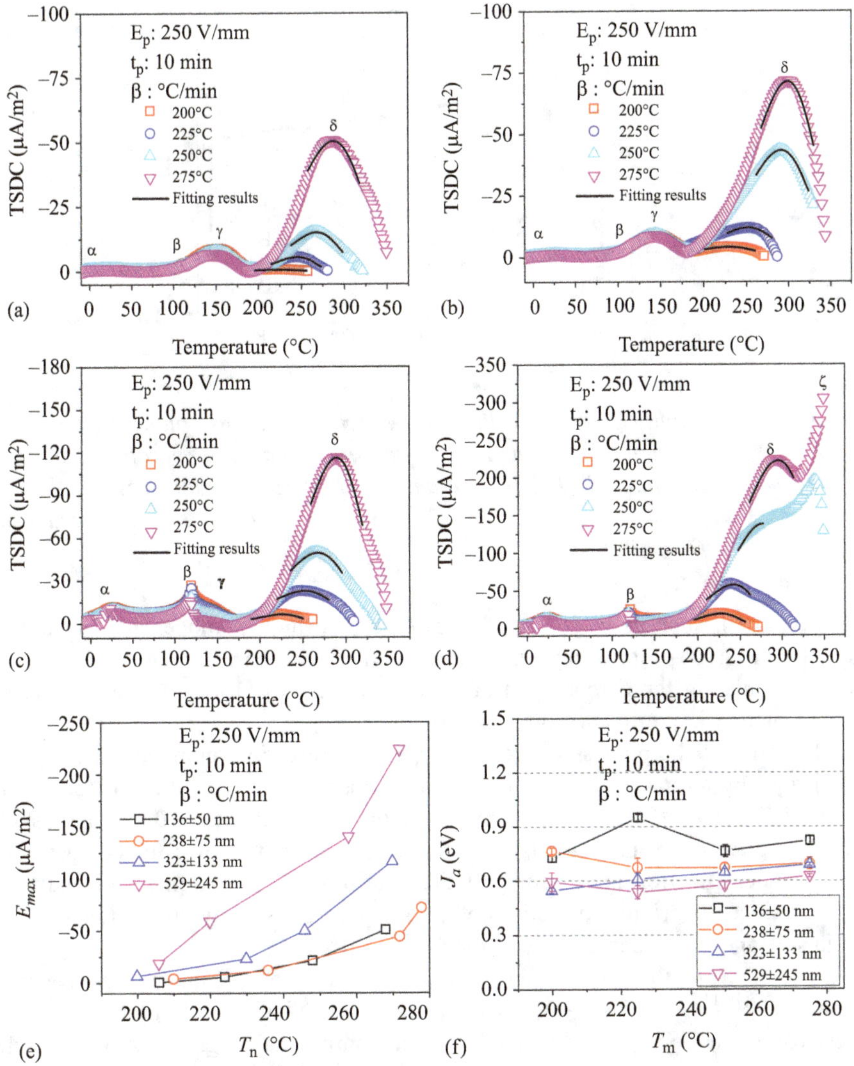

Figure 10.24 TSDC curves for $BaTiO_3$-based ceramics with various grain sizes: (a) BTP1, (b) BTP2, (c) BTP3, and (d) BTP4. (e) Plot of J_{max} of the δ peak versus T_m for the ceramics. (f) Activation energies of the δ peak for the $BaTiO_3$-based ceramics. Reproduced from [92]

grain-size distribution. However, microstructure control in thin-layer multilayer ceramic capacitors (MLCCs) is a major challenge in attempts to improve capacitive volumetric efficiency and high-voltage dielectric properties. The preparation of $BaTiO_3$-based dielectric ceramics with grain sizes below 200 nm using the conventional pressureless sintering method is difficult. While various methods of synthesizing dielectrics with high densification and good grain-size control have

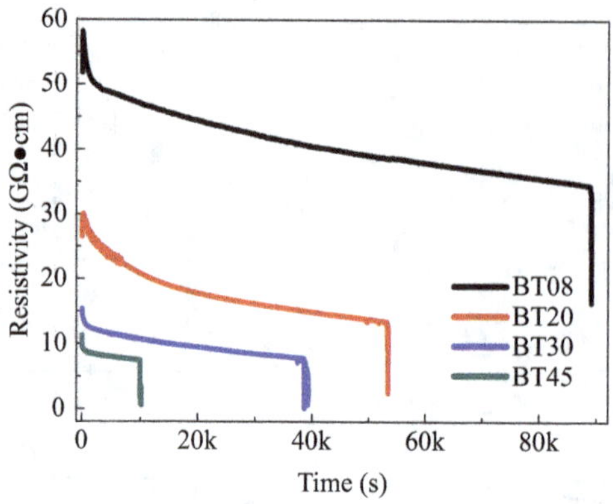

Figure 10.25 Highly accelerated lifetime test (HALT) curves obtained under a sustained DC field (1.2 kV/mm) and at a high temperature (260 °C) for the $BaTiO_3$-based ceramics with various grain sizes (BT08: 118 nm, BT20: 227 nm, BT30: 346 nm, BT45: 462 nm). Reproduced from [99]

been reported, such as preparing BT powders with high homogeneity and tetragonality [103,104], uniformly dispersing additives around BT particles [80,105], and manipulating the composition of dielectric ceramics [106], the sintering process is another key influence on grain growth.

Sintering is the process whereby interparticle pores in a granular material are eliminated by atomic diffusion driven by capillary forces, and the final stage of sintering is always accompanied by rapid grain growth [107,108]. Success at avoiding grain coarsening is related to the degree of control over the competition between densification and grain growth. Good control is extremely difficult to achieve since the driving forces for both processes are inversely proportional to the grain size and hence comparable in magnitude [109].

The TSS method has been successfully utilized to obtain many nanoceramics,[47,48,117] such as Y_2O_3, $BaTiO_3$, ZnO, and Ni–Cu–Zn ferrite. In each of these cases, the TSS method delivers a significantly smaller grain size (not much larger than the size of the particles in the initial powder). Importantly, arresting grain growth, which is fundamentally a random-walk process [110], naturally leaves a more definitive grain-size distribution that mimics the size distribution of the initial powder and suppresses stochastic variations.

In our group, two-step sintering (TSS) in a reducing atmosphere was employed to obtain fine-grain $BaTiO_3$-based ceramics with a core–shell microstructure, a more uniform grain-size distribution, and superior reliability for multilayer ceramic capacitor applications [4,111–113]. Obviously, both the sintering temperature and the soaking time employed during the first and second stages of the sintering process significantly affect the evolution of the microstructure of the ceramic and its associated properties, especially the development of core–shell structure.

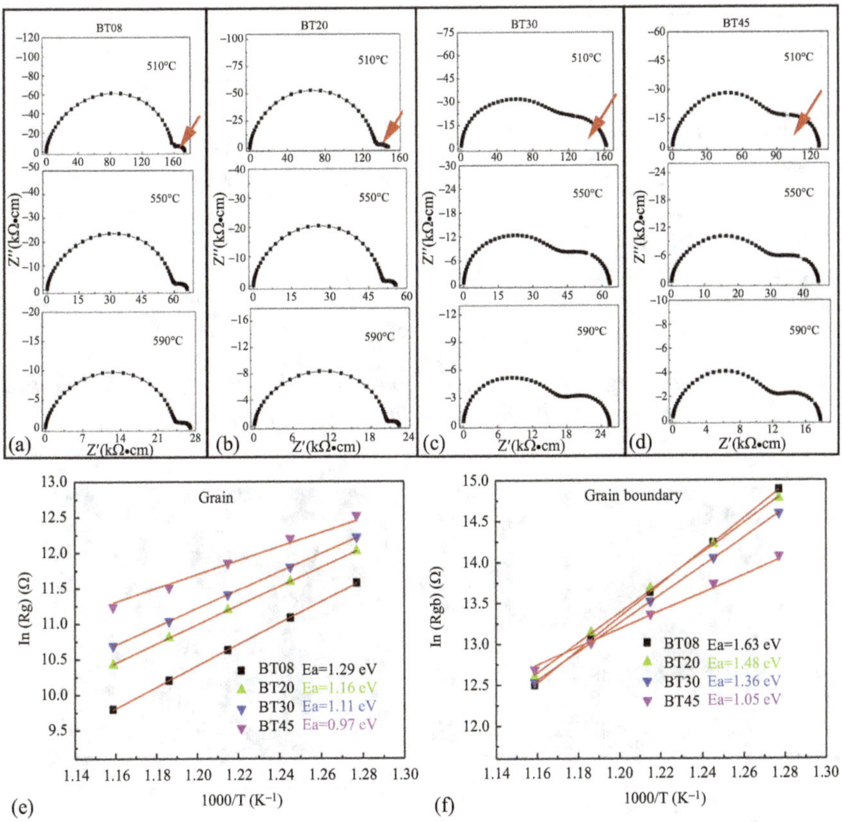

Figure 10.26 Complex impedance data for the ceramic samples: (a) BT08, (b) BT20, (c) BT30, and (d) BT45, obtained at a series of temperatures (the red arrow indicates the Warburg impedance). Also shown are the fitting results for the activation energies of the grains (e) and grain boundaries (f) for grains of different sizes. Reproduced from [99]

Sintering in reducing atmosphere ($H_2/N_2/H_2O$, $P_{O_2} = 10^{-11}$~10^{-13} Pa) was conducted using two schedules of comparable durations sketched in Figure 10.27: (a) conventional sintering (CS) at 1,210 °C for CS-A and CS-B samples and (b) two-step sintering (TSS) with $T_1 = 1250$ °C (1 min) and, after rapid cooling, $T_2 = 1050$ °C (5 h) for TSS-A and TSS-B samples [111]. Before sintering, all samples were held at 600 °C for 2 h to remove binder; after sintering, they were reoxidized under 1~50 Pa O_2 for 2.5 h at 1050 °C and 950 °C, respectively.

10.3.3.1 Microstructure and dielectric properties

TSS led to a smaller average grain size and a narrower grain-size distribution in the sample. TEM images and EDS analysis revealed that each grain exhibited

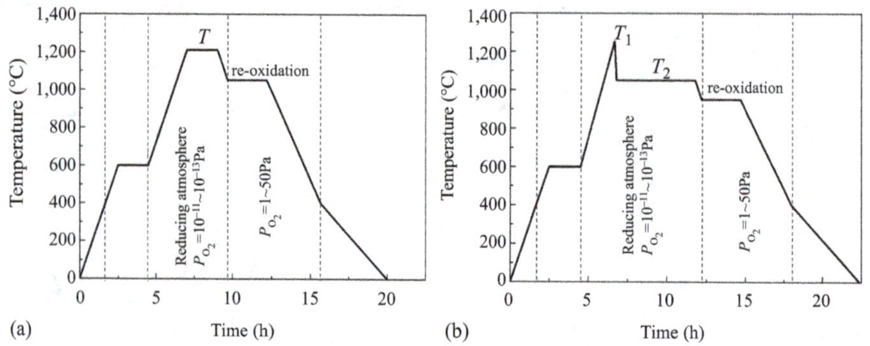

Figure 10.27 Sintering schedules for (a) conventional sintering (CS) and (b) two-step sintering (TSS). Reproduced from [111]

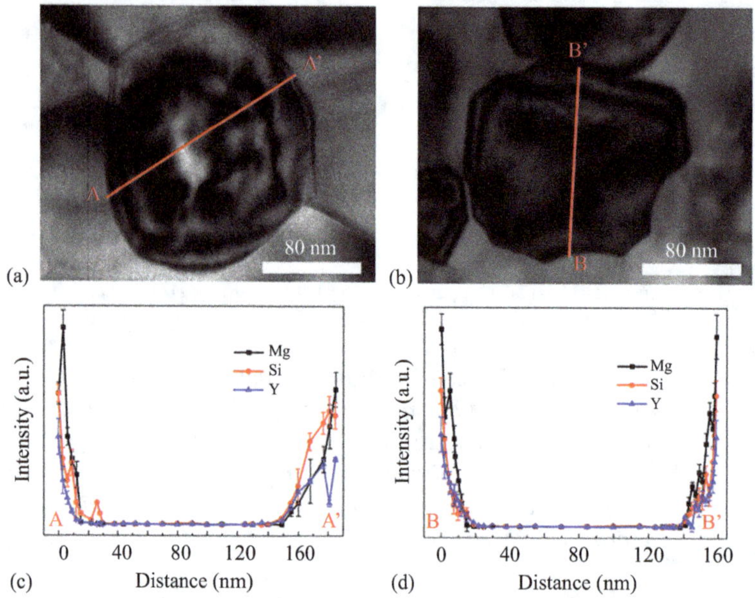

Figure 10.28 TEM images of grains: (a) from CS-generated sample CS-A (note the line AA' across the grain) and (b) from TSS-generated sample TSS-A (note the line BB' across the grain). Also shown are EDS line profiles of Y, Mg, and Si along the lines AA' (c) and BB' (d). Reproduced from [111]

core–shell microstructure, as shown in Figure 10.28. The shell region is enriched in doping elements such as Y, Mg, and Si; a solid solution forms as a paraelectric phase in the shell. The core region is composed of almost pure $BaTiO_3$, which is a ferroelectric phase. The grain from the TSS sample has a thinner shell region than the grain from the CS sample. Thus, considering that the grains shown for the two samples are of a similar size (CS: ~160 nm, TSS: ~145 nm), the grain from the TSS sample must have a larger core region than the grain from the CS sample. Since the Curie temperature for $BaTiO_3$ is about 120 °C, the TSS sample with the higher ratio of core region exhibits better temperature stability at high temperature (see Figure 10.29). The difference between the dielectric properties of the samples prepared using different sintering methods can be explained by the shift of the broaden dielectric peak controlled by the diffuse phase transition (DPT) of the shell region [112]. The dielectric constant is the sum of the core and shell regions. The shell region often shows a diffusive dielectric peak that shifts to a lower temperature as the dopant concentration increases in the shell. A schematic comparing CS and TSS samples in terms of the temperature dependence of the dielectric constant, including the contribution of the diffuse phase transition (DPT: a series of dielectric peaks over a given temperature range), is presented in Figure 10.30. Compared to the CS sample, the TSS sample has a thinner shell with higher dopant concentrations, leading to the movement toward the low temperature of the diffusive dielectric peak corresponding to the shell and thus improving the temperature stability of the TSS sample compared to the CS sample at low temperature. Meanwhile, the higher core:shell ratio of the TSS sample results in enhanced stability at high temperature compared to the CS sample. These results demonstrate that understanding the evolution and regulation of the core–shell structure is beneficial to the further development of nanograin dielectric ceramic materials and next-generation ultrathin MLCC devices.

10.3.3.2 Super-reliability and mechanism analysis

A fundamental reliability concept is that failure is a statistical event, and that breakdown and degradation initiate at the weakest link. Such a link is likely to result from statistically extreme variabilities in grain size and other microstructural/microchemical features on which failure-inducing defects/fields reside, accumulate,

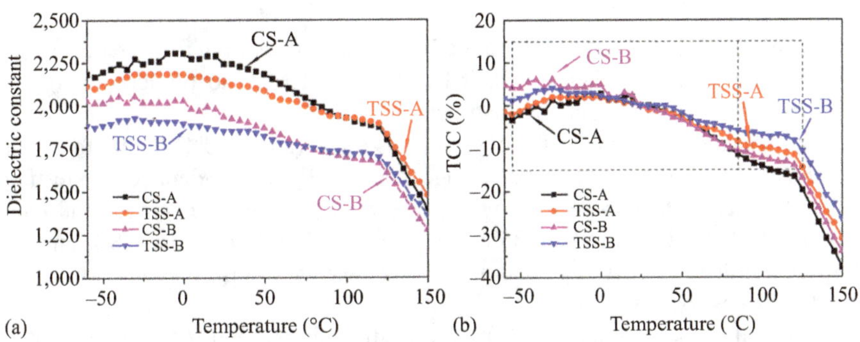

Figure 10.29 Temperature dependence of (a) the relative dielectric constant and (b) the temperature coefficient of capacitance (TCC) for TSS and CS samples. Reproduced from [111]

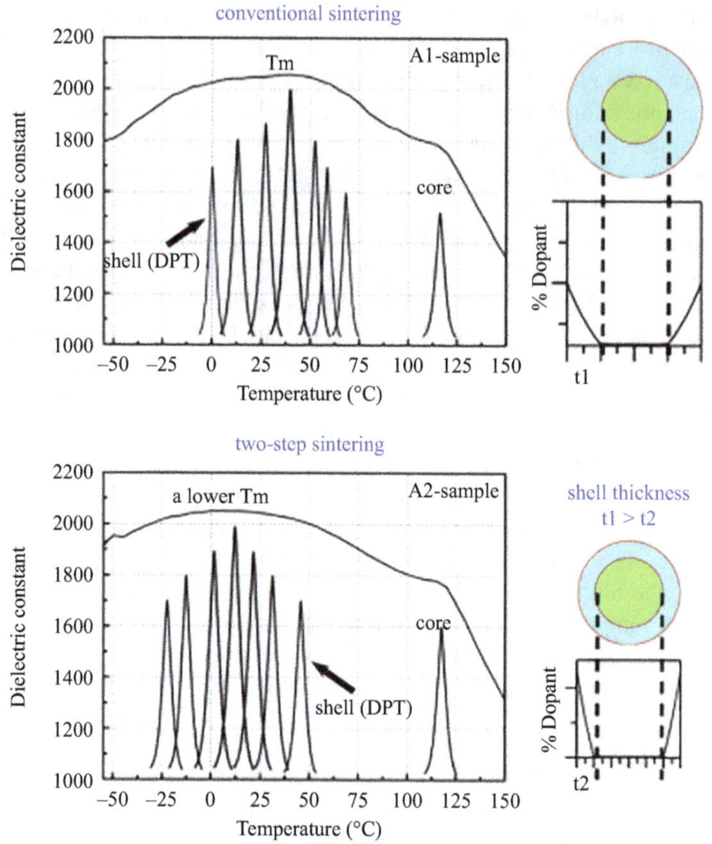

Figure 10.30 The schematic of temperature dependence of the dielectric constant for samples prepared by different sintering methods with diffuse phase transition (DPT) in shell region. Reproduced from [112]

or concentrate. This realization has naturally motivated the search for better ceramics through processing; in other words, it has driven attempts to produce fine-grain ceramics with a narrow grain-size distribution.

The breakdown strength of TSS samples is much (about 50%) higher than that of CS samples, and it has less percentage scatter. Highly accelerated life testing (HALT) indicates that the resistivity of CS samples gradually degrades and then fails catastrophically after about 35 h (see Figure 10.31). In contrast, TSS samples show very little resistivity degradation; the time taken to fail is around four times as long (155 h) as for CS samples.

It is well known that the resistivity degradation observed in HALT is correlated with oxygen vacancy ($V_O^{\cdot\cdot}$) migration in $BaTiO_3$ and related perovskites [114]. Perhaps the best evidence for the presence of $V_O^{\cdot\cdot}$ is provided by the TSDC data in Figure 10.32, which shows an inverse correlation between the strengths of the α and

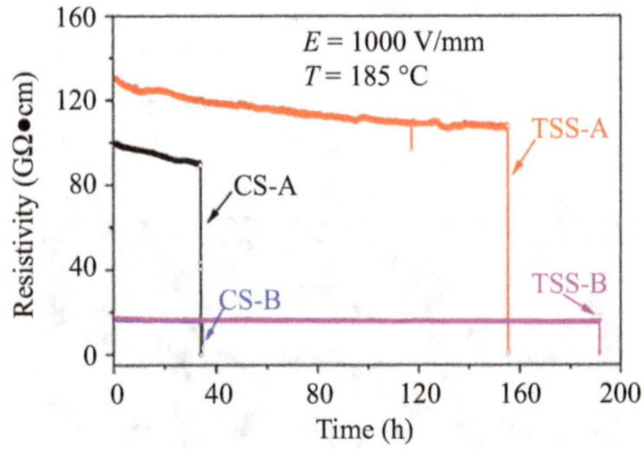

Figure 10.31 Evolution of electrical resistivity under accelerated DC stressing. Reproduced from [111]

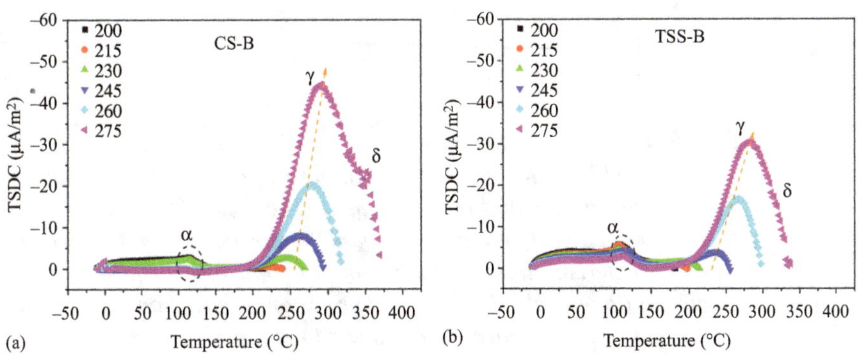

Figure 10.32 Thermally stimulated depolarization current (TSDC) for samples (a) CS-B and (b) TSS-B, which were identically poled and tested except that different poling temperatures were applied (200, 215, 230, 245, 260, and 275 °C). The characteristic α, γ, and δ peaks are also indicated. Reproduced from [111]

γ peaks. Oxygen vacancies cause polarization fatigue in ferroelectrics [115] and suppress dielectric relaxation in relaxors [116]. Hence, based on the inverse α–γ correlation, the γ peak can be associated with $[V_O^{\cdot\cdot}]$, the $V_O^{\cdot\cdot}$ concentration (likewise, the δ peak can be associated with $[V_O^{\cdot\cdot}]$ across grain boundaries). Since the strength of the γ peak is an indicator of $[V_O^{\cdot\cdot}]$, it is clear from Figure 10.32 that TSS leaves fewer oxygen vacancies in the ceramic than CS does. It was speculated that this could be because TSS samples have thiner grain shells, which may alter $[V_O^{\cdot\cdot}]$. To check this possibility and link it to the HALT results, a scaling analysis was performed.

Noting that grain boundaries block carrier migration, these boundaries can be regarded as double Schottky barriers. Further, analogously to electrochemical devices [117], we can view the dielectric–electrode assembly as an RC circuit for which the characteristic time τ for charge (i.e., oxygen) redistribution may be

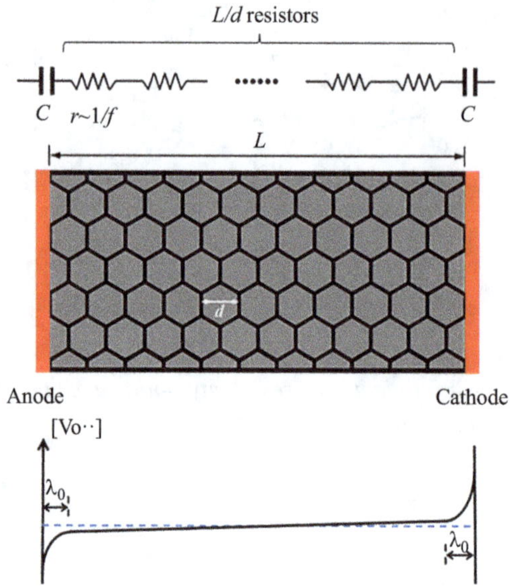

Figure 10.33 Schematic of a device with two oxygen blocking electrodes separated by a distance L that contains a polycrystalline ceramic with a grain size of d (middle panel). The dynamics of oxygen vacancy redistribution (bottom panel) approach a steady state with a characteristic time s, which is the RC time of the equivalent circuit (top panel). R contains L/d resistors r, each controlled by the oxygen vacancy transmission factor f for crossing a set of grain boundaries and adjacent shell regions. The resulting charge accumulation at each electrode is represented by C, which corresponds to a capacitor with (double) layer thickness λ_D *(bottom panel). Reproduced from [111]*

estimated from the RC constant, that is, $\tau \sim RC$. As shown in Figure 10.33, this simple relation allows us to gain useful insight into the HALT lifetime without the need to numerically solve any equation. To estimate the RC time, we first assume that oxygen diffusion across the shell/grain boundary region is the rate-limiting step. The oxygen conductance R^{-1} is thus proportional to (a) the oxygen flux, which scales with $[V_O^{\cdot\cdot}]$ and the vacancy diffusivity D in the core (where the dopants are scarce), and (b) the transmission factor f of $V_O^{\cdot\cdot}$ across the grain boundary [114,118] and the shell region. Since there are approximately L/d grain boundaries/shells to cross in a sample with a thickness L and a grain size d, we can write [117]

$$R \sim \frac{kTL}{dfD[V_O^{\cdot\cdot}]e^2} \tag{10.4}$$

where e is the charge on an electron and kT has its usual meaning. The relevant capacitance C for internal charge (oxygen) accumulation is largely localized within the two serial Debye double-layer capacitors near the two electrodes. Thus, $C \sim \varepsilon/\lambda_D$ in the small voltage limit ($eV \ll kT$), where ε is the dielectric constant and the layer thickness λ_D is given by [117]

$$\lambda_D \sim \left(\frac{\varepsilon kT}{z^2 e^2 [V_O^{\cdot\cdot}]}\right)^{1/2} \tag{10.5}$$

where $z = 2$ is the charge on an oxygen vacancy. This gives

$$\tau \sim \left(\frac{\varepsilon z^2 kT}{e^2 [V_O^{\cdot\cdot}]}\right)^{1/2} \cdot \frac{L}{dfD} \tag{10.6}$$

Although (10.4) is obtained in the small voltage limit, we expect its parametric dependence to hold qualitatively even for large voltages. Since the redistribution dynamics should scale with the characteristic time τ, we can use (10.4) to quantitatively assess the HALT lifetime, which corresponds to a certain oxygen vacancy redistribution state in which there is excessive electronic current. Combining relevant data, we can conclude that neither the modest variation in $[V_O^{\cdot\cdot}]$ (less than three times according to the TSDC data), nor the 10% difference in grain size, nor the likely small variation in the diffusivity D of $V_O^{\cdot\cdot}$ in the core, nor the small variation in the dielectric constant of the ceramic (less than 20%) can account for the four- to five-fold improvement in the HALT lifetimes of the TSS ceramics over the CS ceramics. Hence, the transmission factor must play the dominant role: f must be much smaller in TSS ceramics. This can be attributed to the higher local dopant concentrations in the shell regions of the TSS ceramics compared to the CS ceramics, which imply more dopant–$V_O^{\cdot\cdot}$ associations that immobilize $V_O^{\cdot\cdot}$ and greater dopant segregation at the grain boundary, forming a stronger barrier against $V_O^{\cdot\cdot}$ transmission.

Overall, compared to the CS method, the TSS method leads to smaller and more uniform grains, improves the dielectric temperature stability, minimizes reduction-induced point defects (mainly $V_O^{\cdot\cdot}$ and associated electron carriers), and reduces resistance/strength degradation, implying that it will be an important tool for the development of next-generation ultrathin MLCC devices.

10.3.4 Large-capacity ultrathin-layer BME-MLCCs

High-performance nanograin ceramic dielectric materials and related nanograin control technologies have been applied to prepare large-capacity multilayer ceramic capacitors with ultrathin layers, yielding remarkable results. Our group applied the chemical coating method to develop a high-performance, nonreducible, temperature-stable nanograin (grain size: 120–150 nm) BT-based ceramic, which we used to prepare small-sized, high-capacity, and ultrathin-layer BME-MLCC devices cooperating with Shenzhen Eyang Technology Development Co., Ltd. The resulting three kinds of temperature-stable BMEMLCC devices C0201, B0402, and A0603 exhibit high capacities of 0.22, 2.2, and 10 μF, respectively. The average dielectric layer thicknesses of these three MLCCs are 0.8, 1.0, and 1.2 μm, respectively. All three devices meet the requirements of X5R-type MLCCs. A SEM image of BMEMLCC- C0201 is shown in Figure 10.34(a). Also, since the chemical-coated powders show highly uniform core–shell structure and a very small and uniform grain size, the devices prepared from the chemical-coated powders present excellent performance with similar capacities to commercial devices, but much smaller capacity dispersions. Figure 10.34(b) provides a comparison of C0201 with two commercial products that were developed to the same specification by one of the world's largest electronics companies, Samsung Electro-Mechanics. The capacity dispersion is an important influence on the product qualification rate, and the current pass rate is as high as 90%–95%, especially for large-capacity, ultrathin-layer BME-MLCCs. Due to the excellent microstructure of the chemical-coated powders, the loss characteristics of the resulting devices are much better than those of conventional devices, even the devices from Samsung Electro-Mechanics, as shown in Figure 10.34(c). Tables 10.3 and 10.4 list basic parameters of the X5R-type BME-MLCCs developed in our work and compares them with commercial MLCCs. We can conclude that high-performance, nonreducible, nanograin $BaTiO_3$-based ceramics optimized via material composition and the chemical coating method are highly advantageous for preparing next-generation ultrathin-layer BME-MLCCs.

It is reported that the interfacial properties of MLCCs with Ni electrodes significantly influence the lifetimes of these devices [119]. Thus, microstructure and

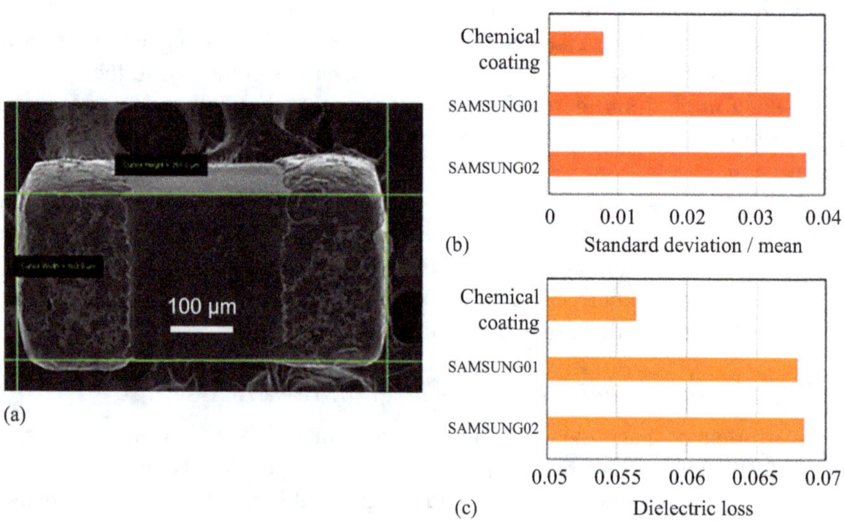

Figure 10.34 (a) SEM image of the BME-MLCC C0201; (b) comparison of the capacity dispersion of devices, and (c) comparison of the loss of devices

Table 10.3 Basic parameters of the X5R-type BME-MLCCs developed in our work

	A0603	**B0402**	**C0201**
Rated voltage	6.3 V	6.3 V	6.3 V
Capacity (μF)	7.241	2.062	0.201
Electrostatic capacity tolerance	1.2~3.2%	0.8~1.9%	2.2~5.2%
Loss	1.56%	2.98%	3.8%
Resistance(MΩ)	3050	2800	2150

Table 10.4 Detailed comparison of the specifications of X5R-type BME-MLCCs developed in our work with those of commercial X5R-type MLCCs

	A0603	**Samsung**	**B0402**	**Murata**
Product code	106	CL10A106KQ8NNNB	225	GRM155R60J225KA01D
Dimensions	1.6 mm × 0.8 mm × 0.8 mm		1.0 mm × 0.5 mm × 0.5 mm	
Rated voltage	6.3 V		6.3 V	
Capacity (μF)	7.241	7.592	2.062	1.702
Loss	1.56%	4.37%	2.98%	6.08%
TCC	X5R(−55 to 85 °C, TCC $\leq \pm 15\%$)		X5R(−55 to 85 °C, TCC $\leq \pm 15\%$)	
Dielectric layer thickness (μm)	1.2	1.1	1.0	1.8
Electrode layer thickness (μm)	0.7	0.8	0.8	0.7
Number of layers	400	400	270	160
Grain size (nm)	140	175	130	240

compositional analyses were carried out to investigate diffusion at the interface between the Ni electrode and the $BaTiO_3$-based dielectric layer in ultrathin-layer BME-MLCCs (layer thickness = 1 μm) [18]. As presented in Figure 10.35(a), there is a transition zone for the Ni, Ba, and Ti contents around the BT/Ni interface, with a thickness of about 15 nm. The amount of diffusion between the Ni electrode to BT is different from the amount between BT and Ni electrode. It is easier for Ni to diffuse into the $BaTiO_3$ dielectric layer than for $BaTiO_3$ to diffuse into the Ni electrode in the samples, which indicates that the diffusion of Ni is the dominant factor in the interfacial diffusion behavior of BME-MLCCs. To check this, HRTEM experiments were conducted. The incorporation of Ni does not change the crystal structure of BT, but it does introduce lattice distortion. According to Figure 10.35(b), the distance between the (011) planes in BT is 0.263 nm and 0.258 nm at different sites to the interface. The slightly larger interplanar distance near the interface is due to the lattice distortion caused by the diffusion of Ni into the BT phase. In contrast (see Figure 10.35(c)), the distance between (1–11) planes in the Ni electrode is virtually the same at different sites to the interface, indicating

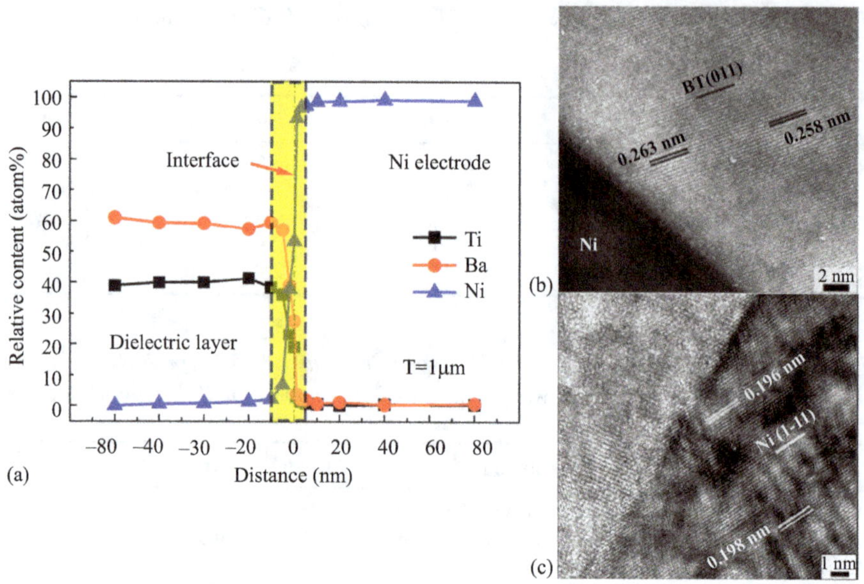

Figure 10.35 Results of an EDS analysis of the interface between the Ni electrode layer and the $BaTiO_3$ dielectric layer (a), and HRTEM images of the interface: (b) dielectric layer and (c) Ni electrode. Reproduced from [18]

that the diffusion of BT into Ni is much weaker. The incorporation of Ni into the BT lattice facilitates the creation of oxygen vacancies, which can decrease electron conduction. However, migration of oxygen vacancies can lead to BME-MLCC degradation under an applied electric field. Therefore, controlling the diffusion of Ni into the ceramic layer is key to the production of reliable BME-MLCCs, especially as the active layer becomes thinner.

10.4 MLCCs for energy-storage applications

10.4.1 Advanced energy-storage MLCCs

New energy technology is undergoing rapid development to overcome the growing environmental problems and meet the idea of sustainable development. As a result, environmentally friendly, and low-cost energy-storage devices are in high demand. In contrast to batteries and electrochemical capacitors, dielectric ceramics possess ultrahigh power density and can release the stored energy very rapidly (within in a microsecond) to create intense power pulses. This fast charge–discharge capability is crucial in pulsed power systems such as femtosecond lasers, medical defibrillators, hybrid electric vehicles, high-voltage pulse-power circuits, and kinetic energy weapons [24–28]. However, the energy density of a dielectric capacitor is much lower than that of an electrochemical capacitor or battery.

In recent decades, much research effort has been directed into improving the energy-storage performance of ferroelectric ceramics. The utilization of MLCCs in energy-storage applications is of high interest since MLCCs possess high volumetric capacitance, and are suitable for surface mount due to their special structures [48–51]. Furthermore, the energy-storage density of MLCCs has been enhanced significantly due to the significantly enhanced breakdown strength, which makes them one of the most promising candidates for high-power energy-storage system.

Relaxor ferroelectrics and antiferroelectrics are considered among the most promising dielectric materials used in the fabrication of MLCCs because of their high energy efficiencies. The dielectric layer thickness of an energy-storage MLCC is always less than 20 μm, so they can be operated under a relatively high applied field. Low energy efficiency may lead to high heat generation, accelerating the failure of MLCCs. In this section, we consider various energy-storage MLCCs, including a relaxor ferroelectric BT-BNZT-based energy-storage MLCC [120], a relaxor antiferroelectric BNT-based energy-storage MLCC [121], and a chemically inhomogeneous but electrically homogeneous BF-BT-based energy-storage MLCC [122], in order to highlight their advantages.

10.4.1.1 BT-BZNT-based energy-storage MLCC

Due to the continuous push for new high energy-density and efficiency dielectrics, $Bi(M)O_3$-modified $BaTiO_3$ (BT-B*M*O) ceramics, where *M* represents a net-trivalent *B* site occupied by one or more species, have recently received a great deal of attention [123]. Dielectrics with this composition exhibit similar weakly coupled relaxor behavior. As an example, the relaxor ferroelectric material $0.87BaTiO_3$–$0.13Bi(Zn_{2/3}(Nb_{1-x}Ta_x)_{1/3})O_3$ (BT-BZNT) was developed in our lab. This material typically presented a high discharge energy density of 1.44 J cm^{-3} and an efficiency of 92.5% at an applied field of 218 kV/cm. Due to its excellent energy-storage performance, we have fabricated BT-BZNT-based energy-storage MLCCs (1210-size: ~3.2 mm × 2.7 mm) via the Roll-to-Roll tape-casting technique [120].

Cross-sectional SEM images of this MLCC are presented in Figure 36(c) and (d), which shows that this MLCC contains two active dielectric layers with a layer thickness of ~ 5 μm. From Figure 10.36(a), it is clear that the BT-BZNT dielectric is densely sintered with an average grain size of ~434 nm. Most importantly, the internal electrodes are 60Ag/40Pd, which is far less expensive than Pt.

The BT-BZNT-based energy-storage MLCC exhibits typical relaxor ferroelectric behavior, as shown by the "slim" hysteresis loops in Figure 10.37(a). The discharge energy density and energy efficiency as functions of the electric field (extracted from the hysteresis loops) are plotted in Figure 10.37(b). As expected, a large discharge energy density of 10.12 J cm^{-3} and a high energy efficiency of 89.4% are obtained at an applied electric field of 104.7 MV m^{-1}. The discharge energy density obtained with this BT-BZNT-based MLCC (>10 J cm^{-3}) is much higher than those of bulk ceramics (< 4 J cm^{-3}) and comparable with those of polymer-based dielectrics (~ 20 J cm^{-3}). As mentioned in the introduction to this chapter, the low polarization of polymer-based dielectrics induces low dielectric

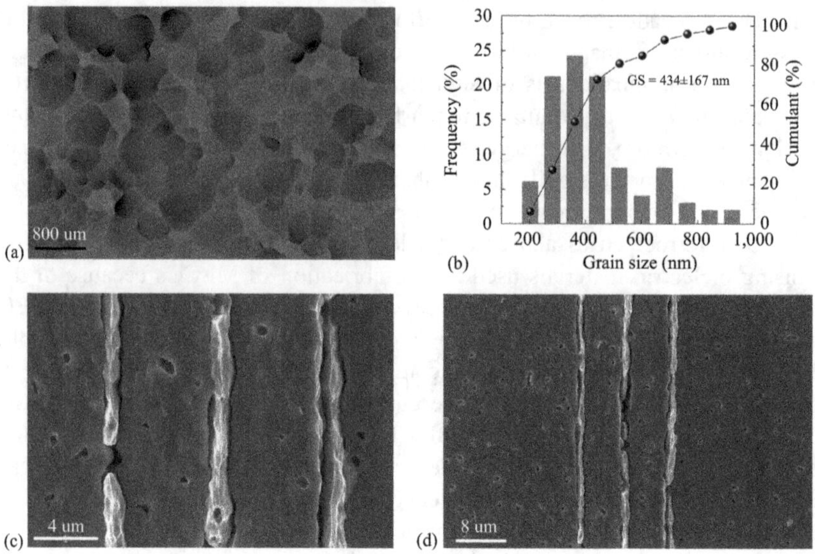

Figure 10.36 (a) SEM image of the surface of the BT-BZNT-based energy-storage MLCC. (b) Grain-size distribution derived from (a). (c–d) SEM images of a cross-section of the BT-BZNT-based energy-storage MLCC. Reproduced from [120]

constant and low energy efficiency, limiting their applicability at high temperatures. They also show reduced breakdown strength and much greater energy loss at high than at low temperatures. In contrast, the BT-BZNT-based energy-storage MLCC based on a relaxor ferroelectric shows superior thermal stability, with a minimal variation ($< \pm 5\%$) in discharge energy density over the broad temperature range -75 °C to 150 °C at 40 MV m^{-1}, as seen in Figure 37(c) and (d).

10.4.1.2 NBT-SBT-based energy-storage MLCC

The utilization of antiferroelectric materials is thought to be an effective way of improving the discharge energy density of dielectric ceramics, since antiferroelectric dielectrics exhibit unique double hysteresis loops with high recoverable polarization during charging and discharging processes, as shown in Figure 10.38 (a). However, an obvious drawback of antiferroelectrics is their high energy dissipation, which results from the large polarization and electric field hysteresis caused by the first-order antiferroelectric/ferroelectric phase transition. Fei Li's group [121] designed a relaxor antiferroelectric material (i.e., a dielectric that combines the features of both relaxor ferroelectrics and antiferroelectrics) with high energy density and high energy efficiency, as shown in Figure 10.38(b).

To realize this relaxor antiferroelectric dielectric, they selected $(Na_{0.5}Bi_{0.5})TiO_3$ (NBT) as the base material, as it exhibits both relaxor ferroelectric

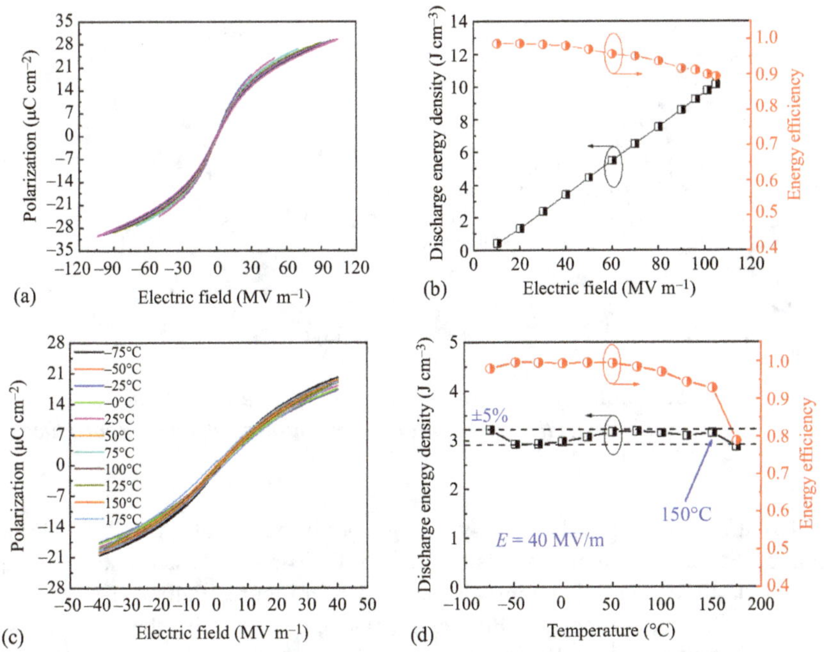

Figure 10.37 (a) Hysteresis loops of the BT-BZNT-based energy-storage MLCC measured at different electric fields. (b) Calculated discharge energy density and efficiency of the BT-BZNT-based energy-storage MLCC measured at room temperature and 10 Hz. (c) Hysteresis loops of the BT-BZNT-based energy-storage MLCC measured at a maximum field of 40 MV m^{-1} and various temperatures. (d) Temperature stability of the BT-BZNT-based energy-storage MLCC measured at a maximum electric field of 40 MV m^{-1} and 10 Hz. Reproduced from [120]

characteristics and antiferroelectric characteristics. The relaxor ferroelectric $(Sr_{0.7}Bi_{0.2})TiO_3$ (SBT) was added to NBT to decrease the polarization hysteresis. The *P*–*E* loops for the resulting NBT-*x*SBT ceramics with $x = 0.30$–0.50 are presented in Figures 39(a)–(e). Under an applied electric field of 10 MV m^{-1}, both the maximum polarization and the hysteresis decrease with increasing SBT content. When $x \geq 0.45$, the *P*–*E* loops show almost no hysteresis. Considering the need for both low hysteresis and high polarization, the NBT-0.45SBT ceramic was chosen for further study. The unipolar hysteresis loop and the current–electric field curve for the NBT-0.45SBT ceramic is presented in Figure 10.39(f). This ceramic shows an enhanced breakdown strength of up to 20 MV m^{-1}, a high energy-storage density of 2.5 J cm^{-3}, and an excellent energy efficiency of 95%, suggesting that this material would be very useful for the fabrication of energy-storage MLCCs.

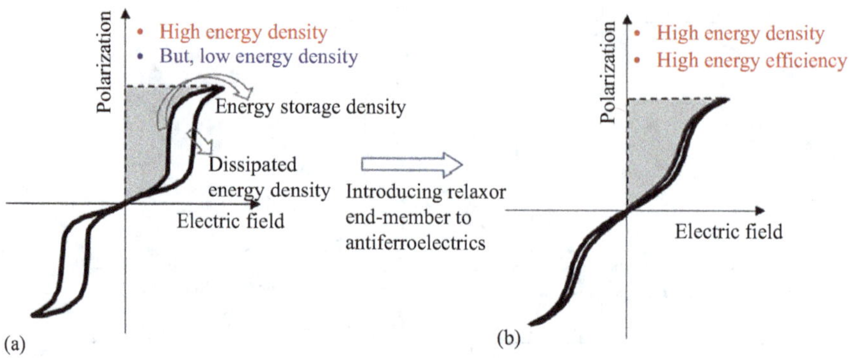

Figure 10.38 Schematic of the energy-storage properties of (a) an antiferroelectric and (b) a relaxor antiferroelectric. Reproduced from [121]

To further enhance the energy-storage performance of the NBT-0.45SBT ceramic, Fei Li's group also fabricated NBT-0.45SBT-based energy-storage MLCCs via the tape casting approach, as shown in Figure 10.40(a). Based on the SEM and EDX images shown in Figures 40(b)–(d), it is clear that the thicknesses of the ceramic layers and the Pt electrodes are ~20 μm and 1 μm, respectively.

The energy-storage performance of the NBT-0.45SBT-based MLCCs is shown in Figure 10.41. Due to the significant decrease in dielectric layer thickness, the breakdown strength is enhanced by up to 72 MV m^{-1}, leading to a high energy density of 9.5 J cm^{-3} and a corresponding energy efficiency of 92%. The temperature stability of the energy density and efficiency of the NBT-0.45SBT-based MLCC was measured under an applied electric field of 35 MV m^{-1}, approximately half the breakdown strength, to ensure that the MLCC did not break down at high temperature. The energy density and efficiency are displayed as functions of temperature in Figure 10.41(b); the variations in energy density and efficiency are observed to be less than 10% over the temperature range from −60 °C to 120 °C, indicating satisfactory thermal stability.

10.4.1.3 BF-BT-based energy-storage MLCC

Aside from the improvements in the *P–E* relation achieved using a relaxor ferroelectric or relaxor antiferroelectric as the dielectric, Dawei Wang's group [122] targeted high dielectric breakdown strength by designing ceramics with chemically inhomogeneous but electrically homogeneous microstructure: $(0.7-x)BiFeO_3$–$0.3BaTiO_3$ $x Nd(Zn_{0.5}Zr_{0.5})O_3$ (BF-BT-*x*NZZ). Electrical homogeneity (but not chemical homogeneity) helps to optimize the dielectric breakdown strength by suppressing the conductive pathways that lead to breakdown at high field.

Impedance spectroscopy of these BF-BT-*x*NZZ ceramics ($0.00 \leq x \leq 0.10$) was instigated to study their electrical properties and in particular the role of the electrical microstructure in optimizing energy storage. Combined plots of Z'' and M'' are often more instructive than Z^* plots when probing electrical heterogeneity

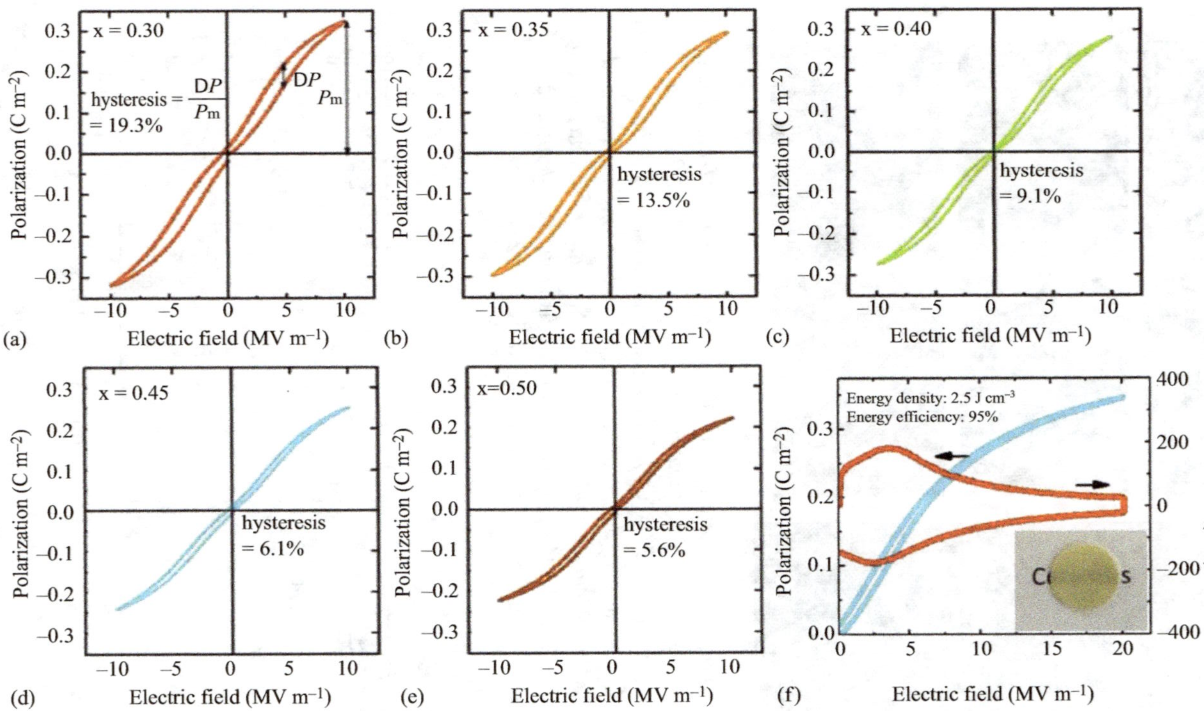

Figure 10.39 *(a–e) Bipolar P–E loops for NBT-xSBT ceramics with x = 0.30–0.50, as measured at 10 Hz. (f) The unipolar hysteresis loop and the current–electric field curve for the NBT-xSBT ceramic with x = 0.45, as measured at 10 Hz. A photograph of the NBT-0.45SBT ceramic sample is also shown in the bottom-right corner. Reproduced from [121]*

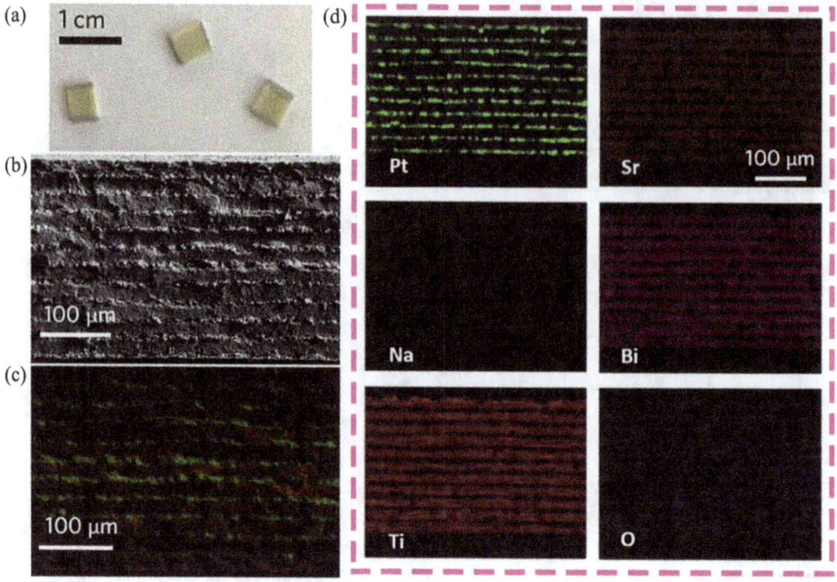

Figure 10.40 *Photograph and microstructure of the NBT-0.45SBT-based energy-storage MLCC. (a) Photograph of some NBT-0.45SBT-based energy-storage MLCCs following fabrication. (b) SEM image of the MLCC, showing the ceramic and electrodes. (c) Superimposed SEM/EDS images of the elements Pt, Sr, Na, Bi, Ti, and O in the MLCC. (d) SEM/EDX images of the distributions of specific elements in the MLCC. Reproduced from [121]*

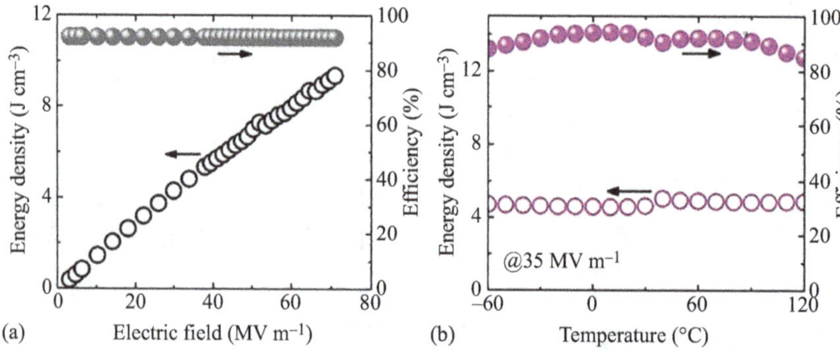

Figure 10.41 *(a) Energy density and efficiency of the NBT-0.45SBT-based energy-storage MLCC, as extracted from P–E loops. (b) Energy density and efficiency of the NBT-0.45SBT-based energy-storage MLCC as functions of temperature. Reproduced from [121]*

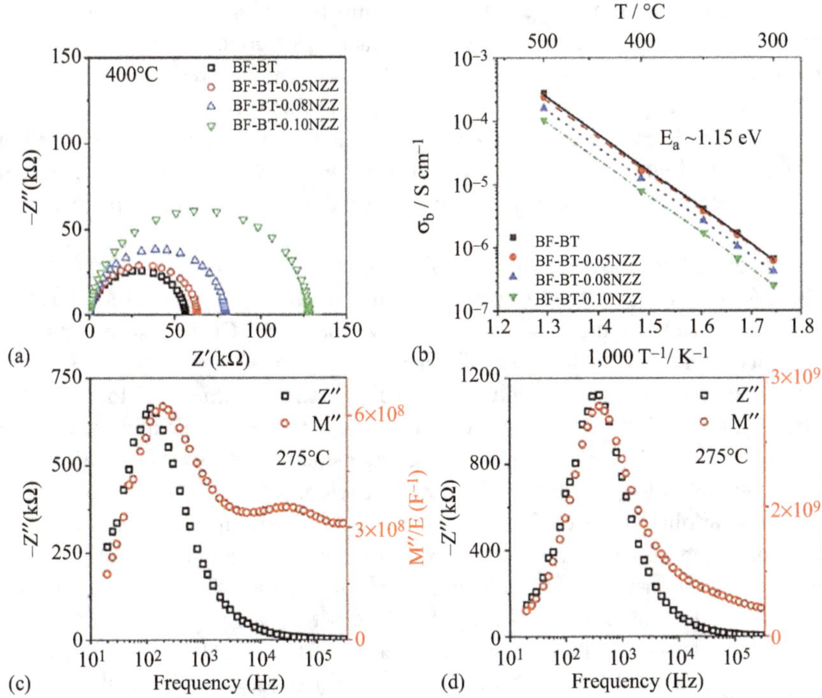

Figure 10.42 (a) Z plots for BF-BT-xNZZ ceramics at 400 °C. (b) Arrhenius plots of total electrical conductivity for BF-BT-xNZZ ceramics. Also shown are combined Z″ and M″ spectra obtained at 275 °C for (c) x = 0.00 and (d) x = 0.08. Reproduced from [122]*

in ceramics. The M'' spectra are dominated by electroactive regions with small capacitances (typically grain-type responses), whereas the Z'' spectra (and Z^* plots) are dominated by the most resistive electroactive region(s) in the ceramic (typically grain-boundary-type responses or resistive shell regions in core–shell microstructures). In general, each Debye peak in the Z'' and M'' spectra can be considered to be a parallel resistor–capacitor (RC) element. If multiple Debye peaks are observed in the Z'' and M'' spectra, they can be modeled using an equivalent circuit based on two parallel RC elements connected in series.

Figure 10.42(a) shows complex impedance plane (Z^*) plots for various BF-BT-xNZZ ceramics at 400 °C, in which a single arc is observed for each composite. As x increases, the total resistivity of the ceramic (extracted from the low-frequency intercept of the Z' axis) gradually increases, indicating improved intrinsic breakdown strength. Arrhenius plots of the total electrical conductivity for BF-BT-xNZZ ceramics are shown in Figure 10.42(b), with an activation energy for conduction (E_a) of ~1.15 eV observed for all compositions. Two peaks are observed in the M'' spectrum for $x = 0.00$, while only a single peak is observed in the Z'' and M'' spectra for $x = 0.08$ at 275 °C, as shown in Figure 10.42(c) and (d), which means that the electrical homogeneity of the grains is much greater at $x = 0.08$ than at $x = 0.0$. The presence of a large volume fraction of conductive cores in these ceramics is undesirable for high-voltage applications, as this results in a stronger effective electrical field across the insulating grain shells, making electrical breakdown more likely than in insulating grains that contain few or no conducting cores (for equivalent grain-sized ceramics). For $x = 0.08$, the improved electrical homogeneity coupled with higher resistivity provides a more difficult pathway for conduction at high field.

Considering the above results for BF-BT-xNZZ ceramics, BF-BT-0.08NZZ-based MLCCs were fabricated to explore their potential for energy-storage applications. As shown in Figure 10.43, these energy-storage MLCCs consisted of seven active dielectric layers, each with a thickness of ~16 μm.

The energy density and efficiency, as calculated from the *P–E* loops of the $x = 0.08$ MLCC in Figure 10.44(a), are shown as functions of the applied electric field in Figure 10.44(b). The optimal discharge energy density (W_{dis}) of 10.5 J cm^{-3} is obtained, along with a high energy efficiency of 87%, for the $x = 0.08$ MLCC at an applied field of 700 kV cm^{-1}. The temperature dependence of the energy-storage properties of this MLCC was also evaluated, and the variation in the discharge energy density across the temperature range from 25 to 150 °C at 300 kV cm^{-1} was found to be less than ±15%, which is critical for practical applications. However, the energy efficiency is much lower than 70% at 150 °C. The discharge energy density was still lower than that reported for oxide thin films ($W_{dis} > 20$ J cm^{-3} at > 2 MV cm^{-1}), although we note that for an equivalent electric field (0.7 MV cm^{-1}), the 0.62BF-0.3BT-0.08NZZ-based energy-storage MLCC performs better. Mass market production of MLCCs, however, favors thick over thin film technology due to the ease with which it can be scaled up and its significantly lower cost.

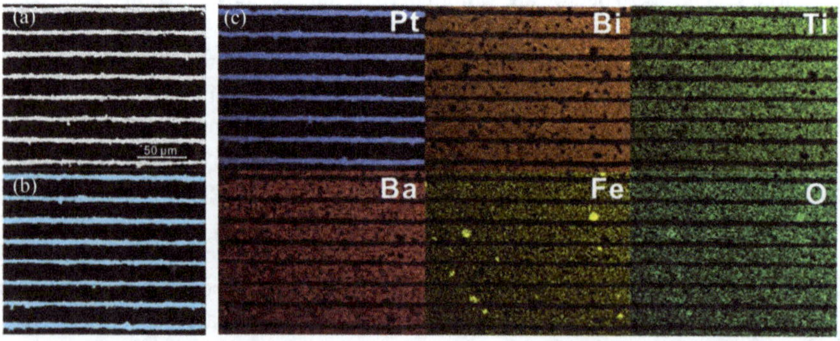

Figure 10.43 (a) Backscattered electron (BSE) cross-section micrographs of a BF-BT-xNZZ-based energy-storage MLCC with x = 0.08. (b) Superimposed EDS mapping images of the elements Pt, Bi, Ti, Ba, Fe, and O. (c) EDS mapping images of individual elements. Reproduced from [122]

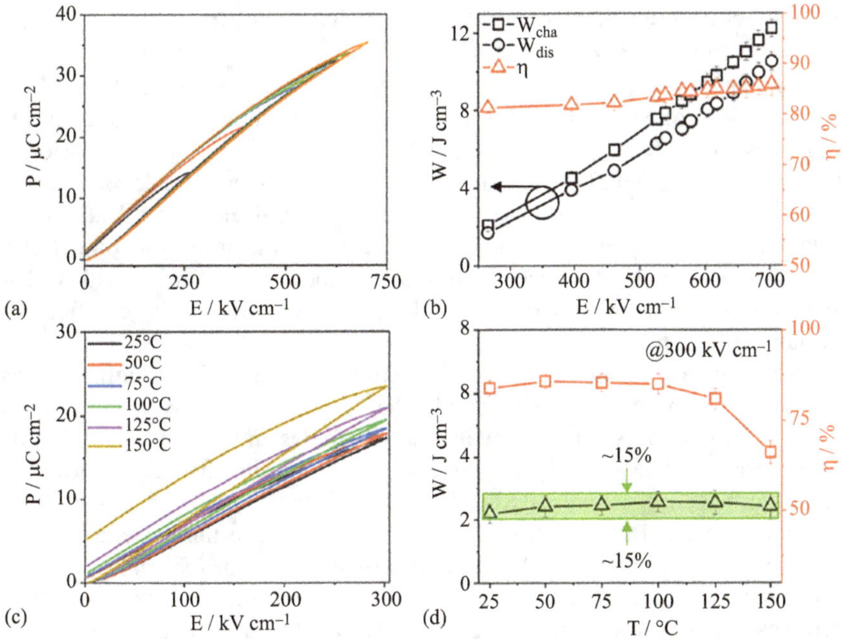

Figure 10.44 *Unipolar P–E loops obtained (a) at RT and (c) in the temperature range from 25 to 150 °C for the BF-BT-0.08NZZ-based energy-storage MLCC. Also shown are the calculated energy storage properties for the BF-BT-0.08NZZ-based MLCC (b) at 25 °C and (d) in the temperature range 25–150 °C. Reproduced from [122]*

10.4.2 High-temperature energy-storage MLCC

The energy-storage performance of MLCCs has improved considerably, such that discharge energy densities of >10 J cm^{-3} are now attainable. However, the high loss observed for energy-storage MLCCs at high temperatures is still a huge challenge that limits their usefulness at such temperatures. For example, the energy-storage efficiency of the BT-BF-based MLCC at 150 °C was found to be lower than 70%. This low energy-storage efficiency leads to high heat production, accelerating the failure of the MLCC. In addition, it wastes energy. On the other hand, the rapid development of the avionics and automotive industries, underground oil and gas exploration, advanced propulsion systems, and hybrid electric vehicles has resulted in an urgent need for high-temperature dielectric capacitors [124–126]. For example, in hybrid electric vehicles, the underhood temperature is always more than 140 °C. Therefore, we now introduce an optimized two-step sintering method for the fabrication of high-temperature energy-storage MLCCs with high energy efficiency. A X9R-type energy-storage MLCC is also presented that possesses high-temperature stability.

10.4.2.1 Sintering optimization for high-temperature energy-storage MLCCs

The bonding at the interfaces between the internal metal electrodes and the dielectric layers of MLCCs and the continuity of these internal electrodes are the key factors that affect the dielectric breakdown strength and the energy-storage efficiency of MLCCs [127,128]. The existence of pores in the electrodes and at dielectric layer interfaces and discontinuities in internal electrodes lead to concentrations of the local electric field, greatly increasing the probability of dielectric breakdown. Additionally, concentrations of the local electric field will lead to enhanced leakage current and further decrease the energy-storage efficiency, especially at high temperature.

Based on our advanced two-step sintering method [47], we utilize a fast heating rate (such as 40 °C/min) during the first sintering step to improve the quality of the interfaces between the internal electrodes and dielectric layers of the energy-storage MLCCs [125]. As an example, the high-performance lead-free relaxor ferroelectric ceramic $0.87BaTiO_3$–$0.13Bi(Zn_{2/3}(Nb_{0.85}Ta_{0.15})_{1/3})O_3$ (BT-BZNT) was selected as the dielectric layer material, since this material exhibits very high energy efficiency.[31] Meanwhile, 60Ag/40Pd was the material selected for the internal electrodes to permit the co-sintering of ceramic dielectrics and metal electrodes and thus lower the cost. Each MLCC (1210 type) contained ten active dielectric layers with a single-layer thickness of 11 μm and an effective single-layer electrode area of ~4.8 mm^2.

Three heating rates were trialed during the first sintering step (4, 20, and 40 °C/min) for comparison purposes. The corresponding fabricated energy-storage MLCCs were denoted MLCC-4, MLCC-20, and MLCC-40, respectively. Surface and cross-sectional SEM images of these MLCCs are shown in Figure 10.45. It is clear from those images that as the heating rate during the first step increases, the grain size of the dielectric material (BT-BZNT) decreases significantly, causing the dielectric layers to become more dense. This is also confirmed by the cross-sectional SEM images shown in Figure 10.45(d)–(f). The dielectric layers of MLCC-4 show some pores, whereas there are no obvious pores in the dielectric layers of MLCC-40. Notably, the morphologies of the internal electrodes of the three MLCCs are quite different. When the heating rate is slow during the first sintering step (e.g., for MLCC-4), a large number of pores can be observed in the internal electrodes, together with some discontinuities. However, in MLCCs sintered at a high heating rate (e.g., MLCC-40), barely any pores between the internal electrodes and the dielectric layers and almost no discontinuities in the internal electrodes are observed.

A comparison of the energy-storage properties of MLCCs sintered at various first-step heating rates is provided in Figure 10.46. The temperature-dependent dielectric constant and loss are displayed in Figure 10.46(a). All of the MLCCs sintered with various first-step heating rates present similar dielectric losses, indicating that the grain size has only a minor influence on the dielectric properties of this relaxor ferroelectric material. Differences in dielectric constant are mainly

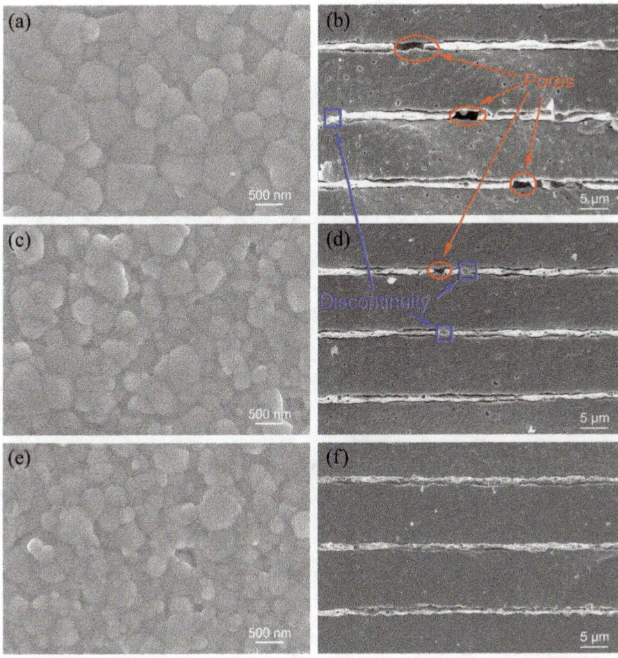

Figure 10.45 Surface SEM images (a)–(c) and cross-sectional SEM images (d)–(f) of the MLCCs sintered at first-step heating rates of (a) 4 °C/min, (b) 20 °C/min, and (c) 40 °C/min. Note that pores and discontinuities in internal electrodes are clearly visible in (d) and (e), but not in (f). Reproduced from [125]

caused by different area losses from internal electrodes. There are almost no pores in the internal electrodes of MLCCs sintered at a high heating rate during the first step, so there is only minimal area loss from internal electrodes, which accounts for the higher dielectric constant of the MLCCs sintered at a high heating rate. To explore the energy-storage performance, the *P–E* loops of MLCCs sintered at various first-step heating rates were measured at 1 Hz. A typical comparison of *P–E* loops obtained under an electric field of 500 kV/cm is presented in Figure 10.46(b). The polarization of the MLCC sintered at the fast heating rate under a field of 500 kV/cm is shown to be slightly enhanced due to the reduced loss from internal electrodes. The discharge energy density and energy efficiency as functions of the applied electric field were calculated and the results are plotted in Figure 10.46(c). When the applied electric field is lower than 250 kV/cm, the energy-storage performance was almost the same for all the MLCCs. However, as the applied electric field is increased, the MLCC sintered at the fast heating rate (MLCC-40) displays a slightly higher discharge energy density than those of MLCC-4 and MLCC-20 due to its greater energy efficiency. Under a high applied field, energy loss is mainly caused by an increased leakage current. This increased

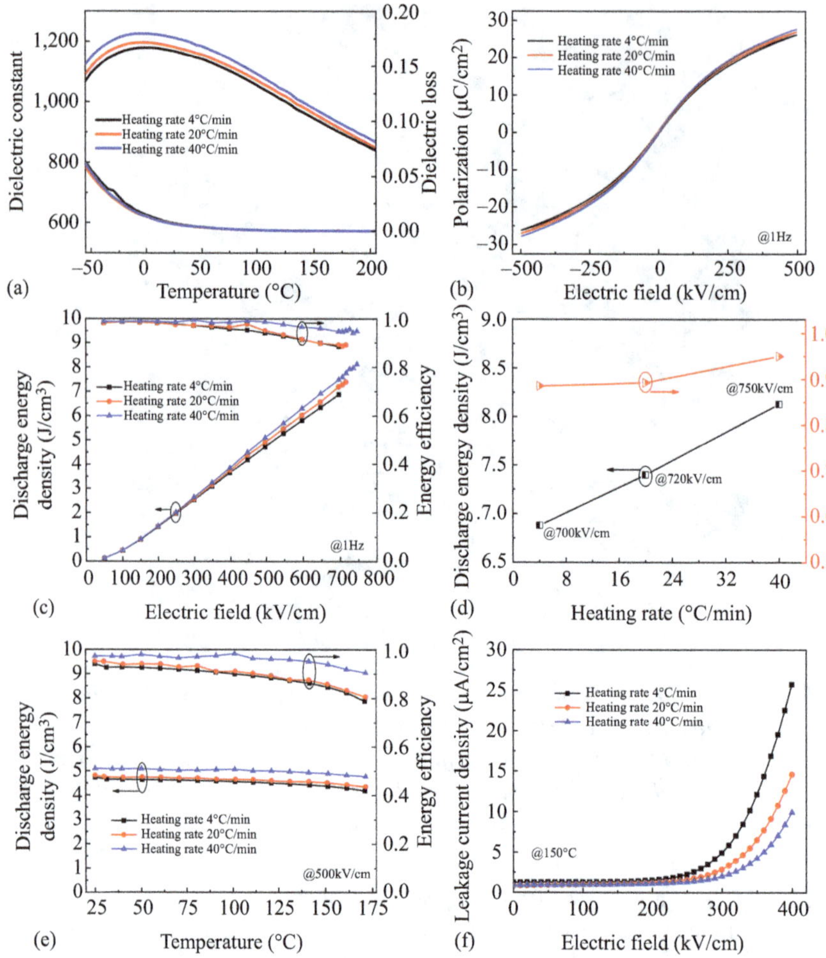

Figure 10.46 *(a) Plot of temperature-dependent dielectric constant and loss for the MLCCs sintered at various first-step heating rates. (b) Hysteresis loops of these MLCCs measured under an electric field of 500 kV/cm at 1 Hz. (c) Calculated discharge energy densities and energy efficiencies of the MLCCs sintered at various heating rates as a function of the applied electric field, as measured at 1 Hz. (d) Discharge energy densities and energy efficiencies of the MLCCs sintered at various heating rates, as measured under the maximum applied field. (e) Plot of temperature-dependent discharge energy density and energy efficiency for the MLCCs sintered at various heating rates, as measured under an applied field of 500 kV/cm at 1 Hz. (f) Plot of MLCC leakage current density MLCC as a function of the applied electric field at 150 °C. Reproduced from [125]*

leakage current is much more serious in MLCCs with poor bonding at the interfaces between the internal metal electrodes and the dielectric layers, as seen for MLCC-4 and MLCC-20. Therefore, the energy efficiencies of MLCC-4 and MLCC-20 gradually drop below 90% under a high applied field, limiting the applicability of these MLCCs to high-voltage systems. On the other hand, a high energy density of 8.13 J cm^{-3} and an ultrahigh efficiency of 95% are achieved under an applied electric field of 750 kV/cm when the MLCC is sintered at the fast heating rate using the TSS method (i.e., for MLCC-40), implying that MLCC-40 is a particularly promising candidate for high-voltage applications. Figure 10.46(d) summarizes the maximum discharge energy densities of MLCCs sintered at various first-step heating rates. Due to improved bonding at the interfaces between its internal metal electrodes and dielectric layers, the MLCC sintered at the fast heating rate by TSS exhibits the best breakdown strength and discharge energy density as well as an ultrahigh energy efficiency.

The temperature-dependent discharge energy densities and energy efficiencies of the MLCCs sintered at various heating rates, as measured under an applied field of 500 kV/cm at 1 Hz, are shown in Figure 10.46(e). It is clear that the energy efficiency of MLCC-4 or MLCC-20 gradually decreases with temperature, dropping below 80% at 170 °C. The leakage current is the main reason for the energy loss at high temperature and high applied field. A lower energy efficiency implies greater heat generation, accelerating the failure of the MLCC. However, for an MLCC sintered at the fast heating rate by the TSS method, an ultrahigh energy efficiency of 91% can be maintained at a high temperature of 170 °C and a high applied field of 500 kV/cm, with the variation in discharge energy density observed to be less than 7% over the temperature range 25–170 °C. To further confirm this leakage-current-induced energy loss, the leakage current densities of the MLCCs are shown as a function of the applied electric field at 150 °C in Figure 10.46(f). It is clear that the leakage current density of MLCC-4 is nearly triple that of MLCC-40.

Given these results, it is expected that this TSS approach with a fast first-step heating rate will be applied not only to BTBZNT-based energy-storage MLCCs but to all kinds of energy-storage MLCCs in order to improve the bonding at the interfaces between their internal electrodes and dielectric layers and to enhance their high-voltage and high-temperature energy-storage performance.

10.4.2.2 X9R-type energy-storage MLCC

The temperature stability of its capacitance is one of the most important characteristics of an MLCC from a practical perspective. The internationally used Electronic Industries Alliance standard EIA-198-1-F-2002 proposes that every capacitor should be assigned a characteristic code that denotes the temperature stability of the capacitor. These codes comprise two letters and one number; examples include X7R (indicating temperature stability from −55 to 125 °C, $\Delta C/C_{25\,°C} \leq \pm 15\%$) and X8R (stable from −55 °C to 150 °C, $\Delta C/C_{25\,°C} \leq \pm 15\%$). As mentioned at the beginning of Section 10.4.2, the rapid development of the avionics and automotive industries, underground oil and gas exploration,

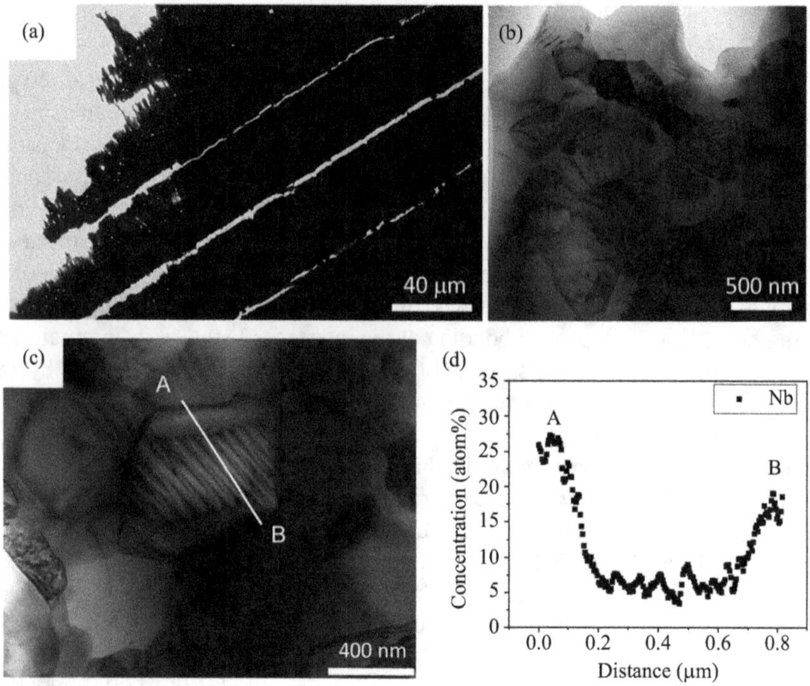

Figure 10.47 TEM images of the BTBNT-Nb-Mn-based MLCC: (a) microstructure of the dielectric layers and internal electrodes and (b, c) the core–shell structure of grains. (d) The distribution of Nb along the line AB in (c). Reproduced from [129]

advanced propulsion systems, and hybrid electric vehicles has resulted in an urgent demand for dielectric capacitors that are stable above 150 °C. Therefore, X9R-type (i.e., stable from −55 to 200 °C, $\Delta C/C_{25°C} \leq \pm 15\%$) MLCCs are in urgent demand, and research into these MLCCs for energy storage applications is underway.

X9R specification energy-storage MLCCs (1210 size: ~3.2 mm × 2.7 mm) based on Nb_2O_5- and MnO_2-modified $0.9BaTiO_3$–$0.1(Bi_{0.5}Na_{0.5})TiO_3$ (BTBNT-Nb-Mn) ceramics with core–shell structure have been fabricated in our lab [129]. The room-temperature capacitance of this MLCC is higher than 20 nF. Each MLCC contains ten active dielectric layers. The thicknesses of the internal electrodes and the dielectric layers in these MLCCs are approximately 1–2 μm and 30 μm, respectively, as shown in Figure 10.47(a). The grains inside the dielectric layer show core–shell structure and are approximately 500 nm in size, as shown in the TEM images in Figure 10.47(b)–(d), where the ferroelectric domains are clearly discernible in the grain core and Nb is virtually nonexistent inside the grain core.

This core–shell structure enhances not only the temperature stability of the capacitance but also the dielectric breakdown strength.

The temperature dependence of the capacitance, the loss tangent, and the temperature coefficient of the capacitance (TCC $= \Delta C/C_{25°C}$) measured at frequencies of 100 Hz, 1 kHz, 10 kHz, and 100 kHz in the temperature range from −100 to 250 °C are presented in Figure 10.48. Two dielectric peaks can be seen. The dielectric peak at low temperature corresponds to the grain shell and is dependent on the frequency, whereas the dielectric peak at high temperature corresponds to the grain core and does not exhibit relaxor behavior. The variation of the TCC curves of the BTBNT-Nb-Mn-based MLCC, which is within ±15% from −90 to 225 °C at 1 kHz, satisfies the X9R specification (indicated by the dashed box in Figure 10.48(b)). The TCC curves are almost linear from 15 to 140°C at 1 kHz to within ±1%.

The *P–E* loops of the BTBNT-Nb-Mn-based MLCC at a fixed electric field of 360 kV/cm, as measured at 1 Hz and different temperatures, are plotted in Figure 10.49(a). As the temperature increases, the maximum polarization decreases. The remanent polarization initially decreases and then increases dramatically above the Curie temperature. The difference between the maximum polarization and the remanent polarization increases before the Curie temperature is reached, resulting in enhanced energy efficiency with increasing temperature under the same electric field. The discharge energy density and the energy efficiency as functions of the electric field at various temperatures were calculated from the *P–E* loops and are presented in Figure 10.49(b). When the electric field is lower than 125 kV/cm, the discharge energy density curves almost overlap, suggesting that the temperature has little effect on the discharge energy density under a low electric field. When the electric field reaches 150 kV/cm, the curves of the discharge energy density begin to separate. At room temperature, the energy efficiency increases as the applied

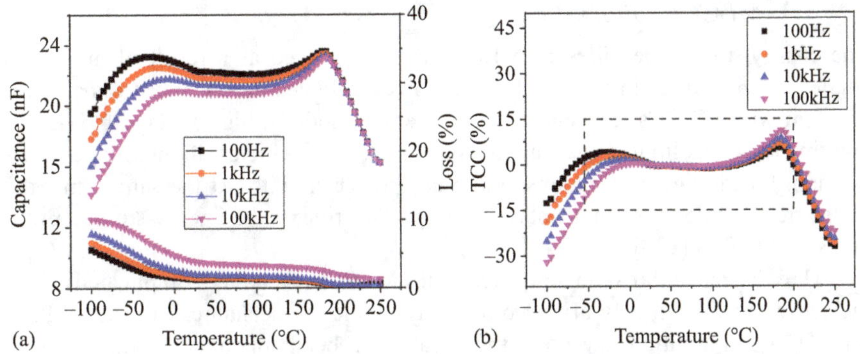

Figure 10.48 Temperature dependence of (a) the capacitance and dielectric loss and (b) the TCC for the BTBNT-Nb-Mn-based MLCC at different frequencies from −100 to 250 °C. Reproduced from [129]

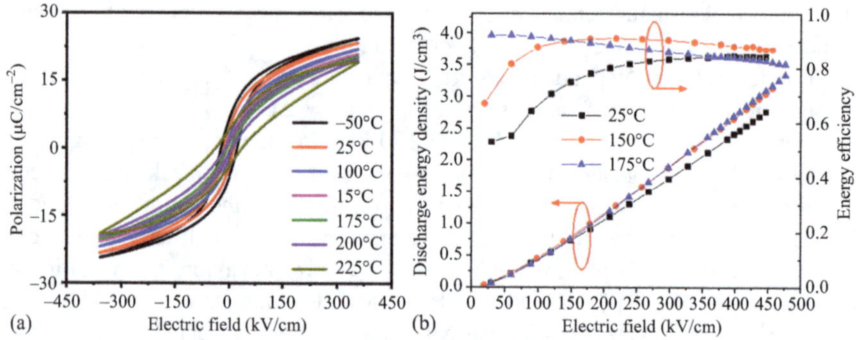

Figure 10.49 (a) P–E loops of the BTBNT-Nb-Mn-based MLCC at a fixed electric field of 360 kV/cm. (b) Discharge energy density and energy efficiency as functions of the electric field at various temperatures. Reproduced from [129]

electric field increases, but at a temperature of 150 °C, the energy efficiency first increases but then decreases due to a gradual rise in the leakage current. At high temperature (175 °C), the energy efficiency decreases as the applied electric field increases. This phenomenon is related to the field-induced transition from a ferroelectric to a paraelectric state [130].

A high discharge energy density of 3.33 J/cm^3 is obtained with this X9R-type MLCC at an applied electric field of 480 kV/cm and 175 °C, together with a high energy efficiency of 81.6%. These results demonstrate that the X9R-type BTBNT-Nb-Mn-based MLCC is a promising candidate for high-temperature applications due to the superb temperature stability of its capacitance.

10.4.3 Advanced design strategy for energy-storage MLCCs: phase-field simulations

The energy-storage densities of MLCCs have improved as a result of numerous studies, as mentioned in Section 10.4.1. However, dielectric breakdown is a major problem with MLCCs, especially those operated under a high voltage [131,132], and this issue has limited their application in electric vehicles, high-voltage transmission, kinetic energy weapons, and pulse-power circuits. At the same time, the dielectric breakdown strength plays an important role in the energy-storage properties of MLCCs [133].

The dielectric breakdown process in MLCCs is quite complicated since dielectric breakdown may arise from electrical, electromechanical, or thermal stimuli [134]. In addition, the dielectric breakdown behavior may be influenced by defects, grain size, and the second phase of the dielectric material [135–137]. Various physical models have been proposed to improve the understanding of dielectric breakdown, but those models focus mainly on electric stimuli, ignoring

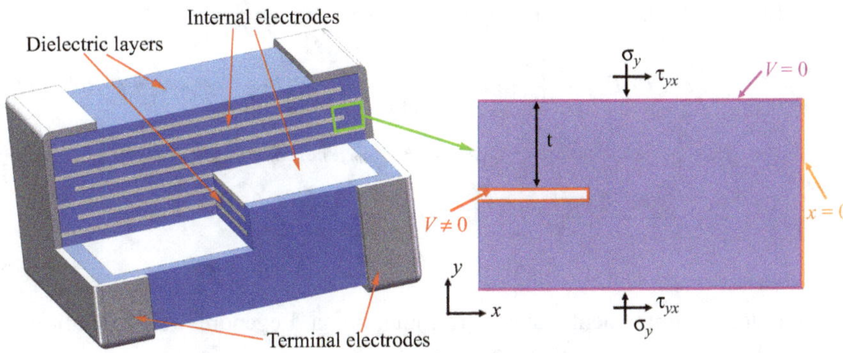

Figure 10.50 Schematic of a multilayer ceramic capacitor (MLCC), and the loading setup employed in a phase-field electromechanical breakdown model of a selected typical region of the MLCC. Reproduced from [146]

electromechanical effects [138–141]. However, electromechanical effect is a crucial factor in dielectric breakdown of MLCC, as MLCCs always suffer from compressive or shear stresses induced by the sintering process or normal operation. Also, the dielectric materials used in MLCCs are mainly ferroelectrics, such as $BaTiO_3$, $BiFeO_3$, and BT-B*M*O. Piezoelectricity, an electromechanical coupling property, is intrinsically displayed by these ferroelectric materials when they are operated under an applied voltage. Therefore, studying electromechanical breakdown is key to understanding the breakdown behavior of MLCCs. In this section, we introduce a phase-field model that we have used to simulate the electromechanical breakdown behavior of MLCCs. The electromechanical coupling properties of ferroelectrics are described by the Ginzburg–Landau kinetic equation [142]. In contrast to the existing dielectric breakdown model, the need to address the intricate microscopic details in the process zone can be avoided by employing a Griffith-type energy criterion. Advantageously, the evolution of the polarization vectors and the breakdown path can be visualized using this model, and it can be used to calculate the nominal breakdown strength.

The local electric field and stress are concentrated around the tips of the internal electrodes, so a 2D region of the MLCC containing an internal electrode tip is selected MLCC to serve as the computational region. The loading setup of this region in our phase-field electromechanical breakdown model is shown in Figure 10.50. Inspired by the phase-field model of brittle fracture, a scalar spatially and temporally dependent damage field $s(x, t)$ is introduced to characterize the dielectric breakdown process in the MLCC [143–145]. The value of s varies continuously from the intact state, $s = 1$, to the fully damaged state, $s = 0$. A large but finite dielectric constant κ_0/η is assumed for the material, where η is a sufficiently small number. Therefore, the modified Landau–Ginzburg–Devonshire energy density function during the breakdown process can be written as [146]

$$\begin{aligned}\tilde{\psi} &= \frac{1}{2}a_{ijkl}P_{i,j}P_{k,l} + \left\{\frac{1}{2}\bar{a}_{ij}P_iP_j + \frac{1}{4}\bar{\bar{a}}_{ijkl}P_iP_jP_kP_l + \frac{1}{6}\bar{\bar{\bar{a}}}_{ijklmn}P_iP_jP_kP_lP_mP_n\right\} \\ &+ \left\{b_{ijkl}\varepsilon_{ij}P_kP_l + \frac{1}{2}c^H_{ijkl}\varepsilon_{ij}\varepsilon_{kl} + f_{ijklmn}\varepsilon_{ij}\varepsilon_{kl}P_mP_n + g_{ijklmn}\varepsilon_{ij}P_kP_lP_mP_n\right\} \\ &+ \frac{f(s)+\eta}{2\kappa_0}(D_i - P_i)(D_i - P_i)\end{aligned} \tag{10.7}$$

In order to implement such a formulation, a Legendre transformation is introduced to derive the electrical enthalpy h:

$$h(P_i, P_{i,j}, E_i) = \tilde{\psi} - E_iD_i \tag{10.8}$$

The constitutive relation between D and P is given by $D_i = \kappa_0 E_i + P_i$. Finally, the total potential energy including polarization evolution and dielectric breakdown can be written as

$$\Pi[s, P_i, \phi] = \int_\Omega [h + W_\mathrm{d}(s) + W_\mathrm{i}(\nabla s)]\mathrm{d}V \tag{10.9}$$

where $W_\mathrm{d}(s) = W_\mathrm{c}[1 - f(s)]$ is the breakdown energy function, with W_c representing the critical density of electrostatic energy, and $W_\mathrm{i}(\nabla s) = \frac{\Gamma}{4}\nabla s \cdot \nabla s$ is the gradient energy term to regulate sharp phase boundaries. Notably, the material parameter Γ is approximately the breakdown energy.

Considering the time-dependent procedure, we take the polarization P and the damage field s as the order parameters of the system. As the system evolves toward equilibrium, its total energy continues to decrease. In the simplest case, the evolution of P can be taken to be proportional to its driving force, according to

$$\frac{\partial P_i(x,t)}{\partial t} = -L_{ij}\frac{\delta\Pi}{\delta P_j} \tag{10.10}$$

where L_{ij} are the components of the positive-definite kinetic coefficients. Equation (10.4) is the well-known time-dependent Ginzburg–Landau kinetic equation. We can therefore define the criterion for dielectric breakdown as follows: breakdown will occur if the process decreases the total potential energy of the system. Thus, s evolves according to the linear kinetic law

$$\partial s(x,t)/\partial t = -m\delta\Pi/\delta s \tag{10.11}$$

where, damage mobility m is a material parameter that incorporates the speed of breakdown path propagation. The modeling was performed on a COMSOL Multiphysics platform.

As an example, we first performed this phase-field simulation in an MLCC with a dimensionless dielectric layer thickness of 40 under stress-free conditions. The nominal field–charge density relation is shown in Figure 10.51. The

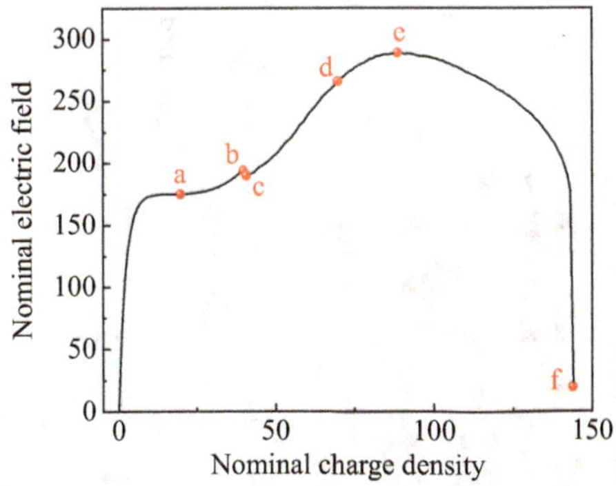

Figure 10.51 Nominal field–charge density relation for the selected region of the MLCC with a dimensionless dielectric layer thickness of 40. Reproduced from [146]

field–charge density relation shows typical ferroelectric character with increasing charge density before point e. However, when the charge density exceeds that for point e, the MLCC cannot sustain a higher electric field due to partial breakdown, as seen in Figure 10.52(e). Finally, full electromechanical breakdown occurs and the electric field reduces to zero at point f. The detailed evolution of the polarization vectors and that of the breakdown path are shown in the six snapshots of Figure 52(a)–(f). It is clear that the breakdown initiates at the sharp tip of the internal electrode, where the local electric field is most concentrated. These results agree with experimental observations [131]. Following the initiation of the breakdown path, the local electric field is always concentrated around the tip of the breakdown path, which further promotes the propagation of the path. As a result, as the controlled charge increases and the breakdown path propagates to the negative electrode. Total breakdown occurs when the positive electrode and the negative electrode are bridged by one of the breakdown paths, as seen in Figure 10.52(f) with the final breakdown path showing a tree-like pattern with branches.

In line with current technological trends for miniaturization, high portability, networking, and multimedia, the dielectric layer thicknesses of MLCCs are decreasing to ~1 μm. The thickness-dependent dielectric breakdown strength of MLCCs has become an area of great research interest, because it may provide useful guidance for the design of new MLCCs with improved performance [136,147,148] The thickness dependence of the breakdown strength is generally reported to follow an inverse power law, $E_b \propto t^{-n}$, where E_b and t are the dielectric breakdown strength and the thickness of the dielectric material, respectively. n is a constant associated with dielectric testing that ranges from 0.4 to 0.5. Our phase-field electromechanical breakdown model was used to comprehensively simulate the thickness-dependent breakdown strength, and the results were found to agree well with existing experiments [149].

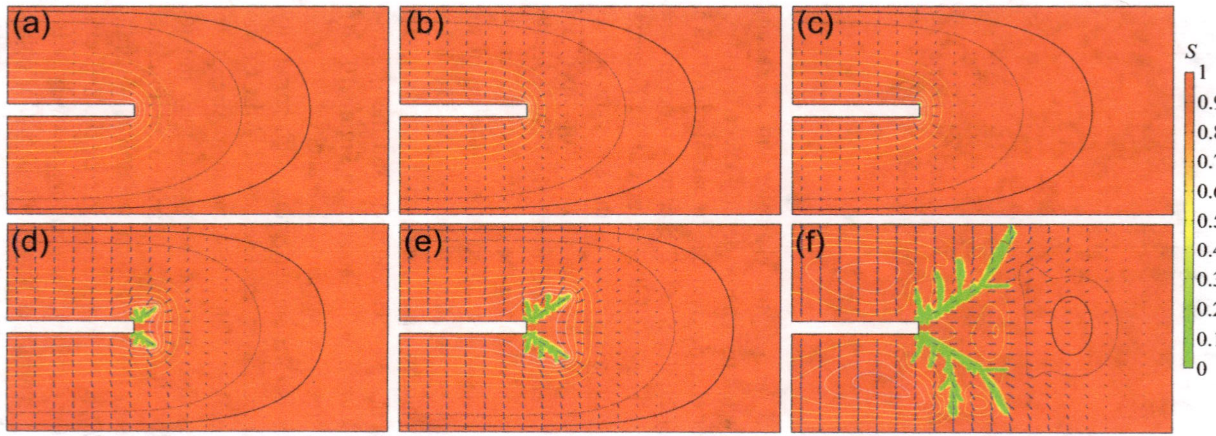

Figure 10.52 *Snapshots of the electromechanical breakdown process for the selected region of the MLCC, showing the coupled evolution of the polarization vectors and the breakdown path. The six snapshots (a)–(f) correspond to the six dots a–f in Figure 10.51. The shading reflects the value of the damage variable s, with s = 1 indicating the intact state and s = 0 the fully damaged state. The arrows represent the polarization vector P and the contour lines are the equipotential lines for the then-current electric field. Reproduced from [146]*

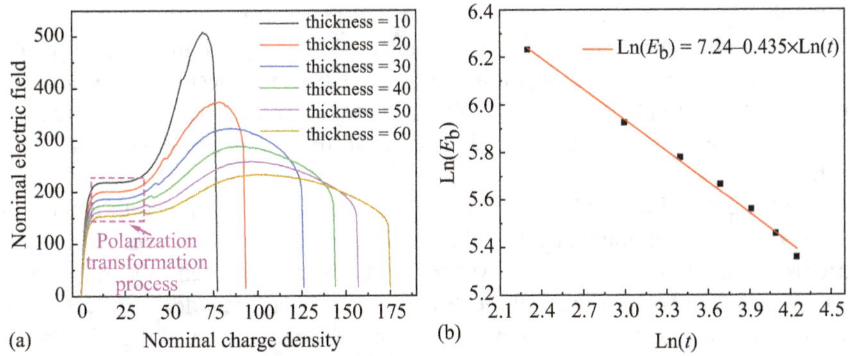

Figure 10.53 (a) Nominal field–charge density relations for MLCCs with various dimensionless dielectric layer thicknesses. (b) Nominal breakdown strength as a function of the dimensionless dielectric layer thickness for MLCCs. The solid red line, with a slope of –0.435, is the best fit to the numerical data points. Reproduced from [146]

The electromechanical breakdown model was then applied to MLCCs with various dielectric layer thicknesses using the nominal field–charge density relation plotted in Figure 10.53(a). As we can see, all of the samples present a polarization transformation at nominal charge densities ranging from ~5 to ~35, where the electric field remains almost unchanged while the polarization surges. The typical polarization–field relation (i.e., a hysteresis loop) is observed around the coercive field. We denote this unchanged electric field as the polarization transformation field (PTF). The PTF is seen to increase to some extent with decreasing dielectric layer thickness. A higher PTF means that it is harder for polarization to transform, and such a transformation will dissipate more energy, implying enhanced dielectric breakdown strength. Further, a thinner dielectric layer corresponds to fewer defects, which also contributes to the enhanced dielectric strength of MLCCs with thin dielectric layers. In addition, it is worth noting that an MLCC with thicker dielectric layers can hold more charge density before suddenly breaking down than an MLCC with thinner dielectric layers. This is because there is enough distance across a layer in an MLCC with thicker dielectric layers to allow the formation of extra branches of the breakdown path. As a whole, an MLCC with thicker dielectric layers is more stable with respect to degradation than an MLCC with thinner dielectric layers, although its effective breakdown strength is lower. The relation between the nominal dielectric breakdown strength and the dimensionless dielectric layer thickness is plotted in Figure 10.53(b). The data are well fitted by the inverse power law $E_b \propto t^{-n}$ with $n = 0.435$. The numerical results for the thickness-dependent breakdown strength agree well with the results from theoretical analysis and experimental observations [136,148].

MLCCs have tens of thin dielectric layers arranged in parallel between internal electrodes. Under a given applied voltage, the local electric field is concentrated around the tips of the internal electrodes, which will favor the initiation and propagation of dielectric breakdown paths inside MLCCs and ultimately greatly reduce their dielectric breakdown strength. The structural design of the internal electrodes, for example the margin length, is quite important since different margin lengths will lead to different degrees of local field concentration around the electrode tips.

We therefore simulated MLCCs with various dimensionless margin lengths of 50, 75, and 100 using the phase-field breakdown model. The final breakdown path patterns inside the MLCCs are shown in Figures 54(c)–(e). Each breakdown path initiates at the end of the internal electrode tip, and total breakdown occurs when the electrodes are bridged by the breakdown path. Generally speaking, the breakdown paths all display similar tree-like branched patterns. The nominal field–charge density relation during the breakdown process is presented in Figure 10.54 (a). At nominal charge densities ranging from ~10 to ~40, the applied electric field remains almost unchanged, indicating that the dielectric ceramics undergo a polarization transformation process. Note that the polarization transformation inside the MLCC with the smallest margin length requires the lowest applied field. This is because the local electric field is especially concentrated around the tips of the internal electrodes in the MLCC with the smallest margin length. The highest applied electric field that the MLCC can sustain during the whole breakdown process is then defined as the nominal breakdown strength, which is plotted for all the MLCCs in Figure 10.54(b). It is quite clear that as the margin length increases, the nominal breakdown strength of the MLCC is significantly enhanced.

To further check and apply our phase-field breakdown model, 1,210-type MLCCs (~3.2 mm × 2.7 mm) with internal 60Ag/40Pd electrodes and ten dielectric layers with a single-layer thickness of 11 μm were fabricated via the tape-casting technique. Different margin lengths were designed on printing screens. The high-performance, lead-free, and relaxor ferroelectric ceramic BT-BZNT was chosen as the dielectric layer material. Various margin lengths of 100, 200, and 400 μm are clearly apparent in the SEM images of Figure 10.55(a)–(c). Continuous internal electrodes and dense dielectric layers can also be observed. The direct current (DC) characteristic dielectric breakdown strengths of the MLCCs with margin lengths of 100, 200, and 400 μm were calculated to be 783, 820, and 895 kV/cm, respectively, based on the Weibull plots shown in Figure 10.55(f). The Weibull distribution function is $P(E_b) = 1 - \exp\left(\left(-E_b/\alpha\right)^\beta\right)$, where $P(E_b)$ is the cumulative probability of dielectric failure, E_b is the measured DC breakdown strength, α is the characteristic breakdown strength corresponding to a ~63.2% probability of dielectric failure, and β is the Weibull shape parameter [140]. This relation between MLCC breakdown strength and margin length MLCCextracted from experimental measurements agrees well with the above results from our phase-field simulations. Increasing the margin length leads to greater dielectric breakdown strength.

To further evaluate the energy-storage performance of MLCCs with various margin lengths, *P–E* loops were measured at an applied electric field of 600 kV/cm and at their alternating current (AC) breakdown strengths and 1 Hz, with the results

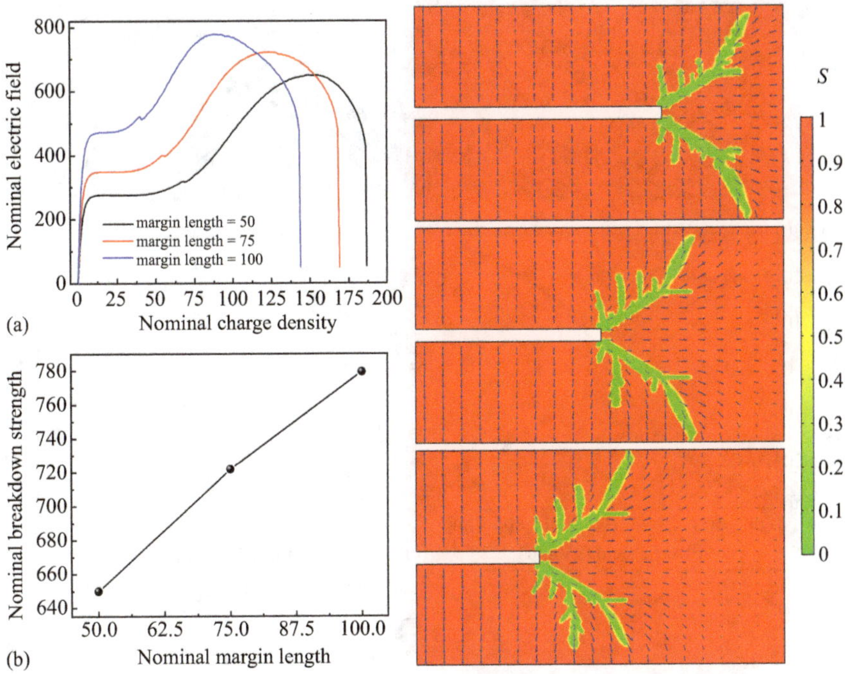

Figure 10.54 Nominal field–charge density relations of MLCCs with various margin lengths (a), nominal breakdown strength as a function of the margin length of the MLCC (b), and final breakdown path patterns obtained in the selected MLCC region with various nominal margin lengths: (c) 50, (d) 75, and (e) 100. The shading reflects the value of the damage variable s and the arrows indicate the polarization vector P. Reproduced from [149]

shown in Figure 10.55(d) and (e), respectively. Although their margin lengths are different, the hysteresis loops of these MLCCs under the same applied field are almost the same, since the dielectric layer material in each MLCC is the same BTBZNT ceramic. However, due to differences in breakdown strength, the maximum measured polarizations differ considerably. MLCCs with a margin length of 400 μm achieve a maximum polarization of ~30 μC/m^2 at a field of 790 kV/cm, while MLCCs with a margin length of 100 μm attain a maximum polarization of ~27.5 μC/m^2 at a field of 640 kV/cm. The discharge energy densities and energy efficiencies at 1 Hz for the MLCCs with various margin lengths are plotted as functions of the applied electric field in Figure 10.55(f). This plot shows that the discharge energy densities and energy efficiencies of the various MLCCs are almost the same below an applied electric field of 640 kV/cm. This is a good finding, as it indicates that the MLCCs fabricated in our lab exhibit stable and reliable performance. A large discharge energy density of 7.8 J cm^{-3} was achieved under an applied electric field of 790 kV/cm in the MLCC with a margin length of

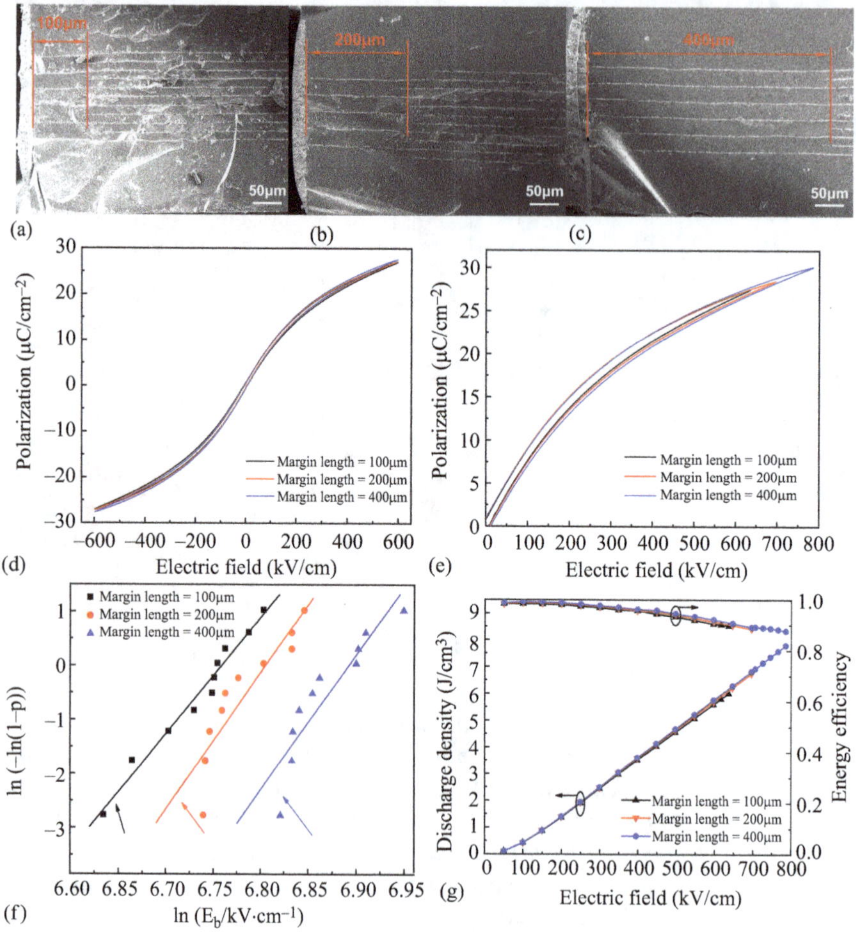

Figure 10.55 (a–c) Cross-sectional SEM images of MLCCs with different margin lengths: (a) 100, (b) 200, and (c) 400 μm; note that the ceramics and electrodes are clearly apparent. (d) P–E loops of MLCCs with various margin lengths measured at an applied electric field of 600 kV/cm and 1 Hz. (e) Unipolar hysteresis loops of MLCCs with various margin lengths, obtained at their AC breakdown strengths and 1 Hz. (f) Weibull plots of the DC dielectric breakdown strengths of MLCCs with various margin lengths. (g) Discharge energy densities and energy efficiencies at 1 Hz for MLCCs with various margin lengths as functions of the applied electric field. Reproduced from [149]

400 μm, which is 30% higher than that attained using the MLCC with a margin length of 100 μm, due to the enhanced breakdown strength of the MLCC with a margin length of 400 μm. This MLCC presented high energy efficiencies (> 88%) across the whole range of field strengths applied and is therefore a promising candidate for high-voltage energy-storage applications.

The experimental results for the margin-length dependence of the dielectric breakdown strength of MLCC agree very well with the predictions obtained from our phase-field model. Therefore, the model not only allows us to deepen our understanding of electromechanical breakdown in MLCCs, but it can also guide the design of novel high-voltage and high-energy-density MLCCs.

10.5 Conclusion and prospects

In this chapter, we first discussed the size effect for BT ceramics in detail. This is an important issue in the development of MLCCs because of the demand for thinner and thinner dielectric layers. In 2006, a dense BT ceramic with an average GS as small as 8 nm was prepared in our lab using a two-step sintering method, and microscopic characterization indicated that this ceramic exhibited a ferroelectric polydomain state. This ceramic also showed robust *K*-temperature (*T*) performance, high dielectric constant (~1400) at room temperature, low dielectric loss, and multiple dielectric/ferroelectric transitions. These results imply that a ceramic capacitor with a layer thickness of 40–50 nm (corresponding to five to seven 8-nm grains) is still viable.

Dielectric ceramics with grain sizes of < 200 nm are currently in high demand in the MLCCs industry as a means to meet the requirement for dielectric layer thicknesses of < 1 μm, since each dielectric layer must be at least five grains thick to ensure reliability. Chemical coating, grain refinement, and two-step sintering methods have been introduced as they facilitate the preparation of high-performance and highly reliable nanocrystalline BT-based dielectric layer ceramics. The reliability of BT ceramics prepared by the chemical coating method and sintered via the two-step method is much better than that of BT ceramics produced by the solid-state method and sintered via the conventional method. However, given the industrial drive to decrease dielectric layer thicknesses even further, the mass production of high-performance BT ceramics with grain sizes of < 100 nm requires further research.

Some high-performance MLCCs that incorporate relaxor ferroelectric, relaxor antiferroelectric, and chemically inhomogeneous but electrically homogeneous ferroelectric ceramics as dielectric layer materials were also reviewed in this chapter. The energy-storage densities of these MLCCs (up to > 10 J cm^{-3}) are significantly enhanced compared to bulk ceramics; indeed, they are comparable to or even larger than those of thin-film capacitors. Given the requirement for operation at high temperatures, we introduced an advanced sintering technique based on the two-step sintering method that yields MLCCs characterized by high energy density and efficiency at high temperature. In addition, a special X9R-type MLCC with capacitance that exhibits high-temperature stability was discussed. We also showed how advanced MLCC design is facilitated by the use of phase-field simulations, as they not only permit a greater understanding of electromechanical breakdown in MLCCs, but they can also provide guidance when designing high-voltage and high-energy-density MLCCs.

The utilization of MLCCs for energy-storage applications shows great promise, as the mass production of MLCCs is easy to achieve through the advanced tape casting technique. However, the internal electrodes used in the MLCCs reported are all based on noble metals; for example, the internal electrodes for NBT-SBT-based MLCC and BT-BF-based MLCC are made from Pt, which is too expensive for mass production. Even the 60Ag/40Pd internal electrodes employed in the BT-BZNT-based MLCC fabricated in our lab are still relatively expensive. From the perspective of minimizing costs during mass production, base metals such as Ni or Cu are preferable internal electrode materials, but this necessitates the use of a dielectric layer that is nonreducible to ensure the cosintering of the base metal with the dielectric material. Therefore, nonreducible dielectric ceramics with high energy storage performance clearly requires further research. Finally, the reliability of MLCCs will play a crucial role in their real-industrial applications, so the reliability issues associated with current MLCCs will also need to be resolved.

Acknowledgments

The work was supported by the Ministry of Sciences and Technology of China through National Basic Program of China (973 Program Nos. 2002CB613301, 2009CB623301, and 2015CB654604) and the National Key Research and Development Program of China (No. 2017YFB0406302).

References

[1] Love, G.R. "Energy-storage in ceramic dielectrics." *Journal of the American Ceramic Society*. 1990;**73**(2): 323–328.

[2] Kishi, H., Mizuno, Y., Chazono, H. "Base-metal electrode-multilayer ceramic capacitors: past, present and future perspectives." *Japanese Journal of Applied Physics Part 1-Regular Papers Short Notes & Review Papers*. 2003;**42**(1): 1–15.

[3] Sakabe, Y., Reynolds, T. "Base-metal electrode capacitors." *American Ceramic Society Bulletin*. 2002;**81**(10): 24–26.

[4] Tian, Z.B., Wang, X.H., Lee, S., Hur, K.H., Li, L.T. "Microstructure evolution and dielectric properties of ultrafine grained $BaTiO_3$-based ceramics by two-step sintering." *Journal of the American Ceramic Society*. 2011;**94**(4): 1119–1124.

[5] Hoshina, T., Kigoshi, Y., Hatta, S., Takeda, H., Tsurumi, T. "Domain contribution to dielectric properties of fine-grained $BaTiO_3$ ceramics." *Japanese Journal of Applied Physics*. 2009;**48**(9): 09KC01.

[6] Frey, M.H., Xu, Z., Han, P., Payne, D.A. "The role of interfaces on an apparent grain size effect on the dielectric properties for ferroelectric barium titanate ceramics." *Ferroelectrics*. 1998;**206**(1–4): 337–353.

[7] Kinoshita, K., Yamaji, A. "Grain-size effects on dielectric properties in barium-titanate ceramics." *Journal of Applied Physics*. 1976;**47**(1): 371–373.

[8] Polotai, A.V., Ragulya, A.V., Randall, C.A. "Preparation and size effect in pure nanocrystalline barium titanate ceramics." *Ferroelectrics*. 2003;**288**(1): 93–102.

[9] Wang, X.-H., Chen, I.-W., Deng, X.-Y., Wang, Y.-D., Li, L.-T. "New progress in development of ferroelectric and piezoelectric nanoceramics." *Journal of Advanced Ceramics*. 2015;**4**(1): 1–21.

[10] Wang, L.M., Deng, X.Y., Zhang, H.T., *et al.* "Effect of grain size on phase transitions and dielectric properties of nano-crystalline barium titanate ceramics." *Advanced Materials Research*. 2012;**602–604**: 192–196.

[11] Wang, X.H., Deng, X.Y., Wen, H., Li, L.T. "Phase transition and high dielectric constant of bulk dense nanograin barium titanate ceramics." *Applied Physics Letters*. 2006;**89**(16): 797–783.

[12] Sun, T.Y., Wang, X.H., Wang, H., *et al.* "A phenomenological model on phase transitions in nanocrystalline barium titanate ceramic." *Journal of the American Ceramic Society*. 2010;**93**(9): 2571–2573.

[13] Zhang, H., Wang, X., Tian, Z., *et al.* "Fabrication of monodispersed 5-nm $BaTiO_3$ nanocrystals with narrow size distribution via one-step solvothermal route." *Journal of the American Ceramic Society*. 2011;**94**(10): 3220–3222.

[14] Pan, M.J., Randall, C.A. "A brief introduction to ceramic capacitors." *IEEE electrical insulation magazine*. 2010;**26**(3): 44–50.

[15] Yang, G.Y., Dickey, E.C., Randall, C.A., *et al.* "Oxygen nonstoichiometry and dielectric evolution of $BaTiO_3$. Part I—Improvement of insulation resistance with reoxidation." *Journal of Applied Physics*. 2004;**96**(12): 7492–7499.

[16] Hong, K., Lee, T.H., Suh, J.M., Yoon, S.H., Jang, H.W. "Perspectives and challenges in multilayer ceramic capacitors for next generation electronics." *Journal of Materials Chemistry C*. 2019;**7**(32): 9782–9802.

[17] C. Pithan, D.H. "Progress in the synthesis of nanocrystalline $BaTiO_3$ powders for MLCC." *International Journal of Applied Ceramic Technology*. 2005;**2**(1): 1–14.

[18] Gong, H.L., Wang, X.H., Tian, Z.B., Zhang, H., Li, L.T. "Interfacial diffusion behavior in Ni- $BaTiO_3$ MLCCs with ultra-thin active layers." *Electronic Materials Letters*. 2014;**10**(2): 417–421.

[19] Zhao, Q.C., Wang, X.H., Gong, H.L., Liu, B.B., Luo, B.C., Li, L.T. "The properties of Al_2O_3 coated fine-grain temperature stable $BaTiO_3$-based ceramics sintered in reducing atmosphere." *Journal of the American Ceramic Society*. 2018;**101**(3): 1245–1254.

[20] Gong, H.L., Wang, X.H., Zhang, S.P., Li, L.T. "Synergistic effect of rare-earth elements on the dielectric properties and reliability of $BaTiO_3$-based ceramics for multilayer ceramic capacitors." *Materials Research Bulletin*. 2016;**73**: 233–239.

[21] Park, K.J., Kim, C.H., Yoon, Y.J., Song, S.M., Kim, Y.T., Hur, K.H. "Doping Behaviors of dysprosium, yttrium and holmium in $BaTiO_3$ ceramics." *Journal of the European Ceramic Society*. 2009;**29**(9): 1735–1741.

[22] Gong, H.L., Wang, X.H., Zhang, S.P., Tian, Z.B., Li, L.T. "Electrical and reliability characteristics of Mn-doped nano $BaTiO_3$-based ceramics for ultrathin multilayer ceramic capacitor application." *Journal of Applied Physics*. 2012;**112**(11): 114119.

[23] Xie, J., Hao, H., Yao, Z.H., *et al*. "Energy storage properties of low concentration Fe-doped barium strontium titanate thin films." *Ceramics International*. 2018;**44**(6): 5867–5873.

[24] Zhang, X., Shen, Y., Zhang, Q., *et al*. "Ultrahigh energy density of polymer nanocomposites containing $BaTiO_3$@TiO_2 nanofibers by atomic-scale interface engineering." *Advanced Materials*. 2015;**27**(5): 819–824.

[25] Zou, K.L., Dan, Y., Xu, H.J., *et al*. "Recent advances in lead-free dielectric materials for energy storage." *Materials Research Bulletin*. 2019;**113**: 190–201.

[26] Yang, L.T., Kong, X., Li, F., *et al*. "Perovskite lead-free dielectrics for energy storage applications." *Progress in Materials Science*. 2019;**102**: 72–108.

[27] Hu, Q., Bian, J., Zelenovskiy, P.S., *et al*. "Symmetry changes during relaxation process and pulse discharge performance of the $BaTiO_3$–$Bi(Mg_{1/2}Ti_{1/2})O_3$ ceramic." *Journal of Applied Physics*. 2018;**124**(5): 054101.

[28] Li, F., Yang, K., Liu, X., *et al*. "Temperature induced high charge–discharge performances in lead-free $Bi_{0.5}Na_{0.5}TiO_3$-based ergodic relaxor ferroelectric ceramics." *Scripta Materialia*. 2017;**141**: 15–19.

[29] Zhang, X., Jiang, J., Shen, Z., *et al*. "Polymer nanocomposites with ultrahigh energy density and high discharge efficiency by modulating their nanostructures in three dimensions." *Advanced Materials*. 2018;**30**(16): e1707269.

[30] Jiang, J., Shen, Z., Qian, J., *et al*. "Ultrahigh discharge efficiency in multilayered polymer nanocomposites of high energy density." *Energy Storage Materials*. 2019;**18**: 213–221.

[31] Wang, Y.F., Wang, L.X., Yuan, Q.B., *et al*. "Ultrahigh electric displacement and energy density in gradient layer-structured $BaTiO_3$/Pvdf nanocomposites with an interfacial barrier effect." *Journal of Materials Chemistry A*. 2017;**5**(22): 10849–10855.

[32] Luo, B.C., Wang, X.H., Wang, H.X., Cai, Z.M., Li, L.T. "P(Vdf-Hfp)/Pmma flexible composite films with enhanced energy storage density and efficiency." *Composites Science and Technology*. 2017;**151**: 94–103.

[33] Guan, J.P., Xing, C.Y., Wang, Y.Y., Li, Y.J., Li, J.Y. "Poly (Vinylidene Fluoride) Dielectric composites with both ionic nanoclusters and well dispersed graphene oxide." *Composites Science and Technology*. 2017;**138**: 98–105.

[34] Yuan, Q., Yao, F., Wang, Y., Ma, R., Wang, H. "Relaxor ferroelectric $BaTiO_3$–$0.1Bi(Zn_{0.5}Zr_{0.5})O_3$ ceramic capacitors with high energy density and temperature stable energy storage properties." *Journal of Materials Chemistry C*. 2017;**5**(37): 9552–9558.

[35] Jia, W., Hou, Y., Zheng, M., Zhu, M. "High-temperature dielectrics based on $(1-X)(0.94Bi_{0.5}Na_{0.5}TiO_3$–0.06 $BaTiO_3)$–$X$$NaNbO_3$ system." *Journal of Alloys and Compounds*. 2017;**724**: 306–315.

[36] Hou, D., Usher, T.-M., Zhou, H., *et al.* "Temperature-induced local and average structural changes in $BaTiO_3-XBi(Zn_{1/2}Ti_{1/2})O_3$ solid solutions: The origin of high temperature dielectric permittivity." *Journal of Applied Physics*. 2017;**122**(6): 064103.

[37] Muhammad, R., Iqbal, Y., Reaney, I.M., Randall, C. " $BaTiO_3$–$Bi(Mg_{2/3}Nb_{1/3})O_3$ ceramics for high-temperature capacitor applications." *Journal of the American Ceramic Society*. 2016;**99**(6): 2089–2095.

[38] Hao, X. "A review on the dielectric materials for high energy-storage application." *Journal of Advanced Dielectrics*. 2013;**03**(01): 1330001.

[39] Shen, Z., Wang, X., Luo, B., Li, L. " $BaTiO_3$–$BiYbO_3$ perovskite materials for energy storage applications." *Journal of Materials Chemistry A*. 2015;**3**(35): 18146–18153.

[40] Wang, T., Jin, L., Li, C.C., Hu, Q.Y., Wei, X.Y. "Relaxor ferroelectric $BaTiO_3$–$Bi(Mg_{2/3}Nb_{1/3})O_3$ ceramics for energy storage application." *Journal of the American Ceramic Society*. 2015;**98**(2): 559–566.

[41] Yuan, Q., Li, G., Yao, F.-Z., *et al.* "Simultaneously achieved temperature-insensitive high energy density and efficiency in domain engineered $BaTiO_3$–$Bi(Mg_{0.5}Zr_{0.5})O_3$ lead-free relaxor ferroelectrics." *Nano Energy*. 2018;**52**: 203–210.

[42] Liu, B.B., Wang, X.H., Zhao, Q.C., Li, L.T. "Improved energy storage properties of fine-crystalline $BaTiO_3$ ceramics by coating powders with Al_2O_3 and SiO_2." *Journal of the American Ceramic Society*. 2015;**98**(8): 2641–2646.

[43] Wang, J.W., Xu, C., Shen, B., Zhai, J.W. "Enhancing energy storage density of (Ba, Sr)TiO_3 ceramic particles by coating with Al_2O_3 and SiO_2". *Journal of Materials Science-Materials in Electronics*. 2013;**24**(9): 3309–3314.

[44] Chen, I.W., Wang, X.H. "Sintering Dense Nanocrystalline Ceramics Without Final-Stage Grain Growth." *Nature*. 2000;**404**(6774): 168–171.

[45] Zou, T., Wang, X., Zhao, W., Li, L. "Preparation and properties of fine-grain $(1-X)BiScO_3-XPbTiO_3$ ceramics by two-step sintering." *Journal of the American Ceramic Society*. 2007;**91**(1): 121–126.

[46] Wang, X.H., Chen, P.L., Chen, I.W. "Two-step sintering of ceramics with constant grain-size, I. Y_2O_3." *Journal of the American Ceramic Society*. 2006;**89**(2): 431–437.

[47] Wang, X.H., Deng, X.Y., Bai, H.L., *et al.* "Two-Step Sintering of ceramics with constant grain-size, II: $BaTiO_3$ and Ni-Cu-Zn ferrite." *Journal of the American Ceramic Society*. 2006;**89**(2): 438–443.

[48] Groh, C., Kobayashi, K., Shimizu, H., *et al.* "High-temperature multilayer ceramic capacitors based on $100-X(94Bi_{1/2}Na_{1/2}TiO_3$–6 $BaTiO_3)$–$XK_{0.5}Na_{0.5}NbO_3$." *Journal of the American Ceramic Society*. 2016;**99**(6): 2040–2046.

[49] Bultitude, J., Jones, L., Xu, B., Magee, J., Phillips, R. "An evaluation of BME C0G multilayer ceramic capacitors as building blocks for DC-link capacitors in 3-D power electronics." in *International Symposium on 3D Power Electronics Integration and Manufacturing (3D-PEIM)* (IEEE, 2016).

[50] Xu, X.L., Gurav, A.S., Lessner, P.M., Randall, C.A. "Robust BME Class-I MLCCs for harsh-environment applications." *IEEE Transactions on Industrial Electronics*. 2011;**58**(7): 2636–2643.

[51] Sun, Z., Ma, C., Liu, M., *et al.* "Ultrahigh energy storage performance of lead-free oxide multilayer film capacitors via interface engineering." *Advanced Materials*. 2017;**29**(5).

[52] Mao, Y., Banerjee, S., Wong, S.S. "Hydrothermal synthesis of perovskite nanotubes." *Chem Commun (Camb)* . 2003;**3**(3): 408–409.

[53] Boulos, M., Guillemet-Fritsch, S., Mathieu, F., Durand, B., Lebey, T., Bley, V. "Hydrothermal synthesis of nanosized $BaTiO_3$ powders and dielectric properties of corresponding ceramics." *Solid State Ionics*. 2005;**176**(13–14): 1301–1309.

[54] Wang, D.L., Zhu, K.J., Ji, H.L., Qiu, J.H. "Two-step sintering of the pure $K_{0.5}Na_{0.5}NbO_3$ lead-free piezoceramics and its piezoelectric properties." *Ferroelectrics*. 2009;**392**: 120–126.

[55] Mazaheri, M., Zahedi, A.M., Haghighatzadeh, M., Sadrnezhaad, S.K. "Sintering of titania nanoceramic: Densification and grain growth." *Ceramics International*. 2009;**35**(2): 685–691.

[56] Maca, K., Pouchly, V., Zalud, P. "Two-step sintering of oxide ceramics with various crystal structures." *Journal of the European Ceramic Society*. 2010;**30**(2): 583–589.

[57] Wang, X., Chen, I.-W. "Sintering of nanoceramics." *Nanomaterials Handbook*. New York: Taylor Francis; 2006. pp. 359–382.

[58] Kim, H.D., Han, B.D., Park, D.S., Lee, B.T., Becher, P.F. "Novel two-step sintering process to obtain a bimodal microstructure in silicon nitride." *Journal of the American Ceramic Society*. 2002;**85**(1): 245–252.

[59] Chen, I.W., Wang, Y., Sun, T., Wang, X. "Size effect on ferroelectric transitions in nanograin barium titanate polycrystals." in *Bulletin of the American Physical Society (APS Meeting)*, (2007)

[60] Huan, Y., Wang, X.H., Fang, J., Li, L.T. "Grain size effects on piezoelectric properties and domain structure of $BaTiO_3$ ceramics prepared by two-step sintering." *Journal of the American Ceramic Society*. 2013;**96**(11): 3369–3371.

[61] Huan, Y., Wang, X.H., Fang, J., Li, L.T. "Grain size effect on piezoelectric and ferroelectric properties of $BaTiO_3$ ceramics." *Journal of the European Ceramic Society*. 2014;**34**(5): 1445–1448.

[62] Burns, G., Scott, B.A. "Raman studies of underdamped soft modes in PbTi." *Physical Review Letters*. 1970;**25**(3): 167–170.

[63] Fu, D., Suzuki, H., Ishikawa, K. "Size-induced phase transition in $PbTiO_3$ nanocrystals: Raman scattering study." *Physical Review B*. 2000;**62**(5): 3125–3129.

[64] Pirc, R., Blinc, R. "Off-center Ti model of barium titanate." *Physical Review B*. 2004;**70**(13): 134107.

[65] Shirane, G., Frazer, B.C., Minkiewicz, V.J., Leake, J.A., Linz, A. "Soft optic modes in barium titanate." *Physical Review Letters*. 1967;**19**(5): 234.

[66] DiDomenico, M., Wemple, S.H., Porto, S.P.S., Bauman, R.P. "Raman spectrum of single-domain $BaTiO_3$." *Physical Review*. 1968;**174**(2): 522–530.

[67] Zhu, J.L., Han, W., Zhang, H., *et al.* "Phase coexistence evolution of nano $BaTiO_3$ as function of particle sizes and temperatures." *Journal of Applied Physics*. 2012;**112**(6): 484–748.

[68] Aneggi, E., de Leitenburg, C., Trovarelli, A. "On the role of lattice/surface oxygen in ceria–zirconia catalysts for diesel soot combustion." *Catalysis Today*. 2012;**181**(1): 108–115.

[69] Buscaglia, M.T., Buscaglia, V., Viviani, M., *et al.* "Ferroelectric properties of dense nanocrystalline $BaTiO_3$ ceramics." *Nanotechnology*. 2004;**15**(9): 1113–1117.

[70] Kishi, H., Kohzu, N., Iguchi, Y., *et al.* "Occupational sites and dielectric properties of rare-earth and Mn substituted $BaTiO_3$." *Journal of the European Ceramic Society*. 2001;**21**(10–11): 1643–1647.

[71] Buscaglia, M.T., Buscaglia, V., Alessio, R. "Coating of $BaTiO_3$ crystals with TiO_2: Versatile approach to the synthesis of $BaTiO_3$ tetragonal nano-particles." *Chemistry of Materials*. 2007;**19**(4): 711–718.

[72] Makovec, D., Samardzija, Z., Drofenik, M. "Solid solubility of holmium, yttrium, and dysprosium in $BaTiO_3$." *Journal of the American Ceramic Society*. 2004;**87**(7): 1324–1329.

[73] Kishi, H., Okino, Y., Honda, M., *et al.* "The Effect of MgO and rare-earth oxide on formation behavior of core-shell structure in $BaTiO_3$." *Japanese Journal of Applied Physics Part 1-Regular Papers Short Notes & Review Papers*. 1997;**36**(9b): 5954–5957.

[74] Kaneda, K., Lee, S., Donnelly, N.J., Qu, W., Randall, C.A., Mizuno, Y. "Kinetics of oxygen diffusion into multilayer ceramic capacitors during the re-oxidation process and its implications on dielectric properties." *Journal of the American Ceramic Society*. 2011;**94**(11): 3934–3940.

[75] Albertsen, K. "Re-oxidation of Ni-MLCS." *Journal of the European Ceramic Society*. 2004;**24**(6): 1883–1887.

[76] Kishi, H., Mizuno, Y., Hagiwara, T., Orimo, H., Ohsato, H. "Effect of re-oxidation on dielectric properties in Ni-MLCC." *Journal of Electroceramics*. 2007;**21**(1–4): 22–28.

[77] Ferkel, H., Hellmig, R.J. "Effect of nanopowder deagglomeration on the densities of nanocrystalline ceramic green bodies and their sintering behaviour." *NanoStructured Materials*. 1999;**11**: 617–622.

[78] Lee, J.Y., Lee, J.H., Hong, S.H., Lee, Y.K., Choi, J.Y. "Coating $BaTiO_3$ nanolayers on spherical Ni powders for multilayer ceramic capacitors." *Advanced Materials*. 2003;**15**(19): 1655.

[79] Yaseen, H., Baltianski, S., Tsur, Y. "Effect of incorporating method of niobium on the properties of doped barium titanate ceramics." *Journal of the American Ceramic Society*. 2006;**89**(5): 1584–1589.

[80] Tian, Z.B., Wang, X.H., Shu, L.K., *et al*. "Preparation of nano $BaTiO_3$-based ceramics for multilayer ceramic capacitor application by chemical coating method." *Journal of the American Ceramic Society*. 2009;**92**(4): 830–833.

[81] Zhang, Y.C., Wang, X.H., Tian, Z.B., Hur, K.H., Li, L.T. "Preparation of BME MLCC powders by aqueous chemical coating method." *Journal of the American Ceramic Society*. 2011;**94**(10): 3286–3290.

[82] Zhang, Y.C., Wang, X.H., Kim, J.Y., *et al*. "High performance $BaTiO_3$-based BME-MLCC nanopowder prepared by aqueous chemical coating method." *Journal of the American Ceramic Society*. 2012;**95**(5): 1628–1633.

[83] Gong, H.L., Wang, X.H., Zhang, S.P., Li, L.T. "Sintering behavior and reliability characteristics of $BaTiO_3$-based ceramics prepared by different methods." *Journal of Materials Science*. 2015;**50**(10): 3523–3530.

[84] Tian, Z.B., Wang, X.H., Zhang, Y.C., Song, T.H., Hur, K.H., Li, L.T. "Fabrication of $BaTiO_3$-based dielectrics for ultrathin-layer multilayer ceramic capacitor application by a modified coating approach." *Japanese Journal of Applied Physics*. 2011;**50**(2).

[85] Hennings, D., Rosenstein, G. "Temperature-stable dielectrics based on chemically inhomogeneous $BaTiO_3$." *Journal of the American Ceramic Society*. 1984;**67**(4): 249–254.

[86] Yang, G.Y., Lian, G.D., Dickey, E.C., *et al*. "Oxygen nonstoichiometry and dielectric evolution of $BaTiO_3$. Part II - Insulation resistance degradation under applied DC bias." *Journal of Applied Physics*. 2004;**96**(12): 7500–7508.

[87] Polotai, A.V., Fujii, I., Shay, D.P., Yang, G.-Y., Dickey, E.C., Randall, C.A. "Effect of heating rates during sintering on the electrical properties of ultra-thin Ni–$BaTiO_3$ multilayer ceramic capacitors." *Journal of the American Ceramic Society*. 2008;**91**(8): 2540–2544.

[88] Chang, C.u., Wang, W.i., Huang, C.u., Alford, N. "Effect of MgO and Y_2O_3 doping on the formation of core–shell structure in $BaTiO_3$ ceramics." *Journal of the American Ceramic Society*. 2013;**96**(8): 2570–2576.

[89] Donnelly, N.J., Randall, C.A. "Mixed conduction and chemical diffusion in a $Pb(Zr_{0.53},Ti_{0.47})O_3$ buried capacitor structure." *Applied Physics Letters*. 2010;**96**(5): 1060.

[90] Gong, H.L., Wang, X.H., Zhao, Q.C., Li, L.T. "Effect of Mg on the dielectric and electrical properties of $BaTiO_3$-based ceramics." *Journal of Materials Science*. 2015;**50**(21): 6898–6906.

[91] Ghayour, H., Abdellahi, M. "A brief review of the effect of grain size variation on the electrical properties of $BaTiO_3$-based ceramics." *Powder Technology*. 2016;**292**: 84–93.

[92] Zhu, C.Q., Wang, X.H., Zhao, Q.C., Cai, Z.M., Cen, Z.Y., Li, L.T. "Effects of grain size and temperature on the energy storage and dielectric tunability

of non-reducible $BaTiO_3$-based ceramics." *Journal of the European Ceramic Society*. 2019;**39**(4): 1142–1148.

[93] Fang, C., Zhou, D.X., Gong, S.P. "Core-shell structure and size effect in barium titanate nanoparticle." *Physica B-Condensed Matter*. 2011;**406**(6–7): 1317–1322.

[94] Badapanda, T., Rout, S.K., Panigrahi, S., Sinha, T.P. "Phase formation and dielectric study of Bi doped $BaTi_{0.75}Zr_{0.25}O_3$ ceramic." *Current Applied Physics*. 2009;**9**(4): 727–731.

[95] Chen, F.J., Pan, Y.N., Lee, C.Y., Lin, C.S. "Internal stress control of nickel-phosphorus electrodeposits using pulse currents." *Journal of the Electrochemical Society*. 2010;**157**(3): D154-D158.

[96] Xu, Q., Zhang, X.F., Huang, Y.H., *et al.* "Effect of MgO on structure and nonlinear dielectric properties of $Ba_{0.6}Sr_{0.4}TiO_3$/MgO composite ceramics prepared from superfine powders." *Journal of Alloys and Compounds*. 2009;**488**(1): 448–453.

[97] Yoon, S.H., Randall, C.A., Hur, K.H. "Correlation between resistance degradation and thermally stimulated depolarization current in acceptor (Mg)-doped $BaTiO_3$ submicrometer fine-grain ceramics." *Journal of the American Ceramic Society*. 2010;**93**(7): 1950–1956.

[98] Yoon, S.H., Randall, C.A., Hur, K.H. "Effect of acceptor (Mg) concentration on the resistance degradation behavior in acceptor (Mg)-doped $BaTiO_3$ bulk ceramics: II. Thermally stimulated depolarization current analysis." *Journal of the American Ceramic Society*. 2009;**92**(8): 1766–1772.

[99] Gong, L., Wang, X.H., Zhang, S.P., Wen, H., Li, L.T. "Grain size effect on electrical and reliability characteristics of modified fine-grained $BaTiO_3$ ceramics for MLCCs." *Journal of the European Ceramic Society*. 2014;**34** (7): 1733–1739.

[100] Oyama, T., Wada, N., Takagi, H., Yoshiya, M. "Trapping of oxygen vacancy at grain boundary and its correlation with local atomic configuration and resultant excess energy in barium titanate: A systematic computational analysis." *Physical Review B*. 2010;**82**(13): 134107.

[101] Schaffrin, C. "Oxygen diffusion in $BaTiO_3$ ceramic." *Physica Status Solidi (a)*. 1976;**25**(1): 79–88.

[102] Loh, E. "A model of DC leakage in ceramic capacitors." *Journal of Applied Physics*. 1982;**53**(9): 6229–6235.

[103] Zhu, W.D., Akbar, S.A., Asiaie, R., Dutta, P.K. "Sintering and dielectric properties of hydrothermally synthesized cubic and tetragonal $BaTiO_3$ powders." *Japanese Journal of Applied Physics Part 1-Regular Papers Short Notes & Review Papers*. 1997;**36**(1a): 214–221.

[104] Mizuno, Y., Hagiwara, T., Kishi, H. "Microstructural design of dielectrics for Ni-MLCC with ultra-thin active layers." *Journal of the Ceramic Society of Japan*. 2007;**115**(1342): 360–364.

[105] Tian, Z.B., Wang, X.H., Zhang, Y.C., *et al.* "Formation of core-shell structure in ultrafine-grained $BaTiO_3$-based ceramics through nanodopant method." *Journal of the American Ceramic Society*. 2010;**93**(1): 171–175.

[106] Han, F., Bai, Y., Qiao, L.J., Guo, D. "A systematic modification of the large electrocaloric effect within a broad temperature range in rare-earth doped $BaTiO_3$ ceramics." *Journal of Materials Chemistry C*. 2016;**4**(9): 1842–1849.

[107] Jung, Y.I., Choi, S.Y., Kang, S.J.L. "Grain-growth behavior during stepwise sintering of barium titanate in hydrogen gas and air." *Journal of the American Ceramic Society*. 2003;**86**(12): 2228–2230.

[108] Hotta, Y., Duran, C., Sato, K., Nagaoka, T., Watari, K. "Densification and grain growth in $BaTiO_3$ ceramics fabricated from nanopowders synthesized by ball-milling assisted hydrothermal reaction." *Journal of the European Ceramic Society*. 2008;**28**(3): 599–604.

[109] Binner, J., Vaidhyanathan, B. "Processing of bulk nanostructured ceramics." *Journal of the European Ceramic Society*. 2008;**28**(7): 1329–1339.

[110] Chen, I.W. "A stochastic-theory of grain-growth." *Acta Metallurgica*. 1987;**35**(7): 1723–1733.

[111] Zhao, Q., Gong, H., Wang, X., Chen, I.W., Li, L. "Superior reliability via two-step sintering: Barium titanate ceramics." *Journal of the American Ceramic Society*. 2016;**99**(1): 191–197.

[112] Tian, Z.B., Wang, X.H., Gong, H.L., Song, T.H., Hur, K.H., Liz, L.T. "Core-shell structure in nanocrystalline modified $BaTiO_3$ dielectric ceramics prepared by different sintering methods." *Journal of the American Ceramic Society*. 2011;**94**(4): 973–977.

[113] Gong, H.L., Wang, X.H., Kim, J.Y., Tian, Z.B., Yao, G.F., Li, L.T. "Preparation of nano $BaTiO_3$-based dielectric ceramics by different sintering methods." *High-Performance Ceramics VII: Parts 1 and 2*. 2012;**512–515**: 1136–1141.

[114] Baiatu, T., Waser, R., Hardtl, K.-H. "DC electrical degradation of perovskite-type titanates: III. A model of the mechanism." *Journal of the American Ceramic Society*. 1990;**73**(6): 1663–1673.

[115] Du, X., Chen, I.-W. "Fatigue of $Pb(Zr_{0.53}Ti_{0.47})O_3$ ferroelectric thin films." *Journal of Applied Physics*. 1998;**83**(12): 7789.

[116] Chen, I.W. "Structural origin of relaxor ferroelectrics—Revisited." *Journal of the Physics and Chemistry of Solids*. 2000;**61**(2): 197–208.

[117] Bazant, M.Z., Thornton, K., Ajdari, A. "Diffuse-charge dynamics in electrochemical systems." *Physical Review E: Statistical, Nonlinear, Biological, and Soft Matter Physics*. 2004;**70**(2 Pt 1): 021506.

[118] Rodewald, S., Fleig, J., Maier, J. "Resistance degradation of iron-doped strontium titanate investigated by spatially resolved conductivity measurements." *Journal of the American Ceramic Society*. 2000;**83**(8): 1969–1976.

[119] Wang, Y.L., Li, L.T., Qi, J.Q., Ma, Z.W., Cao, J.L., Gui, Z.L. "Nickel diffusion in base-metal-electrode MLCCs." *Materials Science and Engineering B-Solid State Materials for Advanced Technology*. 2003;**99** (1–3): 378–381.

[120] Zhao, P., Wang, H., Wu, L., *et al.* "High-performance relaxor ferroelectric materials for energy storage applications." *Advanced Energy Materials.* 2019;**9**(17): 1803048.

[121] Li, J., Li, F., Xu, Z., Zhang, S. "Multilayer lead-free ceramic capacitors with ultrahigh energy density and efficiency." *Advanced Materials.* 2018;**83**(8): 1802155.

[122] Wang, G., Li, J.L., Zhang, X., *et al.* "Ultrahigh energy storage density lead-free multilayers by controlled electrical homogeneity." *Energy & Environmental Science.* 2019;**12**(2): 582–588.

[123] Beuerlein, M.A., Kumar, N., Usher, T.M., *et al.* "Current understanding of structure–processing–property relationships in $BaTiO_3$–$Bi(M)O_3$ dielectrics." *Journal of the American Ceramic Society*. 2016;**99**(9): 2849–2870.

[124] Li, Q., Yao, F.Z., Liu, Y., Zhang, G.Z., Wang, H., Wang, Q. "High-temperature dielectric materials for electrical energy storage." *Annual Review of Materials Research*. 2018;**48**(1): 219–243.

[125] Cai, Z.M., Zhu, C.Q., Wang, H.X., *et al.* "High-temperature lead-free multilayer ceramic capacitors with ultrahigh energy density and efficiency fabricated via two-step sintering." *Journal of Materials Chemistry A.* 2019;**7**(24): 14575–14582.

[126] Jia, W., Hou, Y., Zheng, M., *et al.* "Advances in lead-free high-temperature dielectric materials for ceramic capacitor application." *IET Nanodielectrics.* 2018;**1**(1): 3–16.

[127] Samantaray, M.M., Gurav, A., Dickey, E.C., Randall, C.A. "Electrode defects in multilayer capacitors part II: Finite element analysis of local field enhancement and leakage current in three-dimensional microstructures." *Journal of the American Ceramic Society*. 2012;**95**(1): 264–268.

[128] Samantaray, M.M., Gurav, A., Dickey, E.C., Randall, C.A. "Electrode defects in multilayer capacitors part I: Modeling the effect of electrode roughness and porosity on electric field enhancement and leakage current." *Journal of the American Ceramic Society*. 2012;**95**(1): 257–263.

[129] Chen, L.L., Wang, H.X., Zhao, P.Y., *et al.* "Multifunctional $BaTiO_3$–$(Bi_{0.5}Na_{0.5})TiO_3$-based MLCC with high-energy storage properties and temperature stability." *Journal of the American Ceramic Society*. 2019;**102** (7): 4178–4187.

[130] Wang, D.W., Fan, Z.M., Zhou, D., *et al.* "Bismuth ferrite-based lead-free ceramics and multilayers with high recoverable energy density." *Journal of Materials Chemistry A.* 2018;**6**(9): 4133–4144.

[131] Young, A.L., Hilmas, G.E., Zhang, S.C., Schwartz, R.W. "Mechanical vs. electrical failure mechanisms in high voltage, high energy density multilayer ceramic capacitors." *Journal of Materials Science*. 2007;**42**(14): 5613–5619.

[132] Cai, Z., Wang, X., Luo, B., Hong, W., Wu, L., Li, L. "Multiscale design of high-voltage multilayer energy-storage ceramic capacitors." *Journal of the American Ceramic Society*. 2018;**101**(4): 1607–1615.

[133] Luo, Q., Li, X., Yao, Z., *et al.* "The role of dielectric permittivity in the energy storage performances of ultrahigh-permittivity $(Sr_xBa_{1-x})(Ti_{0.85}Sn_{0.15})O_3$ ceramics." *Ceramics International.* 2018;**44**(5): 5304–5310.

[134] Shen, Z.H., Wang, J.J., Lin, Y., Nan, C.W., Chen, L.Q., Shen, Y. "High-throughput phase-field design of high-energy-density polymer nanocomposites." *Advanced Materials.* 2018;**30**(2): 1704380.

[135] Liebault, J., Vallayer, J., Goueriot, D., Treheux, D., Thevenot, F. "How the trapping of charges can explain the dielectric breakdown performance of alumina ceramics." *Journal of the European Ceramic Society*. 2001;**21**(3): 389–397.

[136] Neusel, C., Jelitto, H., Schmidt, D., Janssen, R., Felten, F., Schneider, G.A. "Thickness-dependence of the breakdown strength: analysis of the dielectric and mechanical failure." *Journal of the European Ceramic Society*. 2015;**35**(1): 113–123.

[137] Grunebohm, A., Nishimatsu, T. "Influence of defects on ferroelectric and electrocaloric properties of $BaTiO_3$." *Physical Review B*. 2016;**93**(13): 12.

[138] Champion, J.V., Dodd, S.J., Stevens, G.C. "Analysis and modeling of electrical tree growth in synthetic resins over a wide-range of stressing voltage." *Journal of Physics D-Applied Physics*. 1994;**27**(5): 1020–1030.

[139] Dissado, L., Sweeney, P., Fothergill, J. "The field dependence of electrical tree growth." in *Sixth International Conference on Dielectric Materials, Measurements and Applications, 1992* (IET, 1992).

[140] Weibull, W. "A statistical distribution function of wide applicability." *Journal of Applied Mechanics-Transactions of the ASME*. 1951;**18**(3): 293–297.

[141] Niemeyer, L., Pietronero, L., Wiesmann, H.J. "Fractal dimension of dielectric-breakdown." *Physical Review Letters*. 1984;**52**(12): 1033–1036.

[142] Su, Y., Liu, N., Weng, G.J. "A phase field study of frequency dependence and grain-size effects in nanocrystalline ferroelectric polycrystals." *Acta Materialia*. 2015;**87**: 293–308.

[143] Hong, W., Pitike, K.C. "Modeling breakdown-resistant composite dielectrics." *Iutam Symposium on Mechanics of Soft Active Materials*. 2015;**12**: 73–82.

[144] Pitike, K.C., Hong, W. "Phase-field model for dielectric breakdown in solids." *Journal of Applied Physics*. 2014;**115**(4): 8.

[145] Cai, Z.M., Wang, X.H., Luo, B.C., Hong, W., Wu, L.W., Li, L.T. "Dielectric response and breakdown behavior of polymer-ceramic nanocomposites: The effect of nanoparticle distribution." *Composites Science and Technology*. 2017;**145**: 105–113.

[146] Cai, Z., Wang, X., Li, L. "Phase-field modeling of electromechanical breakdown in multilayer ceramic capacitors." *Advanced Theory and Simulations*. 2018;**2**(4): 1800179.

[147] Chen, G., Zhao, J.W., Li, S.T., Zhong, L.S. "Origin of thickness dependent DC electrical breakdown in dielectrics." *Applied Physics Letters*. 2012;**100**(22): 222904.

[148] Milliken, A.D., Bell, A.J., Scott, J.F. "Dependence of breakdown field on dielectric (interelectrode) thickness in base-metal electroded multilayer capacitors." *Applied Physics Letters*. 2007;**90**(11): 112910.

[149] Cai, Z.M., Wang, H.X., Zhao, *et al.* "Significantly enhanced dielectric breakdown strength and energy density of multilayer ceramic capacitors with high efficiency by electrodes structure design." *Applied Physics Letters*. 2019;**115**(2): 023901.

Chapter 11
Future prospects: polymer part

Minhao Yang[1] and Zhi-Min Dang[2]

11.1 Introduction

Driven by an urgent need for the development of high-power energy storage devices, both scientists and engineers pay increasing attention to the electrostatic capacitor, which possesses the highest power density among various types of energy storage technologies [1–5]. As the passive electronic component, electrostatic capacitor stores and releases electrical energy through rapid electric field-induced polarization and depolarization. In its simplest form, electrostatic capacitor consists of two electrically conductive plates and a dielectric layer. Among various types of electrostatic capacitors, the film capacitor using an insulating polymeric film as the dielectric layer is highly desirable in the electric power systems due to the high electrical breakdown strength (E_b) of the polymeric film. However, the miniaturization of the power system and the realization of compact energy storage technologies are severely hindered by the low volumetric energy density and operating temperature of the polymeric film. For example, the energy density of biaxially oriented polypropylenes (BOPP), the best commercially available film with a market share of 50%, is merely 1-2 J/cm^3, which is at least one order of magnitude lower than those of electrochemical supercapacitors and batteries. Additionally, the highest operating temperature for BOPP film is limited to 105 °C due to the low melting temperature of PP resin. As a result, an extra cooling system is required to decrease the environmental temperature so that the BOPP film capacitor can be operated at a safe temperature, accordingly leading to the dramatic increases in the weight, volume and cost of the power system. Therefore, improvements in the energy density and operating temperature of the polymeric film have become essential for the development of advanced film capacitors.

In the last two decades, significant advances have been achieved in improving the energy storage density and operating temperature of dielectric films through employing polymer-based nano-dielectric composites (PNCs) [6], molecularly designed polymers (MDPs) [7], multi-layered hierarchical polymer composites

[1]Institute of Advanced Materials, North China Electric Power University, Beijing, China
[2]State Key Laboratory of Power System and Department of Electrical Engineering, Tsinghua University, Beijing, China

(MHPCs) [8] and surface coated polymers (SCPs) [1,9]. Among them, PNCs formed by incorporating inorganic ceramic fillers into an organic polymer matrix have attracted tremendous attention because it combines the high dielectric constant (ε) of inorganic dielectrics with the high E_b of organic dielectrics. Ceramic fillers with perovskite structure such as $BaTiO_3$ (BT) provides PNCs with higher polarization intensity while polymer matrix offers PNCs with higher E_b. However, the energy storage density of PNCs is also determined by other critical factors associated with PNCs including the filler features (e.g. shape, size, concentration and spatial distribution), polymer/filler interface and polymer features (e.g. molecular structure, crystalline structures and processability) [6]. Using MDPs to improve the energy storage density and temperature stability of dielectric films is supposed to be one of the most fundamental methods because of its innovation on the molecular chain level. The incorporation of polar functional groups into polymer chains can increase the dielectric constant and the moderate cross-linking of polymer chains can not only produce reversible polarization but also enhance the breakdown strength. Besides, the glass transition temperature (T_g) of the polymer can also be moved to a high-temperature region after introducing some specific functional groups into the polymer main or side chains. However, the issues of scaling up of polymers and the film processability should be carefully considered. Integrating several hierarchical polymer composites layers into a single MHPCs layer is another popular approach to improve the dielectric and energy storage performance of dielectric film [8]. The dielectric constant and breakdown strength of dielectric film can be easily modulated through either changing the spatial position of different polymer composites layers or changing the composition of the individual layer. Yet not all polymer pairs can be integrated into a multi-layer film [10]. Poor compatibility between the adjacent layers will cause the delamination, accordingly leading to the mechanical failure of multi-layer films. Employing the SCPs with an organic or inorganic thin layer to improve the dielectric constant and breakdown strength of dielectric films has also attracted considerable attention. The introduction of an organic layer can reduce the surface defects of the pristine film, thus improving the breakdown strength. The introduction of an inorganic thin layer can hinder the charge injection from the electrodes, accordingly suppressing the leakage current and hysteresis. Nevertheless, the issue of surface affinity on both inorganic and organic SCPs should be carefully taken into consideration.

This short prospect aims to provide some insights on how to improve the dielectric and energy storage performances of dielectric films through rational selection, design, combination and treatment of polymers from the following PNCs, MDPs, MHPCs and SCPs directions.

11.2 Polymers based nano-dielectric composites

Serving as the host matrix in the PNCs, the dielectric and energy storage performances and high-temperature stability of dielectric films are strongly related to the selected polymer. Polymers with high breakdown strength, high dielectric constant, low dielectric loss and high-temperature tolerance characteristics are highly desired for the film capacitor. However, the scalable production of a uniform and high-quality thin film is a major consideration for the PNCs due to the sharply increased viscosity, voids and defects after the incorporation of 'hard' ceramic fillers into the 'soft' polymer matrix. Additionally, it is difficult to acquire the accurate multi-scale structure of the composite film by controlling the processing conditions during the scalable production. More importantly, higher loading of the 'hard' phase will greatly weaken the breakdown strength of PNCs films. Therefore, lower loading of ceramic fillers in the PNCs should be a prerequisite.

Both linear and nonlinear polymers have been extensively used as the host polymer for PNCs. Typically, linear polymers exhibit low dielectric loss and high breakdown strength. But the improvement of energy storage density of linear polymers is greatly limited owing to its low dielectric constant. For this reason, nonlinear polymers with polar groups in their molecular chains have been widely used for the fabrication of PNCs in the last two decades because of their higher dielectric constant compared to the linear one. Poly (vinylidene fluoride) (PVDF)-based polymers are one family of the most frequently used nonlinear polymers in the PNCs due to their high dielectric constant (>10). Moreover, their solvent dis-solvability makes them compatible with the solution blending process for the incorporation of various fillers into the polymer matrix. Although the energy storage density has been significantly improved by employing nonlinear polymers, it is usually accompanied by a sharp decrease in discharged efficiency (η) due to their high intrinsic ferroelectric losses and remnant polarization. The dissipated energy will be partially converted into the heat, accordingly giving rise to concerns about the safe operation of the film capacitor. Additionally, it is more difficult to process the nonlinear polymer pellets into a thin film and the applied melt-stretching or shearing force will inevitably induce the partial conversion of non-polar α phase into the polar β phase. Such a conversion will subsequently cause an increase of ferroelectric loss and remnant polarization, accordingly further decreasing the discharged efficiency. Compared with the non-polar α and polar β phases in the semi-crystalline PVDF, the γ phase offers PVDF a high breakdown strength, a paraelectric like D-E loop with relatively low remnant polarization and high discharge energy efficiency. Therefore, purposely inducing the formation of γ phase in the PVDF matrix would be a feasible strategy to enhance the dielectric and energy storage performances of PVDF. Apart from controlling the crystalline phase of PVDF, introducing some monomers including trifluoroethylene (TrFE), chlorotrifluoroethylene (CTFE), chlorofluoroethylene (CFE) and hexafluoropropylene (HFP) into the PVDF main chains to fabricate copolymers is another effective approach to improve the dielectric and electrical properties of PVDF [6]. The introduced TrFE can act as defects to enhance the crystallinity and stabilize the β phase. Besides, applying the electron irradiation or introducing a second monomer (CTFE or CFE) into P (VDF-TrFE) can further break up the large ferroelectric domains into nano-domains, consequently leading to a relaxor ferroelectricity behaviour with suppressed remnant polarization. However, their maximum discharged energy density is still limited due to the early saturation behaviour and the electrical displacement exhibits no further increase with continuously increasing the electric field. Compared with ferroelectric and relaxor polymers, the CTFE and HFP monomer-modified PVDF co-polymers hold weaker ferroelectricity and lower remnant polarization. Yet their maximum polarization and discharged energy density are limited due to the suppressed ferroelectric property. Although the dielectric and energy storage properties of PVDF can be easily modified through the irradiation or introduction of some monomers, the produced remnant polarization, maximum polarization and breakdown strength should be seriously considered to achieve an optimum energy storage performance.

Although substantial efforts have been made in the nonlinear systems (PVDF and its copolymers and terpolymers), polypropylene (PP), which is a non-polar linear polymer, holds its position firmly as the industrial resin for the fabrication of BOPP film. However, PP usually suffers from low dielectric constant and low working temperature, accordingly leading to a low volumetric energy storage density and narrow operating temperature range. In order to increase the energy storage density and broaden the operation temperature range of PP, specially designed PP molecular structure is highly desired to modify its intrinsic polarization and breakdown behaviours. Some promising routes for the synthesis of molecularly designed PP will be discussed in the Section 11.3.

In addition to the PVDF-based polymers and PP, various polymers including polystyrene (PS), poly (ethylene terephthalate) (PET), poly (methyl methacrylate) (PMMA), polycarbonate (PC), polytetrafluoroethylene (PTFE), polyetheretherketone (PEEK), polyphenylene sulfide (PPS), polyethylene naphthalate (PEN), polyuria, polythioureas, polyimide (PI) and polyetherimide (PEI) have been widely used as the host for fillers. For the high-temperature application, PEEK, PPS, PEN and PEI are commonly used as the host polymers due to their higher T_g. However, not all of them were widely adopted for the industrial application due to some tough issues including high dielectric loss, low operating temperature, lack of self-healing and uneasy conversion into a thin film. Additionally, the method of composites processing should be properly adapted according to the selected host polymer. Melt blending is greatly appreciated for the industrial application due to its solvent-free process. Despite a homogenous distribution of fillers in the solution blended composites, the use and disposal of environmentally hazardous solvents are of great concern for the solution blending process. In order to reduce the agglomeration of fillers in the melt blended composites, the prolongation of blending time and increase of blending temperature are strongly suggested. For the film formation process, the formation of pinhole and defects and thickness variation should be avoided no matter what kind of film processing methods are adopted including solution casting, hot compression moulding and heated stretching.

11.3 Molecularly designed polymers

Employing MDPs is regarded as one of the most promising approaches to solve the low energy density and narrow working temperature range of dielectric films [7].

This route is especially suitable for the nonlinear polymer system due to its controlled polymerization process. In addition to the synthesis of PVDF copolymers and terpolymers as discussed previously, the synthesis of grafted PVDF based polymers has been proved to be one of the effective approaches to suppress the remnant polarization and hysteresis of nonlinear polymers. The nano-scale ferroelectric crystals can be separated by the introduced block-phase serving as the confining interfacial layer to suppress the remnant polarization.

For the linear polymer systems, the categories of MDPs is greatly limited due to the lack of appropriate chemical modification methods. Nevertheless, incorporating some high polarity groups such as -CN and -OH into the side chains of linear polymers has been experimentally demonstrated as a feasible strategy to modify the dielectric and energy storage performances of linear polymer systems. The introduced polar groups not only offers the linear polymers a high dielectric constant but also maintains a reversible polarization especially for the non-polar PP. As for the high temperature-tolerance behaviour, some high steric hindrance groups should be incorporated into the polymeric main or side chains to increase the T_g of the selected polymers. However, the effects of the density and distribution of these functional groups on the dielectric and energy storage properties of dielectric polymers are still unclear. More importantly, novel MDPs with high dielectric constant and high breakdown strength should be further designed and synthesized. Thus, further experimental and theoretical investigations should be carried out to reveal the unsolved mysteries.

11.4 Multi-layered hierarchical polymer composites

More recently, designing MHPCs has become an emerging strategy to resolve the paradox of simultaneous improvement of dielectric constant and breakdown strength in a single-layered composite film [8]. MHPCs are usually fabricated by a coextrusion process or a layer-by-layer stacking procedure. Among various MHPCs, the sandwich-structured composite as the simplest model of MHPCs has aroused significant interest over the last decade due to its simple structure and tunable property. The sandwiched-structure composite can be mainly divided into two types. The first one consists of two outer layers with high breakdown strength and an inner layer with high dielectric constant. The second type is usually composted of two outer layers with high dielectric constant and an inner layer with high breakdown strength. Each layer in the sandwiched-structure composite can be either a pure polymer or a polymer composite. Placing the high breakdown strength layer near the electrodes as the outer layer can effectively suppress the injection of charges and leakage current and block the growth of electrical trees at the beginning of the electrical breakdown. On the contrary, inserting the high breakdown strength layer as the inner layer into two high dielectric constant layers can not only optimize the electric field distribution but also block the electrical trees crossing the interface between the adjacent layers. The high electrical breakdown strength layer and high dielectric constant layer offer the composites a high breakdown strength and high dielectric constant, respectively, while the special sandwich structure provides the composites with a suppressed hysteresis. Consequently, high energy storage density and discharged efficiency are simultaneously achieved. The finite element analysis of electric field distribution and breakdown process further proves this conclusion. More importantly, the thickness ratio between the outer and inner layers has a significant influence on the polarization and breakdown processes. Apart from these two kinds of sandwiched-structured MHPCs, MHPCs with gradient distribution of fillers have been widely reported due to their high electric displacement and breakdown strength. The layer with higher loading of fillers contributes to high electric displacement while the layer with lower loading of fillers suppresses the growth of electrical trees.

However, not all the polymers can be stacked into a single MHPCs film. The selected polymers should have good adhesion with their neighbouring layers to avoid the delamination but must be simultaneously immiscible. In some special cases of MHPCs, a tie layer should be introduced to improve the interfacial adhesion and the tie layer should be compatible with its adjacent layers. The introduction of tie layer should not weaken the dielectric and breakdown strength of MHPCs. But a huge challenge is how to scalable preparation of large-sized MHPCs films without introducing defects or causing delamination. Therefore, further investigations should be conducted to study the interfacial adhesion at the high operating temperature range and the influence of tie layer on the dielectric and energy storage performances of MHPCs films. MHPCs films with dozens or hundreds of layers are also needed to be further studied due to their multiple tunable parameters.

11.5 Surface coated polymers

Most of polymer films suffer from a low dielectric constant and the charge injection at the electrode-dielectric film is a main concern that causes the early breakdown of the selected films. Using SCPs provides a cost-effective and scalable production approach to improve the dielectric and breakdown strength of dielectric film [9]. The low dielectric constant issue can be resolved by coating an organic layer on the surface of dielectric film. A rough treatment process is usually needed for the commercial dielectric film to improve the adhesion between the dielectric films and metallized electrodes. However, such a treatment will inevitably introduce some surface defects and these defects will lead to a serious deviation of breakdown strength at different locations of the film, thereby causing the deterioration of high electric field performance. The introduced organic coating layer can cover or heal these surface defects and provide a uniform and smooth surface for the dielectric film. As for the issue of charge injection from the electrodes, it can be resolved through coating the dielectric film with the inorganic oxides and nitrides. The inorganic layer is supposed to enhance the breakdown strength of dielectric film. However, its adhesion with the dielectric film should be seriously considered. Based on this, special surface treatment of dielectric film should be carried out before the deposition or coating process. Additionally, substantial efforts should be made to resolve the issues of low dielectric constant and flexibility for the inorganic coated film. The thermal conduction behaviour of the SCPs at high working temperature should also be seriously taken into consideration.

11.6 Summary

In summary, although scientists have made great scientific explorations and progresses in solving the fundamental issues of dielectric polymer films by employing the aforementioned four strategies over the past 20 years, most of these efforts and achievements are made on a lab scale. Huge gap still exists between the innovations of research labs and the application of commercial products, which is especially true in the film capacitor field. The manufacture of a capacitor film capable of being commercially used usually undergoes a complicated process, which is mainly composed of film scale-up, film treatment, metallization and post-winding procedures. From the view of scalable production of high-quality film, appropriate surface modification of fillers with a polymer shell and low loading of fillers should be ensured for the PNCs films. The interfacial adhesion in the MHPCs films should also be carefully considered to avoid the delamination. As for the film treatment, the introduction of an organic coating layer in the SCPs films is necessary to eliminate the introduced surface defects during the film treatment process. Concerning the metallization process, the adhesion between electrodes and polymer film should be seriously considered. For the post-winding process, the introduced stress should be eliminated especially for the MHPCs films. Apart from the dielectric constant and breakdown strength, the thickness of film is the third critical parameter in determining the volumetric energy density. Therefore, more efforts should be made to reduce the thickness of dielectric film, accordingly realizing the compact energy storage of film capacitor.

The development of novel and high-performance polymeric film involves a variety of disciplines such as chemistry, physics, materials science and engineering and electrical engineering. Chemists should devote their efforts to design and synthesize the novel structured MDPs films. This is the most promising approach to achieve the desirable dielectric and energy storage properties without sacrificing other critical performances. Physicists should propose more accurate and precise physical models and visible polarization images to get a better understanding of the local polarization and breakdown behaviours for the interfacial region of dielectric film. Materials scientists and engineers may try to select some commercially manufacturable polymer films or cost-effective polymer films and improve the performance of these existing films from the SCPs strategy. The strategies employing the films of PNCs and MHPCs should be further optimized due to the huge gap between the lab's products and industrial products. Electrical engineers should try to design some novel and smart instruments that can accurately measure the thickness, dielectric and energy storage properties of the thin polymer film.

There is still a long way to go before all the critical issues of dielectric polymer film are ironed out, but we will never know what we can do till we try.

Acknowledgements

The authors gratefully thank Dr Qi Li for the invitation to give some prospects from the view of polymer part. Besides, the authors gratefully acknowledge the support from State Grid Corporation Science Technology Project (5202011600UK), the National Natural Science Foundation of China (Nos. 51937007 and 51425201) and the National Basic Research Program of China (973 Program) (Grant No. 2015CB654603). This project is also supported by State Key Laboratory of New Ceramic and Fine Processing Tsinghua University (No. KF201907) and China Association for Science and Technology (CAST).

References

[1] Tan D.Q. 'Review of polymer-based nanodielectric exploration and film scale-up for advanced capacitors'. *Advanced Functional Materials*. 2020; 30(18): 1808567

[2] Dang Z.M., Yuan J.K., Zha J.W., Zhou T., Li S.T., Hu G.H. 'Fundamentals, processes and applications of high-permittivity polymer-matrix composites'. *Progress in Materials Science*. 2012; 57(4): 660–723

[3] Dang Z.M., Yuan J.K., Yao S.H., Liao R.J. 'Flexible nanodielectric materials with high permittivity for power energy storage'. *Advanced Materials*. 2013; 25(44): 6334–6365

[4] Dang Z.M., Zheng M.S., Zha J.W. '1D/2D carbon nanomaterial-polymer dielectric composites with high permittivity for power energy storage applications'. *Small*. 2016; 12(13): 1688–1701

[5] Zhong S.L., Dang Z.M., Zhou W.Y., Cai H.W. 'Past and future on nanodielectrics'. *IET Nanodielectrics*. 2018; 1(1): 41–47

[6] Zhang X., Li B.W., Dong L., *et al.* 'Superior energy storage performances of polymer nanocomposites via modification of filler/polymer interfaces'. *Advanced Materials Interfaces*. 2018; 5(11): 1800096

[7] Xia W., Zhang Z. 'PVDF-based dielectric polymers and their applications in electronic materials'. *IET Nanodielectrics*. 2018; 1(1): 17–31

[8] Wang Y., Chen J., Li Y., Niu Y., Wang Q., Wang H. 'Multilayered hierarchical polymer composites for high energydensity capacitors'. *Journal of Materials Chemistry A*. 2019; 7: 2965–2980

[9] Zhou Y., Li Q., Dang B., *et al.* 'A scalable, high-throughput, and environmentally benign approach to polymer dielectrics exhibiting significantly improved capacitive performance at high temperatures'. *Advanced Materials*. 2018; 30(49): 1805672

[10] Fan B., Zhou M., Zhang C., He D., Bai J. 'Polymer-based materials for achieving high energy density film capacitors'. *Progress in Polymer Science*. 2019; 97: 101143

Chapter 12

Future prospects: ceramic part

Letao Yang[1], Xi Kong[1], Zhenxiang Cheng[1] and Shujun Zhang[1]

12.1 Introduction

Dielectric capacitors, featured in high charge–discharge rate, low cost and long cyclic lifetime, play an important role in electrical and electronic applications. Among various dielectrics, ceramics and polymers are mainstream materials dominating the dielectric capacitors. Many modern applications including hybrid vehicles and power converters require the capacitors to work under high temperature, which poses challenges for polymer capacitors with low melting/glass-transition temperature [1], while ceramic capacitors can be operated at much higher temperature than polymer capacitors due to the inorganic nature of ceramics. Although the ceramic capacitors are suitable for high-temperature energy storage applications, there are still a lot of challenges such as low energy density, which will be discussed in details in Section 12.2. The perspectives and future development trends for dielectric capacitors are proposed.

12.2 Present challenges to the dielectrics for electrostatic capacitors

12.2.1 Energy density

Since the low energy density of dielectric capacitors is the main drawback when compared with other energy storage devices, enhancing the energy density is among the most studied topics of dielectrics for energy storage applications. The energy density is related to the permittivity and the applied electric field. In the case of high-permittivity dielectrics such as $BaTiO_3$, the permittivity increases significantly near Curie temperature (T_c), exhibiting a dielectric peak in dielectric permittivity-temperature spectroscopy. One approach of enhancing permittivity at room temperature is shifting and broadening the dielectric peaks by doping or

[1]Institute for Superconducting and Electronic Materials, Australia Institute of Innovative Materials, University of Wollongong, Wollongong, New South Wales, Australia

alloying. For example, the dielectric peak of $BaTiO_3$ lies near 120 °C, after alloying with $SrTiO_3$, the broadened dielectric peak is shifted to low temperature [2]. Similar condition exists in $BaTiO_3$–$BaZrO_3$ and $BaTiO_3$–$BaSnO_3$ systems [3,4]. Since most reported energy densities are obtained at room temperature, shifting the dielectric peak to room temperature may be an effective way of increasing energy density, but this may deteriorate the field stability and temperature stability, which will be discussed in Sections 12.3 and 12.4. In addition, elevated dielectric losses are usually associated with the dielectric peaks, thus decrease the breakdown strength (BDS) and energy efficiency.

Comparing with increasing permittivity, increasing BDS is a more efficient way of enhancing energy density. The BDS of ceramics is inversely proportional to the porosity, grain size and sample thickness [5]. Conventional sintering techniques can hardly obtain fully dense ceramics; instead, some pores exist inside the ceramic body. When applying an electric field on the ceramic, the local electric field near the pores is higher than the external electric field and therefore local breakdowns are likely to occur. Long-term local breakdowns will lead to erosion, electrical treeing and pitting and finally the ceramic will breakdown even under normal working voltage [6]. Therefore highly dense ceramics with minimal porosity are always desired for energy storage applications. In general, thinner ceramics possess higher BDS because fewer defects exist in them. The impact of thickness on the BDS is of particular importance for multi-layer ceramic capacitors (MLCCs) in which the submicron dielectric ceramic layers are expected [7]. Reducing dielectric layer thickness is also helpful for the development towards miniaturization of the MLCCs by increasing the volumetric capacitance. However, preparing uniform and submicron-scale dielectric layers are also a challenge.

Typically, the ceramics with uniform and fine grains possess higher BDS. This can be ascribed to the high fraction of grain boundary in fine-grained ceramics. High-resistivity depletion space charge layers build up at grain boundaries and can be acted as the barriers for the charge transporting across the grain boundaries, leading to reduced leakage current and enhanced BDS. Therefore, reducing grain size is a practical method of enhancing BDS of ceramics. Two-step sintering technique was proposed to sinter ceramics with suppressing the final-stage grain growth [8]. By employing two-step sintering, fully densified Y_2O_3 ceramic with a grain size of 60 nm was reported, which was almost an order of magnitude smaller than the ceramics sintered by conventional sintering [8]. The two-step sintering technique was also employed in preparing the MLCCs with the ceramic composition of $0.87BaTiO_3$–$0.13Bi(Zn_{2/3}(Nb_{0.85}Ta_{0.15})_{1/3})O_3$, a smaller grain size of 434 nm was obtained, leading to a superior recoverable energy density of 10.12 J/cm^3 with a high energy efficiency of 89.4% at an electric field of 1047 kV/cm [9]. The fine-grain ceramics can be also prepared by the liquid phase sintering technique. This can be realized by introducing low-melting-point glasses or oxides which can form a liquid phase coating the grains during sintering [10]. For example, BaO–SiO_2–Al_2O_3–B_2O_3–ZrO_2–SrO glass was used to form a thin layer coating of the $BaTiO_3$ particles at high temperature [11]. The low-melting-point glass additive not only reduces the sintering temperature but also reduces the grain size. As a result, the BDS of the ceramics increases from 226 kV/cm for pure $BaTiO_3$ to 645 kV/cm for the ceramics containing 20 vol% glass where the corresponding grain sizes are 145 μm and 1 μm, respectively [11]. However, some studies have shown that the permittivity was decreased due to introducing low-permittivity glasses/oxides [12,13], which may affect the final energy storage properties.

It should be noted that energy density is not an intrinsic property of the dielectrics. The energy density strongly depends on the applied electric field on the dielectrics so just comparing the energy density of different dielectrics without referring the applied electric field is somewhat meaningless.

In summary, increasing permittivity and BDS of the dielectrics are two fundamental methods of increasing energy density of the electrostatic capacitors. The permittivity at room temperature can be improved by shifting the broadened dielectric peaks by doping or forming relaxor-based solid solutions. However, this method may deteriorate other energy storage properties such as the BDS and temperature stability. On the other hand, increasing BDS is another effective approach of improving the energy density. Since the BDS of ceramics is closely related to the porosity, grain size and thickness, improving the density and reducing the grain size and thickness of the ceramic dielectrics are more practical for preparing high energy density ceramic capacitors.

12.2.2 Energy efficiency

In addition to the energy density, the energy efficiency is also important for energy storage applications because it reflects how much energy can be released in discharging process and how much energy is dissipated in the form of heat. The dissipated energy is closely related to the dielectric loss and leakage current. The domain wall motion of ferroelectrics (FEs) contributes to a large fraction of the dielectric loss, leading to a square *P–E* loop with a large dissipated energy density and a low energy efficiency. While in linear dielectrics, no domain structure exists and the dielectric loss is much lower than that of FEs. Therefore, the energy efficiency of linear dielectrics is higher than that of FEs. The dielectric loss-temperature spectra of FEs can be divided into three regions, as shown in Figure 12.1. Below T_c, the dielectric loss is relatively high, mainly coming from the contributions of domain wall motions. In region II where the temperature is higher than T_c, the dielectric loss is reduced because of the absence of domain walls in cubic phase. At even higher temperature, the dielectric loss ramps up due to the free-carrier conductivity initiated by the decreased resistivity at high temperature, including space charge [14]. High energy efficiency is expected to be obtained in region II. In the case of relaxors, there is a similar region above the loss maxima temperature. Shifting the T_m below room temperature is an effective approach of increasing room temperature energy efficiency [15]. The relaxors with T_m below room temperature are suitable for energy storage applications because they also possess high permittivity associated with the presence of nano-sized polar regions and low remnant polarization due to the absence of macroscopic domain switching [16].

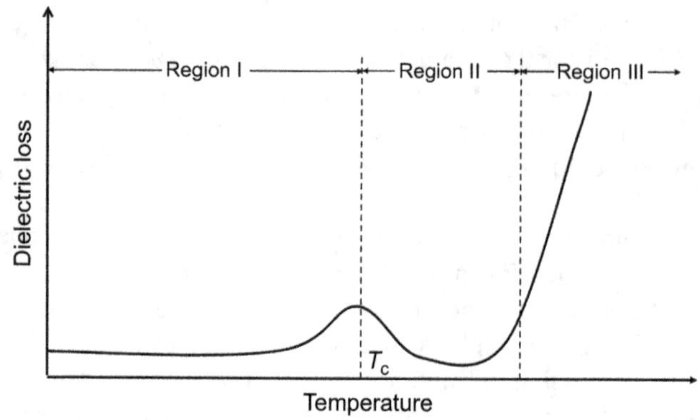

Figure 12.1 Temperature dependence of dielectric loss of ferroelectrics

Polycrystalline ceramics usually possess low conductivity, which contributes little to the loss and dissipated energy density. However, the conductivity increases with the electric field/temperature, leading to an increased leakage current and dissipated energy density. With respect to the dielectric capacitors working under high bias, the conductivity of the dielectrics is required to be low to ensure a satisfactory energy efficiency. It was reported in perovskite ceramics especially those containing volatile elements such as Pb and Bi, the electrical conduction type is p-type where the holes are the dominated charge carriers [17–19]. The holes can be generated by the ionizing process of metal cation vacancies, which are created during high-temperature sintering [19]. Donor dopants, such as Nb^{5+} for $(Zr,Ti)^{4+}$ in $Pb(Zr,Ti)O_3$ are beneficial for reducing the conductivity due to the compensation between the holes and the electrons associated with the donor dopants [18]. Another major defect in perovskite ceramics is oxygen vacancies, which are responsible for high-temperature relaxation and conduction [20–22]. By annealing in an oxygen atmosphere, the oxygen vacancy concentration can be reduced and the insulating properties can be enhanced. As a result, the BDS and energy efficiency of O_2-annealed samples are higher than those of as-sintered and N_2-annealed samples [23].

In summary, in addition to the energy density, the energy efficiency is also important for evaluating the energy storage performance of dielectric capacitors. The dissipated energy mainly comes from the contributions of dielectric loss and conductivity. The energy efficiency of ceramics can be improved either by shifting the loss maxima below room temperature or by doping for reducing the major carrier concentration to decrease the conductivity.

12.2.3 P–E loop shape

Relaxors and anti-ferroelectrics (AFEs) are the two promising dielectric candidates for energy storage applications. Relaxors featured in slim hysteresis with high maximum polarization and low remnant polarization, are expected to yield a high energy density and a high energy efficiency simultaneously at a medium electric field. Under higher electric field, however, the polarization does not increase at the same rate as under low and medium electric fields, which is termed as polarization saturation [24], as shown in Figure 12.2(a). The polarization saturation phenomenon is common in classical FE materials due to the fact that most of the domains being aligned to the electric field direction [25]. After the polarization saturation, the polarization increases linearly as the electric field increases, like the situation in linear dielectrics. That is to say, the permittivity decreases after polarization saturates. When the polarization saturation occurs at the electric field far below the BDS of the ceramics, the energy storage potential will be limited significantly [26,27]. The polarization saturation is similar to the DC bias effect in $BaTiO_3$-based MLCCs, in which the capacitance (permittivity) decreases as applied DC bias increases [28]. This effect is believed to be related to the FE characteristics such as domain switching in $BaTiO_3$-based dielectrics [29]. The polarization saturation in relaxors may come from the FE-active cations, replacing the FE-active cations (such as Ti^{4+}) by FE-inactive cations (such as Sn^{4+}) can mitigate the polarization saturation but decrease zero field permittivity, as shown in Figure 12.2 (b). Considering the applied electric field, the overall energy density is expected to increase at an elevated electric field [24]. On the other hand, it is known that the AFEs experience a phase transition from non-polar AFE phase to polar FE phase under a strong electric field. Adding AFE phase to relaxor phase can be a potential method of increasing the field stability of the dielectrics. For example, by introducing AFE $Pb(Mg_{0.5}W_{0.5})O_3$ to $Pb(Ni_{1/3}Nb_{2/3})O_3$–$PbTiO_3$, the field stability of permittivity under DC bias was markedly improved [29].

Other promising dielectrics for energy storage application are AFEs. Due to the unique double hysteresis loops originated from the reversible AFE-FE phase transition, AFEs can yield higher energy density than other dielectrics. However, the phase transition hysteresis also leads to a large dissipated energy density and a low energy efficiency. Therefore, many AFE ceramics exhibit high recoverable energy density around 4 J/cm^3 but low energy efficiency below 75% [31–33]. A reasonable attempt is to introduce relaxor characteristics into normal AFEs to form so-called 'relaxor anti-ferroelectrics' with a slim double hysteresis loop, accounting for a high energy efficiency, as illustrated in Figure 12.2(c). In addition, a large strain associated with the AFE-FE phase transition usually exists in AFE ceramics, which may result in micro-cracks and mechanical failures of AFE ceramic capacitors in repeated charge/discharge cycles. The relaxor characteristics may be beneficial for smearing the large field-induced strain and the phase transition hysteresis, thus it reduces the mechanical and thermal breakdown of the dielectrics, respectively. The relaxors are featured in structural and chemical heterogeneity, so creating such heterogeneity may be helpful to transform the normal AFEs to relaxor AFEs [16]. For example, 8% La^{3+} was added to $Pb(Zr_{0.91}Ti_{0.09})O_3$ AFE ceramics and a slim *P–E* loop with high energy efficiency of 92% was obtained at 170 kV/cm [34]. Based on the similar idea, a slim double hysteresis loops with both high energy density of 9.5 J/cm^3 and high energy efficiency of 92% was achieved at

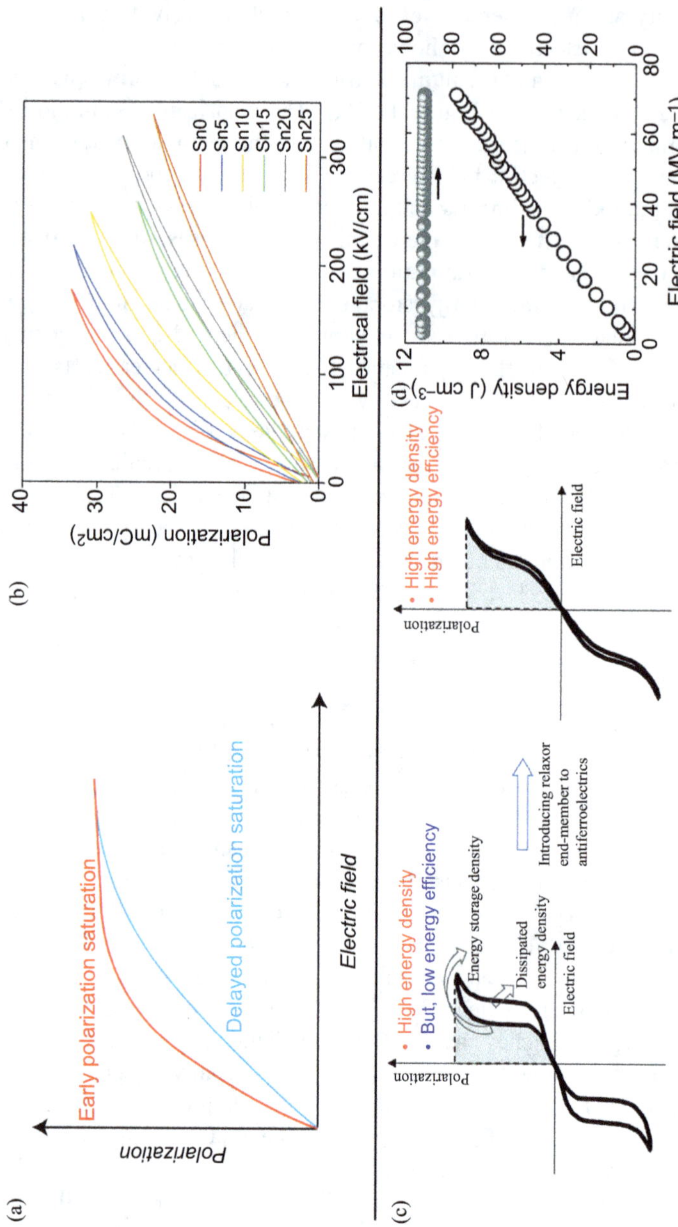

Figure 12.2 Schematic illustration of polarization saturation in relaxors; (b) P–E loops of Sn-doped $0.5(Na_{0.5}Bi_{0.5})TiO_3$–$0.5SrTiO_3$ ceramics. The polarization saturation is delayed as Sn concentration increases. © 2019 Royal Society of Chemistry. Reprinted, with permission, from [24]. (c) Schematic diagram of energy storage properties for AFEs and relaxor AFEs and (d) energy storage properties of relaxor AFE $0.55(Na_{0.5}Bi_{0.5})TiO_3$–$0.45(Sr_{0.7}Bi_{0.2})TiO_3$ MLCC. © 2018 John Wiley and Sons. Reprinted, with permission, from [30]

720 kV/cm for a $0.55(Na_{0.5}Bi_{0.5})TiO_3$–$0.45(Sr_{0.7}Bi_{0.2})TiO_3$ relaxor AFE MLCC, as shown in Figure 12.2(d) [30].

Relaxors and AFEs are two potential candidates for dielectric capacitors. To further improve their energy storage properties, the feature of *P–E* loop can be modified. With respect to the relaxors, the early polarization saturation limits the energy density at high electric field. This can be mitigated by doping FE-non-active ions or adding AFE endmembers. On the other hand, although AFEs can reach high energy density which is benefited from the double hysteresis loops, the large hysteresis in the field-induced FE phase dissipates considerable amount of the stored energy, leading to a low energy efficiency. Analogous to the transformation from FEs to relaxor FEs, the AFEs can be transformed to the 'relaxor anti-ferroelectrics' with both higher energy density and energy efficiency by introducing local heterogeneity.

12.2.4 Temperature stability

Some modern power applications such as power converter and underground gas/oil exploration, require the capacitors operated at high temperature (e.g. >150 °C) [1]. As mentioned above, ceramics usually take advantages over polymers for high-temperature applications due to the inorganic nature. With respect to MLCCs, the weak temperature dependence of permittivity is desired. Ferroelectrics such as $BaTiO_3$ are featured in high permittivity but their permittivity varies rapidly when temperature approaching phase transition temperatures, which is far from the ideal dielectric behaviour of a capacitor. One method of enhancing the temperature stability of MLCCs is adopting relaxor dielectrics. Different from the pure FEs, the phase transitions in relaxors are diffused and smeared, leading to broad dielectric maxima peaks and enhanced temperature stability. Alternatively, the temperature dependence of permittivity for $BaTiO_3$ can be also modified by creating core–shell microstructure, in which the core is almost pure FE $BaTiO_3$ while the shell regions consist of a solid solution of $BaTiO_3$ and additives with diffuse phase transition [7,35]. By introducing the core–shell microstructures, the sharp Curie peak is supressed and another dielectric peak with broadened phase transition being related to the shell appears at low temperature, leading to relatively temperature–insensitive permittivity over a wide temperature range [36]. The permittivity–temperature curves of the ceramics with and without core–shell microstructures are compared in Figure 12.3.

Taking Mg and Ho co-doped $BaTiO_3$ ceramics as an example, the core-shell microstructure formation process is as follows: (1) Mg reacts with $BaTiO_3$ at low temperatures to form the shell phase; (2) Ho reacts with the shell phase and the diffusion of Ho into the core region is suppressed by Mg [37]. The core-shell $BaTiO_3$-based ceramics with improved temperature stability have been widely used as dielectrics in MLCCs. In addition to the formation of core-shell microstructure, the rare earth elements with intermediate ionic radius (Dy, Ho and Er) significantly improve the resistivity and lifetime of $BaTiO_3$-based ceramics. This can be attributed to their amphoteric behaviour as both donor and acceptor dopants in $BaTiO_3$ [38].

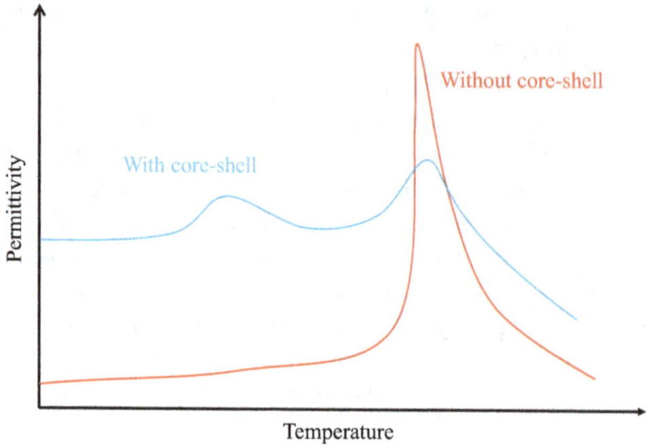

Figure 12.3 Comparison of temperature stability on permittivity for the ceramics with and without core-shell microstructure

At high temperature, the resistance of a ceramic capacitor usually decreases, and the associated leakage current leads to heat inside the ceramic body, which is harmful to the reliability. Therefore, the ideal capacitors for high-temperature applications should possess both high capacitance and high resistance at high temperature. The product of the resistance and capacitance at a given temperature, *RC* time constant, is usually used for evaluating the high-temperature performance of a capacitor [39]. It was known the wide-bandgap materials possess high resistivity at elevated temperature, but their permittivity is usually low [40]. On the other hand, although FE-based dielectrics possess high permittivity and the permittivity can increase several orders of magnitude when approaching phase transition temperature, their resistivity, which is inversely proportional to permittivity, is relatively low. Therefore, there is a trade-off between the *R* and *C* for a given dielectric at high temperature. For high-temperature energy storage capacitors, wide-bandgap dielectrics such as La-doped $BaZrO_3$ whose outstanding insulation properties compensating the relatively low permittivity can yield a high RC constant and suitable for high-temperature applications [39].

High-temperature application is a major development trend for ceramic electrostatic capacitors. Nevertheless, the inferior temperature stability of permittivity usually exists in FE-derived dielectrics, the temperature stability can be improved by introducing the core-shell microstructure and local structure heterogeneity, that is, relaxor component, and this method has been widely used in $BaTiO_3$-based MLCCs. In addition, the *RC* time constant is a useful parameter to evaluate the high-temperature performance of dielectric capacitors, where both high capacitance and resistance are required at high temperature.

12.2.5 Reliability

In practical applications, the capacitors may experience millions of charge–discharge cycles and the reliability of capacitors is important for the safety of the whole system. In addition, modern applications require the capacitors working under harsh environments such as high temperature, humidity and vibration [41]. Because tracking the time-to-failure of each capacitor is impossible, the reliability of capacitors is usually studied by highly accelerated life testing (HALT). The capacitor is tested under various combinations of temperature, humidity and voltage (electric field) and the time-to-failure distributions are recorded. Then the related parameters obtained from the analysis of the distributions can be used to predict the lifetime of the capacitor under normal operating conditions [42].

For the ceramics sintered at high temperature and/or in a reducing atmosphere such as the MLCCs with Ni and Cu electrodes, oxygen vacancies are often generated in the ceramics. The oxygen vacancies are believed to play an important role in the degradation and failure of the capacitors [43,44]. Under the effects of electric field and temperature, the oxygen vacancies migrate and accumulate near the cathode, leading to an abrupt increase of current due to the Schottky effect [44]. Some studies also suggested that the oxygen vacancies are responsible for the degradation of resistivity and capacitance of the capacitors [45–47]. Two common methods of reducing oxygen vacancies in ceramics are post-annealing for reoxidation and doping strategy for suppressing the oxygen vacancies formation during sintering [44].

Other reliability issues for capacitors, especially the ceramic capacitors with inflexible nature, are mechanical and electromechanical failures. Cracks in ceramic capacitors can be formed in many ways, such as rapid cooling when sintering, soldering the MLCCs onto electric board and repetitive charging–discharging cycles. Due to the existence of electrostrictive effect, the high-permittivity ferroelectric ceramics deform during charging–discharging process, leading to internal stresses and micro-cracks in the ceramic capacitors [48]. The micro-cracks may result in degradation of resisitivity and dielectric breakdown strength thus the energy storage properties. For the AFE ceramics with the large strain associated with the AFE-FE or FE-AFE phase transition, the deteriorated ferroelectric properties were also observed and the micro-cracks developed in the AFE ceramic body was attributed to the fatigue phenomenon [49]. One method of relieving the field-induced strain and the associated micro-cracks is enhancing the relaxor characteristics of the ceramics to smear the domain reversals/phase transitions, thus relaxor anti-ferroelectric is a promising material for energy storage applications.

12.3 Perspectives for electrostatic capacitors

Capacitors are one of the most important passive components in electronic and electrical applications. Nowadays, trillions of electrostatic capacitors are manufactured and shipped globally each year [50]. The emerging energy storage and power electronics fields provide a large market for the electrostatic capacitors. However, the requirements for the capacitors become more rigorous. For example, operating temperature above 150 °C is required for the capacitors in hybrid and electrical vehicles [51]. Therefore, novel dielectric formula and device designs have to be explored. The future development trends for ceramic electrostatic capacitors include, but not limited to the following:

1. *Miniaturization and integration*: Similar to many electronic components, the main trend of MLCC is miniaturization and integration in order to save circuit space and power consumption. As the volumetric capacitance of MLCCs increases, each dielectric layer thickness decreases. The dielectric thickness is limited by the particle size of the ceramic powders and preparation techniques such as tape casting. In addition, the applied electric field on the dielectric layers increases significantly for thinner dielectric layer thickness. Therefore, manufacturing thin, dense and uniform dielectric layers is a development trend as well as a challenge for future high volumetric and high-reliability MLCCs. To realize this, advanced manufacture techniques to prepare fine ceramic powders and ultrathin dielectric layers are desired.
2. *Base metal electrodes*: Considering the increased internal electrodes in the MLCCs, the fraction of electrodes in the production cost increased significantly. Replacing the precious metals such as Pt and Pd by base metals such as Ni and Cu can reduce the cost considerably. As mentioned above, the electric properties of $BaTiO_3$-based dielectrics are deteriorated when co-firing with Cu or Ni in a reducing atmosphere. However, some ceramics, such as $(K,Na)NbO_3$ and $Na(Nb,Ta)O_3$, exhibit higher density and resistivity when sintering at a reducing atmosphere [52]. These ceramics are promising candidates for low-cost multi-layer actuators and MLCCs with Ni electrodes.
3. *Low-temperature sintering*: For $BaTiO_3$-based MLCC, high sintering temperature and long duration time are usually required in order to consolidate the ceramic particles and eliminate porosity. However, these sintering conditions often lead to extreme coarsening of ceramic grains and deterioration of mechanical and electrical properties [53]. To solve this problem, various low-temperature sintering techniques are proposed, such as microwave sintering, spark plasma sintering, flash sintering, laser sintering, fast-firing sintering, liquid phase sintering and cold sintering. These novel techniques are expected to save energy consumption, make base metal electrodes compatible and even fabricate flexible MLCC [53,54].
4. *Graceful failure*: The failures of ceramic capacitors are usually catastrophic, that is a short-circuit mode, which leads to a severe damage to the associated components and systems. Therefore, the voltage rating of MLCCs is substantially below the breakdown voltage, limiting the energy density and increasing the cost [39]. Nevertheless, the self-healing mechanisms have been studied intensively in metallized polymer capacitors, the self-healing ability is hard to be realized in ceramic capacitors and MLCCs, and more studies are needed. Some structural designs such as floating electrode can help to reduce the failure possibility of MLCC [55].

References

[1] Tan D., Zhang L., Chen Q., Irwin P. 'High-temperature capacitor polymer films'. *Journal of Electronic Materials*. 2014;**43**(12):4569–4575.

[2] Zhou L., Vilarinho P. M., Baptista J. L. 'Dependence of the structural and dielectric properties of $Ba_{1-x}Sr_xTiO_3$ ceramic solid solutions on raw material processing'. *Journal of the European Ceramic Society*. 1999;**19**(11):2015–2020.

[3] Shvartsman V. V., Zhai J., Kleemann W. 'The dielectric relaxation in solid solutions $BaTi_{1-x}Zr_xO_3$'. *Ferroelectrics*. 2009;**379**(1):77–85.

[4] Lei C., Bokov A. A., Ye Z. G. 'Ferroelectric to relaxor crossover and dielectric phase diagram in the $BaTiO_3$–$BaSnO_3$ system'. *Journal of Applied Physics*. 2007;**101**(8):084105.

[5] Yang L. T., Kong X., Li F., *et al*. 'Perovskite lead-free dielectrics for energy storage applications'. *Progress in Materials Science*. 2019;**102**:72–108.

[6] Morshuis P. H. F., Smit J. J. 'Partial discharges at DC voltage: their mechanism, detection and analysis'. *IEEE Transactions on Dielectrics and Electrical Insulation*. 2005;**12**(2):328–340.

[7] Pan M.-J., Randall C. A. 'A brief introduction to ceramic capacitors'. *IEEE Electrical Insulation Magazine*. 2010;**26**(3):44–50.

[8] Chen I.-W., Wang X.-H. 'Sintering dense nanocrystalline ceramics without final-stage grain growth'. *Nature*. 2000;**404**(6774):168–171.

[9] Zhao P. Y., Wang H. X., Wu L. W., *et al*. 'High-performance relaxor ferroelectric materials for energy storage applications'. *Advanced Energy Materials*. 2019;**9**(17):1803048.

[10] German R. M., Suri P., Park S. J. 'Review: liquid phase sintering'. *Journal of Materials Science*. 2008;**44**(1):1–39.

[11] Young A., Hilmas G., Zhang S. C., Schwartz R. W. 'Effect of liquid-phase sintering on the breakdown strength of barium titanate'. *Journal of the American Ceramic Society*. 2007;**90**(5):1504–1510.

[12] Sarkar S. K., Sharma M. L. 'Liquid phase sintering of $BaTiO_3$ by boric oxide (B_2O_3) and lead borate (PbB_2O_4) glasses and its effect on dielectric strength and dielectric constant'. *Materials Research Bulletin*. 1989;**24**(7):773–779.

[13] Wang X., Zhang Y., Song X., *et al*. 'Glass additive in barium titanate ceramics and its influence on electrical breakdown strength in relation with energy storage properties'. *Journal of the European Ceramic Society*. 2012;**32**(3):559–567.

[14] Härdtl K. H. 'Electrical and mechanical losses in ferroelectric ceramics'. *Ceramics International*. 1982;**8**(4):121–127.

[15] Kong X., Yang L. T., Cheng Z. X., Zhang S. J. 'Bi-modified $SrTiO_3$-based ceramics for high-temperature energy storage applications'. *Journal of the American Ceramic Society*. 2019;**103**(3):1722–1731.

[16] Li F., Zhang S., Damjanovic D., Chen L.-Q., Shrout T. R. 'Local structural heterogeneity and electromechanical responses of ferroelectrics: learning

from relaxor ferroelectrics'. *Advanced Functional Materials*. 2018;**28**(37):1801504.

[17] Masó N., West A. R. 'Electrical properties of Ca-doped $BiFeO_3$ ceramics: from p-type semiconduction to oxide-ion conduction'. *Chemistry of Materials*. 2012;**24**(11):2127–2132.

[18] Dih J. J., Fulrath R. M. 'Electrical conductivity in lead zirconate-titanate ceramics'. *Journal of the American Ceramic Society*. 1978;**61**(9-10):448–451.

[19] Gerson R., Jaffe H. 'Electrical conductivity in lead titanate zirconate ceramics'. *Journal of Physics and Chemistry of Solids*. 1963;**24**(8):979–984.

[20] Kang B. S., Choi S. K., Park C. H. 'Diffuse dielectric anomaly in perovskite-type ferroelectric oxides in the temperature range of 400–700 °C'. *Journal of Applied Physics*. 2003;**94**(3):1904–1911.

[21] Ang C., Yu Z., Cross L. E. 'Oxygen-vacancy-related low-frequency dielectric relaxation and electrical conduction in Bi:$SrTiO_3$'. *Physical Review B*. 2000;**62**(1):228–236.

[22] Choi Y.-K., Hoshina T., Takeda H., Tsurumi T. 'Effect of oxygen vacancy and oxygen vacancy migration on dielectric response of $BaTiO_3$-based ceramics'. *Japanese Journal of Applied Physics*. 2011;**50**:031504.

[23] Zhang L., Yao Z., Lanagan M. T., *et al.* 'Effect of oxygen treatment on structure and electrical properties of Mn-doped $Ca_{0.6}Sr_{0.4}TiO_3$ ceramics'. *Journal of the European Ceramic Society*. 2018;**38**(6):2534–2540.

[24] Yang L. T., Kong X., Cheng Z. X., Zhang S. J. 'Ultra-high energy storage performance with mitigated polarization saturation in lead-free relaxors'. *Journal of Materials Chemistry A*. 2019;**7**(14):8573–8580.

[25] Jin L., Li F., Zhang S. 'Decoding the fingerprint of ferroelectric loops: comprehension of the material properties and structures'. *Journal of the American Ceramic Society*. 2014;**97**(1):1–27.

[26] Zhang X., Shen Y., Shen Z., Jiang J., Chen L., Nan C.-W. 'Achieving high energy density in PVDF-based polymer blends: suppression of early polarization saturation and enhancement of breakdown strength'. *ACS Applied Materials & Interfaces*. 2016;**8**(40):27236–27242.

[27] Chu B., Zhou X., Ren K., *et al.* 'A dielectric polymer with high electric energy density and fast discharge speed'. *Science*. 2006;**313**(5785):334–336.

[28] Seongtae K., Hackenberger W., Alberta E., Furman E., Lanagan M. 'Nonlinear dielectric ceramics and their applications to capacitors and tunable dielectrics'. *IEEE Electrical Insulation Magazine*. 2011;**27**(2):43–55.

[29] Yang G., Yue Z., Zhao J., Wen H., Wang X., Li L. 'Dielectric behaviour of $BaTiO_3$-based ceramic multilayer capacitors under high dc bias field'. *Journal of Physics D: Applied Physics*. 2006;**39**(16):3702–3707.

[30] Li J., Li F., Xu Z., Zhang S. 'Multilayer lead-free ceramic capacitors with ultrahigh energy density and efficiency'. *Advanced Materials*. 2018;**30**(32):1802155.

[31] Zhao L., Liu Q., Gao J., Zhang S., Li J. F. 'Lead-free antiferroelectric silver niobate tantalate with high energy storage performance'. *Advanced Materials*. 2017;**29**(31):1701824.

[32] Han K., Luo N., Mao S., *et al.* 'Ultrahigh energy-storage density in A-/B-site co-doped $AgNbO_3$ lead-free antiferroelectric ceramics: insight into the origin of antiferroelectricity'. *Journal of Materials Chemistry A.* 2019;**7**(46):26293–26301.

[33] Yang X. M., Zhuo F. P., Wang C. X., *et al.* 'High energy storage density and ultrafast discharge in lead lutetium niobate based ceramics'. *Journal of Materials Chemistry A.* 2019;**7**(14):8414–8422.

[34] Jo H. R., Lynch C. S. 'A high energy density relaxor antiferroelectric pulsed capacitor dielectric'. *Journal of Applied Physics*. 2016;**119**(2):024104.

[35] Kishi H., Mizuno Y., Chazono H. 'Base-metal electrode-multilayer ceramic capacitors: past, present and future perspectives'. *Japanese Journal of Applied Physics*. 2003;**42**(Part 1: 1):1–15.

[36] Jeon S.-C., Yoon B.-K., Kim K.-H., Kang S.-J. L. 'Effects of core/shell volumetric ratio on the dielectric-temperature behavior of $BaTiO_3$'. *Journal of Advanced Ceramics*. 2014;**3**(1):76–82.

[37] Kishi H., Okino Y., Honda M., *et al.* 'The effect of MgO and rare-earth oxide on formation behavior of core-shell structure in $BaTiO_3$'. *Japanese Journal of Applied Physics*. 1997;**36**(Part 1: 9B):5954–5957.

[38] Kishi H., Kohzu N., Sugino J., Ohsato H., Iguchi Y., Okuda T. 'The effect of rare-earth (La, Sm, Dy, Ho and Er) and Mg on the microstructure in $BaTiO_3$'. *Journal of the European Ceramic Society*. 1999;**19**(6):1043–1046.

[39] Furman E., Zhang S., Kim N., *et al.* 'High-temperature, high-power capacitors: the assessment of capabilities'. *SAE International Journal of Aerospace*. 2008;**1**(1):822–831.

[40] Robertson J. 'High dielectric constant gate oxides for metal oxide Si transistors'. *Reports on Progress in Physics*. 2005;**69**(2):327–396.

[41] Sarjeant W. J., Zirnheld J., MacDougall F. W. 'Capacitors'. *IEEE Transactions on Plasma Science*. 1998;**26**(5):1368–1392.

[42] Paulsen J. L., Reed E. K. 'Highly accelerated lifetesting of base-metal-electrode ceramic chip capacitors'. *Microelectronics Reliability.* 2002;**42**(6):815–820.

[43] Hayashi H., Kawada S., Kimura M., *et al.* 'Reliability of nickel inner electrode lead-free multilayer piezoelectric ceramics'. *Japanese Journal of Applied Physics*. 2012;**51**(9S1):09LD01.

[44] Yamamatsu J., Kawano N., Arashi T., Sato A., Nakano Y., Nomura T. 'Reliability of multilayer ceramic capacitors with nickel electrodes'. *Journal of Power Sources*. 1996;**60**(2):199–203.

[45] Loh E. 'A model of dc leakage in ceramic capacitors'. *Journal of Applied Physics*. 1982;**53**(9):6229–6235.

[46] Nomura T., Kawano N., Yamamatsu J., Arashi T., Nakano Y., Sato A. 'Aging behavior of Ni-electrode multilayer ceramic capacitors with X7R characteristics'. *Japanese Journal of Applied Physics*. 1995;**34**(Part 1: 9B):5389–5395.

[47] Chazono H., Kishi H. 'DC-electrical degradation of the BT-based material for multilayer ceramic capacitor with Ni internal electrode: impedance

analysis and microstructure'. *Japanese Journal of Applied Physics*. 2001;**40** (Part 1: 9B):5624–5629.

[48] Jiang Q., Cao W., Cross L. E. 'Electric fatigue in lead zirconate titanate ceramics'. *Journal of the American Ceramic Society*. 1994;**77**(1):211–215.

[49] Pan W., Zhang Q., Bhalla A., Cross L. E. 'Field-forced antiferroelectric-to-ferroelectric switching in modified lead zirconate titanate stannate ceramics'. *Journal of the American Ceramic Society*. 1989;**72**(4):571–578.

[50] Ho J., Jow T. R., Boggs S. 'Historical introduction to capacitor technology'. *IEEE Electrical Insulation Magazine*. 2010;**26**(1):20–25.

[51] Ho J. S., Greenbaum S. G. 'Polymer capacitor dielectrics for high temperature applications'. *ACS Applied Materials & Interfaces*. 2018;**10**(35): 29189–29218.

[52] Kobayashi K., Randall C. A., Ryu M., Doshida Y., Mizuno Y. 'New opportunity in alkali niobate ceramics processed in low oxygen partial pressure'. *Proceedings of ISAF-ECAPD-PFM 2012*; Aveiro, Portugal, July 2012. IEEE; 2012. pp. 1–4

[53] Ibn-Mohammed T., Randall C. A., Mustapha K. B., *et al.* 'Decarbonising ceramic manufacturing: a techno-economic analysis of energy efficient sintering technologies in the functional materials sector'. *Journal of the European Ceramic Society*. 2019;**39**(16):5213–5235.

[54] Baker A., Guo H., Guo J., Randall C. 'Utilizing the cold sintering process for flexible–printable electroceramic device fabrication'. *Journal of the American Ceramic Society*. 2016;**99**(10):3202–3204.

[55] Sloka B. Flex crack mitigation [online]. 2008. Available from http://www.kemet.com/Lists/FileStore/Flex%20MItigation%20Solutions%20Rev%203%20%20PHub%20copy%207.30.14.pdf [Accessed 11 Nov 2019].

Index

www.ingramcontent.com/pod-product-compliance
Lightning Source LLC
LaVergne TN
LVHW011558200926
848484LV00001B/330